Encapsulation in Food Processing and Fermentation

Books Published in *Food Biology* series

Food Biology Series

Encapsulation in Food Processing and Fermentation

Editors

Steva Lević

Department of Food Technology and Biochemistry
University of Belgrade – Faculty of Agriculture
Belgrade, Serbia

Viktor Nedović

Department of Food Technology and Biochemistry
University of Belgrade – Faculty of Agriculture
Belgrade, Serbia

Branko Bugarski

Department of Chemical Engineering
University of Belgrade – Faculty of Technology and Metallurgy
Belgrade, Serbia

CRC Press
Taylor & Francis Group
Boca Raton London New York

CRC Press is an imprint of the
Taylor & Francis Group, an **informa** business

A SCIENCE PUBLISHERS BOOK

First edition published 2022
by CRC Press
6000 Broken Sound Parkway NW, Suite 300, Boca Raton, FL 33487-2742

and by CRC Press
4 Park Square, Milton Park, Abingdon, Oxon, OX14 4RN

Library of Congress Cataloging-in-Publication Data

Names: Lević, Steva, 1981- editor. | Nedović, Viktor, editor. | Bugarski, Branko, editor.
Title: Encapsulation in food processing and fermentations / editors, Steva Lević, Department of Food Technology and Biochemistry, University of Belgrade, Belgrade, Serbia, Viktor Nedović, Department of Food Technology and Biochemistry, University of Belgrade, Belgrade, Serbia, Branko Bugarski, Department of Chemical Engineering, University of Belgrade, Belgrade, Serbia.
Description: First edition. | Boca Raton : CRC Press, Taylor & Francis Group, [2022] | Includes bibliographical references and index. | Summary: "The book Encapsulation in food processing and fermentations provides new insights into food encapsulation processes. The chapters are organized in such manner, so even readers who are not familiar with encapsulation could easily educate and find information of interest. Through the chapters, the book offers the most important aspects of encapsulation in general as well as state-of-the-art of modern industrial encapsulation procedures. In the specialized chapters, the topics such as encapsulation of food additives, probiotics, microorganisms for fermentations and nanoencapsulation are discussed"-- Provided by publisher.
Identifiers: LCCN 2021045213 | ISBN 9780367258313 (hbk) | ISBN 9781032160269 (pbk) | ISBN 9780429324918 (ebk)
Subjects: LCSH: Coated foods. | Food--Packaging. | Fermented foods. | Dietary supplements--Preservation. | Extracts--Preservation. | Microencapsulation.
Classification: LCC TP374 .E54 2022 | DDC 664/.09--dc23/eng/20211109
LC record available at https://lccn.loc.gov/2021045213

ISBN: 978-0-367-25831-3 (hbk)
ISBN: 978-1-032-16026-9 (pbk)
ISBN: 978-0-429-32491-8 (ebk)

DOI: 10.1201/9780429324918

Typeset in Times New Roman
by Innovative Processors

Preface to the Series

Food is the essential source of nutrients (such as carbohydrates, proteins, fats, vitamins, and minerals) for all living organisms to sustain life. A large part of daily human efforts is concentrated on food production, processing, packaging and marketing, product development, preservation, storage, and ensuring food safety and quality. It is obvious therefore, our food supply chain can contain microorganisms that interact with the food, thereby interfering in the ecology of food substrates. The microbe-food interaction can be mostly beneficial (as in the case of many fermented foods such as cheese, butter, sausage, etc.) or in some cases, it is detrimental (spoilage of food, mycotoxin, etc.). The *Food Biology* series aims at bringing all these aspects of microbe-food interactions in form of topical volumes, covering food microbiology, food mycology, biochemistry, microbial ecology, food biotechnology and bio-processing, new food product developments with microbial interventions, food nutrification with nutraceuticals, food authenticity, food origin traceability, and food science and technology. Special emphasis is laid on new molecular techniques relevant to food biology research or to monitoring and assessing food safety and quality, multiple hurdle food preservation techniques, as well as new interventions in biotechnological applications in food processing and development.

The series is broadly broken up into food fermentation, food safety and hygiene, food authenticity and traceability, microbial interventions in food bio-processing and food additive development, sensory science, molecular diagnostic methods in detecting food borne pathogens and food policy, etc. Leading international authorities with background in academia, research, industry and government have been drawn into the series either as authors or as editors. The series will be a useful reference resource base in food microbiology, biochemistry, biotechnology, food science and technology for researchers, teachers, students and food science and technology practitioners.

Ramesh C Ray
Series Editor

Preface

As a result of constant market demands and changes in consumers habits, the food industry has been engaged in a plethora of multidisciplinary researches in order to replace unhealthy and potentially dangerous ingredients with natural and more sustainable compounds; implementation of new strategies for improving the product shelf life; replacement of synthetic materials with biodegradable packaging films; use of the food industry byproducts as sources of new ingredients. Within all of these applications, encapsulation has been used as a tool for protecting of active compounds, providing controlled delivery and maintaining overall quality.

Encapsulation has been investigated as an alternative to conventional food processes, providing some benefits but also making the production process more complex and costlier. However, constant improvements and new carrier materials and encapsulation techniques offer critical impulse for establishing food products based on encapsulation technologies. Hence, it is not an exaggeration to claim that encapsulation technologies have become an integral part of the modern food sector. In order to fulfil very diverse demands, encapsulation technologies adopted new tools, materials and practices. The main directions in developing food encapsulation technologies are toward natural and renewable carrier materials, elimination or reduction of pollution and establishing more economically sustainable processes. However, the development and commercialization of encapsulation technologies have been a long process and included many trials and optimizations for any particular encapsulation techniques or carrier materials. For example, just twenty years ago, it was really challenging to purchase new encapsulation equipment, and prices were high as a result of the limited number of suppliers. Today, due to very intensive research and knowledge transfer, food companies and laboratories have numerous opportunities to obtain high-quality encapsulation equipment and materials. Thus, encapsulation is no longer unknown for the food sector, and consequently, numerous encapsulated food products are available for individual consumers and industry.

Considering the number of published papers and patents, it could be concluded that encapsulation is widely recognized and accepted in the food sector. Active compounds such as food flavors, colors, probiotic cells, mineral compounds, vegetable oils, antioxidants and many others are now available in encapsulated forms, optimized for a particular purpose. The main advantages of encapsulates, such as easy handling and protection of active compounds, controlled release, control of sensorial properties and specific marketing effects, are recognized and accepted as benefits for consumers and the overall product quality. On the other hand, encapsulation must be economically sustainable and should not become a burden to food production. Further, the health and ecological impact of encapsulated active compounds, especially those in the form of nanoparticles, are now a major topic in the scientific community.

The aim of this book is to provide an up-to-date overview of the main subjects related to encapsulation technologies in the food sector. Our authors present useful information for those who are in the field of encapsulation but also for those who are beginners in this field. We tried to summarize the main points in the development of encapsulated food compounds and to emphasize some limitations of encapsulation technology. The book contains chapters that review the newest encapsulation methods and carrier materials for the encapsulation of food compounds. Also, we describe general methods for the characterization of encapsulates as well as analytical methods for the analysis of the controlled release of active compounds. Further, the chapters related to specific topics such as the implementation of encapsulation technology in the production of beverages, dairy and meat products as well as the encapsulation of food supplements and plant extracts offer valuable results that may be of interest for academia but also for the broader community. Due to the significance of nanotechnology, we dedicated two chapters to this important topic.

The editors would like to thank and acknowledge all authors for their participating in the creation of this book in these challenging times. It was our great pleasure to be a part of the team, and we look forward to new joint projects in the future.

Steva Lević
Viktor Nedović
Branko Bugarski

Contents

Introduction to Encapsulation Processes

Steva Lević[1]*, Branko Bugarski[2] and Viktor Nedović[1]

[1] University of Belgrade, Faculty of Agriculture, Department of Food Technology and Biochemistry, Nemanjina 6, 11 080 Belgrade-Zemun, Serbia
[2] University of Belgrade, Faculty of Technology and Metallurgy, Department of Chemical Engineering, Karnegijeva 4, 11 120 Belgrade, Serbia

1. Introduction

In industry, particularly the food industry, encapsulation has been generally accepted as a beneficial technology that provides numerous advantages over conventional processes. Encapsulation is now considered as a solution to problems, such as degradation of active compounds, controlled delivery and protection of biocatalysts (Zuidam and Shimoni, 2010; Nedovic *et al.*, 2011; Lević *et al.*, 2016).

Encapsulation processes cover a large number of various active chemical compounds and biological systems, directly or indirectly used in the food industry. Encapsulation has been used for protection of various active food compounds:

- Food additives and ingredients (vitamins, minerals, flavor compounds, colorants, etc.);
- Microorganisms (live cells that produce various food products, such as beer, wine, fermented milk, etc.);
- Enzymes (used in specific processes that require enzyme encapsulation as a method for sustainable production);
- Other food-related active compounds (antimicrobial or antioxidant components of active packaging, etc).

Protection of an active compound is generally based on creating active protective layer(s) around it, making a physical and/or chemical barrier towards the environment. Usually such a system is defined as an *encapsulate* (Fig. 1). Encapsulates can be made into various shapes and sizes, depending on the applied encapsulation technique and properties of the active compound and carrier materials. Encapsulation systems are

*Corresponding author: slevic@agrif.bg.ac.rs

generally designed in accordance with specific applications and preferable properties of the final food product.

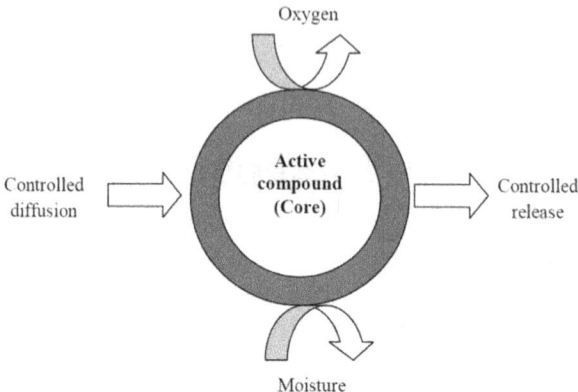

Fig. 1: General scheme of an encapsulate

For production of encapsulates with preferable properties, numerous techniques and industrial processes have been developed. These techniques are based on one or more phenomena that will shape the encapsulate into an adequate form and size (Nedović *et al.*, 2001; Nedović *et al.*, 2002; Thies, 2005; Prüsse *et al.*, 2008; Zuidam and Shimoni, 2010). The size of encapsulates is a particularly important property regarding further usage in food products and food processes. According to Zuidam and Shimoni (2010) the size of encapsulates is mainly controlled by the applied encapsulation techniques. For example, spray-drying as the most frequently used industrial encapsulation technique provides encapsulates in the range 10-400μm; various extrusion techniques can be used for production of capsules with a size up to 5000μm or more.

The size of encapsulates has recently become even more important due to concerns regarding usage of nanoparticles in food industry and the necessary standards and safety procedures.

Carrier materials are of enormous importance in the development of adequate encapsulates as they primarily define the further fate of the encapsulated active compound. In this regard, in encapsulation processes, various materials have been adopted with more or less success. The two main groups of carrier materials are those from natural sources and synthetic materials. Further, the carrier materials can be divided into hydrophilic and hydrophobic groups. This is important since generally hydrophobic active compounds are encapsulated (coated) with a hydrophilic carrier and vice versa. The main groups of carrier materials used in food applications are carbohydrates (sugars, starch, cellulose, gums, dextrins), proteins (soy proteins, corn proteins, gelatin, casein, whey protein isolates), lipids (hard fat, fatty acids, waxes, phospholipids), synthetic polymers and modified natural carrier materials (Sobel *et al.*, 2014).

The selection of appropriate carrier materials in the encapsulation processes remains the crucial point of every encapsulation since there is no universal carrier material that can fulfill all requirements. Also, many carrier materials exhibit some level of biological activity; hence there is an interest in using such materials in food-related encapsulations.

2. The Goals of Encapsulation in Food Technology

The main goals of encapsulation in food processes differ between various applications and types of active compounds, but generally, there are several main goals of encapsulation:

- Protection of an active compound against unfavorable environmental conditions;
- Controlled delivery of an active compound;
- Improved handling of an active compound;
- Masking of an unpleasant odor and taste associated with an active compound.

The list is a simplified presentation of encapsulation possibilities that may also include reuse of encapsulated/immobilized biocatalysts and creation of certain visual and textural effects upon consumption (Zuidam and Shimoni, 2010).

Benefits of introducing encapsulation into production processes are potentially significant. The main role of encapsulation in food industry remains protection of an active compound against the negative influence of environmental factors. Among these factors, high temperatures and oxygen-related degradation processes are main reasons for the loss of an active compound. For example, improving thermal stability of aromas is critical in many food thermal processes. Hence, protection of thermal-sensitive compounds by encapsulation is sometimes the only solution for their application in food (Kayaci and Uyar, 2012). Besides protection, encapsulation simultaneously could be used for production of encapsulates that transform active compounds into easier-to-use forms. Good examples are liquid aroma compounds for which easy handling and thermal protection could be achieved in a single encapsulation process (Lević *et al.*, 2015).

Controlled delivery (or controlled release) of active compound is one of the main benefits of encapsulation and one of the main reasons for the great progress of encapsulation in recent decades. The most important step in the development of a controlled delivery system is the recognition of place of delivery (i.e. place of release) of the active compound and consequently selection of an adequate carrier material and encapsulation technique. According to Huynh and Lee (2014) the mechanisms of active compound release depend on physical and chemical properties of carrier materials, but can also be considered a stimuli-controlled process. Therefore, the projection of active compound controlled release is critical for successful application of encapsulate in real food products. For this purpose, *in vitro* tests have been developed in order to simulate specific conditions of human gastrointestinal system (Minekus *et al.*, 2014). In the food industry, controlled release encapsulation systems are usually implemented in food products as carrier systems for probiotics (Burgain *et al.*, 2011; Dimitrellou *et al.*, 2016; Dimitrellou *et al.*, 2019). Other applications include thermal protection (and release) and facilitating handling of additives, such as flavors (Milanovic *et al.*, 2010; Kayaci and Uyar, 2012; Lević *et al.*, 2015), edible oils (Beindorff and Zuidam, 2010; Stajić *et al.*, 2014; Stajić *et al.*, 2018), micronutrients (Zimmermann and Windhab, 2010), etc.

On the other hand, the goals of biocatalyst encapsulation are mainly related to the protection against stress conditions, such as extreme temperature and pH and inhibitory effects of substrate or products. Also, encapsulation of biocatalysts is performed in order to ensure reuse of expensive compounds (i.e. biocatalysts), such as enzymes, or in continuous bioreactor processes to prevent or minimize loss of biocatalysts as a result of intense substrate flow (Lević *et al.*, 2016).

Introduction of encapsulation into the food production process is a complex task and requires knowledge of technological details of encapsulation, management of processes as well as precise economic analysis. For example, sometimes encapsulation is the only way to overcome technical problems, such as inhibition of a biocatalyst during a biotechnological process (Lalou *et al.*, 2013), or to contain active compound that in the free form may be lost during food processing (Stajić *et al.*, 2014; Stajić *et al.*, 2018). However, encapsulation is sometimes a complex process and a broad analysis is needed prior to encapsulation of active compound and introduction into a food product. Moreover, the fate of encapsulated active compound must be also projected as well as its storage stability, storage conditions and package options. For these reasons, establishing of food encapsulation processes on an industrial level is a multidisciplinary task as well as a significant financial effort.

3. Historical Background of Encapsulation

Historically, encapsulation processes are relatively well established in industry and science. The successful applications of encapsulation processes could be found in various industrial sectors and usually depend on current needs of the market.

There is a general consensus among the authors that the first encapsulation technologies started with first patent applications and reported research in the 19th century (Sobel *et al.*, 2014; Tucker *et al.*, 2012). This is not a surprise considering the fact that at the same time many countries entered the Industrial Revolution, a period of history with never-before-seen level of inventions and ideas.

The practices that involve some form of protection of active ingredients or biocatalysts are recorded through the history, emphasizing the needs and problems of a particular epoch. A good example for this is the solution for production of copy paper that comes from the 1930s. The basis for this invention was encapsulated color that was released as result of pressure during writing (Fanger, 1974). The needs sometimes dictate the unusual solutions and sometimes could be considered as true masterpieces, and encapsulation technologies are not exempt from this. A good example is the development of coacervates that were first used in the paper industry (Sobel *et al.*, 2014), but later were frequently studied as tools for encapsulation of food active compounds (Zuidam and Shimoni, 2010). Some important points in the development of encapsulation processes are presented in Table 1.

Development of some encapsulation processes is really an interesting story. For example, production of materials consisting of nanofibers has a long history. The first results of using electrostatic force for formation of nanomaterials (i.e. by using an electrospinning configuration) were published (in the form of a patent) in 1900. Further development led to establishment of the first factories for production of electrospun materials for various applications (Tucker *et al.*, 2012). This inevitably led to adoption of electrospun fibers in food-related studies; namely, electrospun fibers containing natural compounds could be a basis for production of new types of packaging materials, optimized for reduction of chemical food preservatives in food products (Vafania *et al.*, 2019). This approach in encapsulation of active ingredients is

Table 1: The Historical Development of Encapsulation Processes

Year	Encapsulation Process	Application	Comments	References
1872	The first spray drying patent.	Milk powder production.	Proposed technique was also evaluated for drying of dextrin, starches and gelatin.	Sobel *et al.*, 2014
1900	The first electrospinning patent.	Model system for electrospinning (nitrocellulose).	Coaxial needle system as support for electrospinning process.	Tucker *et al.*, 2012
1938	Electrospun fibers as basis for gas masks' filters.	Basis for cellulose acetate filters in gas masks.	Known as 'Petryanov filters', they are basis for filters in the nuclear industry.	Tucker *et al.*, 2012
1957	Development of the technique for oil encapsulation that later will become known as coacervation.	Basis for first generation of carbonless paper.	The new technique of encapsulation will become an impulse for development of new encapsulation techniques.	Sobel *et al.*, 2014; Zuidam and Shimoni, 2010
1958	Electrostatic droplet extrusion.	Preparation of emulsions.	One of the first applications of electrostatic force for generation of uniform droplets.	Kim *et al.*, 2019
1960s	Established the basis for liposome encapsulation.	Basic studies on chemistry of new structures that will lead to further development and applications of liposome technology.	The studies were oriented toward analysis of surface and functional properties of liposome.	Sobel *et al.* (2014)
1990	Development of microfluidic devices.	The new process for generation of cells encapsulates.	The development of microfluidic devices provided basis for cells encapsulation and study in many area of biotechnology and medicine.	Kim *et al.*, 2019
1980s and 1990s	Development of numerous encapsulates based on cyclodextrins as carrier materials.	These processes are commonly known as inclusion (or molecular inclusion).	These processes are important for food industry due to the fact that cyclodextrins may provide efficient encapsulation and controlled release of aromas.	Crini (2014)
1980s and 1990s	Many patent applications regarding protection of food compounds.	The focus of these patents was mostly on protection of volatile compounds.		Sobel *et al.*, 2014

promising since various carrier materials are suitable for processing into the form of fine fibers and films (Salević *et al.*, 2019; Dehcheshmeh and Fathi, 2019).

Besides encapsulation techniques, at the same time numerous carrier materials have been developed too, providing possibilities for development of new encapsulation techniques. A good example for this is development of cyclodextrins. The first descriptions of cyclodextrins in literature could be found in 1891, however, the development of encapsulates using these molecules with exceptional properties as carriers did not start until their structure was closely studied. So, as development and availability of analytical methods progressed, more data about cyclodextrins were collected, providing the basis for publishing numerous papers and patents targeting molecular inclusion of various compounds into cyclodextrins structure. Today, cyclodextrins are frequently studied as carriers for encapsulation of active compounds in various industrial applications (Crini, 2014).

4. Encapsulation: State-of-the-Art

Encapsulation is gaining attention in many technological processes and research. Currently, encapsulation systems are mainly applied in pharmacy (medicine) and the food sector. The relevance of encapsulation in the food sector could be measured by using available scientific databases. According to Web of Science and Scopus (accessed on 11 November, 2020), and using 'encapsulation, food' as search criteria, the number of published papers is above 4,000 (Fig. 2a).

Moreover, the success of encapsulation could be also measured by the sum of citations by year (Fig. 2b). Regarding global position of encapsulation in the food sector and science, the main national contributors are China, USA, Brazil, Iran, India, Spain, etc. (Scopus database, accessed on 11 November, 2020; keywords 'encapsulation, food'). As can be seen, this field is dominated by the large world economies and countries with a huge number of research centers. However, due to increased knowledge transfers and more available scientific data, more countries and research facilities have been included in encapsulation of food components and their evaluation.

At the industrial level, spray drying remains the most frequently used method for production of food encapsulates. The main advantage of spray-drying is rapid solvent evaporation and encapsulate formation in a continuous operation (Zuidam

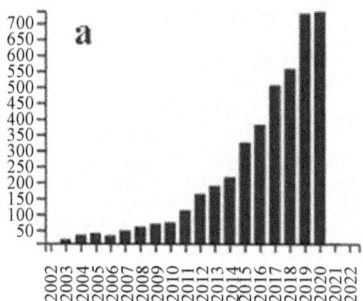 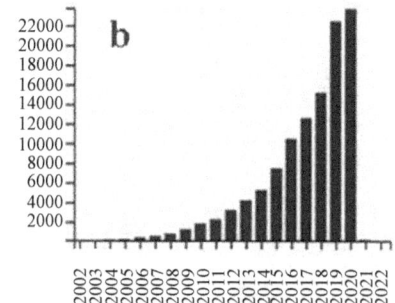

Fig. 2: Total publications by year (a) and sum of times cited by year (b) since 2002 obtained from online database Web of Science using 'encapsulation, food' for search (accessed on 11 November, 2020)

solvent evaporation and encapsulate formation in a continuous operation (Zuidam and Shimoni, 2010). However, to produce encapsulates with satisfactory properties, relatively high concentrations of carrier materials are required (Kalušević *et al.*, 2017b). Further improvements of spray-drying technology could be in the direction of new dispersion systems and solvent evaporation under an inert atmosphere. Liquid dispersion into small droplets (of few microns) using modified spray dryers has been recently adopted as a suitable method for production of nanoparticles and nanoencapsulates. This method is suitable for encapsulation of plant extracts (Del Gaudio *et al.*, 2017, Kyriakoudi and Tsimidou, 2018) or food ingredients, such as salts (Moncada *et al.*, 2015). Besides improved bioavailability, nano spray-drying provides particles that could be more easily dissolved and hence leads to increase in taste perception. Consequently, this could lead to a decrease in concentration of some food ingredients which are generally recognized as causes of some modern human diseases (Moncada *et al.*, 2015).

Spray-drying in an inert atmosphere is necessary when flammable solvents are used for extraction of active compounds. To proceed further with conventional spray drying (i.e. using hot air as drying medium), flammable solvents must be removed by evaporation. However, this step could be avoided by using an inert gas as the drying medium (e.g. nitrogen), which lowers the possibility for explosion during spray-drying. Also, this approach is suitable for application of conventional carrier materials used in spray-drying (Vázquez-León *et al.*, 2020). Inert gases used in spray-drying may cause changes in morphology and structural properties of obtained powders. Depending on the applied gas, a more amorphous powder can be obtained (Islam and Langrish, 2010). Besides solvent regeneration and improved safety, usage of closed-loop spray-drying has become a promising solution in ongoing energy regulations and need for reduction of pollution and establishing of more sustainable drying processes (Moejes *et al.*, 2018).

Other encapsulation techniques may also cause changes in the crystal structure of active compounds, consequently leading to modification of some important properties, such as melting point. This could be explained by the influence of carrier materials and specific conditions during encapsulate formation (Lević *et al.*, 2014).

Emulsions are important systems for the food sector and many food products depend on emulsion stability. Since stability of emulsions is affected by numerous factors, production of successful emulsion-based products is challenging (Dickinson, 2010). Emulsions are very suitable delivery systems for controlled release of food flavors (Mao *et al.*, 2017), some nutritionally valuable compounds (Matos *et al.*, 2018) and probiotics (Su *et al.*, 2021). Although high-energy processes are still the main methods for emulsions preparation, procedures that involve low-energy consumption have become more popular, especially when sensitive active compounds are included in formulations. Santana *et al.* (2013) reviewed these methods in more detail, pointing out the importance of surfactant properties for formation of stable emulsion using low-energy emulsification processes. Also, membrane emulsification is another approach that could be an alternative for conventional emulsions and particles preparation. The emulsion is formed by passing the dispersed phase into the continuous phase, using specially designed membranes and operating under low pressure. These processes are especially suitable for production of encapsulated food active compounds in the form of liposomes (Charcosset, 2009).

Besides food additives, the second important area for application of encapsulation in food production is protection of biocatalysts. The goal of biocatalysts encapsulation/ immobilization is usually oriented toward reuse of enzyme/cells in consecutive production cycles, with additional protection against environmental stress (i.e. high concentration of nutritional medium or inhibitory effects of products). In the case of expensive biocatalysts, encapsulation/immobilization is a solution for easy manipulation, especially in the processes that use enzymes and where, due to their small sizes, it is difficult to separate them and reuse in the next cycle. A good example of a highly successful application of immobilized biocatalysts is the production of high fructose syrup using immobilized enzymes (Lević *et al.*, 2016).

Conventional encapsulation of biocatalysts is usually based on application of adsorption methods, where cells or enzymes are linked to carrier's surface via specific physical/chemical interactions or biocatalysts are entrapped into the structure (i.e. pore) of carrier materials (Verbelen *et al.*, 2010; Costa *et al.*, 2004). The efficiency of adsorption depends on many factors, such as the type of biocatalyst (i.e. enzyme or cells), structure and chemical property of materials, porosity, etc. (Costa *et al.*, 2004; Reddy *et al.*, 2011). Other suitable methods of biocatalyst encapsulation usually include the application of some porous matrix that covers the biocatalyst and provides satisfactory diffusion properties regarding nutrients and reaction products. For this purpose, numerous porous carriers based on natural or synthetic materials have been developed. Their shape and size could be regulated in order to obtain carriers with optimal performance. Also, such carriers must be optimized for application under specific bioreactor conditions (Verbelen *et al.*, 2010; Lević *et al.*, 2016).

The main problem regarding introduction of immobilized biocatalysts in production of food products is the potential negative influence of the encapsulation procedure, particularly the influence of carrier materials on sensorial properties of the final products. To avoid this, research is focused on inert carrier materials, such as natural polysaccharides (e.g. alginate, chitosan and pectin), proteins (e.g. gelatin, collagen) or synthetic polymers, such as polyvinylalcohol (PVA) (Verbelen *et al.*, 2010). Inorganic carrier materials are also suitable for this purpose, especially due to their high chemical, thermal and mechanical stability. For example, special glasses (Marchis *et al.*, 2012) or specially synthesized zeolites (Kumari *et al.*, 2015) may be used for enzyme immobilization. In order to reduce the application of synthetic carrier materials, some studies have been focused on usage of food processed byproducts or low processed natural carriers for encapsulation/immobilization of enzymes and cells. Some examples include use of sugarcane pieces (Reddy *et al.*, 2011), watermelon pieces (Reddy *et al.*, 2008) or pear pieces (Mallios *et al.*, 2004) for yeast cells immobilization. Other renewable resources of carrier materials could be obtained as results of microbial metabolic processes. For example, bacterial cellulose could be used as solid support for yeast cells immobilization (Ton and Le, 2011). Also, active cells, such as yeast can be co-immobilized using filamentous fungi, providing promising support for wine fermentation (Puig-Pujol *et al.*, 2013).

Another interesting concept in cell encapsulation is the creation of protective organic or inorganic layers around individual cells. This concept is described in the literature as 'cyborg cells' (Fakhrullin *et al.*, 2012). This approach offers numerous possibilities, especially for cell protection in those processes that require application of stressful environmental conditions and where conventional encapsulation procedures and materials cannot provide an adequate barrier.

The biotechnological processes that involve use of live cells as biocatalysts are influenced by numerous factors that may make these processes economically unsustainable; namely, cells usually require complex substrates for maintaining cellular functions. During the process, substrate is partially used by the cells for production of products and one significant amount is utilized for the cell's metabolism. Also, cells may undergo the change in its genetic material (i.e. DNA), which further may result in decrease of productivity. The potential solution for this is usage of enzymes. However, complex cellular processes require numerous enzymes and their balanced activity. To overcome these problems and at the same time to exclude the cells from the process, enzymes involved in production of targeted compound are joined by encapsulation in the form of an enzyme cascade. Patterson *et al.* (2014) described one such system where several enzymes were encapsulated into small particles, mimicking cellular conditions. According to Chen *et al.* (2018), an enzyme cascade could be constructed into the form of nano-bioreactors, consisting of several necessary enzymes involved in the desired transformation of substrate. Besides, such an approach may lead to better protection of enzymes and even partial suppression of the formation of undesirable products.

Although the current application of nano-systems in the food sector is limited, further development and potential use are to be expected. Currently, these systems are mainly oriented towards encapsulation/immobilization of enzymes. Inorganic materials based on silica particles, combined with adequate polymers may result in stable enzymatic carriers. Also, materials based on carbon nanotubes are promising enzyme carriers in development of biosensors. Another major group of nano-systems with huge potential in enzyme encapsulation are magnetic nano-particles. The major advantage of magnetic carriers is their easy recovery and control inside bioreactors using a magnetic field. This opens numerous possibilities for process control as well as for a new design of bioreactors (Lević *et al.*, 2016). Also, magnetized cells could be applied with the same goals as enzymes. Cell magnetization could be performed by specially designed magnetic nanoparticles that are positive charged and easily applied on to negative charged cells. Such cells showed satisfactory results when applied in fermentation procedures (Dušak *et al.*, 2016).

5. Health and Environmental Impact of Encapsulation Processes

As pointed out above, the protection of active compounds against degradation or reduction of their nutritional value is one of the main reasons for the introduction of encapsulation in the food sector. Also, by using encapsulation, problems related to dosage and standardization of health beneficial compounds can be solved. According to available literature data, the main application of encapsulation in the food sector is protection of probiotic cells and their controlled delivery. Besides probiotics, one of the most frequently studied group of encapsulated active compounds are plant-based antioxidants (Belščak-Cvitanović *et al.*, 2011; Belščak-Cvitanović *et al.*, 2015; Kalušević *et al.*, 2017a,b; Estakhr *et al.*, 2020). This is not a surprise since plants are a traditional source of nutritionally valuable chemicals, with a generally well-known composition and defined safety risks. However, some plant products, such as some extracts or essential oils have strong (usually negative) sensorial effects on consumers. To overcome these problems, encapsulation could be used for masking the unpleasant taste and to control the sensorial profile of products. Plant essential oils are well known for their medicinal effects, but due to the intense fragrance they require masking and controlled delivery in order to be incorporated into foods. Although essential oils are complex mixtures of various volatile compounds, their encapsulation is relatively simple and could be based on food-grade carriers suitable for incorporation into food products (Yilmaztekin *et al.*, 2019).

Encapsulation could be also important as a tool for prevention of undesirable chemical reactions and formation of harmful compounds. For example, reducing the formation of Maillard reaction products in thermally-processed foods could be achieved by encapsulation of sodium chloride prior to product baking. The prevention of Na^+ cations contact with precursors of Maillard reaction products was found to be critical in reduction of 5-hydroxymethylfurfural and acrylamide content in baked products. Moreover, encapsulation offers possibilities for use of various lipid-based carriers for this purpose (Fiore *et al.*, 2012).

The food industry produces a significant amount of waste that needs to be processed. This is especially important in the case of wastewater generated during food production that could contain a significant concentration of organic materials. The other problem is that the total amount of wastewater generated during food processing can be significant and could be a large financial burden. Hence, intensification of water treatment processes using encapsulated biocatalysts has attracted great attention in the recent decades. Some specific applications of immobilized cells and enzyme technology for wastewater treatments could be removal of cyanide (Chen *et al.*, 2007), endocrine-disrupting chemicals (Maryšková *et al.*, 2020), nitrogen (Dolejš *et al.*, 2019) and pesticides (Lin *et al.*, 2020).

Besides the numerous advantages of encapsulation systems applied in water and waste treatment, we should mention some potential problems related to encapsulation and their potential influence on ecosystems.

There are two major issues that need to be addressed regarding encapsulation processes and ecology. First, the applied carrier materials and active ingredients that form capsules should be from renewable resources. This requirement arises from recent strategies adopted by many countries with the goal of reducing emissions of harmful compounds and making production more energy efficient.

The second issue is closely related to concerns that many materials (mainly synthetic) formed in various industrial processes remain in ecosystems for prolonged periods of time. Browne *et al.* (2011) analyzed pollution by microplastic and concluded that plastic materials tend to accumulate in marine ecosystems and could potentially cause health problems for aquatic organisms. This also raises the question of encapsulates' fate if they are released in the environment without previous knowledge of potential threats to environment.

The food industry usually uses widely approved components that are recognized as safe for nutritional applications. However, in the last several decades, numerous additives and food-related materials have been banned or their usage was significantly restricted. Introduction of new forms of additives and active compounds, i.e. in the form of encapsulates, could become a challenge for existing waste-treatment facilities and processes. Also, biodegradation of encapsulated materials must be thoroughly investigated in order to prevent future problems, such as those with generated microplastics.

6. Economy of Encapsulation Processes

As pointed out above, encapsulation provides numerous advantages and opens new possibilities for development of new food products. However, according to Zuidam and Shimoni (2010) the main obstacles that must be taken into consideration during planning of food encapsulation processes are investment costs; specially introduction of encapsulation into production could be a costly operation which requires a new encapsulation unit (i.e. production line) or use of commercially available encapsulated active compounds. The first approach is more complex and during calculation of investment costs, the following economical prices must be taken into consideration (Dimitrellou *et al.*, 2019):

- Variable costs (cost of carrier materials and active compound, energy costs, etc.);
- Fixed costs (these costs may include depreciation, interest, etc.).

These are the basic inputs for calculating the economical sustainability of the encapsulation process. Further, the structure of costs depends on the applied encapsulation process and all preparation and finishing steps after the main operation (i.e. encapsulation). Figure 3 summarizes the steps in the production of encapsulates that contain living cells. This scheme is, in our opinion, the most complex in the production of encapsulate and includes many steps that critically influence the properties of encapsulated cells as well as economical sustainability of encapsulation.

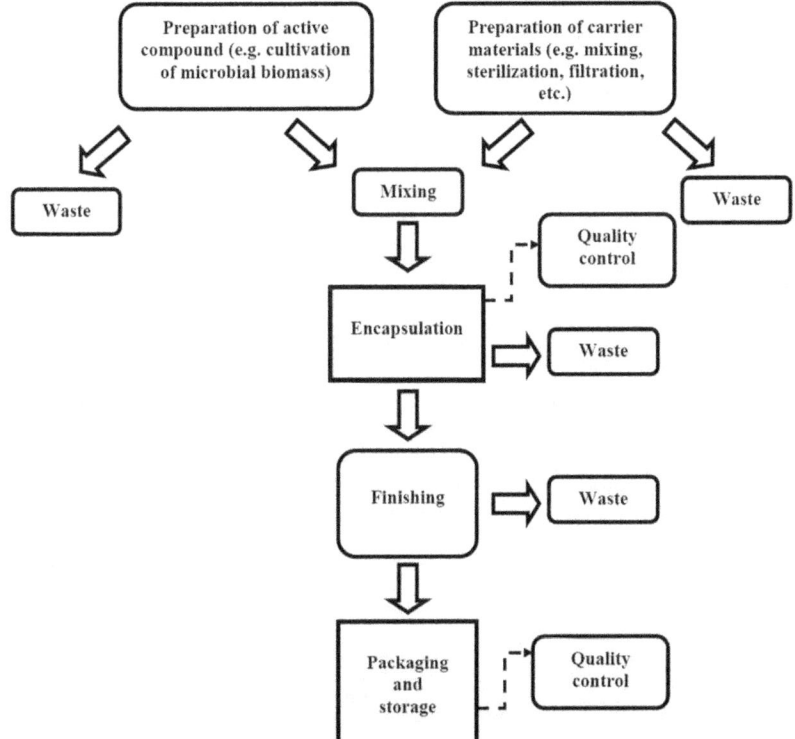

Fig. 3: Schematic representation of main phases of encapsulation process
for production of cells encapsulates

Further, the scheme doesn't include steps like water preparation, storage of materials, manipulative operations, and sanitation. A significant problem is the management of waste created in almost all the main production steps. The structure of waste depends on the encapsulation procedure, but generally may include wastewater, dust, solid particles, biomass, etc. Depending on the structure and local environmental laws, management of the waste could be a costly operation that must be considered during calculation of the encapsulation investment.

Dimitrellou *et al.* (2019) analyzed the economic sustainability of introducing an encapsulation process into the production of fermented milk. The analysis, from the technical point of view, was based on usage of probiotic cells encapsulated into Ca-alginate capsules. Also, the analysis included the projections of fermented milk production capacity and the prices in the regions of interest. According to the same authors, the economic analysis showed that over a period of eight years, and by managing dairy capacities between 70-100 per cent, it is possible to maintain the production of encapsulates and consequently fermented milk, using encapsulated probiotic cells.

Various factors may influence the price of encapsulation process. Some of these factors are costs of drying operations, particles size, type of carrier materials, types of industrial process (i.e. pharmaceutical industry or other industrial branches), etc. On these topics, we recommend the chapter published by Veršič (2014), where many useful data regarding encapsulation process costs are available.

Other limitations regarding economical sustainability of food encapsulation processes may be expressed by the complex influence of climate and political factors, as well as local regulation. The first two factors are closely connected and could have devastating effects on product supply. These factors may influence supply of active ingredients as well as carrier materials. One good example for the complex influence of climate and political factors on encapsulation is availability of gum arabic (gum acacia). According to Wandrey *et al.* (2010) gum arabic is an excellent carrier material that has numerous advantages, such as preferable viscosity (even at high concentrations), good protection of active ingredients against oxidation and negative influence of moisture from the air. For example, gum arabic was found to be an excellent carrier material for encapsulation of plant active compounds via spray-drying (Kalušević *et al.*, 2017a,b). However, the main production region for gum arabica, i.e. for the acacia plant from which this material is obtained is located in Africa, more precisely in the ecoclimatic region of Sahel (Wandrey *et al.*, 2010). The acacia tree is generally more resistant toward arid climate compared to other commercially important tree species of Sahel region (Gonzalez *et al.*, 2012). However, overall climate changes in the region may be sources of further conflicts that could disrupt supply of gum arabica (Arcanjo, 2019). Also, disruption of supply of natural active compounds may consequently lead to a blockade of the whole production process of encapsulation. Hence, alternatives should be closely analyzed and included in the production plans. Some potential strategies regarding potential substitution of active compounds and carrier materials are shown in Table 2.

Another potential alternative to conventional sources of active compounds and carrier materials could be plant cells cultivated in bioreactors. By using this approach many problems related to collected or cultivated plants, such as destruction of environment, pollution, influence of climate and climate changes and political insecurity could be potentially solved. Currently, there are reports of plant cells cultivation for

Table 2: Some Examples of Potential Substitutions of Active Compounds and Carrier Material in the Case of Disruption of Supply Chain

	Used material	Alternative	Comments	References
Carrier materials	Gum arabica	Maltodextrin, skimmed milk	All three materials showed good properties in process of spray drying of plant extracts.	Kalušević *et al.*, 2017a,b
	Alginate (or polyvinyl alcohol)	Combined carrier materials for cell immobilization.	Both the materials mix well and can protect cells. The materials are templates and their share can be changed.	Bezbradica *et al.*, 2004; Radosavljević *et al.*, 2020
	Natural waxes	Various fatty acids	Results depend on potential applications. All materials have good barrier properties but depend on thermal properties of carrier materials.	Fiore *et al.*, 2012
	Native starch	Modified starches	Based on their properties, modified starches could improve properties of encapsulates.	Zhu, 2017
Active compounds	Carotenoids	Alternatives include various plants sources such as red pepper, carrot, some vegetables processing waste.	Regarding their chemical properties, plant carotenoids from various sources could be encapsulated using similar processes and carrier materials.	Vulić *et al.* (2019) ; Šeregelj *et al.*, 2018
	Resveratrol	Acetylresveratrol	Acetylresveratrol is more stable and bioactive compared to resveratrol.	Su *et al.*, 2020
	Vanillin	Ethyl vanillin	Natural vanillin could be replaced even in the encapsulated forms with synthetic ethyl vanillin.	Lević *et al.*, 2014
	Edible oils	Other, more appropriate and more available vegetable oils, but with similar bioactive properties.	Some oils are more acceptable regarding sensorial properties or they are more available than others.	Stajić *et al.*, 2014, 2018

production of proteins (Dörnenburg, 2010), plants secondary metabolites (Marchev and Georgiev, 2020) and microalgae biomass and active compounds (Xu *et al.*, 2009).

Encapsulation in biotechnology offers new possibilities for investors and academia to develop new useful products through cooperation and knowledge transfer. However, encapsulation should first overcome one barrier, namely, according to Veršič (2014), the perception of encapsulation is still mainly focused on it as a service. However, in recent years, there is a trend for new encapsulated products which may change perception on encapsulation and open new possibilities for industry and research.

7. Conclusion and Future Perspectives

In the past several decades, encapsulation has shown its potential to solve many problems related to the preservation of food active compounds, controlled delivery, food packaging, etc. It can be expected that encapsulation will remain a powerful tool in many aspects of food industry, providing a basis for more efficient and innovative processes and products. Also, encapsulation could find place in the newly established concepts and strategies, such as personalized nutrition and the circular economy.

Personalized nutrition is a concept where real nutritional needs are managed in accordance with each person. Under this concept, each individual is analyzed regarding his/her own nutritional needs, especially considering personal health condition, genetics, lifestyle, etc. (Betts and Gonzalez, 2016; McClements, 2020). Encapsulation could be used in personalized nutrition where specific controlled delivery is required for both pure or a mixture of active compounds.

The other important aspect of modern society strongly connected to the food industry is a circular economy. A circular economy is a concept of sustainable use of natural resources via recycling/reuse of available resources and reduction in applying new (natural) resources, reduction or elimination of waste and pollution. One of the main concerns regarding this concept is plastic pollution, particularly plastics that originate from food packaging (Black *et al.*, 2019). Encapsulation has been recognized as a suitable solution in production of active packaging systems, where active ingredients are encapsulated and incorporated into packaging material (preferably natural or biodegradable), and provide a new line in defense of food products.

We hope that readers of this book will find the contents interesting and informative. The authors of the book's chapters cover all major aspects of encapsulation processes in food industry with a critical analysis of available literature data. Editors are grateful to authors for their time and energy invested in this book.

Acknowledgment

Ministry of Education, Science and Technological Development of the Republic of Serbia (Contract No. 451-03-9/2021-14/200116).

References

Arcanjo, M. (2019). Risk and Resilience: Climate Change and Instability in the Sahel, Climate Institute, New York, US; http://climate.org/risk-and-resilience-climate-change-and-instability-in-the-sahel/.

Beindorff, C.M. and Zuidam, N.J. (2010). Microencapsulation of fish oil, pp. 161-186. *In*: N.J. Zuidam and V. Nedovic (Eds.), *Encapsulation Technologies for Food Active Ingredients and Food Processing*, Springer, Dordrecht.

Belščak-Cvitanović, A., Stojanović, R., Manojlović, V., Komes, D., Juranović Cindrić, I., Nedović, V. and Bugarski, B. (2011). Encapsulation of polyphenolic antioxidants from medicinal plant extracts in alginate-chitosan system enhanced with ascorbic acid by electrostatic extrusion, *Food Research International*, 44: 1094-1101.

Belščak-Cvitanović, A., Lević, S., Kalušević, A., Špoljarić, I., Đorđević, V., Komes, D., Mršić, G. and Nedović, V. (2015). Efficiency assessment of natural biopolymers as encapsulants of green tea (*Camellia sinensis* L.) bioactive compounds by spray-drying, *Food Bioprocess Technology*, 8(12): 2444-2460.

Betts, J.A. and Gonzalez, J.T. (2016). Personalized nutrition: What makes you so special? *Nutrition Bulletin*, 41: 353-359.

Bezbradica, D., Matić, G., Nedović, V., Čukalović-Leskošek, I. and Bugarski, B. (2004). Immobilization of brewing yeast in PVA/alginate microbeads using electrostatic droplet generation, *Chemical Industry*, 58: 118-120.

Black, J.E., Kopke, K. and O'Mahony, C. (2019). Towards a circular economy: Using stakeholder subjectivity to identify priorities, consensus, and conflict in the Irish EPS/XPS market, *Sustainability*, 11: 6834.

Browne, M.A., Crump, P., Niven, S.J., Teuten, E., Tonkin, A., Galloway, T. and Thompson, R. (2011). Accumulation of microplastic on shorelines woldwide: Sources and sinks, *Environmental Science & Technology*, 45: 9175-9179.

Burgain, J., Gaiani, C., Linder, M. and Scher, J. (2011). Encapsulation of probiotic living cells: From laboratory scale to industrial applications, *Journal of Food Engineering*, 104: 467-483.

Charcosset, C. (2009). Preparation of emulsions and particles by membrane emulsification for the food processing industry, *Journal of Food Engineering*, 92: 241-249.

Chen, C.Y., Kao, C.M., Chen, S.C., Chien, H.Y. and Lin, C.E. (2007). Application of immobilized cells to the treatment of cyanide wastewater, *Water Science & Technology*, **56**(7): 99-107.

Chen, W.H., Vázquez-González, M., Zoabi, A., Abu-Reziq, R. and Itamar Willner, I. (2018). Biocatalytic cascades driven by enzymes encapsulated in metal-organic framework nanoparticles, *Nature Catalysis*, 1: 689-695.

Costa, S.A., Azevedo, H.S. and Reis, R.L. (2004). Enzyme immobilization in biodegradable polymers for biomedical applications, pp. 301-324. *In*: R.L. Reis and S.J. Román (Eds.), *Biodegradable Systems in Tissue Engineering and Regenerative Medicine*, CRC Press, Boca Raton, Florida.

Crini, G. (2014). Review: A history of cyclodextrins, *Chemical Reviews*, **114**(21): 10940-10975.

Dehcheshmeh, M.A. and Fathi, M. (2019). Production of core-shell nanofibers from zein and tragacanth for encapsulation of saffron extract, *International Journal of Biological Macromolecules*, 122: 272-279.

Del Gaudio, P., Sansone, F., Mencherini, T., De Cicco, F., Russo, P. and Aquino, R.P. (2017). Nanospray drying as a novel tool to improve technological properties of soy isoflavone extracts, *Planta Medica*, **83**(05): 426-433.

Dickinson, E. (2010). Food emulsions and foams: Stabilization by particles, *Current Opinion in Colloid & Interface Science*, 15: 40-49.

Dimitrellou, D., Kandylis, P., Lević, S., Petrović, T., Ivanović, S., Nedović, V. and Kourkoutas, Y. (2019). Encapsulation of *Lactobacillus casei* ATCC 393 in alginate capsules for probiotic fermented milk production, *LWT – Food Science and Technology*, 116: 108501.

Dimitrellou, D., Kandylis, P., Petrović, T., Dimitrijević-Branković, S., Lević, S., Nedović, V. and Kourkoutas, Y. (2016). Survival of spray dried microencapsulated *Lactobacillus casei* ATCC 393 in simulated gastrointestinal conditions and fermented milk, *LWT – Food Science and Technology*, 71: 169-174.

Dolejš, I., Stloukal, R., Rosenberg, M. and Rebroš, M. (2019). Nitrogen removal by co-immobilized anammox and ammonia-oxidizing bacteria in wastewater treatment, *Catalysts*, 9: 523.

Dörnenburg, H. (2010). Cyclotide synthesis and supply: From plant to bioprocess, *Biopolymers*, **94**(5): 602-610.

Dušak, P., Benčina, M., Turk, M., Bavčar, D., Košmerl, T., Berovič, M. and Makovec, D. (2016). Application of magneto-responsive *Oenococcus oeni* for the malolactic fermentation in wine, *Biochemical Engineering Journal*, 110: 134-142.

Estakhr, P., Tavakoli, J., Beigmohammadi, F., Alaei, S. and Mousavi Khaneghah, A. (2020). Incorporation of the nanoencapsulated polyphenolic extract of Ferula persica into soybean oil: Assessment of oil oxidative stability, *Food Science & Nutrition*, 00: 1-10.

Fakhrullin, R.F., Zamaleeva, A.I., Minullina, R.T., Konnova, S.A. and Paunov, V.N. (2012). Cyborg cells: Functionalization of living cells with polymers and nanomaterials, *Chemical Society Reviews*, 41: 4189-4206.

Fanger, G.O. (1974). Microencapsulation: A brief history and introduction, pp. 1-20. *In:* J.E. Vandegaer (Ed.), *Microencapsulation*, Springer, Boston, MA.

Fiore, A., Troise, A.D., Mogol, B.A., Roullier, V., Gourdon, A., Jian, S.A.M., Hamzalıoğlu, B.A., Gökmen, V. and Fogliano, V. (2012). Controlling the Maillard reaction by reactant encapsulation: Sodium chloride in cookies, *Journal of Agricultural and Food Chemistry*, 60: 10808-10814.

Gonzalez, P., Tucker, C.J. and Sy, H. (2012). Tree density and species decline in the African Sahel attributable to climate, *Journal of Arid Environments*, 78: 55-64.

Huynh C.T. and Lee D.S. (2014). Controlled release. *In:* S. Kobayashi and K. Müllen (Eds.), *Encyclopedia of Polymeric Nanomaterials*, Springer, Berlin, Heidelberg. https://doi.org/10.1007/978-3-642-36199-9_314-1

Islam, M.I.U. and Langrish, T.A.G. (2010). The effect of different atomizing gases and drying media on the crystallization behavior of spray-dried powders, *Drying Technology*, **28**(9): 1035-1043.

Kalušević, A.M., Lević, S.M., Čalija, B.R., Milić, J.R., Pavlović, V.B., Bugarski, B.M. and Nedovic, V.A. (2017a). Effects of different carrier materials on physicochemical properties of microencapsulated grape skin extract, *Journal of Food Science and Technology*, **54**(11): 3411-3420.

Kalušević, A., Lević, S., Čalija, B., Pantić, M., Belović, M., Pavlović, V., Bugarski, B., Milić, J., Žilić S. and Nedović, V. (2017b). Microencapsulation of anthocyanin-rich black soybean coat extract by spray drying using maltodextrin, gum arabic and skimmed milk powder, *Journal of Microencapsulation*, **34**(5): 475-487.

Kayaci, F. and Uyar, T. (2012). Encapsulation of vanillin/cyclodextrin inclusion complex in electrospun polyvinyl alcohol (PVA) nanowebs: Prolonged shelf-life and high temperature stability of vanillin, *Food Chemistry*, 133: 641-649.

Kim, H., Bae, C., Kook, Y.M., Koh, W.G., Lee, K. and Park, M.H. (2019). Mesenchymal stem cell 3D encapsulation technologies for biomimetic microenvironment in tissue regeneration, *Stem Cell Research & Therapy*, 10:51.

Kumari, A., Kaur, B., Srivastava, R. and Sangwan, S.R. (2015). Isolation and immobilization of alkaline protease on mesoporous silica and mesoporous ZSM-5 zeolite materials for improved catalytic properties, *Biochemistry and Biophysics Reports*, 2: 108-114.

Kyriakoudi, A. and Tsimidou, M.Z. (2018). Properties of encapsulated saffron extracts in maltodextrin using the Büchi B-90 nano spray-dryer, *Food Chemistry*, 266: 458-465.

Lalou, S., Mantzouridou, F., Paraskevopoulou, A., Bugarski, B., Levic, S. and Nedovic, V. (2013). Bioflavor production from orange peel hydrolysate using immobilized *Saccharomyces cerevisiae*, *Applied Microbiology and Biotechnolgy*, 97: 9397-9407.

Lević, S., Obradović, N., Pavlović, V., Isailović, B., Kostić, I., Mitrić, M., Bugarski, B. and Nedović, V. (2014). Thermal, morphological, and mechanical properties of ethyl vanillin immobilized in polyvinyl alcohol by electrospinning process, *Journal of Thermal Analysis and Calorimetry*, 118(2): 661-668.

Lević, S., Lijaković, I.P., Đorđević, V., Rac, V., Rakić, V., Knudsen, T.Š., Pavlović, V, Bugarski, B. and Nedović, V. (2015). Characterization of sodium alginate/D-limonene emulsions and respective calcium alginate/D-limonene beads produced by electrostatic extrusion, *Food Hydrocolloids*, 45: 111-123.

Lević, S., Đorđević, V., Knežević-Jugović, Z., Kalušević, A., Milašinović, N., Branko Bugarski, B. and Nedović, V. (2016). Encapsulation technology of enzymes and applications in food processing, pp. 469-502. *In*: R.C. Ray and C.M. Rosell (Eds.), *Microbial Enzyme Technology in Food Applications*, CRC Press, Taylor & Francis Group, Boca Raton, US.

Lin, S., Lu, Y., Ye, B., Zeng, C., Liu, G., Li, J., Luo, H. and Zhang, R. (2020). Pesticide wastewater treatment using the combination of the microbial electrolysis desalination and chemical-production cell and Fenton process, *Frontiers of Environmental Science & Engineering*, 14: 12.

Mallios, P., Kourkoutas, Y., Iconomopoulou, M., Koutinas, A.A., Psarianos, C., Marchant, R. and Banat, I.M. (2004). Low-temperature wine-making using yeast immobilized on pear pieces, *Journal of the Science of Food and Agriculture*, 84: 1615-1623.

Mao, L., Roos, Y.H., Biliaderis, C.G. and Miao, S. (2017). Food emulsions as delivery systems for flavor compounds: A review, *Critical Reviews in Food Science and Nutrition*, 57(15).

Marchev, A.S. and Georgiev, M.I. (2020). Plant in vitro systems as a sustainable source of active ingredients for cosmeceutical application, *Molecules*, 25(9): 2006.

Marchis, T., Cerrato, G., Magnacca, G., Crocella, V. and Laurenti, E. (2012). Immobilization of soybean peroxidase on aminopropyl glass beads: Structural and kinetic studies, *Biochemical Engineering Journal*, 67: 28-34.

Maryšková, M., Schaabová, M., Tománková, H., Novotný, V. and Rysová, M. (2020). Wastewater Treatment by Novel Polyamide/Polyethylenimine Nanofibers with Immobilized Laccase, *Water*, 12: 588.

Matos, M., Gutiérrez, G., Martínez-Rey, L., Iglesias, O. and Pazos, C. (2018). Encapsulation of resveratrol using food-grade concentrated double emulsions: Emulsion characterization and rheological behavior, *Journal of Food Engineering*, 226: 73-81.

McClements, D.J. (2020). Nano-enabled personalized nutrition: Developing multicomponent-bioactive colloidal delivery systems, *Advances in Colloid and Interface Science*, 282: 102211.

Milanovic, J., Manojlovic, V., Levic, S., Rajic, N., Nedovic, V. and Bugarski, B. (2010). Microencapsulation of flavors in carnauba wax, *Sensors*, 10(1): 901-912.

Minekus, M., Alminger, M., Alvito, P., Ballance, S., Bohn, T., Bourlieu, C., Carrière, F., Boutrou, R., Corredig, M., Dupont, D., Dufour, C., Egger, L., Golding, M., Karakaya, S., Kirkhus, B., Le Feunteun, S., Lesmes, U., Macierzanka, A., Mackie, A., Marze, S., McClements, D.J., Menard, O., Recio, I., Santos, C.N., Singh, R.P., Vegarud, G.E., Wickham, M.S.J., Weitschies, W. and Brodkorb, A. (2014). A standardised static in vitro digestion method suitable for food – An international consensus, *Food & Function*, 5: 1113-1124.

Moejes, S.N., Visser, Q., Bitter, J.H. and van Boxtel, A.J.B. (2018). Closed-loop spray drying solutions for energy efficient powder production, *Innovative Food Science & Emerging Technologies*, 47: 24-37.

Moncada, M., Astete, C., Sabliov, C., Olson, D., Boeneke, C. and Aryana, K.J. (2015). Nano spray-dried sodium chloride and its effects on the microbiological and sensory characteristics of surface-salted cheese crackers, *Journal of Dairy Science*, 98: 5946-5954.

Nedović, V.A., Obradović, B., Leskošek-Čukalović, I., Trifunović, O., Pešić, R. and Bugarski, B. (2001). Electrostatic generation of alginate microbeads loaded with brewing yeast, *Process Biochemistry*, 37: 17-22.

Nedović, V.A., Obradović, B., Poncelet, D., Goosen, M.F.A., Leskošek-Čukalović, I. and Bugarski, B. (2002). Cell immobilization by electrostatic droplet generation, *Landbauforschung Volkenrode* SH, 241: 11-17.

Nedovic, V., Kalusevic, A., Manojlovic, V., Levic, S. and Bugarski, B. (2011). An overview of encapsulation technologies for food applications, 11th International Congress on Engineering and Food (ICEF11), Athens, Greece, *Procedia Food Science*, 1: 1806-1815.

Patterson, D.P., Schwarz, B., Waters, R.S., Gedeon, T. and Douglas, T. (2014). Encapsulation of an enzyme cascade within the bacteriophage P22 virus-like particle, *ACS Chemical Biology*, 9(2): 359-365.

Prüsse, U., Bilancetti, L., Bučko, M., Bugarski, B., Bukowski, J., Gemeiner, P., Lewinska, D., Manojlovic, V., Massart, B., Nastruzzi, C., Nedovic, V., Poncelet, D., Siebenhaar, S., Tobler, L., Tosi, A., Vikartovská, A. and Vorlop, K.D. (2008). Comparison of different technologies for alginate beads production, *Chemical Papers*, 62: 364-374.

Puig-Pujol, A., Bertran, E., García-Martínez, T., Capdevila, F., Mínguez, S. and Mauricio, J.C. (2013). Application of a new organic yeast immobilization method for sparkling wine production, *American Journal of Enology and Viticulture*, 64: 386-394.

Radosavljević, M., Lević, S., Belović, M., Pejin, J., Djukić-Vuković, A., Mojović, Lj. and Nedović, V. (2020). Immobilization of *Lactobacillus rhamnosus* in polyvinyl alcohol/ calcium alginate matrix for production of lactic acid, *Bioprocess and Biosystems Engineering*, 43: 315-322.

Reddy, L., Reddy, Y., Reddy, L. and Reddy, O. (2008). Wine production by novel yeast biocatalyst prepared by immobilization on watermelon (*Citrullus vulgaris*) rind pieces and characterization of volatile compounds, *Process Biochemistry*, 43: 748-752.

Reddy, L.V., Reddy, L.P., Wee, Y.J. and Reddy, O.V.S. (2011). Production and characterization of wine with sugarcane piece immobilized yeast biocatalyst, *Food Bioprocess Technology*, 4: 142-148.

Salević, A., Prieto, C., Cabedo, L., Nedović, V. and Lagaron, J.M. (2019). Physicochemical, antioxidant and antimicrobial properties of electrospun poly(ε-caprolactone) films containing a solid dispersion of sage (*Salvia officinalis* L.) extract, *Nanomaterials*, 9: 270.

Santana, R.C., Perrechil, F.A. and Cunha, R.L. (2013). High- and low-energy emulsifications for food applications: A focus on process parameters, *Food Engineering Reviews*, 5: 107-122.

Šeregelj, V., Tumbas-Šaponjac, V., Mandić, A., Ćetković, G., Čanadanović-Brunet, J., Vulić, J. and Stajčić, S. (2018). Accelerated solvent extraction of bioactive compounds from carrot – Optimization of response surface methodology, *Journal of Serbian Chemical Society*, 83(11): 1223-1228.

Sobel, R., Versic, R. and Gaonkar, A.G. (2014). Introduction to microencapsulation and controlled delivery in foods, pp. 3-12. *In*: A. Gaonkar, N. Vasisht, A. Khare and R. Sobel (Eds.), *Microencapsulation in the Food Industry: A Practical Implementation Guide*, Academic Press (Elsevier Inc.), San Diego, USA.

Stajić, S., Stanišić, N., Lević, S., Tomović, V., Lilić, S., Vranić, D., Jokanović, M. and Živković, D. (2018). Physico-chemical characteristics and sensory quality of dry fermented sausages with flaxseed oil preparations, *Polish Journal of Food and Nutrition Sciences*, 68: 367-375.

Stajić, S., Živković, D., Tomović, V., Nedović, V., Perunović, M., Kovjanić, N., Lević, S. and Stanišić, N. (2014). The utilization of grapeseed oil in improving the quality of dry fermented sausages, *International Journal of Food Science & Technology*, **49**(11): 2356-2363.

Su, J., Cai, Y., Tai, K., Guo, Q., Zhu, S., Mao, L., Gao, Y., Yuan, F. and Van der Meeren, P. (2021). High-internal-phase emulsions (HIPEs) for co-encapsulation of probiotics and curcumin: Enhanced survivability and controlled release, *Food & Function* (in press).

Su, Y., Sun, C., Sun, X., Wu, R., Zhang, X. and Tu, Y. (2020). Acetylresveratrol as a potential substitute for resveratrol dragged the toxic aldehyde to inhibit the mutation of mitochondrial DNA, *Applied Biochemistry and Biotechnology*, 191: 1340-1352.

Thies, C. (2005): *Microencapsulation. Kirk-Othmer Encyclopedia of Chemical Technology*, 4th edition, 16: 317-327.

Ton, N.M.N. and Le, V.V.M. (2011). Application of immobilized yeast in bacterial cellulose to the repeated batch fermentation in wine-making, *International Food Research Journal*, 18: 983-987.

Tucker, N., Stanger, J.J., Staiger, M.P., Razzaq, H. and Hofman, K. (2012). The History of the Science and Technology of Electrospinning from 1600 to 1995, *Journal of Engineered Fibers and Fabrics*, special edition, 7: 63-73.

Vafania, B., Fathi, M. and Soleimanian-Zad, S. (2019). Nanoencapsulation of thyme essential oil inchitosan-gelatin nanofibers by nozzle-less electrospinning and their application to reduce nitrite in sausages, *Food and Bioproducts Processing*, 116: 240-248.

Vázquez-León, L.A., Olguín-Rojas, J.A., Páramo-Calderón, D.E., Gerardo, F. Barbero, Salgado-Cervantes, M.A., Miguel Palma, García-Alvarado, M.A. and Rodríguez-Jimenes, G.C. (2020). Closed-loop spray drying with N₂ of *Moringa oleifera* leaf ethanolic extracts: Effects on bioactive compounds and antiradical activity, *Drying Technology*, DOI: 10.1080/07373937.2020.1753764

Verbelen, P.J., Nedović, V.A., Manojlović, V., Delvaux, F.R., Laskošek-Čukalović, I., Bugarski, B. and Willaert, R. (2010). Bioprocess intensification of beer fermentation using immobilized cells, pp. 303-326. *In*: N.J. Zuidam and V.A. Nedovic (Eds.), *Encapsulation Technologies for Food Active Ingredients and Food Processing*, Springer, Dordrecht.

Veršič, J.R. (2014). The economics of microencapsulation in the food industry, pp. 409-420. *In*: A. Gaonkar, N. Vasisht, A. Khare and Sobel R. (Eds.), *Microencapsulation in the Food Industry: A Practical Implementation Guide*, Academic Press (Elsevier Inc.), San Diego, USA.

Vulić, J., Šeregelj, V., Kalušević, A., Lević, S., Nedović, V., Tumbas-Šaponjac, V., Čanadanović-Brunet, J. and Ćetković, G. (2019). Bioavailability and bioactivity of encapsulated phenolics and carotenoids isolated from red pepper waste, *Molecules*, 24: 2837.

Wandrey, C., Bartkowiak, A. and Harding, E.S. (2010). Materials for Encapsulation, pp.

31-100. *In*: N.J. Zuidam and V.A Nedovic (Eds.), *Encapsulation Technologies for Food Active Ingredients and Food Processing*, Springer, Dordrecht.

Xu, L., Weathers, P.J., Xiong, X.R. and Liu, C.Z. (2009). Microalgal bioreactors: Challenges and opportunities, *Engineering* in *Life Sciences,* **9**(3): 178-189.

Yilmaztekin, M., Lević, S., Kalušević, A., Cam, M., Bugarski, B., Rakić, V., Pavlović, V. and Nedović, V. (2019). Characterization of peppermint (*Mentha piperita* L.) essential oil encapsulates, *Journal of Microencapsulation*, **36**(2): 109-119.

Zhu, F. (2017). Encapsulation and delivery of food ingredients using starch based systems, *Food Chemistry,* 229: 542-552.

Zimmermann, M.B. and Windhab, E.J. (2010). Encapsulation of iron and other micronutrients for food fortification, pp. 187-210. *In*: N.J. Zuidam and V.A. Nedovic (Eds.), *Encapsulation Technologies for Food Active Ingredients and Food Processing*, Springer, Dordrecht.

Zuidam, N.J. and Shimoni, E. (2010). Overview of microencapsulates for use in food products or processes and methods to make them, pp. 3-29. *In*: N.J. Zuidam and V.A. Nedovic (Eds.), *Encapsulation Technologies for Food Active Ingredients and Food Processing*, Springer, Dordrecht.

Carrier Materials for Encapsulation

Ch.V.K. Sudheendra[1] and Ami Patel[2]*

[1] Department of Dairy Microbiology, College of Dairy Science,
  Kamdhenu University, Amreli – 365601, Gujarat, India
[2] Dairy Microbiology Dept., Mansinhbhai Institute of Dairy & Food Technology –
  MIDFT, Dudhsagar Dairy Campus, Mehsana – 384002, Gujarat, India

1. Introduction

Encapsulation may be defined as 'a technology for packaging small solid particles, liquid droplets, or gas molecules in a form that can release the contents at controlled rates under specific conditions and/or upon receiving a certain stimulus' (Desai and Park, 2005; Picot and Lacroix, 2003). Encapsulation technology is evolving and many materials are encapsulated to protect them from various harmful factors, as well as to deliver specific materials to the targeted areas without getting effected through their journey. Among different encapsulation techniques, microencapsulation is used widely in food industry, where solids, liquid or gaseous material are packed into microcapsules, using polymers as coating material (Gharsallaoui *et al.*, 2007; Wandrey *et al.*, 2010).

In the ideal case, the shell material should have the following properties:

1. It should have good rheological properties and easy to handle during encapsulation process.
2. It should be soluble in solvents acceptable in the food industry, e.g. water, ethanol, etc.
3. It should have the capability to hold the active material in stable emulsion form during processing and storage.
4. It should not react with active material during processing or storage.
5. It should completely release the active material.
6. It should provide maximum protection to the active material against environmental conditions (e.g. heat, light, humidity).
7. It should be economically sustainable and food-grade substance (Shahidi and Han, 1993).

*Corresponding author: amiamipatel@yahoo.co.in, ami@midft.com

Ever since Barrett Green developed the encapsulation technique using gelatin as carrier material (Fanger, 1974), many carrier materials have been tried and tested. A plethora of substances (of different types, origins and properties), including natural or synthetic, are available and could be employed as microcapsule shell material, but most of the commercially prepared microcapsules employ a rather small number of shell materials. The reason for this is that only materials certified as 'generally recognized as safe' (i.e. under the GRAS standard) should be used in the food encapsulation processes. The precise safety requirements for the quality of shell materials are defined by the agencies, such as the European Food Safety Authority (EFSA) or Food and Drug Administration (FDA) in the USA (Wandrey *et al.*, 2010; Gibson *et al.*, 2017).

The majority of materials used for microencapsulation in the food sector are bio-based materials, such as carbohydrates, proteins, lipids and miscellaneous materials (Mishra, 2015; Chaudhary and Patel, 2019). A brief introduction of various materials that are used as carrier materials is given below.

1.1   Polysaccharides

The most widely used encapsulating materials in the food industry belong to the group of polysaccharides. These polysaccharides can be divided into plant-based and animal-based materials, as given in Table 1. Their wide usage in food industry is due the fact that they have desirable physicochemical properties, such as solubility, melting, phase change and can be used to form different shapes in a wide range of sizes. Another important aspect in their selection is that they are very cost-effective materials (Brownlie, 2007).

1.2   Proteins

Proteins are natural biomolecules made up of linear chains of amino acids and are used in encapsulation and other applications in the food industry. There are several types of proteins available. Application of proteins has some advantages, such as biocompatibility, biodegradability, good water solubility, emulsifying properties and foaming capacity. Application of vegetable proteins as shell material in microencapsulation is a trend in the pharmaceutical, cosmetics and food industries. This is because vegetable proteins are identified to be less allergenic when compared to animal proteins (Modi and Seth, 2010). Proteins can encapsulate hydrophobic and hydrophilic compounds alone, but also can be mixed with polysaccharides or synthetic polymers. Disadvantage of proteins as encapsulating agents is due to their low solubility in cold water, probability to react with carbonyls and above all, an important factor is cost, which is higher when compared with polysaccharides and this limits proteins application in food encapsulation processes (Shahidi and Han, 1993).

Proteins which are employed as shell materials can also be divided into two groups, based on their origin (Table 2).

1.3   Lipids

Lipids, such as fats, fatty acids, waxes, and phospholipids are used as edible coating materials. Among the lipids, those that exhibit amphiphilic properties, i.e. the molecules which reduce the surface tension of the medium or reduce interfacial tension between phases at which they are adsorbed, are very useful for encapsulation of active compounds. Due to low polarity of lipids, they obstruct the moisture

Table 1: Polysaccharides Used as Shell Materials for Encapsulation

Plant Origin	Animal and Microbial Origin	Marine Origin
Starch derivatives: Amylose, Amylopectin, Dextrins, Maltodextrins, Polydextrose	Xanthan	Alginate
Octenyl succinic anhydride modified starch	Gellan	Carrageenan
Sodium starch octenyl succinate,	Dextran	
Pearl millet starch	Chitosan	
Arrowroot starch	Curdlan	
Sorghum starch	Pullulan	
Cellulose derivatives: Methylcellulose, Hydroxypropyl cellulose Hydroxypropyl methylcellulose Carboxymethyl cellulose		
Plant exudates: Gum arabic Gum Tragacanth Gum Karaya Mesquite Gum Gum Angum		
Plant extracts: Galactomannans Pectins Soluble soybean polysaccharides Mucilage		
Prebiotics polysaccharides: Inulin Fructooligosaccharides (FOS) Galactooligosaccharides (GOS) Xylooligosaccharides (XOS)		

Table 2: Plant- and Animal-based Proteins Used as Shell Materials for Encapsulation

	Plant Based	Animal Based
Cereal	Oat protein	Casein and caseinates
	Gluten and gliadins – wheat	Whey proteins
	Barley protein	Gelatin
	Zein (corn)	Collagen
	Rice bran	Elastin
Soy and Pulses	Soy protein	Albumin
	Pea protein	
	Chickpea protein	
	Lentil protein	
Miscellaneous	Sunflower protein	
	Gliadin	
	Lectin	
	(Martins *et al.*, 2018)	

transport. However, their hydrophobic nature confers fragility to the formed coatings. Thus, lipids should be melded/mixed with proteins or polysaccharides, in order to improve their coating characteristics. Some of the lipid materials that are utilized in food encapsulation are partial acylglycerols, phospholipids, glycolipids, aminolipids and lipopeptides, phytosterol surfactant, and antioxidant esters, tristearic acid, diglycerides, monoglycerides, oils, fats, hardened oils, wax, paraffin, natural waxes, such as beeswax, paraffin waxes and oxidized polyethylene waxes (synthetic waxes).

1.4   Synthetic Polymers

Natural polymers have certain disadvantages, such as rigid structure that limits potential chemical modifications; also, during extraction of natural polymers, unwanted remnants (e.g. impurities) may remain. The quality of the extracted natural polymers may also vary from batch to batch. To overcome these disadvantages, synthetic polymers, like PVA, PEG, polycaprolactone, PLGA, isocyanates, polyamide, polyurea, polyurethane, melamine formaldehyde, poly (vinyl pyrrolidone) and poly (vinyl acetate-co-crotonic acid) are usually employed for encapsulation of active compounds. They possess greater mechanical and chemical stability, increased reproducibility due to the minimized batch-to-batch variation, reduced non-specific protein binding and could be easily modified, providing carrier materials and encapsulates with tunable properties (Lu *et al.*, 2007; Young *et al.*, 2012).

There are numerous books and copious publications dealing comprehensively with different biopolymers, polysaccharides, hydrocolloids, or gums, and their use in encapsulation process. The aim of the current chapter is to present and update the existing knowledge about different encapsulating materials either already in use or with the potential for application in food sector. Considering that other authors of this book will cover various aspects of encapsulation in the food sector, in this chapter, we primarily focus on the shell materials reviewing and current state-of-the art in the area of probiotics encapsulation, especially focusing on prebiotics as carrier materials for encapsulation.

2.   Prebiotics

Prebiotics are materials that are actually used to enhance the growth of probiotics; in other words, prebiotics serve as probiotics' food. Prebiotics are defined as 'a selectively fermented ingredient that allows specific changes, both in the composition and/or activity in the gastrointestinal microflora that confer benefits upon the host's well-being and health' (Roberfroid, 2007). Much accepted definition of prebiotics is given by the International Scientific Association for Probiotics and Prebiotics (ISAPP), where they are defined as 'a substrate that is selectively utilized by host microorganisms conferring a health benefit'. The term 'prebiotic' may also encompass non-carbohydrates' varied classes other than foods which can act on the entire body of both humans and animals by managing the microorganisms for maintaining health and prevent diseases (Gibson *et al.*, 2017). These prebiotic compounds are usually relatively short chained, possess low molecular-weight carbohydrates that are non-active food constituents and are selectively fermented in the colon, especially by bifidobacteria and lactic acid bacteria. These bacteria utilize the prebiotic compounds, providing essential nutrients and energy (Gourineni *et al.*, 2011; Salvini *et al.*, 2011). They improve the host's health by improving the survival, growth, metabolism and beneficial health activities of probiotics in the digestive system. They are antagonistic to pathogenic organisms, limiting their proliferation (Yee *et al.*, 2019). Regarding the effects of prebiotics on probiotic microorganisms, the logical approach in the development of new food formulations would be to combine these two beneficial nutritional components via encapsulation. In literature, a term *synbiotic encapsulation* (combination of prebiotic carrier(s) and probiotic cells) is used to define an efficient system for targeted delivery of probiotics (Wu and Zhang, 2018).

Addition of prebiotic components (e.g. 'Raftilose P95'and polydextrose) was reported to offer protection to probiotics with improved stability (Capela *et al.*, 2006; Riaz and Masud, 2013; Martinez *et al.*, 2015). Specific prebiotic compound has the ability to augment a particular group of probiotic bacteria in certain intestinal regions (Okuro *et al.*, 2013; Sathyabama *et al.*, 2014). Using this beneficial aspect, utilization of prebiotic for micro capsule preparation could help the probiotic organism to survive under unfavorable conditions; also, encapsulation into prebiotics may help probiotics to grow better in the GI tract where they get additional advantage to compete and grow in the complex environment (Okuro *et al.*, 2013; Sathyabama *et al.*, 2014). Various prebiotic compounds, like inulin, FOS, GOS, inulooligosaccharides (IOS) and XOS have been used in preparation of microcapsules, which could be further applied for preparation of fermented dairy products, like yogurt, without reducing the required probiotic count. Various prebiotics used as carrier material for probiotics encapsulation are diagrammatically represented in Fig. 1. These encapsulated probiotics could also survive simulated gastrointestinal conditions (Li *et al.*, 2020). Additionally, co-encapsulation of probiotic isolate can be a new trend, which might reduce the cost of using prebiotics (Chen *et al.*, 2015).

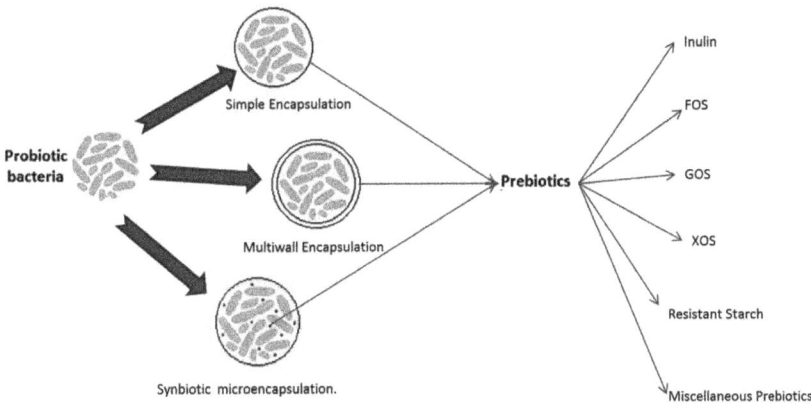

Fig. 1: Prebiotics used as carrier material for probiotics encapsulation

There are some reports which suggest that application of prebiotics as encapsulating material or as co-encapsulant can improve the probiotics' survivability in extreme conditions prevailing in acidic products and frozen products as shown in Table 3.

2.1   Inulin

Inulins are a group of naturally occurring non-structural, storage carbohydrates formed by many types of plants (Abed *et al.*, 2016). The term inulin refers to a heterogeneous blend of fructose polymers, comprising of β-d-fructosyl which residues with (2→1) linkages and generally has a (1→2) linked α-d-glucosyl terminal residue. Inulins are present in commonly consumed fruits and vegetables, like onion, leek, banana, wheat, garlic and rye (Vijn and Smeekens, 1999; Niness 1999; Roberfroid, 2005; Kumar *et al.*, 2015). Survival of probiotics in harsh acidic environments was enhanced in the presence of inulin (Yee *et al.*, 2019). Microcapsules containing inulin have the ability

Table 3: Studies Dealing with the Microencapsulation of Probiotics with Prebiotics as Coating Material Applied in Food Products

Product	Prebiotic and/or Probiotic Types	Major Outcome(s)	Reference
Yogurt	Galacto-oligosaccharide (GOS) and lactitol (LAC)	Microencapsulation and addition of LAC help to enhance the survival of probiotic strains in yogurt during storage, but had a negative relationship between the addition of synbiotic microcapsules and the improvement of yogurt quality.	Li *et al.*, 2020
	Carrageenan	Encapsulation improved the probiotics viability in the prepared yogurt and gastrointestinal tract. In the case of encapsulated bacteria, only 3 logs while for free cells, 7 log reductions were recorded. However, sodium alginate microcapsules exhibited better release profile than carrageenan.	Afzaal *et al.*, 2019
	Inulin in free form or in the form of nanoparticles encapsulated bifidobacterium cultures	The viability of *Bifidobacterium* cultures was enhanced when double-coated microcapsule was applied compared to the free one. Moreover, the enhancement was boosted when inulin was incorporated in either free form or in the form of nanoparticles.	Fayed *et al.*, 2019
Synbiotic diet mousse	Inulin *Lactobacillus acidophilus* La-5	After 6 h of the *in vitro* assays, the lowest reduction of cell counts occurred for mousse with microencapsulated cells (1.3 log cycles), followed by 30 microencapsulated cells (2.0 log cycles), mousse with free cells (3.0 log cycles), and free cells (7.4 log cycles).	dos Santos *et al.*, 2019
Symbiotic chewing gum	Alginate, inulin (0–1%) and lecithin (0–1%). *Lactobacillus reuteri*	The viability of the probiotic in encapsulated samples was retained after 21 days unlike control.	Qaziyani *et al.*, 2019
Functional blueberries	Alginate-based coatings enriched with inulin and oligofructose *Lactobacillus rhamnosus* CECT 8361	All the prebiotics showed enhancement of probiotic viability with counts above 6.2 log CFU/g for the entire period, whereas counts of native microbiota remained under safe levels. Overall visual quality, odor and flavor were acceptable up to storage of 14 days.	Bambace *et al.*, 2019

of self-aggregation and form an insoluble mass inside Ca-alginate matrix, delaying the penetration of H+ ions into the microcapsules by blocking the pores (Zaeim *et al.*, 2019). Atia *et al.* (2017, 2018) reported that co-encapsulation of *lactobacilli* with 50mg/g inulin maintained viability of bacteria in acidic conditions. Encapsulation with inulin increased the encapsulation efficiency of probiotic *Lactobacillus acidophilus*. Microcapsule showed good resistance to simulated gastrointestinal conditions and viability of probiotics was found to be stable even at −18°C for 120 days (Poletto *et al.*, 2019). Another study carried out by Krasaekoopt and Watcharapoka (2014) showed improved viability of the probiotic *Lb. acidophilus* under simulated GI conditions when co-encapsulated with inulin and chitosan coated with Ca-alginate.

When *Lb. casei* was mixed with inulin and then encapsulated with 2 per cent sodium alginate, 98 per cent encapsulation efficiency was observed, while the synbiotic microcapsules' size was around 24μm. Encapsulated probiotic showed good resistance to gastrointestinal liquids *in vitro* (Banerjee *et al.*, 2017). In another study, *Lactobacillus rhamnosus* GG was encapsulated with inulin, alginate and chitosan; encapsulation contributed to a significant improvement in the survival of the probiotic organism under simulated gastrointestinal conditions. Such encapsulates when added to apple juice, improved its sensory properties during 90 days of storage at 4 and 25°C (Gandomi *et al.*, 2016).

Lb. plantarum and *Bifidobacterium lactis*, when co-encapsulated with inulin in calcium alginate and chitosan, exhibited encapsulation efficiency up to 78.9 per cent. Inulin is found to leech out of the matrix, but it also showed that chitosan reduced the release of inulin from the matrix. The presence of inulin ensured that cells have good resistance toward harsh gastrointestinal conditions as compared to free probiotic organisms. However, the efficiency of encapsulation process also depends on the characteristics of probiotic strains as well as on storage conditions. Under the same conditions, *B. lactis* was found to be highly sensitive to environment and even co-encapsulation with inulin did not improve *B. lactis* survival during storage at 25°C, but survivability was enhanced when this encapsulated strain was stored at 4°C and −18°C for 90 days. *Lb. plantarum* showed little loss of viability when stored at 4°C and −18°C, and satisfactory viability after storage at 25°C for 90 days (Zaeim *et al.*, 2019).

Encapsulated *Lactobacillus plantarum* with inulin-sodium alginate coated with skim milk showed good resistance towards freeze-drying and greater resistance to simulated gastric fluid. Loss of viability was also found to be less after 7 weeks of storage period. Gene expression related to probiotic functionality was also found to be unaffected when inulin was incorporated as an encapsulant (Wang *et al.*, 2016).

Microcapsules prepared with sweet whey and inulin provided good protection to *Bifidobacterium* BB-12 (Pinto *et al.*, 2015). *Bifidobacterium* BB-12 was also found to survive in simulated gastrointestinal conditions and was able to show antagonistic activity towards *E.coli* when inulin and goat milk powder were used as carrier materials (Verruck *et al.*, 2020). *Saccharomyces cerevisiae* var. *boulardii* encapsulated in inulin, xanthan gum and alginate survived and grew during 4 weeks of storage at 4°C in berry juice (Fratianni *et al.*, 2014).

Researchers established that the survivability of the encapsulated *Bifidobacterium* in the simulated gastric solution could be significantly higher than the survivability of the free bacteria when alginate and arabic gum are used in combination with inulin nanoparticles as carriers (Fayed *et al.*, 2018). Further, other investigations also confirmed the significance of applying multi-particulate microcapsule for encapsulating probiotic and prebiotic in the same formulations (Cook *et al.*, 2014; Fayed *et al.*, 2018).

Apart from improving stability and survivability of probiotics, addition of prebiotics in encapsulates could also improve their properties, such as sensory effects and antimicrobial activity. Probiotic *Lb. casei* 359, when microencapsulated with inulin and whey protein concentrate (WPCP) along with gum acacia, helped probiotic cells to survive for 4 weeks of storage at refrigerated temperatures in litchi juice, and the juice acceptability was also higher even after storage period (Prakash *et al.*, 2017). In a recent study by Cui and co-workers (2018), it was observed that phthalyl inulin nanoparticle (PIN) internalized with *Pediococcus acidilactici*, a probiotic, had a higher antimicrobial activity than those formulations composed of *Pediococcus acidilactici* in alginate/chitosan/alginate (ACA microcapsules).

2.2   Fructooligosaccharides (FOS)

Fructooligosaccharides (FOS) are inulin-type oligosaccharides, also known as oligofructans or oligofructose. They are simple carbohydrates and small dietary fibers with low sweetness, low caloric value and prebiotic effects (Dominguez *et al.*, 2012; Picazo *et al.*, 2019). Fructooligosaccharides are composed of D-fructose residues linked by $\beta(2 \rightarrow 1)$ bonds with a terminal α-$(1 \rightarrow 2)$ linked D-glucose (Wang *et al.*, 2019). Two types of variants are available in FOS, inulin type with β-1, 2 linkages between fructose units and other type is levan type with β-2,6 linkages between fructosyl units in the main chain. Degree of polymerization (DP) of these variants ranges between 3 and 10. The oligofructose group further comprises short-chain fructooligosaccharides (scFOS) (Picazo *et al.*, 2019). The commercially available prebiotic FOS is inulin type with low DP, ranging from 3 to 9 (Wang, 2015). Neither pancreas nor small intestines produce enzymes that can cleave β-$(1 \rightarrow 2)$ fructosylfructose linkages which allow FOS to reach the large intestine (Aziani *et al.*, 2012). Fructooligosaccharides are used as an encapsulating material for probiotics, but their low glass transition temperature leads to sticky nature which is a major drawback. In order to overcome these limitations, FOSs are used in combination with other materials. Chen *et al.* (2005) reported co-encapsulation of probiotics with fructooligosaccharides, which increased cell survivability by 3 log cycles. FOSs in combination with denatured whey protein isolate showed good encapsulation capability with better storage stability and protection of *Lactobacillus plantarum* in simulated gastric, as well as intestinal conditions in comparison to the other tested encapsulates (Rajam and Anandharamakrishnan, 2015). Also, FOSs showed better protection of *Lb. fermentum* L7 in simulated gastrointestinal conditions and selective release compared to other oligosaccharides, like XOS, IOS and GOS (Liao *et al.*, 2019). When *Lactobacillus plantarum* was microencapsulated using litchi juice mixed with maltodextrin and FOS as carrier material, it showed good resistance to simulated gastric and intestinal conditions. This combination of carrier materials was found to protect *L. plantarum* against harsh environmental conditions and also spray-drying conditions (Kalita *et al.*, 2018).

2.3   Resistant Starch

Resistant starch (RS) is defined as a portion of starch that cannot be digested by amylases in the small intestine and passes to the colon to be fermented by microbiota (Englyst and Cummings, 1985). Resistant starch is a broad category of compounds that encompass several structurally different starches. While all resist digestion by human enzymes, they differ in their effects on the microbiota (DeMartino and Cockburn, 2020). RS cannot be absorbed in the small intestine but can be fermented in the large intestine to produce various short-chain fatty acids (SCFAs), including acetate, propionate and butyrate, which play a particularly positive role in promoting intestinal health (Yuan *et al.*, 2018). Its average degree of polymerization is 30-200; it has high heat-resistance and low water-holding power, ranging between 1.4-2.8 g. X-ray diffraction types of RS is V-type or B-type with the stable double helix, which has a high degree of order that increases tolerance to α-amylase and thus reduces the digestibility of starch. In addition, almost no losses are observed after high-temperature cooking, with a low caloric content (less than 10.5 kJ/g) (Ranhotra *et al.*, 1996; Khawas and Deka, 2017). RS is generally the largest source of energy for microbial growth in the human colon. Growths of bacterial species vary with different types of RS. RS is classified into five different types (as shown in Table 4) and each type of RS is able to alter composition of the gut's microbial community (Martínez *et al.*, 2010). Also, RS added into the feed containing high crude protein helps to reduce the consequences of exposure to harmful nitrogenous metabolites (Mu *et al.*, 2016; Andriamihaja *et al.*, 2010).

Generally, RS makes a beneficial fermentation substrate to balance the intestinal environment and modify and stabilize the gut microbial community to improve the intestinal health and function (Jiang *et al.*, 2020). The unique helical structure of the amylose molecule makes starch a very efficient vehicle for encapsulating molecules like lipids, flavors and others (Conde-Petit *et al.*, 2006).

Table 4*: Classification of Resistant Starch

Type 1	Physical barrier blocking access to the starch
Type 2	Native starch granules that resist digestion due to the presence of B-type crystallinity
Type 3	Retrograded starch
Type 4	Chemically modified starches
Type 5	Amylose-lipid complexes

*Adapted and modified from DeMartino and Cockburn (2020)

Incorporation of starch as encapsulating material results in the improvement of viability of probiotics (Chan *et al.*, 2011; de Araújo Etchepare *et al.*, 2016). Microencapsulation of *Lactobacillus rhamnosus* GG (LGG) using WPI in combination with a physically-modified resistant starch provides better protection to LGG in apple juice or citrate buffer (pH 3.5) for over 5 weeks' period when stored at 4-25 °C compared to the formulation containing RS alone. This effect is most probably caused by formation of a buffered microenvironment within the hydrated colloid particle surrounding the embedded LGG, thus isolating the bacteria from the stresses of the external environment, particularly for the low pH (Ying *et al.*, 2013).

Recently, resistant starch (Hi-maize) was tested and compared to other prebiotics for encapsulation of *Lactobacillus acidophilus*. It showed promising results. Resistance to simulated gastrointestinal conditions was observed too. Hi-maize treatment helped in maintaining the viability of probiotics when stored at 25°C for 120 days (Poletto *et al.*, 2019).

Lb. plantarum and *Bifidobacterium lactis* when co-encapsulated with resistant starch in calcium alginate/chitosan, showed encapsulation efficiency of 99.2 per cent. Resistant starch remained in the matrix even after 24 h and offered good resistance to gastrointestinal conditions. Also, resistant starch showed good protective properties to *Lb. plantarum* during storage at 4°C, −18°C and 25°C for 90 days' period. But *Bifidobacterium lactis* is sensitive when compared with *Lb. plantarum* and could not survive the storage at 4°C and 25°C for 90 days. Similarly, reduction in the viability of encapsulated *Bifidobacterium lactis* was observed when stored at −18°C (Zaeim *et al.*, 2019). Martin and co-workers (2013) also observed that the survival of *Lb. fermentum* co-encapsulated with starch in Ca-alginate matrix was influenced by temperature of the storage. Recently, Ashwar *et al.* (2018) showed that *Lb. plantarum* could tolerate storage conditions more effectively than *Lb. casei* and *Lb. brevis* in the presence of resistant starch.

2.4   Galactooligosaccharides (GOS)

Galactooligosaccharides (GOS) are composed of different galactosyl residues (from two to nine units) and a terminal glucose linked by β-glycosidic bonds, such as β-(1–2), β-(1–3), β-(1–4), and β-(1–6) (Vera *et al.*, 2016). They naturally occur at low concentrations in the milk of many animals, including humans and cows, but they also can be produced by chemical glycosylation or biocatalysis (Ackerman *et al.*, 2017). Some isolates of *Bifidobacterium* demonstrated the capacity to synthesize prebiotics, such as galactooligosaccharides (Tzortzis *et al.*, 2005; Roy *et al.*, 2002). GOS could provide a good protection to the encapsulated probiotics in simulated digestive system. A study carried out on two probiotics, viz. *Lactobacillus acidophilus* 5 and *Lactobacillus casei* 01, showed that addition of GOS in alginate beads with chitosan enhanced the probiotics' survivability during testing in simulated gastric conditions. Probiotic counts were above the recommended therapeutic level in acidic products like yogurt and orange juice, even after refrigerated storage temperature of 4°C for 4 weeks' period (Krasaekoopt and Watcharapoka, 2014). The co-encapsulation of *Lb. fermentum* L7 with GOS and alginate resulted in good resistance to simulated gastrointestinal conditions and could withstand refrigerated storage study at 4°C for 30 days (Liao *et al.*, 2019).

2.5   Xylooligosaccharides (XOS)

Xylooligosaccharides (XOSs) are sugar oligomers composed of xylose units linked by b-(1, 4) bonds, with DP ranging from 2 to 10. They are named according to the number of xylose residues in their composition, i.e. xylobiose, xylotriose, xylotetraose and so on. XOSs are known to be produced by *Aspergillus, Streptomyces, Trichoderma, Penicillium* and *Bacillus* from xylan-rich materials (Belorkar and Gupta, 2016), although they are naturally found in fruits, vegetables, honey, bamboo and milk (Vázquez *et al.*, 2000; Mandelli *et al.*, 2014; Patel and Prajapati, 2015). When *Lb. fermentum* L7 was co-encapsulated, using XOS and alginate, the obtained encapsulates exhibited good protection against simulated gastrointestinal conditions and endured refrigerated storage conditions (Liao *et al.*, 2019).

2.6   Miscellaneous Prebiotics

The prebiotic arabinoxylan oligosaccharides are reported to enhance the encapsulation efficacy in the case of tested probiotics. When *Lactobacillus plantarum* was encapsulated in microspheres prepared through co-gelation of prebiotic arabinoxylan and sodium alginate, the encapsulation efficiency improved 2.5 fold. Survival rate in acidic conditions and bile salt resistance also improved as well as the storage stability (Wu and Zhang, 2018). Also, *Lb. fermentum* L7 co-encapsulated with IOS and alginate offered good protection from simulated gastrointestinal conditions, and endured refrigerated storage conditions (Liao *et al.*, 2019). When sugar beet and chicory were used as encapsulating material along with alginate for two potential probiotics, namely *Staphylococcus succinus* (MAbB4) and *Enterococcus faecium* (FIdM3), their survivability in acidic environment and bile salts improved and could even resist the simulated gastrointestinal conditions. Their count remained well above 8 log cycles throughout refrigerated storage at 4°C for 30 days (Sathyabama *et al.*, 2014). Encapsulation of *Lactobacillus acidophilus* with rice bran yielded 117.70µm particles which exhibited good resistance to the simulated gastrointestinal conditions. Viability of the probiotic organism was stable when stored at 7°C for 120 days (Poletto *et al.*, 2019).

3. Conclusion and Future Perspective

The majority of materials used for microencapsulation in the food sector are bio-based materials, such as carbohydrate polymers (polysaccharides), proteins and lipids. Even though there are many well-known encapsulants, there is still a need to look for novel polymers and new means to use the old ones in other ways to create ideal capsules with excellent resistance and mechanical properties with wide applicability. Application of different prebiotic oligosaccharides as coating material offers several beneficial effects, such as enhanced viability of probiotics and improvement in the shelf-life of some perishable food products. On the other side, it appeared as one of the costlier approaches and very limited studies suggest that it also demands much attention and research focus in future studies. Only a limited number of encapsulating materials have been approved for food, drug, clinical and cosmetics applications. Selection of appropriate microencapsulation technique and encapsulation material is still a challenging task.

References

Aburto, L.C., Tavares, D.D.Q. and Martucci, E.T. (1998). *Microencapsulação de óleo essencial de laranja, Food Science and Technology*, 18(1): 45-48.

Abed, S.M., Ali, A.H., Noman, A., Niazi, S., Ammar, A.F. and Bakry, A.M. (2016). Inulin as prebiotics and its applications in food industry and human health: A review, *International Journal of Agriculture Innovations and Research*, 5(1): 88-97.

Ackerman, D.L., Craft, K.M. and Townsend, S.D. (2017). Infant food applications of complex carbohydrates: Structure, synthesis, and function, *Carbohydrate Research*, 437: 16-27.

Afzaal, M., Khan, A.U., Saeed, F., Ahmed, A., Ahmad, M.H., Maan, A.A., Tufail, T., Anjum, F.M. and Hussain, S. (2019). Functional exploration of free and encapsulated probiotic bacteria in yogurt and simulated gastrointestinal conditions, *Food Science & Nutrition*, 7(12): 3931-3940.

Andriamihaja, M., Davila, A.M., Eklou-Lawson, M., Petit, N., Delpal, S., Allek, F., Blais, A., Delteil, C., Tomé, D. and Blachier, F. (2010). Colon luminal content and epithelial cell morphology are markedly modified in rats fed with a high-protein diet, *American Journal of Physiology-Gastrointestinal and Liver Physiology*, **299**(5): G1030-G1037.

Ashwar, B.A., Gani, A., Gani, A., Shah, A. and Masoodi, F.A. (2018). Production of RS4 from rice starch and its utilization as an encapsulating agent for targeted delivery of probiotics, *Food Chemistry*, 239: 287-294.

Atia, A., Gomma, A.I., Fliss, I., Beyssac, E., Garrait, G. and Subirade, M. (2017). Molecular and biopharmaceutical investigation of alginate-inulin symbiotic coencapsulation of probiotic to target the colon, *Journal of Microencapsulation*, 34: 171-184.

Atia, A., Gomaa, A., Fernandez, B., Subirade, M. and Fliss, I. (2018). Study and understanding behavior of alginate-inulin synbiotic beads for protection and delivery of antimicrobial-producing probiotics in colonic simulated conditions, *Probiotics & Antimicrobial Proteins*, 10: 157-167.

Aziani, G., Terenzi, H.F., Jorge, J.A. and Souza Guimarães, L.H. (2012). Production of fructooligosaccharides by *Aspergillus phoenicis* biofilm on polyethylene as inert support, *Food Technology and Biotechnology*, **50**(1): 40-45.

Bambace, M.F., Alvarez, M.V. and del Rosario Moreira, M. (2019). Novel functional blueberries: Fructo-oligosaccharides and probiotic *lactobacilli* incorporated into alginate edible coatings, *Food Research International*, 122: 653-660.

Banerjee, D., Chowdhury, R. and Bhattacharya, P. (2017). *In-vitro* evaluation of targeted release of probiotic *Lactobacillus casei* (2651 1951 RPK) from synbiotic microcapsules in the gastrointestinal (GI) system: Experiments and modeling, *LWT – Food Science and Technology*, 83: 243-253.

Belorkar, S.A. and Gupta, A.K. (2016). Oligosaccharides: A boon from nature's desk, *AMB Express*, 6: 82.

Brownlie, K. (2007). Marketing perspective of encapsulation technologies in food applications, pp. 213. *In*: J.M. Lakkis (Ed.). *Encapsulation and Controlled Release Technologies in Food Systems*, Blackwell Publishing.

Capela, P., Hay, T.K.C. and Shah, N.P. (2006). Effect of cryoprotectants, prebiotics and microencapsulation on survival of probiotic organisms in yogurt and freeze-dried yogurt, *Food Research International*, **39**(2): 203-211.

Chan, E.S., Wong, S.L., Lee, P.P., Lee, J.S., Ti, T.B., Zhang, Z., Poncelet, D., Ravindra, P., Phan, S.H. and Yim, Z.H. (2011). Effects of starch filler on the physical properties of lyophilized calcium–alginate beads and the viability of encapsulated cells, *Carbohydrate Polymers*, 83: 225-232.

Chaudhary, H.J. and Patel, A.R. (2019). Microencapsulation technology to enhance the viability of probiotic bacteria in fermented foods: An overview, *International Journal of Fermented Foods*, **8**(2): 63-72.

Chen, J., Wang, Q., Liu, C.M. and Gong, J. (2015). Issues deserve attention in encapsulating probiotics: Critical review of existing literature, *Critical Reviews in Food Science and Nutrition*, **57**(6): 1228-1238.

Chen, K.N., Chen, M.J., Liu, J.R., Lin, C.W. and Chiu, H.Y. (2005). Optimization of incorporated prebiotics as coating materials for probiotic microencapsulation, *Journal of Food Science*, 70: 260-266.

Conde-Petit, B., Escher, F. and Nuessli, J. (2006). Structural features of starch-flavor complexation in food model systems, *Trends in Food Science & Technology*, **17**(5): 227-235.

Cook, M.T., Tzortzis, G., Charalampopoulos, D. and Khutoryanskiy, V.V. (2014). Microencapsulation of a synbiotic into PLGA/alginate multiparticulate gels, *International Journal of Pharmacy*, **466**(1): 400-408.

Cui, L.H., Yan, C.G., Li, H.S., Kim, W.S., Hong, L., Kang, S.K., Choi, Y.J. and Cho, C.S. (2018). A new method of producing a natural antibacterial peptide by encapsulated probiotics internalized with inulin nanoparticles as prebiotics, *Journal of Microbiology &. Biotechnology*, **28**(4): 510-519.

de Araújo Etchepare, M., Raddatz, G.C., Cichoski, A.J., Flores, É.M.M., Barin, J.S., Zepka, L.Q., Jacob-Lopes, E., Grosso, C.R.F. and de Menezes, C.R. (2016). Effect of resistant starch (Hi-maize) on the survival of *Lactobacillus acidophilus* microencapsulated with sodium alginate, *Journal of Functional Foods*, 21: 321-329.

DeMartino, P. and Cockburn, D.W. (2020). Resistant starch: Impact on the gut microbiome and health, *Current Opinion in Biotechnology*, 61: 66-71.

Desai, K.G.H. and Jin Park, H. (2005). Recent developments in microencapsulation of food ingredients, *Drying Technology*, **23**(7): 1361-1394.

Dominguez, A., Nobre, C., Rodrigues, L.R., Peres, A.M., Torres, D., Rocha, I., Lima, N. and Teixeira, J. (2012). New improved method for fructooligosaccharides production by *Aureobasidium pullulans*, *Carbohydrate Polymer*, 89: 1174-1179.

dos Santos, D.X., Casazza, A.A., Aliakbarian, B., Bedani, R., Saad, S.M.I. and Perego, P. (2019). Improved probiotic survival to *in vitro* gastrointestinal stress in a mousse containing *Lactobacillus acidophilus* La-5 microencapsulated with inulin by spray-drying, *LWT*, 99: 404-410.

Englyst, H.N. and Cummings, J.H. (1985). Digestion of the polysaccharides of some cereal foods in the human small intestine, *The American Journal of Clinical Nutrition*, **42**(5): 778-787.

Fanger, G.O. (1974). Microencapsulation: A brief history and introduction, pp. 1-20. *In*: Vandegaer, J.E. (Ed), *Microencapsulation*, Springer, Boston, MA.

Fayed, B., Abood, A., El-Sayed, H.S., Hashem, A.M. and Mehanna, N.S. (2018). A symbiotic multiparticulate microcapsule for enhancing inulin intestinal release and Bifidobacterium gastro-intestinal survivability, *Carbohydrate Polymers*, 193: 137-143.

Fayed, B., El-Sayed, H.S., Abood, A., Hashem, A.M. and Mehanna, N.S. (2019). The application of multi-particulate microcapsule containing probiotic bacteria and inulin nanoparticles in enhancing the probiotic survivability in yogurt, *Biocatalysis and Agricultural Biotechnology*, 22: 101391.

Fratianni, F., Cardinale, F., Russo, I., Tremonte, P., Coppola, R. and Nazzaro, F. (2014). Ability of symbiotic encapsulated *Saccharomyces cerevisiae boulardii* to grow in berry juice and to survive under simulated gastrointestinal conditions, *Journal of Microencapsulion*, 31: 299-305.

Gandomi, H.S., Abbaszadeh, S., Misaghi, A., Bokaie, S. and Noori, N. (2016). Effect of chitosan-alginate encapsulation with inulin on survival of *Lactobacillus rhamnosus* GG during apple juice storage and under simulated gastrointestinal conditions, *LWT – Food Science and Technology*, 69: 365-371.

Gharsallaoui, A., Roudaut, G., Chambin, O., Voilley, A. and Saurel, R. (2007). Applications of spray-drying in microencapsulation of food ingredients: An overview, *Food Research International*, **40**(9): 1107-1121.

Gibson, G.R., Hutkins, R., Sanders, M.E., Prescott, S.L., Reimer, R.A., Salminen, S.J., Scott, K., Stanton, C., Swanson, K.S., Cani, P.D. and Verbeke, K. (2017). Expert consensus document: The International Scientific Association for Probiotics and Prebiotics (ISAPP) consensus statement on the definition and scope of prebiotics, *Nature Reviews Gastroenterology and Hepatology*, **14**(8): 491.

Gourineni, V.P., Verghese, M., Boateng, J., Shackelford, L., Bhat, N.K. and Walker, L.T. (2011). Combinational effects of prebiotics and soybean against azoxymethane-induced colon cancer *in vivo*, *Journal of Nutrition and Metabolism*, 868197.

Jiang, F., Du, C., Jiang, W., Wang, L. and Du, S.K. (2020). The preparation, formation, fermentability, and applications of resistant starch, *International Journal of Biological Macromolecules*, 150: 1155-1161.

Kalita, D., Saikia, S., Gautam, G., Mukhopadhyay, R. and Mahanta, C.L. (2018). Characteristics of synbiotic spray dried powder of litchi juice with *Lactobacillus plantarum* and different carrier materials, *LWT – Food Science and Technology*, 87: 351-360.

Khawas, P. and Deka, S.C. (2017). Effect of modified resistant starch of culinary banana on physicochemical, functional, morphological, diffraction and thermal properties, *International Journal of Food Properties*, **20**(1): 133-150.

Krasaekoopt, W. and Watcharapoka, S. (2014). Effect of addition of inulin and galacto-oligosaccharide on the survival of microencapsulated probiotics in alginate beads coated with chitosan in simulated digestive system, yogurt and fruit juice, *Lebensmittel-Wissenschaftund-Technologie – Food Science and Technology*, 57: 761-766.

Kumar, H., Salminen, S., Verhagen, H., Rowland, I., Heimbach, J., Bañares, S., Young, T., Nomoto, K. and Lalonde, M. (2015). Novel probiotics and prebiotics: Road to the market, *Current Opinion in Biotechnology*, 32: 99-103.

Li, H., Zhang, T., Li, C., Zheng, S., Li, H. and Yu, J. (2020). Development of a microencapsulated synbiotic product and its application in yogurt, *Lebensmittel-Wissenschaftund-Technologie*, 122: 109033.

Liao, N., Luo, B., Gao, J., Li, X., Zhao, Z., Zhang, Y., Ni, Y. and Tian, F. (2019). Oligosaccharides as co-encapsulating agents: Effect on oral *Lactobacillus fermentum* survival in a simulated gastrointestinal tract, *Biotechnology Letters*, **41**(2): 263-272.

Lu, H.F., Targonsky, E.D., Wheeler, M.B. and Cheng, Y.L. (2007). Thermally induced gelable polymer networks for living cell encapsulation, *Biotechnology and Bioengineering*, **96**(1): 146-155.

Mandelli, F., Brenelli, L.B., Almeida, R.F., Goldbeck, R., Wolf, L.D., Hoffmam, Z.B., Ruller, R., Rocha, J.M., Mercadante, A.Z. and Squina, F.M. (2014). Simultaneous production of xylooligosaccharides and antioxidant compounds from sugarcane bagasse via enzymatic hydrolysis, *Industrial Crops Production*, 52: 770-775.

Martin, M.J., Lara-Villoslada, F., Ruiz, M.A. and Morales, M.E. (2013). Effect of unmodified starch on viability of alginate-encapsulated *Lactobacillus fermentum* CECT5716, *Lebensmittel-Wissenschaftund Technologie – Food Science and Technology*, 53: 480-486.

Martínez, I., Kim, J., Duffy, P.R., Schlegel, V.L. and Walter, J. (2010). Resistant starches types 2 and 4 have differential effects on the composition of the fecal microbiota in human subjects, *PloS One*, **5**(11): e15046.

Martinez, R.C.R., Bedani, R. and Saad, S.M.I. (2015). Scientific evidence for health effects attributed to the consumption of probiotics and prebiotics: An update for current perspectives and future challenges, *British Journal of Nutrition*, **114**(12): 1993-2015.

Martins, J.T., Bourbon, A.I., Pinheiro, A.C., Fasolin, L.H. and Vicente, A.A. (2018). Protein-based structures for food applications: From macro to nanoscale, *Frontiers in Sustainable Food Systems*, 2: 77.

Mishra, M. (2015). Materials of natural origin for encapsulation, pp. 493. *In*: M. Mishra (Ed.), *Handbook of Encapsulation and Controlled Release*, Taylor & Francis Group, LLC.

Modi, V.C. and Seth, A.K. (2010). Formulation and evaluation of diltiazem sustained release tablets, *Journal of Global Trends in Pharmaceutical Sciences*, **1**(3): 1-10.

Mu, C., Yang, Y., Luo, Z., Guan, L. and Zhu, W. (2016). The colonic microbiome and epithelial transcriptome are altered in rats fed a high-protein diet compared with a normal-protein diet, *The Journal of Nutrition*, **146**(3): 474-483.

Niness, K.R. (1999). Inulin and oligofructose, *Journal of Nutrition*, 129: 1402-1406.

Okuro, P.K., Thomazini, M., Balieiro, J.C.C., Liberal, R.D.C.O. and Fávaro-Trindade, C.S. (2013). Co-encapsulation of *Lactobacillus acidophilus* with inulin or polydextrose in solid lipid microparticles provides protection and improves stability, *Food Research International*, 53: 96-103.

Patel, A. and Prajapati, J.B. (2015). Biological properties of xylooligosaccharides as an emerging prebiotic and future perspective, *Current Trends in Biotechnology and Pharmacy*, **9**(1): 28-36.

Picazo, B., Flores-Gallegos, A.C., Muñiz-Márquez, D.B., Flores-Maltos, A., Michel-Michel, M.R., de la Rosa, O., Rodríguez-Jasso, R.M., Rodríguez-Herrera, R. and Aguilar-González, C.N. (2019). Enzymes for fructooligosaccharides production: Achievements and opportunities, pp. 303-320. *In: Enzymes in Food Biotechnology*, Academic Press.

Picot, A. and Lacroix, C. (2003). Production of multiphase water-insoluble microcapsules for cell microencapsulation using an emulsification/spray-drying technology, *Journal of Food Science*, **68**(9): 2693-2700.

Pinto, S.S., Verruck, S., Vieira, C.R., Prudêncio, E.S., Amante, E.R. and Amboni, R.D. (2015). Influence of microencapsulation with sweet whey and prebiotics on the survival of *Bifidobacterium*-BB-12 under simulated gastrointestinal conditions and heat treatments, *LWT – Food Science and Technology*, **64**(2): 1004-1009.

Poletto, G., Raddatz, G.C., Cichoski, A.J., Zepka, L.Q., Lopes, E.J., Barin, J.S., Wagner, R. and de Menezes, C.R. (2019). Study of viability and storage stability of *Lactobacillus acidophillus* when encapsulated with the prebiotics rice bran, inulin and Hi-maize, *Food Hydrocolloids*, 95: 238-244.

Prakash, K.S., Bashir, K. and Mishra, V. (2017). Development of symbiotic litchi juice drink and its physiochemical, viability and sensory analysis, *Journal of Food Process & Technology*, **8**(12): 1-6.

Qaziyani, S.D., Pourfarzad, A., Gheibi, S. and Nasiraie, L.R. (2019). Effect of encapsulation and wall material on the probiotic survival and physicochemical properties of synbiotic chewing gum: Study with univariate and multivariate analyses, *Heliyon*, **5**(7): e02144.

Rajam, R. and Anandharamakrishnan, C. (2015). Microencapsulation of *Lactobacillus plantarum* (MTCC 5422) with fructooligosaccharide as wall material by spray drying, *LWT – Food Science and Technology*, **60**(2): 773-780.

Ranhotra, G.S., Gelroth, J.A. and Glaser, B.K. (1996). Energy value of resistant starch, *Journal of Food Science*, **61**(2): 453-455.

Riaz, Q.U. and Masud, T. (2013). Recent trends and applications of encapsulating materials for probiotic stability, *Critical Reviews in Food Science and Nutrition*, **53**(3): 231-244.

Roberfroid, M. (2005). Introducing inulin-type fructans, *British Journal of Nutrition*, **93**(1): S13-S25.

Roberfroid, M. (2007). Prebiotics: The concept revisited, *The Journal of Nutrition*, **137**(3): 830S-837S.

Roy, D., Daoudi, L. and Azaola, A. (2002). Optimization of galacto-oligosaccharide production by *Bifidobacterium infantis* RW-8120 using response surface methodology, *Journal of Industrial Microbiology & Biotechnology*, 29: 281-285.

Salvini, F., Riva, E., Salvatici, E., Boehm, G., Jelinek, J., Banderali, G. and Giovannini, M. (2011). A specific prebiotic mixture added to starting infant formula has long-lasting bifidogenic effects. *The Journal of Nutrition*, **141**(7): 1335-1339.

Sathyabama, S. and Vijayabharathi, R. (2014). Co-encapsulation of probiotics with prebiotics on alginate matrix and its effect on viability in simulated gastric environment, *LWT – Food Science and Technology*, **57**(1): 419-425.

Shahidi, F. and Han, X.Q. (1993). Encapsulation of food ingredients, *Critical Reviews in Food Science & Nutrition*, **33**(6): 501-547.

Tzortzis, G., Goulas, A.K. and Gibson, G.R. (2005). Synthesis of prebiotic galactooligosaccharides using whole cells of a novel strain, *Bifidobacterium bifidum* NCIMB 41171, *Applied Microbiology and Biotechnology*, 68: 412-416.

Vázquez, M.J., Alonso, J.L., Domõnguez, H. and Parajó, J.C. (2000). Xylooligo-saccharides: Manufacture and applications, *Trends in Food Science & Technology*, 11: 387-393.

Vera, C., Córdova, A., Aburto, C., Guerrero, C., Suárez, S. and Illanes, A. (2016). Synthesis and purification of galacto-oligosaccharides: State-of-the-art, *World Journal of Microbiology and Biotechnology*, **32**(12): 197.

Verruck, S., Barretta, C., Miotto, M., Canella, M.H.M., de Liz, G.R., Maran, B.M., Garcia, S.G., da Silveira, S.M., Vieira, C.R.W., da Cruz, A.G. and Prudencio, E.S. (2020). Evaluation of the interaction between microencapsulated *Bifidobacterium* BB-12 added in goat's milk Frozen Yogurt and *Escherichia coli* in the large intestine, *Food Research International*, 127: 108690.

Vijn, I. and Smeekens, S. (1999). Fructan: More than a reserve carbohydrate? *Plant Physiology*, 120: 351-359.

Wandrey, C., Bartkowiak, A. and Harding, S.E. (2010). Materials for encapsulation, pp. 31-100. *In:* N.J Zuidam and V.A. Nedovic (Eds), *Encapsulation Technologies for Food Active Ingredients and Food Processing*, Springer, Dordrecht.

Wang, L., Yu, X., Xu, H., Aguilar, Z. and Wei, H. (2016). Effect of skim milk coated inulin-alginate encapsulation beads on viability and gene expression of *Lactobacillus plantarum* during freeze-drying, *Lebensmittel-Wissenschaftund-Technologie – Food Science and Technology*, 68: 8-13.

Wang, T.H. (2015). Synthesis of neofructooligosaccharides, *Organic Chemistry Insights*, 5: 1-6.

Wang, Y., Guo, Q., Goff, H.D. and LaPointe, G. (2019). Oligosaccharides: Structure, function and application, *Encyclopedia of Food Chemistry*, 202.

Wu, Y. and Zhang, G. (2018). Synbiotic encapsulation of probiotic *Lactobacillus plantarum* by alginate-arabinoxylan composite microspheres, *LWT*, 93: 135-141.

Yee, W.L., Yee, C.L., Lin, N.K. and Phing, P.L. (2019). Microencapsulation of *Lactobacillus acidophilus* NCFM incorporated with mannitol and its storage stability in mulberry tea, *Ciência e Agrotecnologia*, 43.

Ying, D., Schwander, S., Weerakkody, R., Sanguansri, L., Gantenbein-Demarchi, C. and Augustin, M.A. (2013). Microencapsulated *Lactobacillus rhamnosus* GG in whey protein and resistant starch matrices: Probiotic survival in fruit juice, *Journal of Functional Foods*, **5**(1): 98-105.

Young, C.J., Poole-Warren, L.A. and Martens, P.J. (2012). Combining submerged electrospray and UV photopolymerization for production of synthetic hydrogel microspheres for cell encapsulation, *Biotechnology and Bioengineering*, **109**(6): 1561-1570.

Yuan, H.C., Meng, Y., Bai, H., Shen, D.Q., Wan, B.C. and Chen, L.Y. (2018). Meta-analysis indicates that resistant starch lowers serum total cholesterol and low-density cholesterol, *Nutrition Research*, 54: 1-11.

Zaeim, D., Sarabi-Jamab, M., Ghorani, B. and Kadkhodaee, R. (2019). Double layer co-encapsulation of probiotics and prebiotics by electro-hydrodynamic atomization, *Lebensmittel-Wissenschaftund-Technologie – Food Science and Technology*, 110: 102-109.

Encapsulation Techniques for Food Purposes

Laura G. Gómez-Mascaraque[1]**, Bojana Balanč**[2]**, Verica Djordjević**[2]**,
Branko Bugarski**[2]** and Kata Trifković**[1]*

[1] Teagasc Food Research Centre, Moorepark, Fermoy, Co. Cork, Ireland
[2] Department of Chemical Engineering, Faculty of Technology and Metallurgy,
University of Belgrade, Belgrade, Serbia

1. Introduction

The innovation trends in the food sector in the last few decades have been influenced by an increased consumer awareness of the relationship between nutrition and health and, thus, the intensive promotion of a healthier eating and lifestyle (Bleiel, 2010). Consequently, the demand for food products that provide health benefits beyond basic nutrition, i.e. functional foods, has grown in the last years (Bimbo *et al.*, 2017). This has resulted in an estimated increase of the market size of functional foods, beverages and dietary supplements of about 10 per cent per year (Hilton, 2017). Some authors argue that the concept of functional foods is not only appealing for consumers and the food industry, but is also a potential way of reducing the burden on the healthcare system by a continuous preventive mechanism (Gul *et al.*, 2016; Trifković and Benković, 2019).

Microencapsulation is considered a promising approach to overcome some of the challenges faced when developing enriched or fortified functional food products (Gómez-Mascaraque and Lopez-Rubio, 2020), such as potential bioactivity losses of the incorporated bioactive ingredient, its immiscibility with the food matrix, the alteration of the sensorial properties of the food products, and the low bioavailability of some of the active ingredients to be incorporated (Deng *et al.*, 2014; Lafarga and Hayes, 2017). Microencapsulation technologies can be defined as processes in which a compound or ingredient of interest is coated with or embedded within a protective matrix (Jiménez-Martín *et al.*, 2014) generating microstructured or micron-sized materials. The obtained microcapsules can exhibit a variety of structures (Fig. 1), where the encapsulation matrix works as a physical barrier limiting the direct contact of the bioactive ingredients with the food environment (Ye *et al.*, 2009).

---

*Corresponding author: kata.trifkovic@inlecomsystems.com

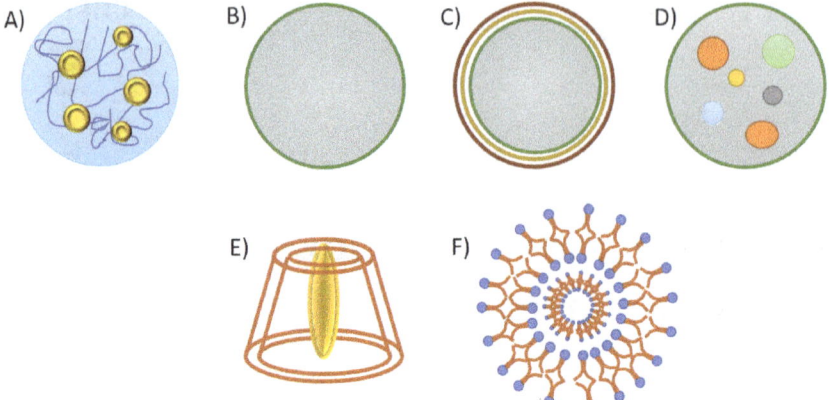

Fig. 1: Schematic illustration of different types of microcapsules: A) matrix; B) reservoir (core/shell); C) multiwall; D) multicore; E) inclusion complexation; F) phospholipid bilayer

Numerous technologies and methods have been developed to date for the microencapsulation of bioactive compounds, many of which can be exploited for food applications. Selection of the most adequate microencapsulation technique has to be made for each specific application, taking into account different factors. The most relevant criteria to be considered include the physico-chemical properties of the bioactive ingredient to be encapsulated, the characteristics of the coating materials to be used, the reproducibility and productivity of the method and the cost and scalability of the process (Gómez-Mascaraque *et al.*, 2016a). Besides, the feasibility of working in food-grade conditions is an unavoidable requirement when the produced microcapsules are intended to be used for food applications. This requisite limits the range of microencapsulation technologies which are acceptable for the food industry to those which do not require the use of toxic solvents or reagents, and can be operated using edible coating materials such as proteins, polysaccharides or lipids (Gómez-Mascaraque *et al.*, 2018). Additionally, cost considerations are more restrictive in the food industry than in other sectors, such as the cosmetics or the pharmaceutical industries, due to its much lower profit margin (Gómez-Mascaraque *et al.*, 2016a).

In this chapter, the different microencapsulation approaches used for food applications will be described and the main advantages and current challenges of each of them will be addressed. Additionally, the latest advances in the application of these technologies for the microencapsulation of food ingredients will be discussed.

1.1   Classification of Microencapsulation Technologies

Although the classification of the available microencapsulation technologies is complex and non-consistent throughout literature, in general, microencapsulation technologies can be classified into chemical, physico-chemical and physical (or physico-mechanical) methods (Gómez-Mascaraque *et al.*, 2016a), according to the nature of the phenomena taking place to form the capsules. Some works have also reported biological methods of microencapsulation, where the bioactive compound is encapsulated within biological entities (i.e. cells) (Ciamponi *et al.*, 2012).

Chemical microencapsulation technologies involve chemical reactions to achieve microencapsulation. Polymerization is the main chemical microencapsulation technology, and exploits the formation of new chemical bonds between reactive monomers to synthetize (i) an encapsulation matrix for a compound of interest *in situ* (Perignon *et al.*, 2015) or (ii) so called *ready-made-support* for the post-loading of compound of interest (Trifković *et al.*, 2014, 2015). Despite being of exceptional interest in diverse fields, including the design of sensors (Barkade *et al.*, 2013), textiles (Tawiah *et al.*, 2017), self-healing materials (He *et al.*, 2017) or aerospace coatings (Guo *et al.*, 2016), this technique entails a critical limitation for its application in food products, as it involves the use of toxic solvents and/or reactants. Therefore, chemical microencapsulation technologies will not be covered in this chapter.

Most of the microencapsulation technologies used for food applications fall into the categories of physical or physico-chemical methods. Physical methods rely on changes in thermodynamic properties (e.g. temperature, pressure) and/or mechanical processing for capsule formation, and include spraying and drying techniques (spray-drying, freeze-drying, spray-chilling, electrospraying), fluidized-bed coating, supercritical fluids and emulsification technologies. Physico-chemical methods, on the other hand, are based on structural changes and/or macromolecular reorganization due to the interaction between components, and they include the formation of microgels, coacervation, liposome encapsulation and inclusion complexation. Due to their relevance for food applications, the various physical and physic-chemical methods available are described in detail in the following sections.

Biological microencapsulation technologies are interesting approaches in which intact cells, usually yeast or plant cells, are used as carriers to stabilize bioactive compounds (Bishop *et al.*, 1998). Especially, the yeast strain *Saccharomyces cerevisiae* has been extensively investigated for this purpose (Ciamponi *et al.*, 2012), as it can be easily produced on a large scale and constitutes a food-grade material with high nutritional value (Pham-Hoang *et al.*, 2013). In the food sector, encapsulation in yeasts has been mainly applied for aromas, flavors and antioxidants (Dardelle *et al.*, 2007; Zuidam and Heinrich, 2010; Hafner *et al.*, 2011). Being a simple and cost-effective process, the main challenge of this technology is to achieve an effective transfer of the active molecules across the cell wall and plasma membrane of the yeast cells, for which different techniques are being developed (da Silva Pedrini *et al.*, 2014). More recently, Gómez-Mascaraque *et al.* (2017a) also suggested the potential of potato cells as natural capsules for polyphenols and demonstrated the ability of different phenolic compounds to bind to starch granules and cell walls in intact potato cells. The use of biological microencapsulation approaches is still in its early stage and only a few works have been conducted to explore their potential.

1.1.1 Spray-drying

Spray-drying is the most commonly employed microencapsulation technology in the food industry (Franco *et al.*, 2017). This drying technique has long been used in the food industry for the production of powder ingredients, such as dairy powders (Drapala *et al.*, 2017), and its use was later extended to the field of microencapsulation. Since then, spray-drying has been used for the encapsulation of a wide range of food ingredients, including carotenoids (Robert *et al.*, 2003; Deng *et al.*, 2014), phenolic compounds (Mahdavi *et al.*, 2014; Paini *et al.*, 2015), oils and polyunsaturated fatty acids (Kolanowski *et al.*, 2004; Klinkesorn *et al.*, 2005; Shaw *et al.*, 2007; Encina *et al.*, 2016) or probiotics (Arslan *et al.*, 2015; Liu *et al.*, 2015), among others. It is estimated that about 80-90 per cent of the microcapsules produced in the food industry are obtained by spray-drying (Nedović *et al.*, 2013; Anandharamakrishnan and Ishwarya, 2015a).

Spray-drying allows production of matrix type (Fig. 1A) encapsulation structures in the form of fine powders with low water activity, which is very convenient for handling, transport and storage of the ingredients, ensuring microbiological stability (Gouin, 2004). Microencapsulation through spray-drying requires dissolving or dispersing the ingredient of interest in an aqueous solution or dispersion of the wall material prior to the atomization or spraying of the blend through a nozzle. The produced fine droplets are rapidly dried using a hot gas stream, obtaining dry powders (Deshmukh *et al.*, 2016) with sizes generally ranging from 10 μm to 3mm (Đorđević *et al.*, 2015) depending on the feed formulation and the processing conditions. Recently, spray-driers designed to produce particles in the sub-micron range have also been developed (Gómez-Mascaraque *et al.*, 2016b; Assadpour and Jafari, 2019a), often referred to as 'nano spray-driers' (Lee *et al.*, 2011).

Figure 2A shows a simplified scheme of a spray-drying process. The basic principle of spray-drying is heat transfer between the atomized feed droplets and hot air, which results in water evaporation and, thus, dry particles formation. These particles are then collected by using a cyclone separator, an electrostatic precipitator or a bag filter, depending on the equipment configuration and the particle size of the powders (Ishwarya and Anandharamakrishnan, 2017). Several publications have previously reviewed the different equipment configurations available for spray-drying (Anandharamakrishnan and Ishwarya, 2015b; Miller *et al.*, 2016; Ishwarya and Anandharamakrishnan, 2017; Jaskulski *et al.*, 2017; O'Sullivan *et al.*, 2019), to which the readers are directed for more information about nozzle/atomizer and collector types.

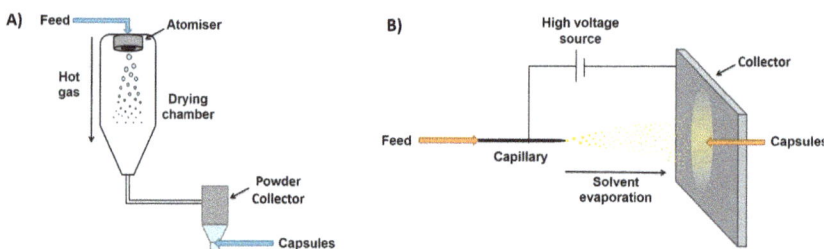

Fig. 2: Simplified schemes of: A) spray-drying and B) electrospraying processes

Some of the advantages of spray-drying for microencapsulation of food ingredients include the high production rates of conventional spray-driers, ease of operation, reproducibility, possibility of continuous operation and automation, low operating costs, and the versatility for processing feeds with variable properties as this allows use of a wide range of encapsulating materials (Vega and Roos, 2006; Đorđević *et al.*, 2015). Its main drawback, however, is the need to use high temperatures for drying, generally between 130-220°C (Zuidam and Heinrich, 2010; Chen *et al.*, 2013a; Jaskulski *et al.*, 2017), which may have a negative impact on thermosensitive bioactive ingredients (Pérez-Masiá *et al.*, 2015a; Gómez-Mascaraque and López-Rubio, 2016). Although the drying of feed droplets itself occurs very fast, in the order of milliseconds to a few seconds, the exposed surface is high, and the obtained microparticles are still subjected to high temperatures during their residence time in the collector. On the other hand, particle agglomeration frequently occurs during spray-drying due to the adhesion and collision forces between droplets and/ or dry particles (Anandharamakrishnan and Ishwarya, 2015c). The thermal efficiency of spray-drying has also been reported to be considerably low, with limited options of heat recovery from the dried materials (Jaskulski *et al.*, 2017).

The encapsulation efficiency of spray-drying is quite variable. Values varying from 10-90 per cent have been reported (Đorđević *et al.*, 2015). Both the properties of the feed fluid to be dried and the process parameters have an impact on the encapsulation efficiency of a particular spray-drying process. The type of wall material selected and its concentration in the feed has a great impact on the size, shape and encapsulation efficiency of the systems. However, there is no universal consensus on what material performs best; a lot of work has been done in an attempt to determine the most suitable wall material or blends for encapsulation (Matsuno and Adachi, 1993; Pérez-Alonso *et al.*, 2003; Jafari *et al.*, 2008; Munin and Edwards-Lévy, 2011; Shi *et al.*, 2013; Muzaffar and Kumar, 2016). In addition, encapsulation efficiency also depends on the nature of the ingredient to be encapsulated. In that respect, efficiency of encapsulation is generally lower for volatile and/or thermosensitive compounds (Rodrigues *et al.*, 2011). Special consideration has to be given to encapsulation of probiotic bacteria through spray-drying, since high temperatures can adversely affect their viability. Nevertheless, bacterial viability of some strains after spray-drying can be enhanced by optimizing the matrix composition and the process parameters, as shown in previous works (Kingwatee *et al.*, 2015; Liu *et al.*, 2015; Khem *et al.*, 2016).

Recent trends in encapsulation through spray-drying include the use of so-called 'nano spray-driers', which allow production of particles with sizes in the submicron range. This is achieved through a different nozzle design based on a vibrating mesh with controlled pore size (Xue *et al.*, 2018) and a collector that applies an electric field to create an electrostatic charge on the dried particles to separate them by deposition on the cathode (Assadpour and Jafari, 2019b). The main advantage of this spray-drying variant is the reduced particle size. However, the productivity is considerably lower than that of the traditional spray-driers. Moreover, the concentration of the feed that can be processed is generally much lower (Anandharamakrishnan and Ishwarya, 2015c; Gómez-Mascaraque *et al.*, 2016b), since too viscous solutions cannot be processed by using meshes with such small-pore sizes. It is also worth mentioning that increase in the size of the powder particles generally improves their storage stability due to reduction in the surface area (Fang and Bhandari, 2017). So even though many recent works tend to produce capsules as small as possible, these considerations should be taken into account.

Another variant of the conventional process is the 'water-free spray-drying'. This refers to spray-drying processes in which instead of water, organic solvents are used to dissolve or disperse the bioactive ingredients and the encapsulation matrix. However, this technique is more commonly used in the pharmaceutical industry; its main drawback in food applications is the toxicity of most of the organic solvents used. Only ethanol is permitted for use in the food industry (Encina *et al.*, 2016). Some potential advantages of not using water, however, include the lower temperatures required for drying, which would decrease oxidation reactions taking place in some sensitive ingredients dispersed in water upon processing at high temperatures (e.g. fish oil). However, other alternatives to reduce the extent of lipid oxidation during spray-drying without the need to use organic solvents have already been proposed. For instance, Serfert *et al.* (2009) reported enhanced oxidative stability of spray-dried fish oil microparticles by using nitrogen instead of air as the drying gas.

1.1.2 Freeze-drying

Freeze-drying is another drying technology suitable for entrapment of bioactive food ingredients within protective matrices. Some experts argue that freeze-drying cannot be considered as a microencapsulation method, since technically there is no capsule formation involved in the process. Instead, a blend of the ingredient(s) of interest and the protective matrix is frozen and subsequently dried in bulk. Usually, the obtained freeze-dried material has to be then crushed, for example, by grinding, to obtain a fine powder. However, according to the generally accepted definition given in the introduction for microencapsulation, freeze-drying meets the criteria, since it allows embedding of an ingredient of interest within a protective matrix.

The concept of freeze-drying is based on drying a frozen formulation by water sublimation under low pressure and temperature (Zuidam and Shimoni, 2010). In principle, the process consists of several stages including freezing, sublimation (often referred to as primary drying), desorption (e.g. secondary drying) and storage stages (Laokuldilok and Kanha, 2015). Together with spray-drying, it is one of the most commonly used microencapsulation technologies in the food industry (Kwak, 2014; Ballesteros *et al.*, 2017). However, it is much more expensive and time consuming than spray-drying (Gharsallaoui *et al.*, 2007). Moreover, the materials obtained are highly porous, which generally results in poor barrier properties (Zuidam and Shimoni, 2010). In addition, due to the fact that upon freeze-drying the material has to be grinded into powder, it is difficult to control particle size of the final particles. Therefore, its application is generally limited to heat-sensitive ingredients that cannot be spray-dried.

Although freeze-drying is conceptually simple, there are a number of considerations that need to be taken into account in order to obtain the best encapsulation performance. For instance, the freezing rate has a great impact on the porosity of the final materials, since a faster freezing rate generally generates smaller and more uniformly distributed ice crystals, which result in smaller pores and better product quality (Ceballos *et al.*, 2012). The size and shape of the ice crystals also determine the rate of drying during sublimation (Hottot *et al.*, 2004) and therefore, the overall energy consumption of the process. Similarly, composition and structure of used wall material influence the encapsulation efficiency, active's retention and release. Inability of the material to dry fast enough increases the content of moisture in freeze-dried encapsulated system, consequently causing the hydration of encapsulated material (Laokuldilok and Kanha, 2015), and possibly the loss of its biological activity. Polymeric materials consisting of short chains of monomers are more prone to structural deformations during the drying process, causing decreased protection of encapsulated material (Ersus and Yurdagel, 2007).

Despite the aforementioned, freeze-drying has been used to encapsulate active compounds of fish oil, phytosterol esters and limonene (Chen *et al.*, 2013a, b); flax oil (Quispe-Condori *et al.*, 2011); fennel oleoresin (Chranioti and Tzia, 2014), curcumin (Cano-Higuita *et al.*, 2015), blackberry anthocyanins (Yamashita *et al.*, 2017), propolis (Šturm *et al.*, 2019), red beet extract (Antigo *et al.*, 2017) and probiotic bacteria (Gul and Atalar, 2019).

Recently, a hybrid microencapsulation technique that combines spraying and freeze-drying has been developed with the aim of producing actual microcapsules while reducing the drying time. This method is referred to as 'spray-freeze-drying' and consists of atomization of a liquid feed followed by freezing the obtained droplets by contact with a cold fluid (e.g. liquid nitrogen) prior to drying by sublimation (Ishwarya *et al.*, 2015). Spray-freeze-drying is about four times less time-consuming than bulk freeze-drying (Shishir *et al.*, 2018), and it has been already applied to microencapsulate food ingredients, such as vitamin E (Parthasarathi and Anandharamakrishnan, 2016), vanillin (Hundre *et al.*, 2015) or ω-3 fatty acids (Karthik and Anandharamakrishnan, 2013), with reported improved oral bioavailability and thermal or oxidative stabilities, respectively.

1.1.3 Electrospraying

Electrospraying, also known as electrohydrodynamic atomization (EHDA), is another drying technique and it is based on the use of high voltage electric fields to spray and dry the feed formulations. The application of electrospraying for microencapsulation was initially proposed for biomedical purposes (Bock *et al.*, 2012), and was extended to food only recently, when the challenges of using edible biopolymers as wall materials began to be addressed (Gómez-Mascaraque *et al.*, 2016a).

Figure 2B shows a simplified scheme of an electrospraying process, in which the feed fluid is slowly pumped through a conductive nozzle connected to a high voltage source. As a result, the drop of fluid which is flowing out of the nozzle is subjected to an electric field that generates electrostatic interactions within the fluid. When these electrical forces overcome the surface tension of the fluid, a charged jet is ejected towards the opposite electrode, where the collector is placed. During the flight from the nozzle to the collector, the solvent evaporates due to the combination of the great surface/volume ratio of the droplets and the repulsive electrical forces, and dry material is finally collected (Chakraborty *et al.*, 2009; Bhardwaj and Kundu, 2010; Bhushani and Anandharamakrishnan, 2014). Depending on the fluid properties and the resulting balance of forces imposed on it, different phenomena may take place. If the cohesion between molecules in the fluid is high enough, the fluid jet is elongated, giving rise to nanofibers upon drying. This process is generally referred to as 'electrospinning' (Kriegel *et al.*, 2008). On the contrary, 'electrospraying' occurs when the intermolecular cohesion is low enough so that the repulsive forces dominate and cause the jet to break into fine droplets, which form dry micro- or nano-particles upon solvent evaporation (Alehosseini *et al.*, 2017).

Both electrospraying and electrospinning, which are based on the same principles and differ only in the morphology of the obtained structures, have been proposed for the encapsulation of food ingredients (Alehosseini *et al.*, 2019a). However, electrosprayed powders are easier to handle and to disperse within food matrices (Gómez-Mascaraque *et al.*, 2015), so electrospraying is generally the preferred mode to produce capsules for functional food applications (Costamagna *et al.*, 2017; Gómez-Mascaraque *et al.*, 2017c) (Fig. 3A), while electrospun fiber mats are mainly proposed as food coatings or active layers for food packaging (Alehosseini *et al.*, 2019b; Moreno *et al.*, 2019) (Fig. 3B). The potential to tailor the morphology of the obtained encapsulation structures for the intended application is one of the advantages of the electrohydrodynamic processing encapsulation techniques.

Generally, matrix type encapsulation structures (Fig. 1A) are obtained through electrospraying. However, it is possible to produce core-shell (Fig. 1B) or even multi-

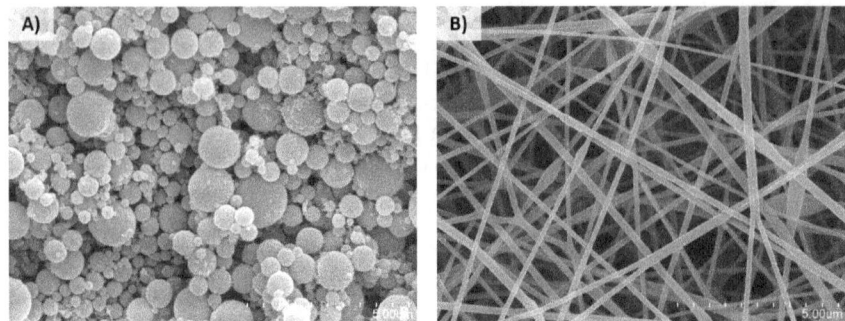

Fig. 3: Scanning electron microscopy images of: A) electrosprayed capsules *vs* B) electrospun fibers (*SEM micrographs are taken by L.G. Gómez-Mascaraque and have not been published elsewhere*)

layered structures (Fig. 1C) by using coaxial or multiaxial configurations, in which the nozzle consists of two or more concentric capillaries through which different fluids are pumped (Zhang *et al.*, 2012; Zhang *et al.*, 2017). Usually, the bioactive ingredient flows through the inner capillary to be located in the core of the capsules and the solution containing the protective matrix is pumped through the external one from independent circuits (Gómez-Mascaraque *et al.*, 2019). The main advantage of multiaxial electrospraying over other techniques to obtain multi-layered encapsulation structures is that they can be obtained in a single step, avoiding the need for subsequent coating steps. Moreover, it offers the possibility of encapsulating lipophilic ingredients within hydrophilic matrices without prior emulsification steps, since both fluids are pumped through independent circuits (Pérez-Masiá *et al.*, 2015a).

Additional advantages of electrospraying include ease of operation, cost-effectiveness (Gómez-Mascaraque and Lopez-Rubio, 2019), high microencapsulation efficiencies (Gómez-Mascaraque *et al.*, 2015, 2017c), high product yields (Jaworek and Sobczyk, 2008) and small particle sizes, often in the submicron range (Gómez-Mascaraque *et al.*, 2015, 2016c), which self-disperse in space due to mutual repulsion of the charged droplets (Zuidam and Velikov, 2018), thus minimizing aggregation. However, its main advantage is the feasibility of working at mild temperature conditions, as it does not require high temperatures nor freezing, making it suitable for both thermo-sensitive and cryo-sensitive ingredients (Gómez-Mascaraque *et al.*, 2016a).

The main drawback of electrospraying is that the properties of feed formulations need to be thoroughly optimized for each particular system in order to be processable by this technique, and only a limited range of concentrations is electrosprayable for each material (Gómez-Mascaraque *et al.*, 2016c; Atay *et al.*, 2018). This is particularly important for food applications, where the use of water as a solvent is almost imperative in order to avoid toxicity issues (López-Rubio and Lagaron, 2012). Water, apart from having a relatively high boiling point compared to other solvents, also has a higher surface tension, which requires higher voltages to obtain stable jets through electrospraying. However, it also has high electrical conductivity; hence, too high voltages would result in corona discharges due to the ionization of water molecules (Pérez-Masiá *et al.*, 2014a). Additionally, many of the edible biopolymers used as encapsulation matrices are polyelectrolytes, with limited chain flexibilities or their molecular weights are too high or too low to provide the required extent of chain entanglements to achieve a stable electrospraying process (Gómez-Mascaraque *et al.*, 2016a). Different approaches have been investigated to enhance the electrosprayability of aqueous biopolymer formulations, such as the addition of food-grade surfactants to reduce surface tension (Pérez-Masiá *et al.*, 2014b) or the denaturation of globular proteins to favor chain entanglements (Nedović *et al.*, 2013).

Electrospraying has been successfully used to encapsulate diverse food ingredients, including curcumin (Gomez-Estaca *et al.*, 2012, 2015, 2017), polyphenols (Atay *et al.*, 2018; Bhushani *et al.*, 2017; Costamagna *et al.*, 2017), vitamins (Pérez-Masiá *et al.*, 2015b), ω-3 fatty acids (Torres-Giner *et al.* 2010; García-Moreno *et al.*, 2017, 2018), carotenoids (Pérez-Masiá *et al.*, 2015a; Gómez-Mascaraque *et al.*, 2017d), probiotic bacteria (López-Rubio *et al.*, 2012; Gomez-Mascaraque *et al.*, 2016d; Moayyedi *et al.*, 2018; Zaeim *et al.*, 2018) and flavors (Khoshakhlagh *et al.*, 2017, 2018). Although most of the research on electrospraying to date has been conducted at lab scale, efforts are currently being made to scale up the process (Zhang *et al.*, 2015).

1.1.4 Spray-chilling and Spray-cooling

Spray-chilling and spray-cooling exploit a similar atomization process as that of spray-drying, but with a different principle for capsule formation. They are low-temperature spray encapsulation methods that consist of solidifying an atomized liquid spray into particles (Oxley, 2012), instead of drying it. Thus, the carriers in this case are generally molten oils which solidify upon contact with cold air. The principles of both spray-chilling and spray-cooling are the same, producing particles as a result of a phase change in the carrier material due to a decrease in temperature. The difference between them is basically the melting point of the carrier material, which is lower in spray-chilling (in the range of 32-42°C) than in spray-cooling (between 45-122°C) (Okuro *et al.*, 2013a). Therefore, mostly polysaccharides, waxes and fats are used as carriers in spray-cooling, while vegetable oils, fatty acids or waxes are mainly used in spray-chilling (Jaskulski *et al.*, 2017).

Like spray-drying, the process involves four stages, i.e. preparation of the feed formulation, spraying, particle solidification and particles collection. However, there are certain particularities of the spray-chilling and spray-cooling processes that differ from spray-drying. In spray-chilling and spray-cooling, the bioactive ingredient is dissolved or dispersed in a molten carrier material. Although one of the advantages of these techniques is that they do not need high temperatures for capsule formation, it is worth mentioning that in order to keep the carrier in its molten state before spraying, the feed needs to be warmed up and kept at a temperature above its melting point before spraying. The spraying stage can be conducted by using the same methods and atomizers as for spray-drying, as long as the pipes and the atomizer are thermo-stated to prevent premature solidification (Jaskulski *et al.*, 2017). The particle formation occurs when the temperature of the sprayed droplets decreases upon contact with the cooling medium, which is usually cold air for spray-cooling, and is further cooled by spraying liquid nitrogen or using solid carbon dioxide baths for spray-chilling (Madene *et al.*, 2006). The collection of particles is also performed in a similar way to spray-drying processes, using cyclones or bag filters.

Spray-cooling and chilling generate matrix type (Fig. 1A) microcapsules, which are generally more spherical, denser, smoother and less porous than spray-dried particles, since no mass transfer (evaporation) occurs in this case (Jaskulski *et al.*,

2017). The size of the particles varies, depending on the type of atomizer used as well as the properties of the feed formulation, and are generally in the range of 500-2000 μm for low-speed rotary atomizers, 150-500 μm for atomization fountain nozzles and 50-150 μm for high-speed rotary atomizers (Đorđević *et al.*, 2015).

Similar to spray-drying, the advantages of these techniques include high production rates, ease of operation and reproducibility. Additionally, spray cooling and chilling are more suitable for heat-sensitive and volatile ingredients. However, cooling the chamber is in some cases more expensive than heating in spray-drying (Jaskulski *et al.*, 2017). Moreover, encapsulation efficiencies are in some cases low due to the crystallization of lipids in the solid particles, that may cause expulsion of the core (Morselli Ribeiro *et al.*, 2012). Also, the obtained capsules need to be stored under very controlled conditions to avoid melting of the wall material and, even doing so, the active ingredients may be lost by diffusion from the lipid matrix during storage (Okuro *et al.*, 2013a). Regarding their application, these solid lipid particles are generally restricted to foods containing high levels of fat (Deshmukh *et al.*, 2016).

Many of the recent works regarding encapsulation, using spray-cooling/chilling, focus on optimization of the wall material composition to avoid or minimize core expulsion and increase the encapsulation efficiency of the method, as well as the shelf-life of the capsules. One strategy consists of adding unsaturated lipids ('liquid lipids') to saturated lipids ('solid lipids'), so that the crystallization kinetics are altered and crystallization defects are created. This less-packed crystalline structure allows better accommodation of the bioactive ingredients to be encapsulated. This approach was studied by Morselli Ribeiro *et al.* (2012), who reduced the migration of the glucose (as a model hydrophilic compound) to the surface of the particles from 22 per cent to 2 per cent by adding oleic acid to the initial stearic acid matrix. The addition of surfactants is another approach that results in better encapsulation efficiency.

Spray-cooling and chilling have been applied for the encapsulation of different food ingredients, such as probiotics (de Lara Pedroso *et al.*, 2012; Okuro *et al.*, 2013b), vitamins (Gamboa *et al.*, 2011; Zoet *et al.*, 2011; Paucar *et al.*, 2016), antioxidants (Consoli *et al.*, 2016; Tulini *et al.*, 2016; de Matos-Jr *et al.*, 2017) and carotenoids (Pelissari *et al.*, 2016). Spray-cooled/chilled products find applications in bakery products, dry soup mixes and other food products containing considerable fat contents (Đorđević *et al.*, 2015).

1.1.5 Emulsification

Emulsions are widely used as an encapsulation technique for food purposes considering they are appropriate for hydrophilic, lipophilic, but also amphiphilic bioactive compounds. In addition, emulsions usually provide high encapsulation efficiency, chemical stability of encapsulated actives and their controlled release (Lu *et al.*, 2016). Emulsions as carriers for active ingredients have an advantage over other encapsulation systems since they are already present in the everyday food we consume (mayonnaise, milk, ice cream, dressings, etc.). Thus their implementation and final application is facilitated.

The process of emulsification is based on dispersing one liquid phase in the other immiscible liquid phase, accompanied by addition of energy to the system. The addition of energy can be low or high (Đorđević *et al.*, 2015). The application of low energy is typical for spontaneous emulsification where the emulsion is made only by mixing the oil phase, water phase, surfactants and co-surfactant at constant temperature. This is convenient for encapsulation of thermo-sensitive actives, but it is challenging due to restricted choice of food-grade surfactants (Santana *et al.*, 2013). On the other hand, utilization of high energy for emulsion preparation is frequently employed in industry. The application of high-pressure homogenization provides emulsions with droplets size between 0.3 and 1 μm and narrow size distribution. The advantage of high energy application lies in the possibility of usage of different emulsifiers and co-emulsifiers (e.g. Generally Recognized As Safe, GRAS – polysaccharides and proteins), but the drawbacks are low energy efficiencies, high production costs and expensive equipment.

There are different ways to categorize emulsions; according to the type of the emulsion structure, they can be classified as conventional emulsions, multiple emulsion (double emulsion), multilayered emulsions, Pickering emulsion and nanoemulsions (Araiza-Calahorra *et al.*, 2018). Conventional emulsions are divided, depending on dispersed and continuous phase to oil-in-water (O/W) emulsions and water-in-oil (W/O) emulsions. Emulsification of O/W and W/O is usually achieved by utilizing some mixing device (colloid mill, sonicator, microfluidizer or high shear mixer). They are mainly thermodynamically unstable and tend to be opaque due to the droplet size (0.2-100μm) (Araiza-Calahorra *et al.*, 2018).

Double emulsions or multiple emulsions are also commonly used for actives encapsulation and they can be water-in-oil-in-water type (W/O/W) or oil-in-water-in-oil (O/W/O) type (Santana *et al.*, 2013). They have two thermodynamically unstable interfaces and thus it is harder to obtain stable multiple emulsions in comparison to conventional ones. Two-step homogenization method is necessary to produce this type of emulsion: during the first step, inner W/O (or O/W) emulsion is formed and then it is further dispersed into the outer phase by using the second homogenization step which is usually conducted in milder conditions (Mao and Miao, 2015). One-step processes were also developed for preparation of multiple emulsions, but they involve a strict selection of emulsifiers, and are still prone to droplet fusion. Multiple emulsions find application mainly in low-calorie foods (Souilem *et al.*, 2014) or as a convenient system for slow and controlled release of actives (Farahmand *et al.*, 2006).

Multilayered emulsions have an additional layer of oppositely-charged emulsifiers on the surface of the emulsion droplet; hence, they are electrostatically stabilized as compared to the conventional ones, especially towards droplets aggregation, freeze-thaw cycling, thermal processing and harsh environmental conditions common in foods (for example, high calcium concentrations) (Mao and Miao, 2015; Trifković *et al.*, 2016). There are many reports about the application of the layer-by-layer (LbL) electrostatic deposition as a method to produce multilayered emulsions (Arnon-Rips and Poverenov, 2018; Bilbao-Sainz *et al.*, 2018; Julio *et al.*, 2019). This method is based on the absorption process of charged polyelectrolyte on to an oppositely-charged droplet surface, forming multilayered structure on the interface (Mao and Miao, 2015). One of the effective applications of LbL method was encapsulation of oils with high ω-3 content (Jiménez-Martín *et al.*, 2016; Fioramonti *et al.*, 2017), but it was also used for microencapsulation of chia seed oil where efficiency achieved was higher than 80 per cent. In general, multilayer emulsions are found to be more efficient in retaining actives (especially flavors) compared to conventional ones (Kaasgaard and Keller, 2010; Benjamin *et al.*, 2013; Xu *et al.*, 2014). In addition, multilayer emulsions are found to be effective in protection of actives from oxidation (i.e. essential oils or polyphenolic compounds) (Maswal and Dar, 2014; Hermund *et al.*, 2016; Acevedo-

Fani *et al.*, 2017). Đorđević *et al.* (2015) reported the mechanisms of actives release from the multilayer emulsions.

In order to avoid the utilization of emulsifiers in the emulsification process, in the last two decades researchers have focused on development of the so-called 'Pickering emulsions'. Here, solid particles partially wetted in both phases (but not soluble in either of them) are employed as stabilizing agents by adsorption on the emulsion droplet interface. In this way, steric barrier is formed, preventing emulsion droplets fusion. It is important to emphasize that solid particles have to be 10-100 times smaller in size than the emulsion droplets in order to be effective as stabilizers (Araiza-Calahorra *et al.*, 2018). Some of the examples of food-grade Pickering particles are cellulose, chitin, nanocrystals, starch granules, insoluble flavonoids, cocoa particles and microgels (Lam *et al.*, 2014; Xiao *et al.*, 2016; Aditya *et al.*, 2017; Winuprasith *et al.*, 2018; Zhou *et al.*, 2018). Recent trends emphasize colloidal particles to be emulsion stabilizers; Xiao *et al.* (2016) discussed in detail five different approaches for fabrication of colloidal particles for Pickering emulsions.

In general, good physical stability of Pickering emulsions designates them as an attractive delivery system. For instance, vitamin D_3 was successfully encapsulated and delivered, using oil-in-water Pickering emulsion (stabilized by nano-fibrillated cellulose), as shown by Winuprasith *et al.* (2018). Furthermore, natural phenolic compound, curcumin, was encapsulated in starch Pickering emulsions by Wang *et al.* (2014), in chitosan-tripolyphosphate stabilized Pickering emulsion by Shah *et al.* (2016) and more recently in Pickering emulsion stabilized by protein nanogel particles by Araiza-Calahorra and Sarkar (2019). However, the advantages offered by Pickering emulsions have to be furthermore explored in order to increase their applicability in the industrially processed foods.

Recent trends in development of nanotechnologies have led to formulation of nanoemulsions with very small droplets, in the size ranging between 50 and 200 nm (Yalçınöz and Erçelebi, 2018). They are mainly transparent and thermodynamically unstable due to their tendency to break down during storage (Trifković *et al.*, 2016). Nanoemulsions are produced via either low- or high-energy processes, usually involving two steps – pre-emulsification and emulsification (Araiza-Calahorra *et al.*, 2018). Nanoemulsions made of food-grade ingredients gained much attention as vehicles to increase bioaccessibility of nutrients as well as their bioavailability (Pool *et al.*, 2013; Guttoff *et al.*, 2015), to deliver bioactive compounds to foods (Donsi *et al.*, 2013; Yalçınöz and Erçelebi, 2018), to enhance antifungal and mycotoxin inhibitory activities and preserve food (Donsi and Ferrari, 2016; Yalçınöz and Erçelebi, 2018; Wan *et al.*, 2019) or to modify the final product structure (McClements, 2017). Recently, Donsi (2018) gave a detailed overview of compounds that have been encapsulated in nanoemulsions in the last few years and reported the main benefits of their utilization along with the problems arising while formulating nanoemulsions. Generally, the application of nanoemulsions in foods is still quite challenging due to limitations like: (i) capital costs of the homogenization systems, (ii) lack of efficient food grade emulsifiers, (iii) customers' concern and global regulations regarding the usage of nano-grade technology, (iv) not entirely revealed interactions of nanoemulsions with food ingredients, and (v) inability to assure stability of nanoemulsion over the period in which food product shelf-life is defined (Donsi, 2018). Nowadays, only few food products enriched with nanoemulsions are available in the food market (Salvia-Trujillo *et al.*, 2017; Yalçınöz and Erçelebi, 2018). Despite the awareness of these obstacles, huge efforts of researchers worldwide to overcome them and an increasingly eager market will definitely lead to an enlarged application of nanoemulsions in foods.

1.1.6 Fluidized Bed Coating

The purpose of fluidized bed-coating technique is to apply an additional layer over previously formed particles. In terms of encapsulation, resulting shell-like capsule can encompass the active compounds either in the particle itself, on particle surface or even in the coating layer (Guignon *et al.*, 2002). Although firstly developed for pharmaceutical purposes, fluidized bed is nowadays increasingly used in the food industry as well.

The process of coating in fluidized bed can be performed in a batch or continuous mode and implies several steps: (i) active ingredients (particles) are suspended in the chamber with the aid of airstream; (ii) solution of coating material is atomized through the nozzle and on to fluidized active ingredients (particles), where the layer of coating is formed; (iii) finally, produced encapsulates are dried and cooled. Usual set-up implies introducing the coating solution either from the bottom (so-called Wurster set-up) or from top of the fluidized chamber; in some cases rotational fluidized chamber can be used. The upward airstream introduced in the chamber counterweights the gravitational force experienced by particles, enabling their fluidization and subsequent coating. Over time, particles are spent in the spraying zone and they become gradually layered with the coating material (Zuidam and Shimoni, 2010). In general, particles obtained by fluidized bed coating are reservoir type (Fig. 1B). Commonly, particles ranging from 100 µm to several mm can be efficiently coated by the fluidized-bed technique. For particles of smaller diameter, other physical forces (e.g. electrostatic forces) may interfere with the successful fluidization of particles, leading to incomplete coating. For purposes of fluidized coating of submicron particles, an optimized system where particles are contained within the rotating perforated drum while airstream is introduced in the chamber tangentially was proposed (Matsuda *et al.*, 2001). Yet another obstacle to overcome when concerning the coating of submicron particles is atomization of coating solution in a particle size range, notably smaller than fluidized particles (usually in the range of 10-100 nm) (Gouin, 2004), which is extremely challenging.

As opposed to most encapsulation techniques, fluidized-bed technique can utilize any kind of wall material being polysaccharides, proteins, fats, powder coatings, emulsifiers, extract of yeast cells (Gouin, 2004). By freely choosing the optimal coating material or mixture of coating materials, it is possible to finely tailor the release of encapsulated materials in the controlled manner. Additionally, it is possible to coat the particles with multiple coating layers, producing particles with multiwall morphology (Fig. 1C) and extended release features. Aqueous solution of polysaccharides (Coronel-Aguilera and San Martín-González, 2015; Pitigraisorn *et al.*, 2017; Pellicer *et al.*, 2019), proteins (Sun *et al.*, 2013; Schell and Beermann, 2014; Benelli and Oliveira 2019) and gums (Benelli and Oliveira, 2019), melted waxes and fats (Müller *et al.*, 2018; Goslinska and Heinrich, 2019) and synthetic polymers (Knezevic *et al.*, 1998) are all used for fluidized-bed encapsulation purposes. However, regardless of the type of coating material used, it is important that particles of coating material are within the size range of 10 µm to ensure retention in the coating layer (Frey, 2016), as well as to achieve complete and homogenous coating over the fluidized particles while evading particle agglomeration (Gouin, 2004). Additionally, during atomization of the coating solution, temperature in the fluid chamber should be kept above the melting point of the coating material so as to avoid particles agglomeration (Abbas *et al.*, 2012). This can pose a challenge for the heat-sensitive bioactive ingredients to be encapsulated.

The main process parameters influencing the efficiency of fluidized-bed coating are flow rate and temperature of the airstream, flow rate and pressure during spraying, as well as nozzle characteristics. Composition and characteristics (e.g. rheology) of the coating solution are important input parameters as well; it is necessary to optimize its solubility, hardness and permeability. Last, but not the least, particles to be coated must possess adequate features: size and size distribution, shape, surface morphology as well as appropriate hydrophilicity/hydrophobicity in regard to coating solution. For a detailed discussion on fluidized-bed coating process parameters, readers are referred to a review by Guignon *et al.* (2002). One drawback of the fluidized-bed technique that has to be taken into consideration is unpredictable trajectory of particles suspended in the airstream; however, there are still many advantages that designate this technique highly desirable for industrial application. In particular, limited pressure drop in the fluid bed chamber, homogeneity of temperature that enables fast heat and consequently mass transfer, as well as ease of control of process parameters, such as flow rate and reaction kinetics, contribute to the applicability of the fluid-bed technique (Teunou and Poncelet, 2002).

Fluidized bed coating is generally considered suitable for encapsulation of water-soluble substances. Different food ingredients have been encapsulated via fluidized bed coating: pigments (Coronel-Aguilera and San Martín-González, 2015), flavors (Sun *et al.*, 2013; Pellicer *et al.*, 2019), antioxidants (Benelli and Oliveira, 2019), probiotics (Schell and Beermann, 2014; Pitigraisorn *et al.*, 2017). Traditionally, fluidized-bed-coated sorbic and lactic acid, calcium propionate and potassium sorbate have been used in the bakery industry (Đorđević *et al.*, 2015). As discussed, fluidized bed is particularly advantageous when tailoring the release of encapsulated material (Sun *et al.*, 2013).

1.1.7 Supercritical Fluids

Supercritical fluid (SCF) is defined as a fluid that is in a specific physical state and is above its critical temperature and critical pressure. This physical state is mainly defined as lying between the gas and the liquid, showing the density similar to the liquid but thermal conductivity, viscosity and diffusivity more similar to those of the gas. This intermediate state is convenient since properties of liquid are useful for active compound solubilization while the transport properties of gas are crucial to improve the mass transfer, which is often the limiting factor of conventional processes, and to enhance selectivity of reactions or extractions (Varona *et al.*, 2016). The most commonly used SCF in the food sector, especially in precipitation and encapsulation processes, is supercritical CO_2 due to its relatively mild critical condition. Furthermore, supercritical CO_2 ($SC\text{-}CO_2$) is nonflammable, environmentally benign, relatively inexpensive, nontoxic, easily removable and chemically stable (Martín *et al.*, 2010; Visentin *et al.*, 2012). $SC\text{-}CO_2$ technology finds many applications in the food industry; some of the most common are aroma extraction, decaffeination, phytoextraction, defatting, and pesticide removal (Lang and Wai, 2001; Kim *et al.*, 2008; Tello *et al.*, 2011; Varona *et al.*, 2016). As an alternative process, $SC\text{-}CO_2$ technology can be also used for particle formation and thus encapsulation. Shortly, the active compound and the carrier are solubilized in SCF and then the solution is sprayed through a nozzle

(Đorđević *et al.*, 2017). In this way, SCF is evaporated, leading to formation of nano- or micro-particles (Fig. 4). Brunner (1994) described two basic configurations that can be used for SC-CO_2 processes, where the pump is used for liquid recirculation while the recirculation of gaseous CO_2 is achieved through a compressor; basically the equipment used is almost the same as in other processes with spray towers and nozzles. The exact conditions must be optimized for each process since the particles formation, as well as particles final properties, depend on the main process conditions. The effect of each process parameter on the particles size is described in detail by Kalani and Yunus (2011) and readers are referred to it.

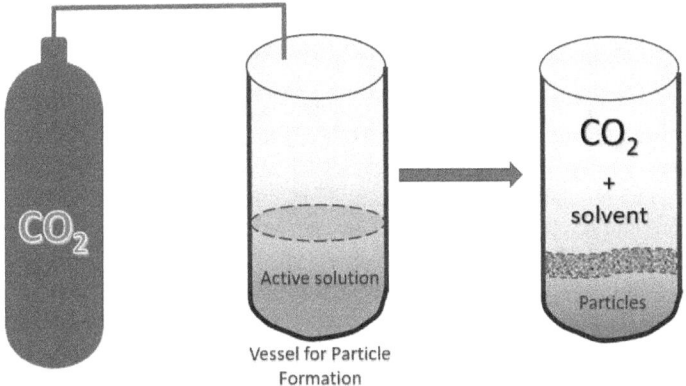

Fig. 4: Simplified scheme of SC-CO_2 process

The role of SC-CO_2 in the process of particles formation classifies the supercritical fluid techniques as solvent (rapid expansion of supercritical solution – RESS); antisolvent (supercritical antisolvent – SAS); cosolvent or solute (particles from gas-saturated solutions – PGSS); nebulization compound (supercritical fluid-assisted atomization – SAA); extractor and antisolvent techniques (supercritical fluid extraction of emulsions – SFEE) (Silva and Meireles, 2014).

RASS processes are based on dissolution of both carrier material and actives to be encapsulated in the SC-CO_2, followed by spraying through nozzle and precipitation. The morphology of the obtained particles is determined by two major factors – the material structure and the process parameters (temperature, pressure drop, size of the atomization vessel, distance of the jet from the surface, nozzle design) (Varona *et al.*, 2016). A few food compounds were encapsulated using RESS process: jaboticaba peel extract (Santos *et al.*, 2013), rutin and anthocyanin-rich extract (with polyethylene glycol) (Santos and Meireles, 2013).

The SAS process can be operated in the batch or semi-continuous mode and the initial concentrations of actives and carrier in this solution are main parameters that affect particle morphology, while the temperature and interactions between carrier and SC-CO_2 may lead to an uneven precipitation. Recently, Arango-Ruiz *et al.* (2018) used SAS process for encapsulation of curcumin and the obtained particles were smaller than 10 µm. Slightly smaller particles (from 0.8 to 3 µm) were obtained when SAS process was used for encapsulation of lycopene with lecithin and α-tocopherol (Cheng *et al.*, 2017). Oliveira *et al.* (2017) applied SAS process to prepare particles of encapsulated passion fruit seed oil with a poly(lactic-co-glycolic acid), PLGA. They varied in processing conditions (temperature: 35 and 45°C, CO_2 mass fraction: 92.5 and 95.0 per cent, pressure: 90 and 110 bar) and obtained particle sized from 721 to 1498nm and the oil encapsulation efficiency from 67.8 to 91 per cent. Similarly, the optimization of SAS was done to obtain riboflavin, δ-tocopherol and β-carotenein-loaded zein microcapsules (Rosa *et al.*, 2019). The mean particle size of the microcapsules was from 8 to 18 μm, depending on encapsulated vitamin, while the precipitation yield varied from 41 to 82g/100g. An interesting study compared three different methods for encapsulation of anthocyanin-rich extract from blackberry residues: spray-drying, freeze-drying and SAS. The SAS process achieved favored precipitation of anthocyanins in comparison to spray-drying and freeze-drying, since $SC-CO_2$ does not have any affinity to such compounds (Da Fonseca Machado *et al.*, 2018).

PGSS can be performed in batch or continuous mode and obtained particles are spherical in shape and usually 10 μm to 100 μm in size (slightly bigger than those obtained when using RESS or SAS processes). The morphology of the particles can be optimized by regulating the process conditions, predominantly the volume of SCF dissolved into the solution and the nozzle geometry. An encapsulation efficiency of 60 per cent was achieved when β-carotene was encapsulated in soy lecithin using PGSS process, producing spherical particles sized from 10 to 500 μm (Paz *et al.*, 2019). Furthermore, PGSS has been applied to encapsulate omega-3 polyunsaturated fatty acids by using non-oxidative conditions and 98 per cent encapsulation efficiency was achieved (Melgosa *et al.*, 2019).

The SAA process enables the usage of either organic solvents or aqueous solutions. Final particle size and particle size distribution can be controlled by the selection of solvent, solute concentration and feed ratio, providing particles with size between 0.05 and 5μm. The SAA was mainly applied for preparation of particles with a variety of pharmaceutical compounds (Silva and Meireles, 2014), but it is lately being used for encapsulation of phenolic compounds in maltodextrin (Aliakbarian *et al.*, 2017).

SFEE can be considered as an evolution of the SAS process where the initial droplet size in the emulsion has a main influence on the final particle size (Mattea *et al.*, 2010). Mendonça *et al.* (2019) encapsulated neem seed oil in poly(3-hydroxybutyrate-co-3-hydroxyvalerate), using SFEE where initial emulsion had average droplets size of 463nm while average particle size after application of the SFEE was 228 nm. SFEE was also employed for encapsulation of fish oil and three different initial emulsion formulations were tested (containing Tween 80, polycaprolactone and acetone). Very small nanoparticles with sizes ranging from 6 to 73 nm (depending on the formulation) were obtained (Prieto and Calvo, 2017). The SFEE process was also used for encapsulation of shrimp-residue extract enriched with astaxanthin or essential oil of laurel leaves, using the modified starch and dichloromethane (Mezzomo *et al.*, 2012; Reis *et al.*, 2018) and lycopene and β-carotene by using OSA modified starch and dichloromethane (Santos *et al.*, 2012).

Generally speaking, the operational expenses of SCF processes have a tendency to be relatively low due to the low energy consumption, but the investment costs can be extremely high. Moreover, investment costs scale-up proportionately as the size of the plant is enlarged (Perrut, 2000). Further studies evaluating food-grade organic solvents and natural carriers may result in food-engineering innovations, possibly leading to cost reduction. In addition, new knowledge and detailed understanding of the interactions between supercritical fluid, carrier and active compound will provide novel opportunities for controlling the properties of encapsulates prepared by SCF processes.

1.1.8 Microgels

Microgels (or micro hydrogels) are considered soft microstructures consisting of cross-linked molecules of polymers. In general, the term 'microgels' encompasses both nanogels, with a diameter below 0.5 μm and microgels that are bigger in diameter (Vinogradov, 2010). Microgels have the unique property of reversible swelling and de-swelling in an adequate solvent, based on the type of polymer and its response to surroundings. This feature is used to tailor the specific delivery systems responsive to environment stimuli, e.g. changes of temperature, pH or ionic strength, in order to deliver and release encapsulated material at targeted place. In addition, by modulating properties of microgel particles, such as the degree of crosslinking, particle size and size distribution, particle surface charge or particle shape (including fibers, rods and spheres), it is possible to produce delivery systems to suit specific need (Shewan and Stokes, 2013).

To produce microgels, either emulsification or extrusion can be used (Đorđević *et al.*, 2015). Extrusion techniques are based on extruding liquid solution of encapsulating material and bioactives through the nozzle and into the gelling solution. Droplet formation on the nozzle tip as well as the jet breaking are governed by different physical forces: gravitational force, surface tension, impulse and frictional forces (Whelehan and Marison, 2011). Consequently, extrusion techniques can be classified as simple dripping, coaxial airflow, vibrating jet/nozzle, spinning disk atomization, jet cutting and electrostatic extrusion (Fig. 5). The characteristics of the techniques as well as features of the produced particles are compared in Table 1.

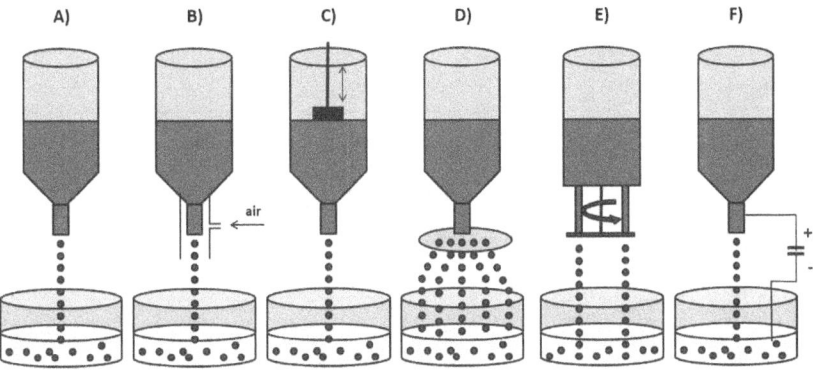

Fig. 5: Extrusion techniques: A) simple dripping; B) coaxial air flow; C) vibrating jet/nozzle; D) spinning disc atomization; E) jet cutting and F) electrostatic extrusion

The physical properties of the produced particles greatly depend on the polymer features, e.g. viscosity, composition and surface tension. By adjusting the concentration of polymer solution, it is possible to impact the shape and sphericity of the particles; highly viscous polymer solutions may result in formation of filaments and irregularly

Table 1: Comparison of Different Extrusion Encapsulation Techniques
(Whelehan and Marison, 2011)

	Simple Dripping	Coaxial Airflow	Vibrating Jet/Nozzle	Spinning Disc Atomization	Jet Cutting	Electrostatic Extrusion
Characteristics of the Technique						
Relatively easy to set-up and operate[a]	√	√	√	√	√	√
Short production time	√	√	√	√	√	√
High production rates[b]	×	×	×	√	×	×
High efficiency[c]	√	√	√	×	×	√
Capacity to extrude viscous solutions	×	×	×	√	√	×
Capacity to produce a range of different size droplets	√	√	√	√	√	√
Possibility to operate under sterile conditions	√	√	√	√	√	√
Features of the Microparticles						
Mono-dispersed	√	√	√	√	√	√
Homogenous with spherical shape	√	√	√	√	√	√
Small size[d]	×	×	√	×	×	√
Narrow size distribution[e]	√	√	√	√	√	√

√ - Technique complies with the requirement; × - Technique doesn't comply with the requirement

[a] - Does not require experts to repeatedly set up, nor extensive training and supervision to operate the process

[b] - No extensive loss of wall or encapsulated material

[c] - Production capacity of tons/day

[d] - Particle size smaller than 100 μm

[e] - Overall deviation of smaller than ± 5 per cent

shaped particles. Additionally, polymers with the ability to form polyelectrolyte complexes with oppositely charged polymers can be utilized to produce microgels with increased mechanical stability, reduced porosity and enhanced actives entrapment (Vilela *et al.*, 2015). Extrusion encapsulation techniques are generally performed under mild environmental conditions (e.g. temperature, solvents); thus they are suitable for encapsulation of living cells. In addition, it is possible to perform the encapsulation under sterile conditions (Heinzen *et al.*, 2004; Obradovic *et al.*, 2020).

Production of microgels by emulsification is a two-step process, entailing, firstly, dispersion of polymer solution into the oil phase (for food purposes, vegetable oils are usually used) by high shear mixing, e.g. high pressure homogenization or colloidal milling; second step involves addition of gelling agent in order to promote particles formation, which can be achieved by introducing heat, salt or acid treatment (Torres *et al.*, 2016). Microgels produced in this way are called emulsion gels (or emulsion hydrogels), where lipid droplets (e.g. fillers) are entrapped within the aqueous phase (e.g. gel matrix). An alternative approach to obtain microgels by emulsification is production of so-called filled hydrogel particles (or emulsion microgel particles), where several emulsion droplets are enclosed within gel shell (Torres *et al.*, 2016). This type of microgels is basically an O/W/W system, since hydrogel particles containing oil droplets are dispersed within the aqueous phase (McClements *et al.*, 2009). The advantage of filled hydrogel particles is that they can be used to encapsulate hydrophobic actives, by adding them to the oil phase prior to O/W emulsion preparation, while hydrophilic actives can be entrapped within the water phase.

The main constraints for application of microgels on industrial scale are related to the productivity issues. Among extrusion techniques, higher capacity of production can be achieved by utilizing the vibrating jet/nozzle, spinning disk atomization or jet cutting techniques (Fig. 5C, D and E) (Whelehan and Marison, 2011). On the contrary, production of microgels by emulsification is easier to scale up; nevertheless, processing costs can be higher than those of extrusion encapsulation given that oil used for emulsification has to be removed from the system (Burey *et al.*, 2008). Emulsification can yield microgels of 1-10 μm size, while extrusion techniques, in particular electrostatic extrusion, enables production of particles down to 50 μm in size (Nedovic *et al.*, 2011).

Microgels are frequently used in food products, such as yogurts, dairy desserts, cheese and sausages (Mun *et al.*, 2015). Commonly, encapsulation in microgels is used to entrap hydrophilic compounds, such as plant extracts (Belščak-Cvitanović *et al.*, 2011; Stojanovic *et al.*, 2012; Trifković *et al.*, 2014; Trifković *et al.*, 2015), probiotics (Dimitrellou *et al.*, 2019; Lee *et al.*, 2019) and food colorants (Pradeep and Nayak, 2019). However, microgels can be used for hydrophobic compounds as well (Levic *et al.*, 2013; Waterhouse *et al.* (2014); Chew and Nyam, 2016; Volić *et al.*, 2018). Additionally, extrusion encapsulation has been utilized for production of immobilized biocatalysts, where yeasts, bacteria and enzymes are encapsulated and used for fermentation purposes (Nedović *et al.*, 2013).

1.1.9 Coacervation

Coacervation processes are essentially liquid-liquid separation processes. Depending on the polymer(s) participating in the reaction and the mechanism of phase separation, two different types of coacervation can be distinguished – simple and complex coacervation. Simple coacervation implies single polymer which under certain environmental stimuli (such as addition of salts) re-assembles its intermolecular structure to form coacervates; usually, polymers involved in simple coacervation carry self-charge, both negative and positive (Mohanty and Bohidar, 2005). In complex coacervation, phase separation is induced by interactions between two (or more) oppositely-charged polymers. Commonly, governing force for formation of complex coacervates is electrostatic interaction among charged polymers. Nevertheless, Van der Waals intermolecular interactions, hydrophobic interactions within proteins or conformation changes in nucleic acids can also contribute to formation of complex coacervates (Turgeon *et al.*, 2007; Schmitt and Turgeon, 2011; Xiao *et al.*, 2014). Type of interaction between polymers determines if the phase separation in the system is segregative type, where repulsive interactions are behind the separation, or associative type, when the cause of phase separation is electrostatic interaction between adversely-charged polymers (Gupta *et al.*, 2012). Usually, to produce coacervates, proteins-polysaccharides mixtures are used although there are studies on protein-protein (Elzoghby *et al.*, 2012), as well as polysaccharide-polysaccharide coacervates (Fathi *et al.*, 2014).

The mechanism of complex coacervation typically involves four steps: (i) polymers (two or more) solutions preparation, which is usually done when the temperature is higher than gelling point of polysaccharide and pH higher than the protein isoelectric point; (ii) dispersion of active compound into respective polymer solution followed by homogenization until stable emulsion is achieved; (iii) coacervation – pH or temperature-induced phase separation, and (iv) hardening of the produced particles either by decrease in temperature or by addition of desolvation agent or cross-linking agent (Ortega-Rivas *et al.*, 2006). Particles produced by complex coacervation are usually core-shell type (Fig. 1B), where the morphology and size of the particles is mainly dependent on the stirring rate (Lemetter *et al.*, 2009). In addition, the type of polymers used and interactions occurring between polymers are significantly influencing the texture, structure and stability of coacervates, as well as the encapsulated material (Timilsena *et al.*, 2019). Another important factor is pH, since it determines the degree of ionization of polymers (both proteins and polysaccharides) and enables the initiation of coacervation process (Weinbreck *et al.*, 2003).

It is claimed that coacervation is one of the most effective techniques for actives encapsulation in the food industry (Golzi *et al.*, 2013; Butstraen and Salaün, 2014; Xiao *et al.*, 2014; Timilsena *et al.*, 2019). Coacervation is a simple method, easily scalable, with low costs and high encapsulation efficiency, where the payloads of up to 99 per cent can be achieved (Gosh, 2006; Drusch *et al.*, 2012). The release of actives encapsulated within coacervates is controlled by mechanical stress, temperature and pH shifts or sustained release (Gouin, 2004). The main drawback of coacervation is that it is more suitable for hydrophobic compounds. In order to encapsulate hydrophilic compounds, several changes in the method have to be introduced (Comunian *et al.*, 2013; Rocha-Selmi *et al.*, 2013).

Coacervation has been used to encapsulate various food components, such as omega-3 rich oils – tuna oil (Bakry *et al.*, 2019), *sacha inchi* oil (da Silva Soares *et al.*, 2019) or fish oil (Mirzaei and Jafarpour, 2019); cinnamon extract (de Souza *et al.*, 2020); garlic extract (Tavares *et al.*, 2019); ascorbic acid (Rodrigues da Cruz *et al.*, 2019), vitamin D (Jannasari *et al.*, 2019) and probiotic cells (de Almeida Paula *et al.*, 2019), to name a few. Recently, an innovative easily scalable approach has been proposed to encapsulate D-limonene, where the multiple steps are collapsed into one, to form dry complex coacervates (Tang *et al.*, 2020).

1.1.10 Liposomes

Liposomes are one of the most widely used encapsulation techniques and thus they are applied in different cosmetic, pharmaceutical, nutraceutical and food products. They are suitable for encapsulation and consequently for protection of both hydrophobic and hydrophilic active compounds; but this is not the only advantage of liposomal technique. Mainly, liposomes can be prepared by using endogenic constituents providing biocompatible, non-immunogenic and non-toxic carrier which can act as target delivery system in the organism (Amjadi *et al.*, 2018). This fact allows quicker and easier implementation of the liposomes in foods and cosmetics, overcoming regulatory barriers (Đorđević *et al.*, 2015).

Liposomal vesicles can be described or defined as closed lipid and/or phospholipid colloidal structures composed of hydrophilic core and lipophilic bilayered shell (Fig. 1F). The bilayered shell is formed by input of energy (e.g. mechanical agitation, homogenization, heating, sonication, etc.) to phospholipids in an aqueous surrounding (Khorasani *et al.*, 2018). The outcome of energy input is the arrangement of the lipid molecules (formation of lipid vesicles-liposomes) in such a manner to achieve a thermodynamic balance in an aqueous surrounding. Available literature suggests several laboratory-scale and a few large-scale techniques for liposome preparation, but only a small number of these are applicable in the food industry. Conventional techniques, like thin-film hydration method, ethanol injection and reverse phase evaporation involve usage of potentially toxic solvents; they are also time consuming and difficult to scale-up. Microfluidization, proliposome technique, membrane contactor technique, freeze drying of double emulsion and Mozafari technique are easily scalable techniques and most of them can use commercial lipid mixtures as a starting material which reduces processing costs (Đorđević *et al.*, 2015; Khorasani *et al.*, 2018). The proper choice of technique to be applied for liposomes formation depends on several factors – type of active compound to be encapsulated; properties (both physical and chemical) of solvents used for dissolution; anticipated size, polydispersity, shelf-life, desired bioactive release profile and last but not the least, adequate (effective) quantity of active compound encapsulated within the liposomes (Khorasani *et al.*, 2018). By using previously mentioned techniques and optimizing them, it is possible to prepare multilamellar liposomes of several hundred nanometers, but if an additional step for size reduction is applied, multilayer liposomes can be downsized even to the nano-scale level.

Currently there is a rising interest in application of liposome vesicles in the food industry (Mozafari *et al.*, 2008). Bioactive compounds (essential oils, antioxidants, fish oils, vitamins, flavors, enzymes, etc.) incorporated in the liposomal structure can be added to functional food to improve nutraceutical value of the food or to extend its shelf-life. In recent literature, numerous information is given about liposomes loaded with a great variety of natural active compounds. For example, an antioxidant peptide fraction from rainbow trout-skin, gelatin was encapsulated in liposomes by Ramezanzade and coworkers (2017). Different polyphenolic compounds were also successfully incorporated in liposomes (Isailović *et al.*, 2013; Balanč *et al.*, 2015; Gibis *et al.*, 2016; Popova and Hincha, 2016; Dag and Oztop, 2017; Pravilović *et al.*, 2017; Jovanović *et al.*, 2019). Equally effective were encapsulations of lipophilic compounds (such as carotenoids or unsaturated fatty acids) in liposomes (Du *et al.*, 2015; Semenova *et al.*, 2016; Tan *et al.*, 2016; Beltrán *et al.*, 2019). Despite all advantages, liposomes as vehicles have some drawbacks, such as possibility of leakage of entrapped compound during storage. They can also be instable in the presence of sugars and salts in real food applications, causing vesicle fusion or aggregation (Frenzel and Steffen-Heins, 2015). Thus, alternative processes to overcome these drawbacks to some extent are proposed. For example, lyophilization can be applied to increase the shelf-life of liposomes by conserving them in a dehydrated state, but formation of ice crystals during this process can damage liposomal vesicles; so the addition of cryoprotectants is recommended (Marín *et al.*, 2018). Another way of liposomes protection is coating of their surface with a biopolymer, applying layer-by-layer-technique. The most commonly used biopolymers for this purpose are chitosan (Gibis *et al.*, 2016; Gültekin-Özgüven *et al.*, 2016; Hao *et al.*, 2017; Jiao *et al.*, 2018; Hasan *et al.*, 2019), whey protein isolate (WPI) (Frenzel and Steffen-Heins, 2015), pectin (Lopes *et al.*, 2017; Feng *et al.*, 2019) or even a combination of two biopolymers, like pectin and WPI (Gomaa *et al.*, 2017).

Up to now different kind of juices were fortified with liposomes encapsulating different actives. For example, ascorbic acid encapsulated in liposomes was added to apple juice (Wechtersbach *et al.*, 2012). A combination of ascorbic acid and alpha-tocopherol incorporated in liposome was applied in orange juice as well as a combination of resveratrol and epigallocatechin gallate-loaded liposomes (Marsanasco *et al.*, 2011; Feng *et al.*, 2019). Quercetin encapsulated in liposomes coated with WPI was added to a functional whey permeate drink (Frenzel *et al.*, 2015). There are also some reports of addition of liposomes to foodstuffs. Addition of liposome loaded with lipase in Cheddar cheese enlarged the production of free fatty acids and improved the flavor profile (Kheadr *et al.*, 2002). In the same way, Rashidinejad *et al.* (2014) described incorporation of liposomes with encapsulated green tea catechin and epigallocatechin gallate into low-fat hard cheese. Aware that gummy candies are main part of the confectionery products, Amjadi *et al.* (2018) successfully fortified them with betanin-loaded liposomes with the idea to deliver natural active compound important for human health of both kids and the elderly. Similarly, black mulberry extract-loaded liposomes, coated with chitosan, were used for fortification of dark chocolate (Gültekin-Özgüven *et al.*, 2016). One of the interesting applications of liposomes is the addition of catechin-loaded liposomes to Chinese dried pork, providing the antioxidant and antibacterial effects on the pork, specifically after 25 days of storage (at room temperature) (Wu *et al.*, 2018). Additionally, lipids (main constituents of liposomes) have no interaction with taste receptors and thus liposomes are extremely favorable for masking the unpleasant taste of bitter active compounds (Sun-Waterhouse and Wadhwa, 2013.). Therefore, liposomes application in the real food systems should be expanded. Researchers are yet to upgrade and promote techniques for production of liposomes on a large scale or innovate new ones in order to encourage food producers to implement liposomes in functional food products.

1.1.11 Inclusion Complexation

Inclusion complexation is a technique where the process of encapsulation happens on a supramolecular level. Active compound (guest molecule or ligand) is entrapped within the host molecule. The process of encapsulation entails non-covalent physicochemical forces, such as hydrogen bonding, hydrophobic interactions or van der Waals forces (Marques, 2010).

Cyclodextrins (CDs) are the most commonly used host molecules. CDs are a group of naturally occurring cyclic oligosaccharides with a truncated cone shape (Cal and Centkowska, 2008) (Fig. 1E). Depending on the number of glucose residues (six, seven or eight) that are forming cone-shaped structure, it is possible to distinguish α-, β- and γ-CDs, respectively (Lucas-Abellan *et al.*, 2008). All three CDs have been approved as additives in the food industry in European Union, at the same time holding GRAS status by FDA (Ciobanu *et al.*, 2013). However, due to the high costs of purification of α- and γ-CDs, β-CD is most widely used for commercial application; up to 97 per cent of CDs used in the market are β-CD (Astray *et al.*, 2009).

The most prominent feature of CDs is their capacity to establish inclusion complexes with a wide range of molecules (solid, liquid or gas) (Hirose and Yamamoto, 2001). CDs can entrap hydrophobic actives within the internal hydrophobic cavity, while hydrophilic actives can be accommodated by the hydrophilic surface of truncated cone. This is due to the presence of the CH_2 and hydroxyl groups in inner and outer side of the CD molecule, respectively (Recharla *et al.*, 2017). By entrapping actives within the CD molecule, it is possible to achieve improvement of solubility of poorly soluble actives, to enhance the stability of actives against degradation caused by oxidation, light or temperature, as well as to control sublimation and volatility of sensitive active compounds, such as flavors. In addition, encapsulation in CDs enables physical separation of incompatible compounds while masking the unpleasant taste, flavor or odor of the encapsulated compound and providing its modified release (Wadhwa *et al.*, 2017). CDs have favorable thermal and chemical stability (Hedges *et al.*, 1995) and they are non-hydrolysable by β-amylases and glucoamylases (Wadhwa *et al.*, 2017). Additionally, it was shown that complexation with CDs increases the bioavailability of flavonoids as confirmed in experiments with rats (dos Santos Lima *et al.*, 2019). The main disadvantage of CDs application is their poor water solubility which can be up to nine times less when compared to linear oligosaccharides (Wadhwa *et al.*, 2017). To solve this, different CD derivatives have been explored, such as methylated, hydroxypropylated or sulfobutylated CDs (Duchene, 2011).

CDs encapsulation methods are generally based on the simple co-precipitation principle (Dias *et al.*, 2010; Duchene, 2011); however, the conditions of inclusion complexation have to be optimized for each guest molecule (Koontz *et al.*, 2009). The active compound dissolved in the compatible solvent is added to CD aqueous solution, under mixing or sonicating, which can be accompanied with heating (Karathanos *et al.*, 2007; Kalogeropoulos *et al.*, 2010; Kfoury *et al.*, 2014) followed with filtration and drying (Marques, 2010). In cases where used CDs have low solubility (i.e. β-CD), precipitation occurs naturally. On the contrary, when CDs used are of higher solubility, precipitation has to be aided via addition of a co-solvent. In these cases, inclusion complex is formed by evaporation of solvent/co-solvent, at the elevated temperature, using either hot air (Wang *et al.* 2011) or vacuum (Duchene, 2011). This is a basic principle of co-evaporation method. Encapsulation within CDs can be achieved by spray-drying and freeze-drying methods as well, where the mixture of CDs and active compounds are sprayed through a nozzle or lyophilized, respectively (Marques, 2010; Recharla *et al.*, 2017). Recently, there have been reports on preparation of CDs complexes using supercritical antisolvent technique, where $SC-CO_2$ is used as antisolvent and as an alternative to organic solvents (Recharla *et al.*, 2017).

The yield and encapsulation efficiency of the inclusion complexation varies due to different factors. Hydrophobic-hydrophilic nature of actives (guest molecules) predominantly affects the efficiency of complexation. However, the type of CDs used, host-to-guest molecule ratio and selected technique to prepare inclusion complex contributes to the inclusion/encapsulation efficiency (Wadhwa *et al.*, 2017). In addition, process parameters, such as temperature, time and rate of mixing or separation conditions can also influence the efficiency of inclusion (Dias *et al.*, 2010; Gomes *et al.*, 2014). In general, efficiency of inclusion has been reported in the range from 30 per cent to 99 per cent (Đorđević *et al.*, 2015).

The cavity in CDs molecule is of hydrophobic nature; hence, CDs are mainly used to encapsulate hydrophobic active compounds. While the depth of the cavity is ~0.8 nm, the diameter depends on the type of CDs, determining as to what molecules can be encapsulated. CDs have been used as a carrier material for essential oils (Wadhwa *et al.*, 2017), plant bioactive compounds (Pinho *et al.*, 2014), linoleic acid (Hadaruga *et al.*, 2006), resveratrol (Lucas-Abellán *et al.*, 2007), red bell pepper carotenoids (De Lima Petito *et al.*, 2016), lycopene (Nerome *et al.*, 2013), to name a few.

2. Conclusion and Future Perspectives

The shift in the functional food demand from the early twentieth century, when functional food products were primarily directed at nutritional deficiency diseases (i.e. iron deficiency, scurvy and avitaminosis in general) to modern demand for foods that support overall health and wellness, or target specific lifestyle (e.g. sports nutrition) or specific age group (infant nutrition, elderly nutrition), created a need to develop technologies to successfully deliver functionality to food products. Natural bioactives and active ingredients in general are sensitive compounds which are prone to degradation under environmental conditions and during food processing. In addition, after oral consumption, bioactives undergo rapid first-pass metabolism, which can lead to modification of their chemical structure and loss of bioactivity. Encapsulation technologies can ensure stability of active ingredients against detrimental effects of light, oxygen, pH and temperature changes during product processing, transport and storage; mask the unpleasant taste or fragrance of actives; improve solubility and bioavailability of actives; as well as enable targeted delivery and prolonged/controlled release of encapsulated actives. Additionally, by adding encapsulated actives to the food products, it is possible to impact physical properties of foods, such as rheological properties or texture.

Wide application of encapsulation at the industrial scale is still facing certain challenges, mainly related to scale-up of encapsulation processes, productivity issues and equipment investment costs. The production of particles of specific size and size distribution, with the well-defined shape and properties on the commercial scale, is still proving difficult. Nevertheless, research within both scientific community and industrial research and development are propelling improvements in encapsulation strategies as to how to successfully overcome challenges associated with nutritional value, color and taste of foods, as well as food preservation. Developing effective encapsulated systems for the delivery of functionality to food is opening enormous business opportunities for food and beverage companies on the functional foods market. Modern-life diseases, such as obesity, especially child obesity, metabolic syndrome, cardiovascular diseases and type 2 diabetes can be tackled by adopting a healthy diet and lifestyle habits. Hence, food companies are emphasizing more on embracing encapsulation technologies and producing functional foods. In 2016 global food encapsulation market was worth over USD 27 billion, and it is predicted to grow over 6 per cent by 2024.

References

Abbas, S., Wei, C.D., Hayat, K. and Xiaoming, Z. (2012). Ascorbic acid: Microencapsulation techniques and trends – A review, *Food Reviews International*, 28: 343-374.

Acevedo-Fani, A., Silva, H.D., Soliva-Fortuny, R., Martín-Belloso, O. and Vicente, A.A. (2017). Formation, stability and antioxidant activity of food-grade multilayer emulsions containing resveratrol, *Food Hydrocolloids*, 71: 207-215.

Aditya, N.P., Espinosa, Y.G. and Norton, I.T. (2017). Encapsulation systems for the delivery of hydrophilic nutraceuticals: Food application, *Biotechnology Advances*, 35: 450-457.

Alehosseini, A., Ghorani, B., Sarabi-Jamab, M. and Tucker, N. (2017). Principles of electrospraying: A new approach in protection of bioactive compounds in foods, *Critical Reviews in Food Science and Nutrition*, 58: 1-18.

Alehosseini, A., Gómez-Mascaraque, L.G., Ghorani, B. and López-Rubio, A. (2019a). Stabilization of a saffron extract through its encapsulation within electrospun/ electrosprayed zein structures, *LWT – Food Science and Technology*, 113: 108280.

Alehosseini, A., Gómez-Mascaraque, L.G., Martínez-Sanz, M. and López-Rubio, A. (2019b). Electrospun curcumin-loaded protein nanofiber mats as active/bioactive coatings for food packaging applications, *Food Hydrocolloids*, 87: 758-771.

Aliakbarian, B., Paini, M., Adami, R., Perego, P. and Reverchon, E. (2017). Use of supercritical assisted atomization to produce nanoparticles from olive pomace extract, *Innovative Food Science & Emerging Technologies*, 40: 2-9.

Amjadi, S., Ghorbani, M., Hamishehkar, H. and Roufegarinejad, L. (2018). Improvement in the stability of betanin by liposomal nanocarriers: Its application in gummy candy as a food model, *Food Chemistry*, 256: 156-162.

Anandharamakrishnan, C. and Ishwarya, S.P. (2015a). Industrial relevance and commercial applications of spray-dried active food encapsulates, pp. 275-284. *In*: C. Anandharamakrishnan and S.P. Ishwarya (Eds.), *Spray-drying Techniques for Food Ingredient Encapsulation*, John Wiley & Sons, New Jersey.

Anandharamakrishnan, C. and Ishwarya, S.P. (2015b). *Spray-drying Techniques for Food Ingredient Encapsulation*, John Wiley & Sons, New Jersey.

Anandharamakrishnan, C. and Ishwarya, S.P. (2015c). Introduction to spray drying, pp. 14-15. *In*: C. Anandharamakrishnan and S.P. Ishwarya (Eds.), *Spray-drying Techniques for Food Ingredient Encapsulation*, John Wiley & Sons, New Jersey.

Antigo, J.L.D., Bergamasco, R.D. and Madrona, G.S. (2017). Effect of pH on the stability of red beet extract (*Beta vulgaris* L.) microcapsules produced by spray-drying or freeze-drying, *Food Science and Technology*, 38: 72-77.

Araiza-Calahorra, A. and Sarkar, A. (2019). Pickering emulsion stabilized by protein nanogel particles for delivery of curcumin: Effects of pH and ionic strength on curcumin retention, *Food Structure*, 21: 100113.

Araiza-Calahorra, A., Akhtar, M. and Sarkar, A. (2018). Recent advances in emulsion-based delivery approaches for curcumin: From encapsulation to bioaccessibility, *Trends in Food Science & Technology*, 71: 155-169.

Arango-Ruiz, Á., Martin, Á., Cosero, M.J., Jiménez, C. and Londoño, J. (2018). Encapsulation of curcumin using supercritical antisolvent (SAS) technology to improve its stability and solubility in water, *Food Chemistry*, 258: 156-163.

Arnon-Rips, H. and Poverenov, E. (2018). Improving food products' quality and storability by using layer by layer edible coatings, *Trends in Food Science & Technology*, 75: 81-92.

Arslan, S., Erbas, M., Tontul, I. and Topuz, A. (2015). Microencapsulation of probiotic *Saccharomyces cerevisiae* var. *Boulardii* with different wall materials by spray drying, *LWT – Food Science and Technology*, 63: 685-690.

Assadpour, E. and Jafari, S.M. (2019a). Nanoencapsulation: Techniques and developments for food applications, pp. 35-61. *In*: A. López Rubio, M.J.F. Rovira, M. Martínez Sanz and L.G. Gómez-Mascaraque (Eds.), *Nanomaterials for Food Applications*, Elsevier, Amsterdam.

Assadpour, E. and Jafari, S.M. (2019b). Advances in spray-drying encapsulation of food bioactive ingredients: From microcapsules to nanocapsules, *Annual Review of Food Science and Technology*, 10: 103-131.

Astray, G., Gonzalez-Barreiro, C., Mejuto, J.C., Rial-Otero, R. and Simal-Gandara, J. (2009). A review on the use of cyclodextrins in foods, *Food Hydrocolloids*, 23: 1631-1640.

Atay, E., Fabra, M.J., Martínez-Sanz, M., Gomez-Mascaraque, L.G., Altan, A. and Lopez-Rubio, A. (2018). Development and characterization of chitosan/gelatin electrosprayed microparticles as food grade delivery vehicles for anthocyanin extracts, *Food Hydrocolloids*, 77: 699-710.

Bakry, A.M., Huang, J., Zhai, Y. and Huang, Q. (2019). Myofibrillar protein with κ- or λ-carrageenans as novel shell materials for microencapsulation of tuna oil through complex coacervation, *Food Hydrocolloids*, 96: 43-53.

Balanč, B.D., Ota, A., Djordjević, V.B., Šentjurc, M., Nedović, V.A., Bugarski, B.M. and Poklar Ulrih, N. (2015). Resveratrol-loaded liposomes: Interaction of resveratrol with phospholipids, *European Journal of Food Science and Technology*, 117: 1615-1626.

Ballesteros, L.F., Ramirez, M.J., Orrego, C.E., Teixeira, J.A. and Mussatto, S.I. (2017). Encapsulation of antioxidant phenolic compounds extracted from spent coffee grounds by freeze-drying and spray-drying using different coating materials, *Food Chemistry*, 237: 623-631.

Barkade, S., Pinjari, D.V., Singh, A.K., Gogate, P.R., Naik, J.B., Sonawane, S.H., Ashokkumar, M. and Pandit, A.B. (2013). Ultrasound-assisted miniemulsion polymerization for preparation of Polypyrrole–Zinc Oxide (PPy/ZnO) functional latex for liquefied petroleum gas sensing, *Industrial & Engineering Chemistry Research*, 52: 7704-7712.

Belščak-Cvitanović, A., Stojanović, R., Manojlović, V., Komes, D., Juranović Cindrić, I., Nedović, V. and Bugarski, B. (2011). Encapsulation of polyphenolic antioxidants from medicinal plant extracts in alginate–chitosan system enhanced with ascorbic acid by electrostatic extrusion, *Food Research Internationl*, 44: 1094-1101.

Beltrán, J.D., Sandoval-Cuellar, C.E., Bauer, K. and Quintanilla-Carvajal, M.X. (2019). *In-vitro* digestion of high-oleic palm oil nanoliposomes prepared with unpurified soy lecithin: Physical stability and nano-liposome digestibility, *Colloids and Surfaces A: Physicochemical and Engineering Aspects*, 578: 123603.

Benelli, L. and Oliveira, W.P. (2019). Fluidized bed coating of inert cores with a lipid-based system loaded with a polyphenol-rich *Rosmarinus officinalis* extract, *Food and Bioproducts Processing*, 114: 216-226.

Benjamin, O., Silcock, P., Beauchamp, J., Buettner, A. and Everett, D.W. (2013). Volatile release and structural stability of b-lactoglobulin primary and multilayer emulsions under simulated oral conditions, *Food Chemistry*, 140: 124-134.

Bhardwaj, N. and Kundu, S.C. (2010). Electrospinning: A fascinating fiber fabrication technique, *Biotechnology Advances*, 28: 325-347.

Bhushani, J.A. and Anandharamakrishnan, C. (2014). Electrospinning and electrospraying techniques: Potential food based applications, *Trends in Food Science & Technology* 38: 21-33.

Bhushani, J.A., Kurrey, N.K. and Anandharamakrishnan, C. (2017). Nanoencapsulation of green tea catechins by electrospraying technique and its effect on controlled release and *in-vitro* permeability, *Journal of Food Engineering*, 199: 82-92.

Bilbao-Sainz, C., Chiou, B.S., Punotai, K., Olson, D., Williams, T., Wood, D., Rodov, V., Poverenov, E. and McHugh, T. (2018). Layer-by-layer alginate and fungal chitosan based edible coating applied to fruit bars, *Journal of Food Science*, 83: 1880-1887.

Bimbo, F., Bonanno, A., Nocella, G., Viscecchia, R., Nardone, G., De Devitiis, B. and Carlucci, D. (2017). Consumers' acceptance and preferences for nutrition-modified and functional dairy products: A systematic review, *Appetite*, 113: 141-154.

Bishop, J., Nelson, G. and Lamb, J. (1998). Microencapsulation in yeast cells, *Journal of Microencapsulation*, 15: 761-773.

Bleiel, J. (2010). Functional foods from the perspective of the consumer: How to make it a success? *International Dairy Journal*, 20: 303-306.

Bock, N., Dargaville, T.R. and Woodruff, M.A. (2012). Electrospraying of polymers with therapeutic molecules: State of the art, *Progress in Polymer Science*, 37: 1510-1551.

Brunner, G. (1994). *Supercritical Fluids: Gas Extraction*, Springer, Berlin.

Burey, P., Bhandari, B.R., Howes, T. and Gidley, M.J. (2008). Hydrocolloid gel particles: Formation, characterization, and application, *Critical Reviews in Food Science and Nutrition*, 48: 361-377.

Butstraen, C. and Salaün, F. (2014). Preparation of microcapsules by complex coacervation of gum arabic and chitosan, *Carbohydrate Polymers*, 99: 608-616.

Cal, K. and Centkowska, K. (2008). Use of cyclodextrins in topical formulations: Practical aspects, *European Journal of Pharmaceutics and Biopharmaceutics*, 68: 467-478.

Cano-Higuita, D., Malacrida, C. and Telis, V.J. (2015). Stability of curcumin microencapsulated by spray and freeze drying in binary and ternary matrices of maltodextrin, gum arabic and modified starch, *Journal of Food Processing and Preservation*, 39: 2049-2060.

Ceballos, A.M., Giraldo, G.I. and Orrego, C.E. (2012). Effect of freezing rate on quality parameters of freeze dried soursop fruit pulp, *Journal of Food Engineering*, 111: 360-365.

Chakraborty, S., Liao, I.C., Adler, A. and Leong, K.W. (2009). Electrohydrodynamics: A facile technique to fabricate drug delivery systems, *Advanced Drug Delivery Reviews*, 61: 1043-1054.

Chen, Q., McGillivray, D., Wen, J., Zhong, F. and Quek, S.Y. (2013a). Co-encapsulation of fish oil with phytosterol esters and limonene by milk proteins, *Journal of Food Engineering*, 117: 505-512.

Chen, Q., Zhong, F., Wen, J.Y., McGillivray, D. and Quek, S.Y. (2013b). Properties and stability of spray-dried and freeze-dried microcapsules co-encapsulated with fish oil, phytosterol esters, and limonene, *Drying Technology*, 31: 707-716.

Cheng, Y.S., Lu, P.M., Huang, C.Y. and Wu, J.J. (2017). Encapsulation of lycopene with lecithin and α-tocopherol by supercritical antisolvent process for stability enhancement, *Journal of Supercritical Fluids*, 130: 246-252.

Chew, S.C. and Nyam, K.L. (2016). Microencapsulation of kenaf seed oil by co-extrusion technology, *Journal of Food Engineering*, 175: 43-50.

Chranioti, C. and Tzia, C. (2014). Arabic gum mixtures as encapsulating agents of freeze-dried fennel oleoresin products, *Food and Bioprocess Technology*, 7: 1057-1065.

Ciamponi, F., Duckham, C. and Tirelli, N. (2012). Yeast cells as microcapsules: Analytical tools and process variables in the encapsulation of hydrophobes in *S. cerevisiae*, *Applied Microbiology and Biotechnology*, 95: 1445-1456.

Ciobanu, A., Landy, D. and Fourmentin, S. (2013). Complexation efficiency of cyclodextrins for volatile flavor compounds, *Food Research International*, 53: 110-114.

Comunian, T.A., Thomazini, M.A., Alves, A.J.G., de Matos Junior, F.E., de Carvalho Balieiro, J.C. and Favaro-Trindade, C.S. (2013). Microencapsulation of ascorbic acid by complex coacervation: Protection and controlled release, *Food Research International*, 52: 373-379.

Consoli, L., Grimaldi, R., Sartori, T., Menegalli, F.C. and Hubinger, M.D. (2016). Acid microparticles produced by spray chilling technique: Production and characterization, *LWT – Food Science and Technology*, 65: 79-87.

Coronel-Aguilera, C.P. and San Martín-González, M.F. (2015). Encapsulation of spray-dried β-carotene emulsion by fluidized bed coating technology, *LWT – Food Science and Technology*, 62: 187-193.

Costamagna, M.S., Gómez-Mascaraque, L.G., Zampini, I.C., Alberto, M.R., Pérez, J., López-Rubio, A. and Isla, M.I. (2017). Microencapsulated chañar phenolics: A potential ingredient for functional foods development, *Journal of Functional Foods*, 37: 523-530.

Da Fonseca Machado, A.P., Rezende, C.A., Rodrigues, R.A., Barbero, G F., de Tarso Vieira e Rosa, P. and Martínez, J. (2018). Encapsulation of anthocyanin-rich extract from blackberry residues by spray-drying, freeze-drying and supercritical antisolvent, *Powder Technology*, 340: 553-562.

da Silva Pedrini, M.R., Dupont, S., de Anchieta Câmara Jr, A., Beney, L. and Gervais, P. (2014). Osmoporation: A simple way to internalize hydrophilic molecules into yeast, *Applied Microbiology and Biotechnology*, 98: 1271-1280.

da Silva Soares, B., Siqueira, R.P., de Carvalho, M.G., Vicente, J. and Garcia-Rojas, E.E. (2019). Microencapsulation of sacha inchi oil (*Plukenetia volubilis* L.) using complex coacervation: Formation and structural characterization, *Food Chemistry*, 298: 125045.

Dag, D. and Oztop, M.H. (2017). Formation and characterization of green tea extract loaded liposomes, *Journal of Food Science*, 82: 463-470.

Dardelle, G., Normand, V., Steenhoudt, M., Bouquerand, P.E., Chevalier, M. and Baumgartner, P. (2007). Flavour-encapsulation and flavour-release performances of a commercial yeast-based delivery system, *Food Hydrocolloids*, 21: 953-960.

de Almeida Paula, D., Furtado Martins, E.M., de Almeida Costa, N., de Oliveira, P.M., de Oliveira, E.B. and Ramos, A.M. (2019). Use of gelatin and gum arabic for microencapsulation of probiotic cells from *Lactobacillus plantarum* by a dual process combining double emulsification followed by complex coacervation, *International Journal of Biological Macromolecules*, 133: 722-731.

de Lara Pedroso, D., Thomazini, M., Heinemann, R.J..B. and Favaro-Trindade, C.S. (2012). Protection of *Bifidobacterium lactis* and *Lactobacillus acidophilus* by microencapsulation using spray-chilling, *International Dairy Journal*, 26: 127-132.

De Lima Petito, N., da Silva Dias, D., Costa, V.G., Falcão, D.Q. and de Lima Araujo, K.G. (2016). Increasing solubility of red bell pepper carotenoids by complexation with 2-hydroxypropyl-b-cyclodextrin, *Food Chemistry*, 208: 124-131.

de Matos-Jr, F.E., Comunian, T.A., Thomazini, M. and Favaro-Trindade, C.S. (2017). Effect of feed preparation on the properties and stability of ascorbic acid microparticles produced by spray chilling, *LWT*, 75: 251-260.

de Souza, V.B., Thomazini, M., Chaves, I.E., Ferro-Furtado, R. and Favaro-Trindade, C.S.

(2020). Microencapsulation by complex coacervation as a tool to protect bioactive compounds and to reduce astringency and strong flavor of vegetable extracts, *Food Hydrocolloids*, 98: 105244.

Deng, X.X., Chen, Z., Huang, Q., Fu, X. and Tang, C.H. (2014). Spray-drying microencapsulation of β-carotene by soy protein isolate and/or OSA-modified starch, *Journal of Applied Polymer Science*, 131: 40399.

Deshmukh, R., Wagh, P. and Naik, J. (2016). Solvent evaporation and spray-drying technique for micro- and nanospheres/particles preparation: A review, *Drying Technology*, 34: 1758-1772.

Dias, H.M.A.M., Berbicz, F., Pedrochic, F., Baesso, M.L. and Matioli, G. (2010). Butter cholesterol removal using different complexation methods with beta-cyclodextrin, and the contribution of photoacoustic spectroscopy to the evaluation of the complex, *Food Research International*, 43: 1104-1110.

Dimitrellou, D., Kandylis, P., Lević S., Petrović, T., Ivanović, S., Nedović, V. and Kourkoutas, Y. (2019). Encapsulation of *Lactobacillus casei* ATCC 393 in alginate capsules for probiotic fermented milk production, *LWT*, 116: 108501.

Donsi, F. (2018). Applications of nanoemulsions in foods, pp. 349-377. *In*: S.M. Jafari and D.J. McClements (Eds.), *Nanoemulsions: Formulation, Applications and Characterization*, Academic Press, UK.

Donsi, F. and Ferrari, G. (2016). Essential oil nanoemulsions as antimicrobial agents in food, *Journal of Biotechnology*, 233: 106-120.

Donsi, F., Sessa, M. and Ferrari, G. (2013). Nanometric-size delivery systems for bioactive compounds for the nutraceutical and food industries, pp. 619-666. *In*: D. Bagchi (Ed.), *Bio-nanotechnology: A Revolution in Food, Biomedical and Health Sciences*, Blackwell Publishing Ltd., Oxford.

Đorđević, V., Balanč, B., Belščak-Cvitanović, A., Lević, S., Trifković, K., Kalušević, A., Kostić, I., Komes, D., Bugarski, B. and Nedović, V. (2015). Trends in encapsulation technologies for delivery of food bioactive compounds, *Food Engineering Reviews*, 7: 452-490.

Đorđević, V., Belščak-Cvitanović, A., Drvenica, I., Komes, D., Nedović, V. and Bugarski, B. (2017). Nanoscale nutrient delivery systems, pp. 87-139. *In*: A.M. Grumezescu (Ed.), *Nutrient Delivery: A Volume in Nanotechnology in the Agri-food Industry*, Academic Press, Elsevier, Cambridge, U.S.A.

dos Santos Lima, B., Shanmugam, S., de Souza Siqueira Quintans, J., Quintans-Junior, L.J. and de Souza Araujo, A.A. (2019). Inclusion complex with cyclodextrins enhances the bioavailability of flavonoid compounds: A systematic review, *Phytochemistry Reviews*, 18: 1337-1359.

Drapala, K.P., Auty, M.A.E., Mulvihill, D.M. and O'Mahony, J.A. (2017). Influence of emulsifier type on the spray-drying properties of model infant formula emulsions, *Food Hydrocolloids*, 69: 56-66.

Drusch, S., Regier, M. and Bruhn, M. (2012). Recent advances in the microencapsulation of oils high in polyunsaturated fatty acids, pp. 159-181. *In*: A. McElhatton and P.J.D.A. Sobral (Eds.), *Novel Technologies in Food Science*, Springer, New York.

Du, H.H., Liang, R., Han, R.M., Zhang, J.P. and Skibsted, L.H. (2015). Astaxanthin protecting membrane integrity against photosensitized oxidation through synergism with other carotenoids, *Journal of Agricultural and Food Chemistry*, 63: 9124-9130.

Duchene, D. (2011). Cyclodextrins and their inclusion complexes, pp. 1-18. *In*: E. Bilensoy (Ed.), *Cyclodextrins in Pharmaceutics, Cosmetics, and Biomedicine: Current and Future Industrial Applications*, John Wiley & Sons, New Jersey.

Elzoghby, A.O., Samy, W.M. and Elgindy, N.A. (2012). Protein-based nanocarriers as promising drug and gene delivery systems, *Journal of Controlled Release*, 161: 38-49.

Encina, C., Vergara, C., Giménez, B., Oyarzún-Ampuero, F. and Robert, P. (2016). Conventional spray-drying and future trends for the microencapsulation of fish oil, *Trends in Food Science & Technology*, 56: 46-60.

Ersus, S. and Yurdagel, U. (2007). Microencapsulation of anthocyanin pigments of black carrot (*Daucus carota* L.) by spray dryer, *Journal of Food Engineering*, 80: 805-812.

Fang, Z. and Bhandari, B. (2017). Spray-drying of Bioactives, pp. 261-284. *In*: Y.H. Roos and Y.D. Livney (Eds.). *Engineering Foods for Bioactives Stability and Delivery*, Springer, New York.

Farahmand, S., Tajerzadeh, H. and Farboud, E.S. (2006). Formulation and evaluation of a vitamin C multiple emulsion, *Pharmaceutical Development and Technology*, 11: 255-261.

Fathi, M., Martin, A. and McClements, D.J. (2014). Nanoencapsulation of food ingredients using carbohydrate based delivery systems, *Trends in Food Science and Technology*, 39: 18-39.

Feng, S., Sun, Y., Wang, P., Sun, P., Ritzoulis, C. and Shao, P. (2019). Co-encapsulation of resveratrol and epigallocatechin gallate in low methoxy pectin-coated liposomes with great stability in orange juice, *International Journal of Food Science & Technology*; doi:10.1111/ijfs.14323

Fioramonti, S.A., Rubiolo, A.C. and Santiago, L.G. (2017). Characterisation of freeze-dried flaxseed oil microcapsules obtained by multilayer emulsions, *Powder Technology*, 319: 238-244.

Franco, D., Antequera, T., de Pinho, S.C., Jiménez, E., Pérez-Palacios, T., Fávaro-Trindade, C.S. and Lorenzo, J.M. (2017). The use of microencapsulation by spray-drying and its application in meat products, pp. 333-362. *In*: J.M.L. Rodriguez and F.J.C. García (Eds.), *Strategies for Obtaining Healthier Foods*, Nova Science Publishers, New York.

Frenzel, M. and Steffen-Heins, A. (2015). Whey protein coating increases bilayer rigidity and stability of liposomes in food-like matrices, *Food Chemistry*, 173: 1090-1099.

Frenzel, M., Krolak, E., Wagner, A.E. and Steffen-Heins, A. (2015). Physicochemical properties of WPI coated liposomes serving as stable transporters in a real food matrix, *LWT – Food Science and Technology*, 63: 527-534.

Frey, C.R. (2016). Encapsulation via fluidized bed coating technology, pp. 111-145. *In*: M. Mishra (Ed.), *Handbook of Encapsulation and Controlled Release*, Taylor and Francis Group, CRC Press, Boca Raton.

Gamboa, O.D., Gonçalves, L.G. and Grosso, C.F. (2011). Microencapsulation of tocopherols in lipid matrix by spray chilling method, *Procedia Food Science*, 1: 1732-1739.

García-Moreno, P.J., Özdemir, N., Stephansen, K., Mateiu, R.V., Echegoyen, Y., Lagaron, J.M., Chronakis I.S. and Jacobsen, C. (2017). Development of carbohydrate-based nano-microstructures loaded with fish oil by using electrohydrodynamic processing, *Food Hydrocolloids*, 69: 273-285.

García-Moreno, P.J., Pelayo, A., Yu, S., Busolo, M., Lagaron, J.M., Chronakis, I.S. and Jacobsen, C. (2018). Physicochemical characterization and oxidative stability of fish oil-loaded electrosprayed capsules: Combined use of whey protein and carbohydrates as wall materials, *Journal of Food Engineering*, 231: 42-53.

Gharsallaoui, A., Roudaut, G., Chambin, O., Voilley, A. and Saurel, R. (2007). Applications of spray-drying in microencapsulation of food ingredients: An overview, *Food Research International*, 40: 1107-1121.

Gibis, M., Ruedt, C. and Weiss, J. (2016). *In vitro* release of grape-seed polyphenols encapsulated from uncoated and chitosan-coated liposomes, *Food Research International*, 88: 105-113.

Golzi, R., Boltri, L. and Stollberg, C. (2013). *Microcapsules by Coacervation Containing a Pharmaceutical Incorporated in the Coating Polymer*, US Patent #20,130,156,934.

Gomaa, A.I., Martinent, C., Hammami, R., Fliss, I. and Subirade, M. (2017). Dual coating of liposomes as encapsulating matrix of antimicrobial peptides: Development and characterization, *Frontiers in Chemistry*, **5**(103): 1-10.

Gomes, L.M., Petito, N., Costa, V.G., Falcao, D.Q. and de Lima Araujo, K.G. (2014). Inclusion complexes of red bell pepper pigments with b-cyclodextrin: Preparation, characterisation and application as natural colorant in yogurt, *Food Chemistry*, 148: 428-436.

Gomez-Estaca, J., Balaguer, M.P., Gavara, R. and Hernandez-Munoz, P. (2012). Formation of zein nanoparticles by electrohydrodynamic atomization: Effect of the main processing variables and suitability for encapsulating the food coloring and active ingredient curcumin, *Food Hydrocolloids*, 28: 82-91.

Gomez-Estaca, J., Gavara, R. and Hernández-Muñoz, P. (2015). Encapsulation of curcumin in electrosprayed gelatin microspheres enhances its bioaccessibility and widens its uses in food applications, *Innovative Food Science & Emerging Technologies*, 29: 302-307.

Gomez-Estaca, J., Balaguer, M.P., López-Carballo, G., Gavara, R. and Hernández-Muñoz, P. (2017). Improving antioxidant and antimicrobial properties of curcumin by means of encapsulation in gelatin through electrohydrodynamic atomization, *Food Hydrocolloids*, 70: 313-320.

Gómez-Mascaraque, L.G., Lagarón, J.M. and López-Rubio, A. (2015). Electrosprayed gelatin submicroparticles as edible carriers for the encapsulation of polyphenols of interest in functional foods, *Food Hydrocolloids*, 49: 42-52.

Gómez-Mascaraque, L.G. and López-Rubio, A. (2016). Protein-based emulsion electrosprayed micro- and submicroparticles for the encapsulation and stabilization of thermosensitive hydrophobic bioactives, *Journal of Colloid and Interface Science*, 465: 259-270.

Gómez-Mascaraque, L.G., Ambrosio-Martín, J., Fabra, M.J., Pérez-Masiá, R. and López-Rubio, A. (2016a). Novel nanoencapsulation structures for functional foods and nutraceutical applications, pp. 373-395. *In*: S. Sen and Y. Pathak (Eds.), *Nanotechnology in Nutraceuticals*, CRC Press.

Gómez-Mascaraque, L.G., Soler, C. and Lopez-Rubio, A. (2016b). Stability and bioaccessibility of EGCG within edible micro-hydrogels, chitosan vs. gelatin: A comparative study, *Food Hydrocolloids*, 61: 128-138.

Gómez-Mascaraque, L.G., Sanchez, G. and López-Rubio, A. (2016c). Impact of molecular weight on the formation of electrosprayed chitosan microcapsules as delivery vehicles for bioactive compounds, *Carbohydrate Polymers*, 150: 121-130.

Gomez-Mascaraque, L.G., Morfin, R.C., Pérez-Masiá, R., Sanchez, G. and Lopez-Rubio, A. (2016d). Optimization of electrospraying conditions for the microencapsulation of probiotics and evaluation of their resistance during storage and *in-vitro* digestion, *LWT – Food Science and Technology*, 69: 438-446.

Gómez-Mascaraque, L.G., Dhital, S., López-Rubio, A. and Gidley, M.J. (2017a). Dietary polyphenols bind to potato cells and cellular components, *Journal of Functional Foods*, 37: 283-292.

Gómez-Mascaraque, L.G., Hernández-Rojas, M., Tarancón, P., Tenon, M., Feuillère, N., Vélez Ruiz, J.F., Fiszman, S. and López-Rubio, A. (2017b). Impact of microencapsulation within electrosprayed proteins on the formulation of green tea extract-enriched biscuits, *LWT – Food Science and Technology*, 81: 77-86.

Gómez-Mascaraque, L.G., Sipoli, C.C., de La Torre, L.G. and López-Rubio, A. (2017c). Microencapsulation structures based on protein-coated liposomes obtained through electrospraying for the stabilization and improved bioaccessibility of curcumin, *Food Chemistry*, 233: 343-350.

Gómez-Mascaraque, L.G., Perez-Masiá, R., González-Barrio, R., Periago, M.J. and López-Rubio, A. (2017d). Potential of microencapsulation through emulsion-electrospraying to improve the bioaccesibility of β-carotene, *Food Hydrocolloids*, 73: 1-12.

Gómez-Mascaraque, L.G., Fabra, M.J., Castro-Mayorga, J.L., Sánchez, G., Martínez-Sanz, M. and López-Rubio, A. (2018). Nanostructuring biopolymers for improved food quality and safety, pp. 33-64. *In*: A.M. Grumezescu and A.M. Holban (Eds.), *Biopolymers for Food Design*, Elsevier.

Gómez-Mascaraque, L.G., Tordera, F., Fabra, M.J., Martínez-Sanz, M. and Lopez-Rubio, A. (2019). Coaxial electrospraying of biopolymers as a strategy to improve protection of bioactive food ingredients, *Innovative Food Science & Emerging Technologies*, 51: 2-11.

Gómez-Mascaraque, L.G. and Lopez-Rubio, A. (2019). Encapsulation of plant-derived bioactive ingredients through electrospraying for nutraceuticals and functional foods applications, *Current Medicinal Chemistry*; doi.org/10.2174/0929867326666191010 115343

Gómez-Mascaraque, L.G. and Lopez-Rubio, A. (2020). Production of food bioactive-loaded nanoparticles by electrospraying, pp. 107-149. *In*: S.M. Jafari (Ed.), *Nanoencapsulation of Food Ingredients by Specialized Equipment*, Academic Press, Elsevier, Cambridge, Massachusetts.

Gosh, S.K. (2006). *Functional Coatings: By Polymer Microencapsulation*, John Wiley & Sons, New Jersey.

Goslinska, M. and Heinrich, S. (2019). Characterization of waxes as possible coating material for organic aerogels, *Powder Technology*, 357: 223-231.

Gouin, S. (2004) Microencapsulation: Industrial appraisal of existing technologies and trends, *Trends in Food Science & Technology*, 15: 330-347.

Guignon, B., Duquenoy, A. and Dumoulin, E.D. (2002). Fluid-bed encapsulation of particles: Principles and practice, *Drying Technology*, 20: 419-447.

Gul, K., Singh, A.K. and Jabeen. R. (2016). Nutraceuticals and functional foods: The foods for the future world, *Critical Reviews in Food Science and Nutrition*, 56: 2617-2627.

Gul, O. and Atalar, I. (2019). Different stress tolerance of spray and freeze dried *Lactobacillus casei* Shirota microcapsules with different encapsulating agents, *Food Science and Biotechnology*, 28: 807-816.

Gültekin-Özgüven, M., Karadağ, A., Duman, Ş., Özkal, B. and Özçelik, B. (2016). Fortification of dark chocolate with spray-dried black mulberry (*Morus nigra*) waste extract encapsulated in chitosan-coated liposomes and bioaccessability studies, *Food Chemistry*, 201: 205-212.

Guo, W., Jia, Y., Tian, K., Xu, Z., Jiao, J., Li, R., Wu, Y., Cao, L. and Wang, H. (2016). UV-triggered self-healing of a single robust SiO_2 microcapsule based on cationic polymerization for potential application in aerospace coatings, *ACS Applied Materials and Interfaces*, 8: 21046-21054.

Gupta, R., Basu, S. and Shivhare, U.S. (2012). A review on thermodynamics and functional properties of complex coacervates, *International Journal of Applied Biology and Pharmaceutical Technology*, 3: 64-86.

Guttoff, M., Saberi, A.H. and McClements, D.J. (2015). Formation of vitamin D nanoemulsion-based delivery systems by spontaneous emulsification: Factors affecting particle size and stability, *Food Chemistry*, 171: 117-122.

Hadaruga, N.G., Hadaruga, D.I., Paunescu, V., Tatu, C., Ordodi, V.L., Bandur, G. and Lupea, A.X. (2006). Thermal stability of the linoleic acid/α- and β-cyclodextrin complexes, *Food Chemistry*, 99: 500-508.

Hafner, V., Dardelle, G., Normand, V. and Fieber, W. (2011). Determination of flavor loading in complex delivery systems by time-domain NMR, *European Journal of Lipid Science and Technology*, 113: 856-861.

Hao, J., Guo, B., Yu, S., Zhang, W., Zhang, D., Wang, J. and Wang, Y. (2017). Encapsulation of the flavonoid quercetin with chitosan-coated nano-liposomes, *LWT – Food Science and Technology*, 85: 37-44.

Hasan, M., Elkhoury, K., Kahn, C.J.F., Arab-Tehrany, E. and Linder, M. (2019). Preparation, characterization, and release kinetics of chitosan-coated nanoliposomes encapsulating curcumin in simulated environments, *Molecules*, 24: 2023.

He, Z., Jiang, S., Li, Q., Wang, J., Zhao, Y. and Kang, M. (2017). Facile and cost-effective synthesis of isocyanate microcapsules via polyvinyl alcohol-mediated interfacial polymerization and their application in self-healing materials, *Composites Science and Technology*, 138: 15-23.

Hedges, A.R., Shieh, W.J. and Sikorski, C.T. (1995). Use of cyclodextrins for encapsulation in the use and treatment of food products, pp. 60-71. *In*: S.J. Risch and G.A. Reineccius (Eds.), *Encapsulation and Controlled Release of Food Ingredients*, ACS Symposium Series, American Chemical Society, Washington, DC.

Heinzen, C., Berger, A. and Marison, I. (2004) Use of vibration technology for jet break-up for encapsulation of cells and liquids in monodisperse microcapsules, pp. 257-275. *In*: V. Nedović and R. Willaert (Eds.), *Fundamentals of Cell Immobilization Biotechnology*, Kluwer Academic Publishers, Dordrecht, The Netherlands.

Hermund, D.B., Karadag, A., Andersen, U., Jónsdóttir, R., Kristinsson, H.G., Alasalvar, C. and Jacobsen, C. (2016). Oxidative stability of granola bars enriched with multilayered fish oil emulsion in the presence of novel brown seaweed based antioxidants, *Journal of Agricultural and Food Chemistry*, 64: 8359-8368.

Hilton, J. (2017). Growth patterns and emerging opportunities in nutraceutical and functional food categories: Market overview, pp. 1-28. *In*: D. Bagchi and S. Nair (Eds.), *Developing New Functional Food and Nutraceutical Products*, Academic Press, San Diego.

Hirose, T. and Yamamoto, Y. (2001). *Hinokitol Containing Cyclo-olefin Polymer Compositions and Their Molding with Excellent Antimicrobial and Gas Barrier Properties*; Japanese Patent JP #55480.

Hottot, A., Vessot, S. and Andrieu, J. (2004). A direct characterization method of the ice morphology: relationship between mean crystals size and primary drying times of freeze-drying processes, *Drying Technology*, 22: 2009-2021.

Hundre, S.Y., Karthik, P. and Anandharamakrishnan, C. (2015). Effect of whey protein isolate and β-cyclodextrin wall systems on stability of microencapsulated vanillin by spray-freeze drying method, *Food Chemistry*, 174: 16-24.

Isailović, B.D., Kostić, I.T., Zvonar, A., Đorđević, V.B., Gašperlin, M., Nedović, V.A. and Bugarski, B. (2013). Resveratrol-loaded liposomes produced by different techniques, *Innovative Food Science and Emerging Technologies*, 19: 181-189.

Ishwarya, S.P. and Anandharamakrishnan, C. (2017). Spray drying, pp. 57-94. *In*: C. Anandharamakrishnan (Ed.), *Handbook of Drying for Dairy Products*, John Wiley & Sons, New Jersey.

Ishwarya, S.P., Anandharamakrishnan, C. and Stapley, A.G.F. (2015). Spray-freeze-drying: A novel process for the drying of foods and bioproducts, *Trends in Food Science & Technology*, 41: 161-181.

Jafari, S.M., Assadpoor, E., He, Y. and Bhandari, B. (2008). Encapsulation efficiency of food flavours and oils during spray drying, *Drying Technology*, 26: 816-835.

Jannasari, N., Fathi, M., Moshtaghian, S.J. and Abbaspourrad, A. (2019). Microencapsulation of vitamin D using gelatin and cress seed mucilage: Production, characterization and *in vivo* study, *International Journal of Biological Macromolecules*, 129: 972-979.

Jaskulski, M., Kharaghani, A. and Tsotsas, E. (2017). Encapsulation methods, spray drying, spray chilling and spray cooling, pp. 67-113. *In*: M. Krokida (Ed.), *Thermal and Nonthermal Encapsulation Methods*, CRC Press, Boca Raton, Florida.

Jaworek, A. and Sobczyk, A.T. (2008). Electrospraying route to nanotechnology: An overview, *Journal of Electrostatics*, 66: 197-219.

Jiao, Z., Wang, X., Yin, Y., Xia, J. and Mei, Y. (2018). Preparation and evaluation of a chitosan-coated antioxidant liposome containing vitamin C and folic acid, *Journal of Microencapsulation*, 35: 272-280.

Jiménez-Martín, E., Antequera Rojas, T., Gharsallaoui, A., Ruiz Carrascal, J. and Pérez-Palacios, T. (2016). Fatty acid composition in double and multilayered microcapsules of ω-3 as affected by storage conditions and type of emulsions, *Food Chemistry*, 194: 476-486.

Jiménez-Martín, E., Gharsallaoui, A., Pérez-Palacios, T., Carrascal, J.R. and Antequera Rojas, T. (2014). Suitability of using monolayered and multilayered emulsions for microencapsulation of ω-3 fatty acids by spray drying: Effect of storage at different temperatures, *Food and Bioprocess Technology*, 8: 100-111.

Jovanović, A., Balanč, B., Djordjević, V., Ota, A., Skrt, M., Šavikin, K.P., Bugarski, B.M., Nedović, V.A. and Poklar Ulrih, N. (2019). Effect of gentisic acid on the structural-functional properties of liposomes incorporating β-sitosterol, *Colloids and Surfaces B: Biointerfaces*, 183: 110422.

Julio, L.M., Copado, C.N., Crespo, R., Diehl, B.W.K., Ixtaina, Y.V. and Tomás, M.C. (2019). Design of microparticles of chia seed oil by using the electrostatic layer-by-layer deposition technique, *Powder Technology*, 345: 750-757.

Kaasgaard, T. and Keller, D. (2010). Chitosan coating improves retention and redispersibility of freeze-dried flavour oil emulsions, *Journal of Agricultural and Food Chemistry*, 58: 2446-2454.

Kalani, M. and Yunus, R. (2011). Application of supercritical antisolvent method in drug encapsulation: A review, *International Journal of Nanomedicine*, 6: 1429-1442.

Kalogeropoulos, N., Yannakopoulou, K., Gioxari, A., Chiou, A. and Makris, P.D. (2010). Polyphenol characterization and encapsulation in b-cyclodextrin of a flavonoid-rich *Hypericum perforatum* (St John's wort) extract, *LWT – Food Science and Technology*, 43: 882-889.

Karathanos, V.T., Mourtzinos, I., Yannakopoulou, K. and Andrikopoulos, N.K. (2007). Study of the solubility, antioxidant activity, and structure of inclusion complex of vanillin with beta-cyclodextrin, *Food Chemistry*, 101: 652-658.

Karthik, P. and Anandharamakrishnan, C. (2013). Microencapsulation of docosahexaenoic acid by spray-freeze-drying method and comparison of its stability with spray-drying and freeze-drying methods, *Food and Bioprocess Technology*, 6: 2780-2790.

Kfoury, M., Auezova, L., Greige-Gerges, H., Ruellan, S. and Fourmentin, S. (2014). Cyclodextrin, an efficient tool for trans-anethole encapsulation: Chromatographic, spectroscopic, thermal and structural studies, *Food Chemistry*, 164: 454-461.

Kheadr, E.E., Vuillemard, J.C. and El Deeb, S. (2002). Acceleration of cheddar cheese lipolysis by using liposome- entrapped lipases, *Journal of Food Science*, 67: 485-492.

Khem, S., Bansal, V., Small, D.M. and May, B.K. (2016). Comparative influence of pH and heat on whey protein isolate in protecting *Lactobacillus plantarum* A17 during spray drying, *Food Hydrocolloids*, 54: 162-169.

Khorasani, S., Danaei, M. and Mozafari, M.R. (2018). Nanoliposome technology for the food and nutraceutical industries, *Trends in Food Science & Technology*, 79: 106-115.

Khoshakhlagh, K., Koocheki, A., Mohebbi, M. and Allafchian, A. (2017). Development and characterization of electrosprayed *Alyssum homolocarpum* seed gum nanoparticles for encapsulation of d-limonene, *Journal of Colloid and Interface Science*, 490: 562-575.

Khoshakhlagh, K., Mohebbi, M., Koocheki, A. and Allafchian, A. (2018). Encapsulation of D-limonene in *Alyssum homolocarpum* seed gum nanocapsules by emulsion electrospraying: Morphology characterization and stability assessment, *Bioactive Carbohydrates and Dietary Fiber*, 16: 43-52.

Kim, W.J., Kim, J.D., Kim, J., Oh, S.G. and Lee, YW. (2008). Selective caffeine removal from green tea using supercritical carbon dioxide extraction, *Journal of Food Engineering*, 89: 303-309.

Kingwatee, N., Apichartsrangkoon, A., Chaikham, P., Worametrachanon, S., Techarung, J. and Pankasemsuk, T. (2015). Spray-drying *Lactobacillus casei* 01 in lychee juice varied carrier materials, *LWT – Food Science and Technology*, 62: 847-853.

Klinkesorn, U., Sophanodora, P., Chinachoti, P., McClements, D.J. and Decker, E.A. (2005). Stability of spray-dried tuna oil emulsions encapsulated with two-layered interfacial membranes, *Journal of Agricultural and Food Chemistry*, 53: 8365-8371.

Knezevic, Z., Gosaki, D., Hraste, M. and Jalsenjak, I. (1998). Fluid-bed microencapsulation of ascorbic acid, *Journal of Microencapsulation*, 15: 237-252.

Kolanowski, W., Laufenberg, G. and Kunz, B. (2004). Fish oil stabilization by microencapsulation with modified cellulose, *International Journal of Food Sciences and Nutrition*, 55: 333-343.

Koontz, J.L., Marcy, J.E., O'Keefe, S.F. and Duncan, S.E. (2009). Cyclodextrin inclusion complex formation and solid-state characterization of the natural antioxidants alpha-tocopherol and quercetin, *Journal of Agricultural and Food Chemistry*, 57: 1162-1171.

Kriegel, C., Arrechi, A., Kit, K., McClements, D.J. and Weiss, J. (2008). Fabrication, functionalization and application of electrospun biopolymer nanofibers, *Critical Reviews in Food Science and Nutrition*, 48: 775-797.

Kwak, H.S. (2014). Overview of nano- and microencapsulation for foods, pp. 1-14. *In*: H.S. Kwak (Ed.), *Nano- and Microencapsulation for Foods*, John Wiley & Sons, New Jersey.

Lafarga, T. and Hayes, M. (2017). Bioactive protein hydrolysates in the functional food ingredient industry: Overcoming current challenges, *Food Reviews International*, 33: 217-246.

Lam, S., Velikov, K.P. and Velev, O.D. (2014). Pickering stabilization of foams and emulsions with particles of biological origin, *Current Opinion in Colloid & Interface Science*, 19: 490-500.

Lang, Q.Y. and Wai, C.M. (2001). Supercritical fluid extraction in herbal and natural products studies – A practical review, *Talanta*, 53: 771-782.

Laokuldilok, T. and Kanha, N. (2015). Effects of processing conditions on powder properties of black glutinous rice (*Oryza sativa* L.) bran anthocyanins produced by spray-drying and freeze-drying, *LWT – Food Science and Technology*, 64: 405-411.

Lee, S.H., Heng, D., Ng, W.K., Chan, H.K. and Tan, R.B.H. (2011). Nano spray drying: A novel method for preparing protein nanoparticles for protein therapy, *International Journal of Pharmaceutics*, 403: 192-200.

Lee, Y.J., Ji, Y.R., Lee, S., Choi, M.J. and Cho, Y. (2019). Microencapsulation of Probiotic *Lactobacillus acidophilus* KBL409 by Extrusion Technology to Enhance Survival under Simulated Intestinal and Freeze-drying Conditions, *Journal of Microbiology and Biotechnology*, 29: 721-730.

Lemetter, C., Meeuse, F. and Zuidam, N. (2009). Control of the morphology and the size of complex coacervate microcapsules during scale-up, *AIChE Journal*, 55: 1487-1496.

Levic, S., Djordjevic, V.B., Rajic, N.Z., Milivojevic, M.M., Bugarski, B.M. and Nedovic, V.A. (2013). Entrapment of ethyl vanillin in calcium alginate and calcium alginate/poly(vinyl alcohol) beads, *Chemical Papers*, 67: 221-228.

Liu, H., Gong, J., Chabot, D., Miller, S.S., Cui, S.W., Ma, J., Zhong, F. and Wang, Q. (2015). Protection of heat-sensitive probiotic bacteria during spray-drying bysodium caseinate stabilized fat particles, *Food Hydrocolloids*, 51: 459-467.

Lopes, N.A., Pinilla, C.M.B. and Brandelli, A. (2017). Pectin and polygalacturonic acid-coated liposomes as novel delivery system for nisin: Preparation, characterization and release behavior, *Food Hydrocolloids*, 70: 1-7.

López-Rubio, A. and Lagaron, J.M. (2012). Whey protein capsules obtained through electrospraying for the encapsulation of bioactives, *Innovative Food Science & Emerging Technologies*, 13: 200-206.

López-Rubio, A., Sanchez, E., Wilkanowicz, S., Sanz, Y. and Lagaron, J.M. (2012). Electrospinning as a useful technique for the encapsulation of living *bifidobacteria* in food hydrocolloids, *Food Hydrocolloids*, 28: 159-167.

Lu, W., Kelly, A.L. and Miao, S. (2016). Emulsion-based encapsulation and delivery systems for polyphenols, *Trends in Food Science & Technology*, 47: 1-9.

Lucas-Abellan, C., Fortea, I., Gabaldon, J.A. and Nunez-Delicado, E. (2008). Encapsulation of quercetin and myricetin in cyclodextrins at acidic pH, *Journal of Agricultural and Food Chemistry*, 56: 255-259.

Lucas-Abellán, C., Fortea, I., López-Nicolás, J.M. and Núñez-Delicado, E. (2007). Cyclodextrins as resveratrol carrier system, *Food Chemistry*, 104: 39-44.

Madene, A., Jacquot, M., Scher, J. and Desobry, S. (2006). Flavour encapsulation and controlled release – A review, *International Journal of Food Science & Technology*, 41: 1-21.

Mahdavi, S.A., Jafari, S.M.., Ghorbani, M. and Assadpoor, E. (2014). Spray-drying microencapsulation of anthocyanins by natural biopolymers: A review, *Drying Technology*, 32: 509-518.

Mao, L. and Miao, S. (2015). Structuring food emulsions to improve nutrient delivery during digestion, *Food Engineering Reviews*, 7: 439-451.

Marín, D., Aleman, A., Montero, P. and Gomez-Guillen, M.C. (2018). Encapsulation of food waste compounds in soy phosphatidylcholine liposomes: Effect of freeze-drying, storage stability and functional aptitude, *Journal of Food Engineering*, 223: 132-143.

Marques, H.M.C. (2010). A review on cyclodextrin encapsulation of essential oils and volatiles, *Flavour and Fragrance Journal*, 25: 313-326.

Marsanasco, M., Marquez, A.L., Wagner, J.R., del Alonso, S. and Chiaramoni, N.S. (2011). Liposomes as vehicles for vitamins E and C: An alternative to fortify orange juice and offer vitamin C protection after heat treatment, *Food Research International*, 44: 3039-3046.

Martín, Á., Varona, S., Navarrete, A. and José Cocero, M. (2010). Encapsulation and co-precipitation processes with supercritical fluids: Applications with essential oils, *The Open Chemical Engineering Journal*, 4: 31-41.

Maswal, M. and Dar, A.A. (2014). Formulation challenges in encapsulation and delivery of citral for improved food quality, *Food Hydrocolloids*, 37: 182-195.

Matsuda, S., Hatano, H., Kuramoto, K. and Tsutsumi, A. (2001). Fluidization of ultrafine particles with high G, *Journal of Chemical Engineering of Japan*, 34: 121-125.

Matsuno, R. and Adachi, S. (1993). Lipid encapsulation technology-techniques and applications to food, *Trends in Food Science & Technology*, 4: 256-261.

Mattea, F., Martín, Á., Schulz, C., Jaeger, P., Eggers, R. and Cocero, M.J. (2010). Behavior of an organic solvent drop during the supercritical extraction of emulsions, *AIChE Journal*, 56: 1184-1195.

McClements, D.J. (2017). The future of food colloids: Next-generation nanoparticle delivery systems, *Current Opinion in Colloid and Interface Science*, 28: 7-14.

McClements, D.J., Decker, E.A., Park, Y. and Weiss, J. (2009). Structural design principles for delivery of bioactive components in nutraceuticals and functional foods, *Critical Reviews in Food Science and Nutrition*, 49: 577-606.

Melgosa, R., Benito-Román, Ó., Sanz, M. T., de Paz, E. and Beltrán, S. (2019). Omega-3 encapsulation by PGSS-drying and conventional drying methods. Particle characterization and oxidative stability, *Food Chemistry*, 270: 138-148.

Mendonça, F.M.R., Polloni, A.E., Junges, A., da Silva, R.S., Rubira, A.F., Borges, G.R., Dariva, C. and Franceschi, E. (2019). Encapsulation of neem (*Azadirachta indica*) seed oil in poly(3-hydroxybutyrate-co-3-hydroxyvalerate) by SFEE technique, *Journal of Supercritical Fluids*, 152: 104556.

Mezzomo, N., de Paz, E., Maraschin, M., Martin, A., Cocero, M.J. and Ferreira S.R.S. (2012). Supercritical anti-solvent precipitation of carotenoid fraction from pink shrimp residue: Effect of operational conditions on encapsulation efficiency, *Journal of Supercritical Fluids*, 66: 342-349.

Miller, D.A., Ellenberger, D. and Gil, M. (2016). Spray-drying technology, pp. 437-525. *In*: R.O. Williams Iii, A.B. Watts and D.A. Miller (Eds.), *Formulating Poorly Water-soluble Drugs*, Springer International, New York.

Mirzaei, F. and Jafarpour, S.A. (2019). Integrated encapsulation of fish oil and vitamin E with complex coacervation technique and its efficiency optimization by Response Surface Method (RSM), *Journal of Research and Innovation in Food Science and Technology*, 8: 53-66.

Moayyedi, M., Eskandari, M.H., Rad, A.H.E., Ziaee, E., Khodaparast, M.H.H. and Golmakani, M.T. (2018). Effect of drying methods (electrospraying, freeze drying and spray drying) on survival and viability of microencapsulated *Lactobacillus rhamnosus* ATCC 7469, *Journal of Functional Foods*, 40: 391-399.

Mohanty, B. and Bohidar, H.B. (2005). Microscopic structure of gelatin coacervates, *International Journal of Biological Macromolecules*, 36: 39-46.

Moreno, M.A., Orqueda, M.E., Gómez-Mascaraque, L.G., Isla, M.I. and López-Rubio, A. (2019). Crosslinked electrospun zein-based food packaging coatings containing bioactive chilto fruit extracts, *Food Hydrocolloids*, 95: 496-505.

Morselli Ribeiro, M.D.M., Barrera Arellano, D. and Ferreira Grosso, C.R. (2012). The effect of adding oleic acid in the production of stearic acid lipid microparticles with a hydrophilic core by a spray-cooling process, *Food Research International*, 47: 38-44.

Mozafari, M.R., Johnson, C., Hatziantoniou, S. and Demetzos, C. (2008). Nanoliposomes and their applications in food nanotechnology, *Journal of Liposome Research*, 18: 309-327.

Müller, M.G., Lindner, J.A., Briesen, H., Sommer, K. and Foerst, P. (2018). On the properties and application of beeswax, carnauba wax and palm fat mixtures for hot melt coating in fluidized beds, *Advanced Powder Technology*, 29: 781-788.

Mun, S., Kim, Y.R., Shin, M. and McClements, D.J. (2015). Control of lipid digestion and nutraceutical bioaccessibility using starch-based filled hydrogels: Influence of starch and surfactant type, *Food Hydrocolloids*, 44: 380-389.

Munin, A. and Edwards-Lévy, F. (2011). Encapsulation of natural polyphenolic compounds: A review, *Pharmaceutics*, 3: 793-829.

Muzaffar, K. and Kumar, P. (2016). Effect of soya protein isolate as a complementary drying aid of maltodextrin on spray-drying of tamarind pulp, *Drying Technology*, 34: 142-148.

Nedovic, V., Kalusevic, A., Manojlovic, V., Levic, S. and Bugarski, B. (2011). An overview of encapsulation technologies for food applications, *Procedia Food Science*, 1: 1806-1815.

Nedović, V., Kalušević, A., Manojlović, V., Petrović, T. and Bugarski, B. (2013). Encapsulation systems in the food industry, pp. 229-253. *In*: S. Yanniotis, P. Taoukis,

N. Stoforos and V.T. Karathanos (Eds.), *Advances in Food Process Engineering Research and Applications*, Springer, New York.

Nerome, H., Machmudah, S., Fukuzato, R., Higashiura, T., Youn, Y.S., Lee, Y.W. and Goto, M. (2013). Nanoparticle formation of lycopene/b-cyclodextrin inclusion complex using supercritical antisolvent precipitation, *Journal of Supercritical Fluids*, 83: 97-103.

Obradovic, N., Pajic-Lijakovic, I., Krunic, T., Belovic, M., Rakin, M. and Bugarski, B. (2020). Effect of encapsulated probiotic starter culture on rheological and structural properties of natural hydrogel carriers affected by fermentation and gastrointestinal conditions, *Food Biophysics*, 15: 18-31.

Okuro, P.K., de Matos Junior, F.E. and Favaro-Trindade, C.S. (2013a). Technological challenges for spray chilling encapsulation of functional food ingredients, *Food Technology and Biotechnology*, 51: 171-182.

Okuro, P.K., Thomazini, M., Balieiro, J.C.C., Liberal, R.D.C.O. and Fávaro-Trindade, C.S. (2013b). Co-encapsulation of *Lactobacillus acidophilus* with inulin or polydextrose in solid lipid microparticles provides protection and improves stability, *Food Research International*, 53: 96-103.

Oliveira, D.A., Mezzomo, N., Gomes, C. and Ferreira, S.R.S. (2017). Encapsulation of passion fruit seed oil by means of supercritical antisolvent process, *Journal of Supercritical Fluids*, 129: 96-105.

Ortega-Rivas, E., Juliano, P. and Yan, H. (2006). *Food Powders: Physical Properties, Processing, and Functionality*, Springer Science & Business Media, New York.

O'Sullivan, J.J., Norwood, E.A., O'Mahony, J.A. and Kelly, A.L. (2019). Atomisation technologies used in spray-drying in the dairy industry: A review, *Journal of Food Engineering*, 243: 57-69.

Oxley, J.D. (2012). Spray cooling and spray chilling for food ingredient and nutraceutical encapsulation, pp. 110-130. *In*: N. Garti and D.J. McClements (Eds.), *Encapsulation Technologies and Delivery Systems for Food Ingredients and Nutraceuticals*, Elsevier, Woodhead Publishing, Cambridge.

Paini, M., Aliakbarian, B., Casazza, A.A., Lagazzo, A., Botter, R. and Perego, P. (2015). Microencapsulation of phenolic compounds from olive pomace using spray drying: A study of operative parameters, *LWT – Food Science and Technology*, 62: 177-186.

Parthasarathi, S. and Anandharamakrishnan, C. (2016). Enhancement of oral bioavailability of vitamin E by spray-freeze drying of whey protein microcapsules, *Food and Bioproducts Processing*, 100: 469-476.

Paucar, O., Tulini, F.L., Thomazini, M., Balieiro, J.C.C., Pallone, E.M.J.A. and Favaro-Trindade, C.S. (2016). Production by spray chilling and characterization of solid lipid microparticles loaded with vitamin D_3, *Food and Bioproducts Processing*, 100: 344-350.

Paz, E., Martín, Á. and Cocero, M.J. (2019). Formulation of β-carotene with soybean lecithin by PGSS (Particles from Gas Saturated Solutions)-drying, *Journal of Supercritical Fluids*, 72: 125-133.

Pelissari, J.R., Souza, V.B., Pigoso, A.A., Tulini, F.L., Thomazini, M., Rodrigues, C.E.C., Urbano, A. and Favaro-Trindade, C.S. (2016). Production of solid lipid microparticles loaded with lycopene by spray chilling: Structural characteristics of particles and lycopene stability, *Food and Bioproducts Processing*, 98: 86-94.

Pellicer, J.A., Fortea, M.I., Trabal, J., Rodríguez-López, M.I., Gabaldón, J.A. and Núñez-Delicado, E. (2019). Stability of microencapsulated strawberry flavor by spray drying, freeze drying and fluid bed, *Powder Technology*, 347: 179-185.

Pérez-Alonso, C., Báez-González, J.G., Beristain, C.I., Vernon-Carter, E.J. and Vizcarra-Mendoza M.G. (2003). Estimation of the activation energy of carbohydrate polymers blends as selection criteria for their use as wall material for spray-dried microcapsules, *Carbohydrate Polymers*, 53: 197-203.

Pérez-Masiá, R., Lagaron, J.M. and López-Rubio, A. (2014a). Surfactant-aided electrospraying of low molecular weight carbohydrate polymers from aqueous solutions, *Carbohydrate Polymers*, 101: 249-255.

Pérez-Masiá, R., Lagaron, J. and López-Rubio, A. (2014b). Development and optimization of novel encapsulation structures of interest in functional foods through electrospraying, *Food and Bioprocess Technology*, 7: 3236-3245.

Pérez-Masiá, R., Lagaron, J.M. and Lopez-Rubio, A. (2015a). Morphology and stability of edible lycopene-containing micro- and nanocapsules produced through electrospraying and spray drying, *Food and Bioprocess Technology*, 8: 459-470.

Pérez-Masiá, R., López-Nicolás, R., Periago, M.J., Ros, G., Lagaron, J.M. and López-Rubio, A. (2015b). Encapsulation of folic acid in food hydrocolloids through nanospray-drying and electrospraying for nutraceutical applications, *Food Chemistry*, 168: 124-133.

Perignon, C., Ongmayeb, G., Neufeld, R., Frere, Y. and Poncelet, Denis. (2015). Microencapsulation by interfacial polymerisation: Membrane formation and structure, *Journal of Microencapsulation*, 32: 1-15.

Perrut, M. (2000). Supercritical fluid applications: Industrial developments and economic issues, *Industrial & Engineering Chemistry Research*, 39: 4531-4535.

Pham-Hoang, B.N., Romero-Guido, C., Phan-Thi, H. and Waché, Y. (2013). Encapsulation in a natural, preformed, multi-component and complex capsule: Yeast cells, *Applied Microbiology and Biotechnology*, 97: 6635-6645.

Pinho, E., Grootveld, M., Soares, G. and Henriques, M. (2014). Cyclodextrins as encapsulation agents for plant bioactive compounds, *Carbohydrate Polymers*, 101: 121-135.

Pitigraisorn, P., Srichaisupakit, K., Wongpadungkiat, N. and Wongsasulak, S. (2017). Encapsulation of *Lactobacillus acidophilus* in moist-heat-resistant multilayered microcapsules, *Journal of Food Engineering*, 192: 11-18.

Pool, H., Mendoza, S., Xiao, H. and McClements, D.J. (2013). Encapsulation and release of hydrophobic bioactive components in nanoemulsion-based delivery systems: Impact of physical form on quercetin bioaccessibility, *Food & Function*, 4: 162-174.

Popova, A.V. and Hincha, D.K. (2016). Effects of flavonol glycosides on liposome stability during freezing and drying, *Biochimica et Biophysica Acta – Biomembranes*, 1858: 3050-3060.

Pradeep, H.N. and Nayak, C.A. (2019). Enhanced stability of C-phycocyanin colorant by extrusion encapsulation, *Journal of Food Science and Technology*, 56: 4526-4534.

Pravilović, R., Balanč, B., Trifkoić, K., Đorđević, V., Bošković-Vragolović, N., Bugarski, B. and Pjanović, R. (2017). Comparative effects of span 20 and span 40 on liposomes release properties, *International Journal of Food Engineering*, 13: 20170339.

Prieto, C. and Calvo, L. (2017). The encapsulation of low viscosity omega-3 rich fish oil in polycaprolactone by supercritical fluid extraction of emulsions, *Journal of Supercritical Fluids*, 128: 227-234.

Quispe-Condori, S., Saldana, M.D.A. and Temelli, F. (2011). Microencapsulation of flax oil with zein using spray and freeze drying, *LWT – Food Science and Technology*, 44: 1880-1887.

Ramezanzade, L., Hosseini, S.F. and Nikkhah, M. (2017). Biopolymer-coated nanoliposomes as carriers of rainbow trout skin-derived antioxidant peptides, *Food Chemistry*, 234: 220-229.

Rashidinejad, A., John Birch, E., Sun-Waterhouse, D. and Everett, D.W. (2014). Delivery of green tea catechin and epigallocatechin gallate in liposomes incorporated into low-fat hard cheese, *Food Chemistry*, 156: 176-183.

Recharla, N., Riaz, M., Ko, S. and Park, S. (2017). Novel technologies to enhance solubility of food-derived bioactive compounds: A review, *Journal of Functional Foods*, 39: 63-73.

Reis, P.M.C.L., Mezzomo, N., Aguiar, G.P.S., Senna, E.M.T.L., Hense, H. and Ferreira, S.R.S. (2018). Ultrasound-assisted emulsion of laurel leaves essential oil (*Laurus nobilis* L.) encapsulated by SFEE, *Journal of Supercritical Fluids*, 147: 284-292.

Robert, P., Carlsson, R.M., Romero, N. and Masson L. (2003). Stability of spray-dried encapsulated carotenoid pigments from *rosa mosqueta* (*Rosa rubiginosa*) oleoresin, *Journal of the American Oil Chemists' Society*, 80: 1115-1120.

Rocha-Selmi, G.A., Bozza, F.T., Thomazini, M., Bolini, H.M.A. and Favaro-Trindade, C.S. (2013). Microencapsulation of aspartame by double emulsion followed by complex coacervation to provide protection and prolong sweetness, *Food Chemistry*, 139: 72-78.

Rodrigues da Cruz, M.C., Andreotti Dagostin, J.L., Perussello, C.A. and Masson, M.L. (2019). Assessment of physicochemical characteristics, thermal stability and release profile of ascorbic acid microcapsules obtained by complex coacervation, *Food Hydrocolloids*, 87: 71-82.

Rodrigues, R.A.F., Rodrigues, M.V.N., Oliveira, T.I.V., Bueno, C.Z., de Oliveira Souza, I.M., Sartoratto, A. and Foglio, M.A. (2011). Docosahexaenoic acid ethyl esther (DHAEE) microcapsule production by spray-drying: Optimization by experimental design, *Food Science and Technology*, 31: 589-596.

Rosa, M.T.M.G., Alvarez, V.H., Albarelli, J.Q., Santos, D.T., Meireles, M.A.A. and Saldana, M.D.A. (2019). Supercritical anti-solvent process as an alternative technology for vitamin complex encapsulation using zein as wall material: Technical-economic evaluation, *Journal of Supercritical Fluids* (in press).

Salvia-Trujillo, L., Soliva-Fortuny, R., Rojas-Graü, M.A., McClements, D.J. and Martín-Belloso, O. (2017). Edible nanoemulsions as carriers of active ingredients: A review, *Annual Review of Food Science and Technology*, 8: 439-466.

Santana, R.C., Perrechil, F.A. and Cunha, R.L. (2013). High- and low-energy emulsifications for food applications: A focus on process parameters, *Food Engineering Reviews*, 5: 107-122.

Santos, D.T. and Meireles, M.A.A. (2013). Micronization and encapsulation of functional pigments using supercritical carbon dioxide, *Journal of Food Process Engineering*, 36: 36-49.

Santos, D.T., Albarelli, J.Q., Beppu, M.M. and Meireles, M.A. (2013). Stabilization of anthocyanin extract from jabuticaba skins by encapsulation using supercritical CO_2 as solvent, *Food Research International*, 50: 617-624.

Santos, D.T., Martín, A., Meireles, M.A.A. and Cocero, M.J. (2012). Production of stabilized sub-micrometric particles of carotenoids using supercritical fluid extraction of emulsions, *Journal of Supercritical Fluids*, 61: 167-174.

Schell, D. and Beermann, C. (2014). Fluidized bed microencapsulation of *Lactobacillus reuteri* with sweet whey and shellac for improved acid resistance and *in-vitro* gastro-intestinal survival, *Food Research International*, 62: 308-314.

Schmitt, C. and S.L. Turgeon. (2011). Protein/polysaccharide complexes and coacervates in food systems, *Advances in Colloid and Interface Science*, 167: 63-70.

Semenova, M.G., Antipova, A.S., Zelikina, D.V., Martirosova, E.I., Plashchina, I.G., Palmina, N.P., Binyukov, V.I., Bogdanova, N.G., Kasparov, V.V., Shumilina, E.A. and Ozerova, N.S. (2016). Biopolymer nanovehicles for essential polyunsaturated fatty acids: Structure-functionality relationships, *Food Research International*, 88: 70-78.

Serfert, Y., Drusch, S., Schmidt-Hansberg, B., Kind, M. and Schwarz, K. (2009). Process engineering parameters and type of n-octenylsuccinate-derivatised starch affect oxidative stability of microencapsulated long chain polyunsaturated fatty acids, *Journal of Food Engineering*, 95: 386-392.

Shah, B.R., Li, Y., Jin, W., An, Y., He, L., Li, Z., Xu, W. and Li, B. (2016). Preparation and optimization of Pickering emulsion stabilized by chitosan-tripolyphosphate nanoparticles for curcumin encapsulation, *Food Hydrocolloids*, 52: 369-377.

Shaw, L.A., McClements, D.J. and Decker, E.A. (2007). Spray-dried multilayered emulsions as a delivery method for ω-3 fatty acids into food systems, *Journal of Agricultural and Food Chemistry*, 55: 3112-3119.

Shewan, H.M. and Stokes J.R. (2013). Review of techniques to manufacture micro-hydrogel particles for the food industry and their applications, *Journal of Food Engineering*, 119: 781-792.

Shi, Q., Fang, Z. and Bhandari, B. (2013). Effect of addition of whey protein isolate on spray-drying behavior of honey with maltodextrin as a carrier material, *Drying Technology*, 31: 1681-1692.

Shishir, M.R.I., Xie, L., Sun, C., Zheng, X. and Chen, W. (2018). Advances in micro and nano-encapsulation of bioactive compounds using biopolymer and lipid-based transporters, *Trends in Food Science & Technology*, 78: 34-60.

Silva, E.K. and Meireles, M.A.A. (2014). Encapsulation of food compounds using supercritical technologies: Applications of supercritical carbon dioxide as an antisolvent, *Food and Public Health*, 4: 247-258.

Souilem, S., Kobayashi, I., Neves, M.A., Sayadi, S., Ichikawa, S. and Nakajima, M. (2014). Preparation of monodisperse food-grade oleuropein-loaded w/o/w emulsions using microchannel emulsification and evaluation of their storage stability, *Food and Bioprocess Technology*, 7: 2014-2027.

Stojanovic, R., Belscak-Cvitanovic, A., Manojlovic, V., Komes, D., Nedovic, V. and Bugarski, B. (2012). Encapsulation of thyme (*Thymus serpyllum* L.) aqueous extract in calcium alginate beads, *Journal of the Science of Food and Agriculture*, 92: 685-696.

Šturm, L., Osojnik Črnivec, I.G., Istenič, K., Ota, A., Megušar, P., Slukan, A., Humar, M., Levic, S., Nedović, V., Kopinč, R., Deželak, M., Pereyra Gonzales, A. and Poklar Ulrih, N. (2019). Encapsulation of non-dewaxed propolis by freeze-drying and spray-drying using gum arabic, maltodextrin and inulin as coating materials, *Food and Bioproducts Processing*, 116: 196-211.

Sun, P., Zeng, M., He, Z., Qin, F. and Chen, J. (2013). Controlled release of fluidized bed-coated menthol powder with a gelatin coating, *Drying Technology*, 31: 1619-1626.

Sun-Waterhouse, D. and Wadhwa, S.S. (2013). Industry-relevant approaches for minimising the bitterness of bioactive compounds in functional foods: A review, *Food and Bioprocess Technology*, 6: 607-627.

Tan, C., Feng, B., Zhang, X., Xia, W. and Xia, S. (2016). Biopolymer-coated liposomes by electrostatic adsorption of chitosan (chitosomes) as novel delivery systems for carotenoids, *Food Hydrocolloids*, 52: 774-784.

Tang, Y., Scher, H.B. and Jeoh, T. (2020). Industrially scalable complex coacervation process to microencapsulate food ingredients, *Innovative Food Science & Emerging Technologies*, 59: 102257.

Tavares, L., Pelayo, C. and Noreña, Z. (2019). Encapsulation of garlic extract using complex coacervation with whey protein isolate and chitosan as wall materials followed by spray drying, *Food Hydrocolloids*, 89: 360-369.

Tawiah, B., Asinyo, B.K., Badoe, W., Zhang, L. and Fu, S. (2017). Phthalocyanine green aluminum pigment prepared by inorganic acid radical/radical polymerization for water-borne textile applications, *International Journal of Industrial Chemistry*, 8: 17-28.

Tello, J., Viguera, M. and Calvo, L. (2011). Extraction of caffeine from Robusta coffee (*Coffea canephora* var. *Robusta*) husks using supercritical carbon dioxide, *Journal of Supercritical Fluids*, 59: 53-60.

Teunou, E. and Poncelet, D. (2002). Batch and continuous fluid bed coating – Review and state-of-the-art, *Journal of Food Engineering*, 53: 325-340.

Timilsena, Y.P., Akanbi, T.O., Khalid, N., Adhikari, B. and Barrow, C.J. (2019). Complex coacervation: Principles, mechanisms and applications in microencapsulation, *International Journal of Biological Macromolecules*, 121: 1276-1286.

Torres, O., Murray, B. and Sarkar, A. (2016). Emulsion microgel particles: Novel encapsulation strategy for lipophilic molecules, *Trends in Food Science & Technology*, 55: 98-108.

Torres-Giner, S., Martinez-Abad, A., Ocio, M.J. and Lagaron, J.M. (2010). Stabilization of a nutraceutical omega-3 fatty acid by encapsulation in ultrathin electrosprayed zein prolamine, *Journal of Food Science*, 75: 69-79.

Trifković, K., Milašinović, N., Djordjević, V., Zdunić, G., Kalagasidis Krušić, M., Knežević-Jugović, Z., Šavikin, K., Nedović, V. and Bugarski, B. (2015). Chitosan crosslinked microparticles with encapsulated polyphenols: Water sorption and release properties, *Journal of Biomaterials Applications*, 30: 618-631.

Trifković, K., Đorđević, V., Balanč, B., Kalušević, A., Lević, S., Bugarski, B. and Nedović, V. (2016). Novel approaches in nanoencapsulation of aromas and flavors, pp. 363-419. *In*: A.M. Grumezescu (Ed.), *Encapsulations*, Academic Press, Cambridge, Massachusetts.

Trifković, K. and Benković M. (2019). Introduction to nutraceuticals and pharmaceuticals, pp. 1-31. *In*: C. Galanakis (Ed.), *Nutraceuticals and Natural Product Pharmaceuticals*, Elsevier Academic Press, Cambridge, Massachusetts.

Trifković, K.T., Milašinović, N.Z., Djordjević, V.B., Kalagasidis Krušić, M.T., Knežević-Jugović, Z.D., Nedović, V.A. and Bugarski, B.M. (2014). Chitosan microbeads for encapsulation of thyme (*Thymus serpyllum* L.) polyphenols, *Carbohydrate Polymers*, 111: 901-907.

Tulini, F.L., Souza, V.B., Echalar-Barrientos, M.A., Thomazini, M., Pallone, E.M.J.A. and Favaro-Trindade, C.S. (2016). Development of solid lipid microparticles loaded with a proanthocyanidin-rich cinnamon extract (*Cinnamomum zeylanicum*): Potential for increasing antioxidant content in functional foods for diabetic population, *Food Research International*, 85: 10-18.

Turgeon, S.L., Schmitt, C. and Sanchez, C. (2007). Protein–polysaccharide complexes and coacervates, *Current Opinion in Colloid & Interface Science*, 12: 166-178.

Varona, S., Martín, Á. and Cocero, M.J. (2016). Encapsulation of edible active compounds using supercritical fluids, pp. 16-40. *In*: J.M. Lakkis (Ed.), *Encapsulation and Controlled Release Technologies in Food Systems*, John Wiley & Sons, New Jersey.

Vega, C. and Roos, Y. (2006). Invited review: Spray-dried dairy and dairy-like emulsions – Compositional considerations, *Journal of Dairy Science*, 89: 383-401.

Vilela, J.A.P., Perrechil, F.D.A., Picone, C.S.F., Sato, A.C.K. and da Cunha, R.L. (2015). Preparation, characterization and in vitro digestibility of gellan and chitosan-gellan microgels, *Carbohydrate Polymers*, 117: 54-62.

Vinogradov, S.V. (2010). Hydrophilic colloidal networks (micro- and nanogels) in drug delivery and diagnostics, pp. 367-386. *In*: R. Hidalgo-Alvarez (Ed.), *Structure and Functional Properties of Colloidal Systems*, CRC Press, Boca Raton, Florida.

Visentin, A, Rodríguez-Rojo, S., Navarrete, A., Maestri, D. and Cocero, M.J. (2012). Precipitation and encapsulation of rosemary antioxidants by supercritical antisolvent process, *Journal of Food Engineering*, 109: 9-15.

Volić, M., Pajić-Lijaković, I., Djordjević, V., Knežević-Jugović, Z., Pećinar, I., Stevanović-

Dajić, Z., Veljović, Đ., Hadnadjev, M. and Bugarski, B. (2018). Alginate/soy protein system for essential oil encapsulation with intestinal delivery, *Carbohydrate Polymers*, 200: 15-24.

Wadhwa, G., Kumar, S., Chhabra, L., Mahant, S. and Rao, R. (2017). Essential oil-cyclodextrin complexes: An updated review, *Journal of Inclusion Phenomena and Macrocyclic Chemistry*, 89: 39-58.

Wan, J., Zhong, S., Schwarz, P., Chen, B. and Rao, J. (2019). Physical properties, antifungal and mycotoxin inhibitory activities of five essential oil nanoemulsions: Impact of oil compositions and processing parameters, *Food Chemistry*, 291: 199-206.

Wang, J., Cao, Y., Sun, B. and Wang, C. (2011). Physicochemical and release characterization of garlic oil-b cyclodextrin inclusion complexes, *Food Chemistry*, 127: 1680-1685.

Wang, M.S., Chaudhari, A., Pan, Y., Young, S. and Nitin, N. (2014). Controlled release of natural polyphenols in oral cavity using starch Pickering emulsion, *MRS Proceedings*, 1688; mrss14-1688-y08-11 doi:10.1557/opl.2014.482

Waterhouse, G.I.N., Wang, W. and Sun-Waterhouse, D. (2014). Stability of canola oil encapsulated by co-extrusion technology: Effect of quercetin addition to alginate shell or oil core, *Food Chemistry*, 142: 27-38.

Wechtersbach, L., Poklar Ulrih, N. and Cigic, B. (2012). Liposomal stabilization of ascorbic acid in model systems and in food matrices, *LWT – Food Science and Technology*, 45: 43-49.

Weinbreck, F., de Vries, R., Schrooyen, P. and de Kruif, C.G. (2003). Complex coacervation of whey proteins and gum arabic, *Biomacromolecules*, 4: 293-303.

Whelehan, M. and Marison, I.W. (2011). Microencapsulation by dripping and jet break up, *BRG Newsletter Bioencapsulation Innovations* (September 2011), pp. 4-10; http://bioencapsulation.net/

Winuprasith, T., Khomein, P., Mitbumrung, W., Suphantharika, M., Nitithamyong, A. and McClements, D.J. (2018). Encapsulation of vitamin D_3 in Pickering emulsions stabilized by nanofibrillated mangosteen cellulose: Impact on *in vitro* digestion and bioaccessibility, *Food Hydrocolloids*, 83: 153-164.

Wu, J., Guan, R., Cao, G., Liu, Z., Wang, Z., Shen, H. and Xia, Q. (2018). Antioxidant and antimicrobial effects of catechin liposomes on Chinese dried pork, *Journal of Food Protection*, 81: 827-834.

Xiao, J., Li, Y. and Huang, Q. (2016). Recent advances on food-grade particles stabilized Pickering emulsions: Fabrication, characterization and research trends, *Trends in Food Science & Technology*, 55: 48-60.

Xiao, Z., Liu, W., Zhu, G., Zhou, R. and Niu, Y. (2014). A review of the preparation and application of flavour and essential oils microcapsules based on complex coacervation technology, *Journal of the Science of Food and Agriculture*, 94: 1482-1494.

Xu, D., Yuan, F., Gao, Y., Panya, A., McClements, D.J. and Decker, E.A. (2014). Influence of whey protein-beet pectin conjugate on the properties and digestibility of β-carotene during *in vitro* digestion, *Food Chemistry*, 156: 374-379.

Xue, J., Wang, T., Hu, Q., Zhou, M. and Luo, Y. (2018). Insight into natural biopolymer-emulsified solid lipid nanoparticles for encapsulation of curcumin: Effect of loading methods, *Food Hydrocolloids*, 79: 110-116.

Yalçınöz, S. and Erçelebi, E. (2018). Potential applications of nano-emulsions in the food systems: An update, *Materials Research Express*, 5: 1-17.

Yamashita, C., Chung, M.M.S., dos Santos, C., Mayer, C.R.M., Moraes, I.C.F. and Branco, I.G. (2017). Microencapsulation of an anthocyanin-rich blackberry (*Rubus* spp.) by-product extract by freeze-drying, *LWT – Food Science and Technology*, 84: 256-262.

Ye, A., Cui, J., Taneja, A., Zhu, X. and Singh, H. (2009). Evaluation of processed cheese fortified with fish oil emulsion, *Food Research International*, 42: 1093-1098.

Zaeim, D., Sarabi-Jamab, M., Ghorani, B., Kadkhodaee, R. and Tromp, R.H. (2018). Electrospray-assisted drying of live probiotics in acacia gum microparticles matrix, *Carbohydrate Polymers*, 183: 183-191.

Zhang, C., Chang, M.W., Ahmad, Z., Hu, W., Zhao, D. and Li, J.S. (2015). Stable single device multi-pore electrospraying of polymeric microparticles via controlled electrostatic interactions, *RSC Advances*, 5: 87919-87923.

Zhang, C., Yao, Z.C., Ding, Q., Choi, J.J., Ahmad, Z., Chang, M.W. and Li, J.S. (2017). Tri-needle coaxial electrospray engineering of magnetic polymer yolk – Shell particles possessing dual-imaging modality, multiagent compartments, and trigger release potential, *ACS Applied Materials & Interfaces*, 9: 21485-21495.

Zhang, L., Huang, J., Si, T. and Xu, R.X. (2012). Coaxial electrospray of microparticles and nanoparticles for biomedical applications, *Expert Review of Medical Devices*, 9: 595-612.

Zhou, Y., Sun, S., Bei, W., Zahi, M.R., Yuan, Q. and Liang, H. (2018). Preparation and antimicrobial activity of oregano essential oil Pickering emulsion stabilized by cellulose nanocrystals, *International Journal of Biological Macromolecules*, 112: 7-13.

Zoet, F., Grandia, J. and Sibeijn, M. (2011). *Encapsulated Fat-soluble Vitamin*, NL patent #50668.

Zuidam, N.J. and Heinrich, E. (2010). Encapsulation of aroma, pp. 127-160. *In*: N.J. Zuidam and V. Nedovic (Eds.), *Encapsulation Technologies for Active Food Ingredients and Food Processing*, Springer, New York.

Zuidam, N.J. and Shimoni, E. (2010). Overview of microencapsulates for use in food products or processes and methods to make them, pp. 3029. *In*: N.J. Zuidam and V.A. Nedović (Eds.). *Encapsulation Technologies for Active Food Ingredients and Food Processing*, Springer, New York.

Zuidam, N.J. and Velikov, K.P. (2018). Choosing the right delivery systems for functional ingredients in foods: An industrial perspective, *Current Opinion in Food Science*, 21: 15-25.

Industrial-scale Encapsulation Processes and Products

Farhad Garavand[1,2]*, **Majid Nooshkam**[3], **Mostafa Aghamirzaei**[4,5]*, **Samira Feyzi**[3], **Leila Nateghi**[6], **Shima Yousefi**[7], **Milad Rouhi**[8] and **Seid Mahdi Jafari**[9]

[1] Department of Food Science & Engineering, University of Tehran, Karaj, Iran
[2] Department of Food Chemistry & Technology, Teagasc Food Research Centre, Moorepark, Fermoy, Co. Cork, Ireland
[3] Department of Food Science and Technology, Faculty of Agriculture, Ferdowsi University of Mashhad (FUM), Mashhad, Iran
[4] Department of Food Science and Technology, Faculty of Agriculture, University of Tabriz, Tabriz, 5166616471, Iran
[5] Vice-Chancellor for Food and Drug Administration, Healthcare Network of Fardis, Alborz University of Medical Sciences, Karaj, 3176794107, Iran
[6] Department of Food Science and Technology, Faculty of Agriculture, Varamin-Pishva Branch, Islamic Azad University, Varamin, Iran
[7] College of Food Science and Technology, Science and Research Branch, Islamic Azad University, Tehran, Iran
[8] Department of Food Science and Technology, School of Nutrition Sciences and Food Technology, Research Center for Environmental Determinants of Health (RCEDH), Health Institute, Kermanshah University of Medical Sciences, Kermanshah, Iran
[9] Department of Food Materials and Process Design Engineering, Gorgan University of Agricultural Sciences and Natural Resources, Gorgan, Iran

1. Introduction

Modern lifestyle and changing human habits have resulted in numerous health conditions and diseases related to inadequate nutrition. This trend has driven people to request for health-promoting foods with better organoleptic features in their diet. Such a demand is recognized by food industry and the efforts are put toward fulfillment of consumers' expectations through development of functional foods. Although addition of functional and flavor compounds could be a solution to this challenge, it is accompanied with subsequent challenges. Functional ingredients and flavor compounds are added to food products in order to improve their quality,

*Corresponding authors: farhadgaravand@ut.ac.ir, farhad.garavand@teagasc.com; aghamirzaei.ma88@gmail.com

but they are prone to degradation, inactivation or rapid loss if they are incorporated directly into food products in their normal state, owing to their susceptibility to the environmental, processing, storage and gastrointestinal conditions (de Vos *et al.*, 2010; Nedovic *et al.*, 2011). A well-developed technology in food science and industry is encapsulation, which provides both the possibility of incorporation of functional and flavor compounds into food products and their protection from undesirable chemical reactions, uncontrolled release and physical changes (de Vos *et al.*, 2010; Burgain *et al.*, 2011; Nedovic *et al.*, 2011; Gaonkar *et al.*, 2014).

Different additives, such as flavors, preservatives, colors, etc. are added to foods during industrial-scale food-manufacturing processes to meet consumer demands for palatable products and to extend their shelf-life. In addition, beneficial microelements and specific micronutrients are usually incorporated into foods to obtain high added value, functional and enriched food products. However, there are certain issues regarding addition of these bioactive compounds, which limit the implementation of the specific nutrients due to the incompatibility with food matrix, storage stability and unpleasant sensory features induced by some active bio-compounds (Katouzian and Jafari, 2017). Encapsulation systems are therefore new and proper candidates to achieve food-fortification goals.

In brief, the highlighted reasons for application of encapsulation technology in food industry are listed below (Pegg and Shahidi, 2007; Zuidam and Heinrich, 2010; Dias *et al.*, 2017):

1. Controlled and prolonged release of volatile components like aroma compounds and preventing their loss during preparation and storage steps.
2. Preservation against oxygen which causes oxidation and undesirable organoleptic features, color and off-taste, thus lowering nutritional value.
3. Separation of core materials from reactive compounds of the food matrix;
4. Controlled release and targeted delivery of bioactive compounds at certain conditions, such as pH (acidic or basic), enzyme activity, temperature, etc.
5. Improvement in bioavailability of active agents, like different hydrophobic vitamins, essential oils, omega 3-fatty acids, nutraceuticals, polyphenols, etc., by increasing their surface area and solubility.
6. Convenience of handling by immobilization in a polymeric matrix, such as conversion of a liquid active agent to a powder.
7. Dilution of strong ingredients through their inclusion by wall materials.
8. Homogenous distribution of core materials which could improve visual and textural features.
9. Improvement of flow and compression attributes.

Herein, the various encapsulation techniques and encapsulation of valuable ingredients at industrial scale will be discussed.

2. Encapsulation: Definition and Classification

Encapsulation is referred to a technology of entrapping and shielding gaseous, liquids and solid materials being known as core, active, encapsulants, fills, or internal phase through a stable and protective food-grade material acting as a sealed capsule, well-known as wall, shell, membrane, carrier, capsule, or matrix. The design of such capsules, with a diameter of a few nm to mm, leads to protection of core ingredient from its surroundings, release at a controlled rate over prolonged time and at specific locations (Augustin and Hemar, 2009; Mozafari *et al.*, 2008; Zuidam and Heinrich, 2010; Sauvant *et al.*, 2012; Anandharamakrishnan and Ishwarya, 2015; Lakkis, 2016).

Encapsulation could be classified on the basis of size and morphology of capsules. The terms macroencapsulation, microencapsulation and nanoencapsulation are employed when the encapsulation process results in particles with size >5000 μm, 1.0-5000 μm and <1.0 μm, respectively (Anandharamakrishnan and Ishwarya, 2015). Directing from macro- to micro-, and micro- to nano-encapsulation is accompanied with more surface area leading to superior solubility, enhanced bioavailability, precise targeting and controlled release, increased wall thickness and better protection. All these functionalities result in less consumption of core materials and this is cost-effective and benefits the food industry (Mozafari *et al.*, 2008; Gaonkar *et al.*, 2014; Anandharamakrishnan and Ishwarya, 2015).

From the morphological point of view, encapsulation concludes with two main morphologies, including microcapsules and microspheres. Classification of various morphologies in encapsulation of ingredients is presented in Fig. 1. Microcapsules are resultant of a distinct wall surrounding the core, which could be irregular shaped, core shell or mononuclear, polynuclear and multi-wall microcapsules. Microspheres are created when active agents in the form of small particles or droplets disperse in the matrix phase. This latter morphology has been sub-classified into insoluble and soluble matrix microsphere (Gaonkar *et al.* 2014; Anandharamakrishnan and Ishwarya, 2015).

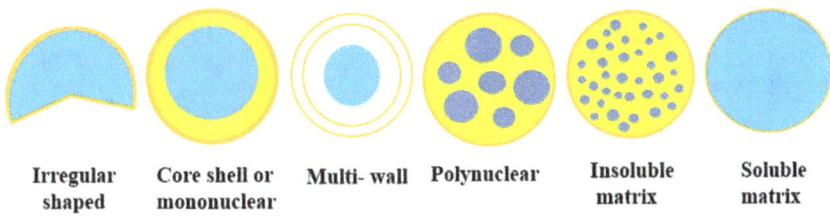

| Irregular shaped | Core shell or mononuclear | Multi- wall | Polynuclear | Insoluble matrix | Soluble matrix |

Fig. 1: Classification of morphologies in encapsulation of various ingredients

3. Importance of Encapsulation for Food Ingredients

Generally, the shell is made of natural and modified polysaccharides (e.g. maltodextrin, cyclodextrins, starch, cellulose), gums (e.g. Na-alginate, carageenan, gum arabic, pectin), proteins (animal and vegetable sources including WPC[*], WPI[†], gelatin, caseinate, casein, and soy proteins, gluten, zein), polysaccharide–protein conjugates/complexes, lipids (including phospholipids, fatty acids, glycerides, plant sterols, sorbitan esters), waxes (e.g. Carnuba wax, microcrystalline wax), and food grade synthetic polymers (e.g. cellulose derivatives, PEG[‡], PVA[§], PVP[¶]) (Mozafari *et al.*, 2008; Marcuzzo *et al.*, 2010; Fathi *et al.*, 2014; Gaonkar *et al.*, 2014; Lakkis, 2016).

---

[*] Whey protein concentrate
[†] Whey protein isolate
[‡] Polyethylene glycol
[§] Polyvinyl acetate
[¶] Polyvinyl pyrrolidone

Knowledge about the nature of core materials would shed light on the role of encapsulation in modern food industry. When active agents or core materials are added in their native- state to food products, they are prone to undesirable reactions before they could play their desired role in food, the consumer's mouth or body. Accordingly, the most common active materials which food industries are interested in are flavors and colorants, minerals, vitamins, peptides, polyphenols, antioxidants, omega-3 fatty acids, essential oils, acidulants, bases, probiotics, sweeteners, enzymes and antimicrobial agents (Marcuzzo *et al.*, 2010; Anandharamakrishnan and Ishwarya, 2015; Lakkis, 2016). Considering the importance of such ingredients in the food industry to fulfill health-conscious consumers' expectations, it is a key factor to increase their life-time in a complex food matrix which is in contact with different environmental conditions (pH, oxygen, temperature, etc.).

It is noteworthy that such advantages are valuable when both the active and shell ingredient are generally recognized as safe (GRAS) according to Food and Drug Administration (FDA) of USA or European Food Safety Authority (EFSA); and when encapsulation does not end in additional costs, unstable matrix during preparation steps and any compliance towards final products through consumers (Zuidam and Heinrich, 2010; Anandharamakrishnan and Ishwarya, 2015).

In order to take the advantages of encapsulation process without any negative effect, it is important to choose the core and wall materials which are compatible with each other in the first place; also with environmental conditions and the target point where delivery is desired (Anandharamakrishnan and Ishwarya, 2015). This final note sheds light on the necessity of having knowledge about the physicochemical properties of desired active agents to be encapsulated, challenges that need to be overcome, product definition and characterization, as well as highlights of the role of encapsulation methods (Table 1).

Table 1: The Main Encapsulation Methods

Active Properties	Product Definition	Encapsulation Method	
		Physical/Mechanical	Chemical
Molecular weight	Appearance	Centrifugal extrusion	Coacervation
Structure	Texture	Electrospraying	Inclusion complexation
Solubility	Mouth feel	Emulsification	Ionotropic gelation
Polarity	Color	Emulsion polymerization	Interfacial cross-linking
Partition coefficient	Flavor	Extrusion	Interfacial polymerization
Chemical reactivity	Shelf-life	Fluidized bed coating	Liposome entrapment
Melting point	pH	Freeze drying	Phase separation
Boiling point	Ionic strength		Self-assembly
Charge	Storage temperature	Molecular encapsulation	Sol-gel process
		Spinning disk	Solvent evaporation
		Spray drying/cooling/ chilling	Solvent exchange
		Supercritical fluid technology	

4. Characteristics of Active Agents

In order to benefit from active agents in improving the food quality and appearance, as well as consumer health and desire, it is necessary to have information about physicochemical properties of active agents, especially the ones leading to challenges in their incorporation into different food formulations. Some of the most common features of active agents which are accompanied with challenges are listed in Table 1. For instance, vitamins A, E and D and β-carotene are hydrophobic compounds with low solubility, leading to their rapid degradation in aqueous matrixes and low bioavailability (Sauvant *et al.*, 2012; McClements, 2018). It has been proved that inclusion of vitamin A within β-cyclodextrin and hydroxypropyl β-cyclodextrins increases its solubility by 100 and 10,000 times, respectively (Lin *et al.*, 2000, Qi and Shieh, 2002). Similarly, nanoencapsulation of vitamin E using medium chain triglycerides in the presence of surfactant mixture of Tween 80/Brij 35 caused high entrapment efficiency by a membrane emulsification method (Laouini *et al.*, 2012). Curcumin and resveratrol, as hydrophobic polyphenols with appreciated antioxidant activity, undergo degradation due to oxidative stress, alkaline conditions and gastrointestinal tract, but nanoencapsulation and co-nanoencapsulation of these two polyphenols improved their *in-vitro* antioxidant activity and photo-stability (Coradini *et al.*, 2014; McClements, 2018).

Moreover, it has been proven that different aroma compounds can be easily released from food matrixes, especially in aqueous and fat-reduced media owing to their relatively high hydrophobicity (log P) and volatility (log K_{aw}) (Weerawatanakorn *et al.*, 2015), and may face flavor-flavor interaction, oxidation, and acid- or light-induced interactions causing off-flavor. Aroma stability (physical and chemical) correlates with food quality and its acceptance (Madene *et al.*, 2006; Weerawatanakorn *et al.*, 2015). Superior binding affinity of WPI towards vanillin in comparison with casein and soy protein isolate (Li *et al.*, 2000), in conjunction with considerable binding affinity of bovine serum albumin (BSA) and WPI towards safranal (Feyzi *et al.*, 2019) led to introduction of whey proteins as an attractive matrix for encapsulation of these aroma compounds (Liu and Mori, 1993; Kanakis *et al.*, 2007).

5. Encapsulation Techniques

Considering the diversity of core and shell/matrix materials, no specific universal encapsulation method could be recommended. The method of encapsulation depends not only on the nature of incorporated materials, but also on the final product characteristics (de Vos *et al.*, 2010). Accordingly, encapsulation methods are classified into two basic chemical and physical or mechanical approaches. Physical or mechanical techniques need some specialized equipment to create and stabilize microencapsulate by controlling the physical conditions under which the process is carried out, usually through precipitation of polymeric solutions. On the other hand, chemical approaches deal with processes which involve different interactions leading to polymerizations, such as adding surfactants, adjusting pH, etc. (Sanguansri and Augustin, 2010).

Table 1 addresses methods/technologies through which the majority of encapsulation processes are done. Also, Table 2 represents particle size range (in μm) that each encapsulation method concludes, giving insights into comparison of

Table 2: Particle Size Range (µm) of Different Encapsulation Techniques

Encapsulation Techniques		Particle Size Range (µm)
Physical methods	Emulsification	
	Spray drying	
	Spray chilling/cooling	
	Freeze drying	
	Fluidized bed coating	
	Melt injection	
	Melt extrusion	
	Centrifugal extrusion	
	Co-extrusion	
	Spinning disk	
	Simple/complex coacervation	
	Liposome entrapment	
	Inclusion complexation	
Chemical methods	Solvent evaporation	
	Interfacial polymerization	
	Phase separation	
	Nanoencapsulation	

0.01   0.1   1   10000

different encapsulation methods (Kheadr *et al.*, 2003; Zuidam and Heinrich, 2010; Gaonkar *et al.*, 2014; Mishra, 2016).

5.1 Emulsification

Emulsion systems (emulsions, nano-emulsions and micro-emulsions) could be classified in different ways. Micro-emulsions are thermodynamically stable systems that can be generated without applying intense forces and have clear appearance due to their small droplet size (>50-100 nm); nano-emulsions are thermodynamically unstable systems needing intense shear force to generate and have droplet size of 100 nm and more (according to literatures); that's why their appearance is clear to somewhat turbid (Rao and McClements, 2011). Smaller size of emulsions, i.e. directing from macro- and nano- toward micro-emulsions (with droplet size of <100 nm) results in more stability and eliminates required surfactants' level from the usual amount of 20 per cent to 10 per cent in nano scales (Laouini *et al.*, 2012; Garavand and Madadlou, 2014; Anandharamakrishnan and Ishwarya, 2015). Generally, encapsulation through emulsification results in a wide range of particle sizes from 0.2 to 5000µm (Zuidam and Heinrich, 2010).

In addition, emulsions could be classified by the nature of two constituent phases – oil in water (O/W), water in oil (W/O), double emulsions (W/O/W or O/W/O) and multilayered emulsions. While, hydrophilic ingredients like probiotics and some polyphenols could act as a core of W/O emulsions, lipophilic ingredients like β-carotene and other carotenoids, aroma compounds, dietary fats and vitamin E could be the core of O/W emulsions (Anandharamakrishnan and Ishwarya, 2015; Lakkis, 2016; Garavand *et al.*, 2018). Double and multilayered emulsions are mainly formed using charged biopolymers; in other words, there is a matrix phase (e.g. W) which contains oil droplets surrounded by a charged biopolymer as an emulsifier. Application of charged emulsifier induces charge on to the droplet surface, preparing them to host the second layer of electrolytes having opposite charge through electrostatic attraction forces, i.e. each layer must have an opposite charge to that of the next depositioning layer. Accordingly, multilayered emulsions, as an extended form of double emulsions, could be considered where oxidation or un-controlled release is possible, similar to what is likely to happen for poly-unsaturated fatty acids, vitamins and aroma compounds (Anandharamakrishnan and Ishwarya, 2015). Directing from mono- and double emulsions toward multilayered emulsions increases thermodynamic stability of emulsions, chemical stability of actives and controls the release of core ingredients as a consequence of greater wall thickness (Güzey and McClements, 2006).

Final emulsion could be converted into powder form by freeze drying or spray drying methods to increase shelf-life of the final product. It has been proven that application of electrostatic layer-by-layer deposition technology on omega-3 fatty acids to create multilayered emulsions using lecithin and chitosan, followed by spray drying in the presence of corn syrup solids, caused considerable oxidative stability (Shaw *et al.*, 2007).

Further studies are required to choose between freeze drying and spray drying as complementary steps of emulsification, since the nature of core and wall materials is important. *Lactobacillus rhamnosus* was stabilized in an emulsion system containing whey protein and starch, followed by spray or freeze drying. While there was no difference in viability of probiotic obtained with the two procedures, more stability was achieved when freeze drying was utilized (Ying *et al.*, 2010).

5.2 Spray Drying/Cooling/Chilling

One of the oldest industrial and most widely used encapsulation technologies is spray drying. The basic principles of this technology is based on dissolving a core material into a matrix phase, in which both are in liquid state, i.e. aqueous or emulsion solution, followed by homogenization (Fig. 2). The resultant dispersion is fed into the hot chamber through a nozzle or spinning wheel which converts the dispersion to droplets. The contact of hot air, passing from the chamber with low humidity and specific flow rate with atomized droplets, evaporates the solvent followed by solidification of shell materials on to the core, resulting in dried/powdered particles. The obtained powders are collected in cyclone. Generally, drying time and final particle size depend on the droplet size. Also, the viscosity and surface tension of the dispersion, flow rate of hot air and pressure drop across the atomizer affect the atomizing droplet size.

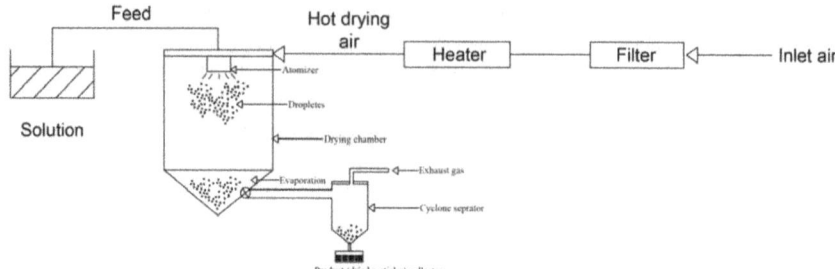

Fig. 2: A schematic diagram of encapsulation using spray drying technology

Spray drying can produce particle sizes with 10-100 μm mean diameter, usually less than 40μm (Madene *et al.*, 2006; Zuidam and Heinrich, 2010). Spray drying is an efficient encapsulation approach for both hydrophilic and hydrophobic agents. However, considering the fact that water-based dispersions are fed into the atomizer, it is necessary that hydrophobic agents are dissolved in an oil phase, followed by formation of O/W emulsions prior to spray drying (de Vos *et al.*, 2010). Although diverse range of active materials, such as essential oils, natural colorants, flavors, vitamins could be applied to spray drying encapsulation, some heat-labile agents, like flavor compounds with low boiling point and probiotics like bifidobacteria (sensitive to temperature higher than 60°C), may be lost or oxidized, and leak into the product upon spray drying (Madene *et al.*, 2006; de Vos *et al.*, 2010).

The most common wall materials for spray drying are dairy- and plant-protein concentrates and isolates and polysaccharides/gums, such as gum arabic, maltodextrin, modified starch, inulin, and cashew gum (de Vos *et al.*, 2010; Dias *et al.*, 2017). In general, spray drying is an economic technique, widespread, continuous and applicable on industrial scale which provides stable products, but the resultant encapsulates release the containing actives upon dissolving in water, which is a disadvantage of this encapsulation technique (Champagne and Fustier, 2007). Indeed, some improvements like employment of more hydrophobic or cross-linked carriers provide gradual release of actives (Zuidam and Heinrich, 2010). Moreover, although spray drying has preservation effect on some probiotic cultures, their activity are dramatically decreased after a few weeks of storage at room temperature owing to the application of high temperature in the process. Desmond *et al.* (2002) showed addition of gum acacia, as a soluble fiber, prior to spray drying of probiotic *Lactobacillus paracasei*, increased its thermo-stability and viability compared with milk powder alone (Desmond *et al.*, 2002). However, it is noteworthy that such thermo-stability and viability-induced effects on probiotics could not be expanded to all soluble fibers. For example, Corcoran *et al.* (2005) observed that inulin and polydextrose could not improve viability of *Lactobacillus rhamnosus*.

In order to overcome the problems related to high inlet temperature used in spray drying, other techniques which are similar in principle but apply lower temperatures, have been developed, like spray cooling and spray chilling (Lakkis, 2016; Dias *et al.*, 2017). In spray cooling and chilling systems, the active agent with matrix, which is usually a high melting-point lipid, is atomized into a lowtemperature chamber at ambient and refrigerator temperature, respectively (de Vos *et al.*, 2010; Anandharamakrishnan and Ishwarya, 2015). As a consequence of encountering low temperature air in the chamber, which must be lower than the solidification point of the molten lipid, droplets would be solidified around the agent. Hydrophilic and heat-labile agents, like ferrous sulfate, acidulants, minerals, enzymes, vitamins and flavors could be encapsulated by spray cooling and chilling methods (de Vos *et al.*, 2010; Anandharamakrishnan and Ishwarya, 2015). Since the matrix is a lipid, the active release occurs after food ingestion (Dias *et al.*, 2017).

5.3 Freeze Drying

In order to perform freeze drying, a solution containing active and carrier materials dissolved in aqueous solvent are first frozen at very low temperature of -40 to -90°C (known as freezing step), followed by drying at low pressure and temperature (-20 to -90°C) (Zuidam and Heinrich, 2010; Anandharamakrishnan and Ishwarya, 2015). This encapsulation process results in matrix morphology with a particle size of about 20-5000µm. It has been reported that freeze drying could be efficient for heat-sensitive agents, like flavors, oils and probiotics (Anandharamakrishnan and Ishwarya, 2015).

There are some disadvantages of freeze drying, like longer process times and high costs; besides that, presence of a considerable amount of open pores in freeze dried product, which is a consequence of application of low pressure in the process, would lead to oxidation (Zuidam and Heinrich, 2010). Similar to this conclusion, it has been reported that encapsulated β-carotene with maltodextrin as a carrier, undergoes faster oxidation and degradation kinetics when spray drying is performed, followed by freeze-drying and drum-drying encapsulation processes; although the initial β-carotene loss (after 24 h) on the powder surface was highest and lowest in drum- and freeze-drying processes, respectively (Desobry *et al.*, 1997).

5.4 Fluidized Bed Coating

This technique is considered as 'modified spray drying' in a batch or continuous setup. It could be performed as a type of top, bottom or tangential spray. In general, the process includes suspension of core material into the air, followed by spraying matrix material on to the core, which becomes a capsule around the core over time. For this purpose, the matrix must be thermally stable with appropriate viscosity to create the film around the particles; thereby it should be aqueous or emulsion in evaporable solvent and should include a wider range of materials, such as proteins, carbohydrates, starch and its derivatives, dextrins and lipids which are molten or dissolved in evaporable solvent (Pegg and Shahidi, 2007; de Vos *et al.*, 2010; Zuidam and Heinrich, 2010; Mishra, 2016). In the case of using molten lipids, including vegetable oils, emulsifiers, fatty acids or waxes, temperature control is important during process in order to prevent lipid solidification prior to contact with particles. Besides, product temperatures, which are close to melting point of lipid, may result in sticky particles and agglomeration phenomenon, which is undesirable (Zuidam and Heinrich, 2010).

Generally, final particle size is usually in the range of 50-350µm, but it might be expanded within 5-5000µm range (Pegg and Shahidi, 2007; Zuidam and Heinrich, 2010). Moreover, fluidized bed coating can be applied on to a spray dried product. In this case, the matrix acts as a second layer on the core, leading to better protection and controlled release (Pegg and Shahidi, 2007).

5.5 Extrusion Technologies

Extrusion is a relatively new encapsulation technology in comparison to spray drying and provides real encapsulation (Pegg and Shahidi, 2007). It could be performed in melt injection, melt extrusion and co-extrusion setups. Both melt injection and melt extrusion are done by melting the coating material usually at temperatures of 110-140°C, dispersing the active in the coating followed by extruding through a filter or twin-screw extruder. The resultant would be hardened through cooling and/or due to contact with dehydrating solvent. In other words, the main difference in these two methods is that melt injection is a vertical and screwless process in which isopropanol or liquid nitrogen is used as dehydrating solvent; but in melt extrusion the screws are horizontal and usually a double screw is used (Zuidam and Heinrich, 2010; Ubbink 2013).

The morphology of resultant particles is dependent on the matrix used for encapsulation. Matrix can be preferably made of polysaccharides, like molten starches, maltodextrin, gums, or proteins, emulsifiers and oils (Zuidam and Heinrich, 2010). Final particle size obtained by melt injection (200-2000 µm) is slightly less than that of melt extrusion (300-5000µm) (Anandharamakrishnan and Ishwarya, 2015; Lakkis, 2016). Encapsulated actives are protected from oxidation during storage and have extended shelf-life, which is really important in the case of flavors and oils (Pegg and Shahidi, 2007; Ubbink 2013; Anandharamakrishnan and Ishwarya, 2015). Although extrusion methods are economic, simple, and generally easy to perform, making them suitable for industries that benefit flavor oils or those seeking efficient incorporation of anaerobic microorganisms in food products, relatively low active load, about 10 per cent in comparison with spray drying technology (20 per cent), may cause some concerns about costs in the use of this method (Pegg and Shahidi, 2007; deVos *et al.*, 2010; Zuidam and Heinrich, 2010).

It has been reported that extrusion prolonged the shelf life of orange peel oil and flavors in the absence of antioxidants up to four years, determined by accelerated shelf life test (Ubbink, 2013).

Centrifugal extrusion, also known as liquid co-extrusion, involves feeding core and wall materials through two different but concentric orifices. Rotation of head causes the co-extrusion of active and wall materials through concentric orifices resulting in a co-extruded rod of core surrounded by a coating material. During rotation, centrifugal force moves the extruded rod outward breaking them into capsules, usually with particle size of 150-2000 µm. This technique has improved the active loading up to 20-80 per cent (Risch and Reineccius, 1995; Pegg and Shahidi, 2007).

5.6 Spinning Disc

Spinning disc, also known as centrifugal suspension separation, is principally similar to centrifugal extrusion. The mixture containing core ingredients dissolved in a wall material is fed into the rotating disc. Rotation action causes droplets' movement towards edge of the disc where the wall solidifies around the core, usually as a result of cooling (Mishra, 2016). Considering that the process takes limited time, from a few seconds to a few minutes, the product particle size of about 30-2000 µm, which could be controlled by feed rate and disk rotation rate and range of lipids and polysaccharides, could be used as wall material, the spinning disc might be an easy and cost-effective encapsulation process (Jenjob *et al.*, 2012; Mishra, 2016).

Optimization of the encapsulation process of lipase with methyl methacrylate chitosan applying disc spinning method resulted in stable immobilized lipase for catalyzing tributyrin, as a reference substrate. So, the process could be suggested as a cost-effective industrial encapsulation method (Jenjob *et al.*, 2012).

5.7   Coacervation

A colloidal phenomenon in which a homogenous polymer solution undergoes phase separation process through creating coacervate, as polymer-rich phase and solvent or polymer-poor phase, is known as coacervation. This chemical encapsulation method could be performed as simple or complex coacervation; the first one is accomplished by employing only one polymer, while the latter requires two or more oppositely-charged polymers. The underlying mechanism of conversion of polymer(s) to a coacervate is based on decreasing the solubility of polyelectrolyte(s), which could be performed by utilizing alcohols (ethanol, propanol), or through addition of salt (sodium sulfate) (Lakkis, 2016; Mishra, 2016). Some of the negatively-charged polysaccharides that are used as a coacervate are gum arabic, which can be replaced by other polysaccharides, like carboxy methyl cellulose, pectin, carrageenan, alginate and alginate derivatives (Zuidam and Heinrich, 2010).

Generally, coacervation includes addition of an ingredient to be encapsulated, to a solution in which polymer(s) have already been converted into coacervate, followed by a change in temperature, pH or salt addition. This latter step causes the coacervate to entrap the core ingredient. Addition of an agent which improves cross-linking in shell, like glutaraldehyde (not allowed for food applications in many countries) or transglutaminase, would help thicken shell using polyelectrolytes in solution. The final microencapsulates can be separated through centrifugation or filtration, followed by solvent removal by freeze or spray drying (Zuidam and Heinrich, 2010; Anandharamakrishnan and Ishwarya, 2015).

Coacervation can be applied to a wide range of actives, like heat- or pH-sensitive ones, since the operational temperature is < 50°C and salt addition could be employed instead of changing pH. Complex coacervation is more efficient and could be a choice for commercial applications. Because a thicker wall provides better protection against oxidation, the payload can be up to 95 per cent, and the formed barrier is not soluble in water, regardless of the method applied for its formation. Limitations in complex coacervation arise from the high polarity of actives; also, usually the resultant single core capsules do not have uniform wall thickness, leading to variable stability and strength of capsules (Mishra, 2016).

Oliveira *et al.* (2007) encapsulated *Bifidobacterium lactis* and *Lactobacillus acidophilus* by applying complex coacervation and using casein-pectin complex followed by spouted bed-drying. They reported that coacervation coupled with spouted bed process was well-organized in microencapsulation of the mentioned bacteria; however, the microencapsulated bacteria were not able to survive in modeled stomach pH condition. The shelf-life test indicated that microencapsulated *Lactobacillus acidophilus* could tolerate ninety days of storage at low temperature (7°C) while its viability significantly reduced at 37°C; on the other hand, microencapsulated *Bifidobacterium lactis* was not able to survive in both low and medium storage temperatures (7°C and 37°C, respectively) over the period of ninety days.

Jun-xia *et al.* (2011) applied soy protein isolate and gum arabic for encapsulation of sweet orange oil through complex coacervation; the best condition was at pH=4 and equimolar ratio of soy protein isolate, gum arabic and sucrose. Complex coacervation showed efficiency in retention of the mix of aroma compounds present in the orange oil, especially D-limonene, as the indicator aroma compound of sweet orange oil of about 90 per cent.

5.8   Liposomal Entrapment

The technology of core entrapment within an aqueous phase through a closed, continuous bilayered structure made of lipids, mainly phospholipids (like lecithin), containing protein and/or carbohydrate, is called liposomal entrapment. The core might be lipid- or aqueous-soluble or amphiphilic ingredient. Final particle size varies from 30nm to a few microns (Mozafari *et al.*, 2008; Anandharamakrishnan and Ishwarya, 2015). Liposomes are classified into different groups, including multilamellar vesicle (MLV) composed of a number of concentric bilayers; unilamellar vesicle (ULV) encompassing one lipidic bilayer with variable size of less or more than 100nm and multivesicular vesicle (MVV) having one lipidic bilayer, which contains numbers of non-concentric vesicles (Zawada, 2004; Fernandez *et al.*, 2009).

Based on the definition, it could be concluded that hydrophobic-hydrophilic interaction of lipid membrane and aqueous phase is the main underlying mechanism of liposomal entrapment. So, the liposome formation requires energy input to make possible such interaction and to stabilize it (Mozafari *et al.*, 2008). Traditional method of liposomal formation is based on the solvation of phospholipid/lipid in a solvent like chloroform or methanol followed by solvent evaporation for formation of a thin film. The procedure is accomplished by addition of aqueous phase to previous hydrophobic phase accompanied by some mechanical (high shear, extrusion, etc.) or thermal energy to reduce the size for separation of bilayer phospholipid membrane from the bulk phase. Application of non-food-grade solvent and high-costs make it incompatible with standards of food industries (Mozafari *et al.*, 2008; Zuidam and Heinrich, 2010; Anandharamakrishnan and Ishwarya, 2015). A modification of liposome process through heating method, based on a one-step procedure in the absence of toxic solvents in one hour, might make it more suitable for food applications (Mozafari *et al.*, 2008).

An important advantage of liposomal entrapment is the targeted release of actives in food stuff or human gut upon specific temperatures, relating to transition temperature of lipid/ phospholipid, which transfers it from gel to liquid form (Anandharamakrishnan and Ishwarya, 2015).

Decrease in the operation steps, diversity in liposome types and compatibility with standards for raw food materials can recommend this technology for food applications, including encapsulation of essential oils, vitamins, colorants, enzymes, microorganisms, antioxidants, antimicrobials, flavors, etc. (Mozafari *et al.*, 2008).

Kheadr *et al.* (2003) applied liposome-loaded enzyme cocktails on Cheddar cheese to accelerate proteolysis and lipolysis. Flavourzyme, a natural bacterial protease, acid fungal protease and lipase were entrapped in liposomes and added to cheese milk prior to renneting. Results showed that cheese with liposome-encapsulated enzyme cocktails had more mature texture and higher flavor intensity in a shorter time as compared with control cheeses.

Yokota *et al.* (2012) applied lyophilized multilamellar liposomes for encapsulation of casein hydrolysate, using soy lecithin. Incorporation of lyophilization with liposome approach required lyoprotectants to protect integrity of bilayer membrane (e.g. sucrose and trehalose) in this study.

5.9   Inclusion Complexation

Interaction of a ligand with a receptor containing a cavity, so-called cavity-based substrate, refers to complex inclusion. There are different bio-polymers, such as proteins and polysaccharides which contain a cavity in their structure, which makes them appropriate substrate for molecular inclusion. β-lactoglobulin (β-Lg) belongs to the lipocalin family which has a calyx encompassing eight anti-parallel β-sheets that act as a specific binding site for small hydrophobic ligands like aroma compounds, colorants like β-carotene, polyphenols and fatty-acids (Feyzi *et al.*, 2019; Allahdad *et al.*, 2018; Kanakis *et al.*, 2011; Le Maux *et al.*, 2014; Tavel *et al.*, 2010). β-Lg is in stable and contact conformation at acidic pH of stomach due to its monomeric state and closing conformation of EF-loop (Qin *et al.*, 1998), suggesting it as an appropriate carrier for small hydrophobic ligands for oral-delivery purposes.

Cyclodextrins, obtained from hydrolysis of starch using cyclodextrin glucosyltransferase which cleavages α (1→4) bonds between glucose subunits, also contain a hydrophobic cavity. There are different types of cyclodextrins, including α-, β-, and γ-cyclodextrin containing six, seven and eight glucose subunits, respectively. The most commonly used cyclodextrin is β-cyclodextrin with internal cavity diameter of 0.65nm. Surface hydrophilic nature of β-cyclodextrin provides it water solubility and presence of the central hydrophobic cavity with specific diameter making it suitable for molecular inclusion of small non-polar compounds, especially flavors, through hydrophobic interaction, hydrogen bonds and Van der Wals forces (Zuidam and Heinrich, 2010; Anandharamakrishnan and Ishwarya, 2015). Generally, inclusion complexation takes place in two steps: (1) mixing carrier, active and water together; (2) followed by incubation and drying, if necessary. It is noteworthy that if a ligand or guest ingredient is insoluble in water at applied concentrations, it is necessary to dissolve it in another solvent, such as alcohol, e.g. ethanol. The first step can be done through stirring and shaking of two phases in water. After inclusion has taken place, filtration and drying can be applied to separate the complex. In order to decrease the amount of water used in the procedure, mixing step could be replaced by kneading of β-cyclodextrin and ligand in a paste form (Pegg and Shahidi, 2007). Reineccius and Risch (1986) reported 0-100 per cent inclusion of β-cyclodextrin with isoeugenol, ethyl hexanoate and linalol, respectively in a model system, indicating the importance of binding affinity of β-cyclodextrin towards active. The advantages of aroma inclusion with β-cyclodextrin are oxidative stability and stability of complex against evaporation (Pegg and Shahidi, 2007).

6. Industrially-implemented Encapsulated Ingredients

6.1 Bioactive Compounds (Phytochemicals, Fatty Acids, Essential Oils (Eos), Herbs and Spices)

6.1.1 Vitamins

Vitamins are an essential part of human diet as they modulate various biochemical reactions in our body. Humans, however, do not have the ability to synthesize vitamins, excepting vitamin B_3 and D. Therefore, these active biomolecules need to be obtained through external sources, i.e. food products and dietary supplements (Azevedo *et al.*, 2014). Furthermore, vitamins are sensitive compounds and can be degraded by harmful agents, such as oxidants and heat. In addition, low absorption and structure breakdown of these compounds during consumption might lead to limited bioavailability (Katouzian and Jafari, 2016). Encapsulation has the potential to improve the intestinal absorption of vitamins, thereby preventing or treating deficiencies (Ruiz Canizales *et al.*, 2018). It can also increase vitamin stability, flow properties and avoid potential off-flavors (Gibbs *et al.*, 1999). Most of the lipophilic vitamins are sold in microencapsulated form in order to protect them from unfavorable surrounding environment (Mougin *et al.*, 2016).

Table 3 summarizes the studies performed on vitamin encapsulation for food applications. By studying the bioavailability of encapsulated vitamin E in food and supplements, Leonard *et al.* (2004) reported that vitamin bioavailability increased in fortified breakfast cereals as compared to supplements which had poor vitamin absorption. Incorporation of liposomes containing vitamins E and C into orange juice provided improved antioxidant potential of vitamins before and after pasteurization. The organoleptic properties of orange juice were not changed by liposomal formulations containing vitamins and the obtained product had enhanced microbial stability after thermal processing and during thirty-seven days of refrigeration period (Marsanasco *et al.*, 2011).

Both children and adults suffer from a worldwide problem of vitamin D_3 deficiency. The major physiological role of vitamin D_3 is to regulate the serum calcium concentration within the normal range. Vitamin D_3 deficiency leads to rickets

Table 3: Examples of Food Products Containing Encapsulated Bioactive Compounds

Bioactive Ingredient	Food Product	Encapsulation Technique	References
Vitamins			
Vitamin E	Breakfast cereal	Emulsification	(Leonard *et al.*, 2004)
Vitamin D	Cheese	Emulsification	(Stratulat *et al.*, 2015)
Vitamin D	Milk	Freeze drying	(Hasanvand *et al.*, 2015)
Vitamin C Vitamin E	Orange juice	Liposome	(Marsanasco *et al.*, 2011)
Vitamin A	Salt	Spray cooling	(Wegmüller *et al.*, 2006)
Vitamin B_{12}	Cheese	Emulsification	(Giroux *et al.*, 2013)
Phytochemicals			
Catechin Curcumin	Beverages	Emulsification	(Aditya *et al.*, 2015)
Curcumin	Ice cream	Emulsification	(Kumar *et al.*, 2016)
Hesperetin	Milk	Solid lipid nanoparticles Nanostructured lipid carriers	(Fathi *et al.*, 2013)
Resveratrol	Yogurt Bars Gummies	Niosome Spray drying	(Koga *et al.*, 2016; Pando *et al.*, 2015)
Catechin Epigallocatechin gallate	Low-fat hard cheese	Liposome	(Rashidinejad *et al.*, 2014; Rashidinejad, 2016)
Fatty Acids			
Flax seed oil	Bread	Spray drying	(Gökmen *et al.*, 2011)

EPA and DHA	Cookies Spaghetti Low fat Iranian UF-Feta cheese	Fluidized bed drying	(Borneo *et al.*, 2007, Verardo *et al.*, 2009; Farbod *et al.*, 2015)
Sardine oil	Cookies	Spray drying Freeze drying	(Taguchi *et al.*, 1992)
Fish oil	Yogurt Sausage Beef burger	Liposome Complex coacervation Emulsification	(Ghorbanzade *et al.*, 2017; Tamjidi *et al.*, 2012; Josquin *et al.*, 2012; Keenan *et al.*, 2015)
Echium oil	Yogurt	Complex coacervation	(Comunian *et al.*, 2017)
Essential Oils (EOs)			
Thyme EOs	Milk Cake Cucumber	Emulsification Complex coacervation Liposome	(Jemaa *et al.*, 2017; Gonçalves *et al.* 2017; Cui *et al.*, 2016)
Oregano EOs	Cheese	Emulsification	(Bedoya-Serna *et al.*, 2018)
Rosemary EOs	Cheese	Spray drying	(Fernandes *et al.*, 2017)
Cinnamon EOs	Beef patties	Emulsion-gelation	(Ghaderi-Ghahfarokhi *et al.*, 2017)
Different EOs	Food packaging	Mainly emulsification	(Ribeiro-Santos *et al.*, 2017)
Natural Colorants			
Curcumin	Yogurt	Solid dispersion	(Almeida *et al.*, 2018)
Lycopene	Cake	Spray drying	(Rocha *et al.*, 2012)
β-carotene	Yogurt	Liposome	(Toniazzo *et al.*, 2014)
Betacyanin	Soft drinks	Spray drying	(Azeredo *et al.*, 2007)
Dinitrosyl ferrohemochrome	Nitrite-free ham	Spray drying	(O'boyle *et al.*, 1992)
Astaxanthin	Yogurt	Complex coacervation	(Taksima *et al.*, 2015)
Flavors			
Limonene and bioactive extracts of lemongrass, red thyme, oregano, and peppermint	Strawberries	Edible coating	(Vu *et al.*, 2011)
Eugenol rich clove extract	Soybean oil	Spray drying	(Chatterjee and Bhattacharjee, 2013)
Trans- cinnamaldehyde	Fresh-cut watermelon	Edible coating	(Sipahi *et al.*, 2013)
Thyme extract	Milk	Emulsification	(Xue *et al.*, 2015)
Cinnamon extract	Beef	Proteoliposomes	(Lin *et al.*, 2017)

in children and osteoporosis in adults. Thus, food fortification with vitamin D_3 is recognized as a crucial nutritional factor, particularly in the absence of sunlight. In this context, vitamin D_3 has been widely incorporated in food products (Ganesan *et al.*, 2011; Stratulat *et al.*, 2015; Kaushik *et al.*, 2017).

Vitamin D_3-fortified milk can be used for cheese making; however, a considerable amount of this vitamin is reduced and transferred into cheese whey. In one study, vitamin D_3 was encapsulated in sodium caseinate-stabilized emulsions to increase its stability during cheese making and subsequent storage period (ninety days at 4°C). The encapsulated vitamin had a manifest retention (91 per cent) and stability in the curd and positively influenced the yield, composition, and chemical stability of the obtained cheese (Stratulat *et al.*, 2015). In another study conducted by Banville *et al.* (2000), Cheddar cheese was fortified with three types of vitamin D_3: a commercial water-soluble emulsion of vitamin D_3 (Vitex D), water-soluble vitamin D_3 entrapped in multilamellar liposomes and crystalline liposoluble vitamin D_3 homogenized in cream. The recovery of vitamin D_3 was remarkably higher in cheese fortified with liposome-entrapped vitamin D_3 than in other fortified Cheddar cheeses. However, it showed a lower vitamin concentration after seven months of ripening, probably due to vitamin destruction during cheese making (e.g. fermentation process by lactic acid bacteria, oxidation and acidification). In addition, the obtained cheese flavor was not influenced by the fortification.

Vitamin A is a lipophilic molecule susceptible to rapid degradation or inactivation, particularly in aqueous media. Encapsulation procedures should therefore increase its solubility, reduce degradation processes and provide a controlled release (Sauvant *et al.*, 2012). There are three commercially available microencapsulated dry vitamin A acetate products 'Beadlets' in the market for food and pharmaceutical applications, which are produced by DSM Nutritional Products Ltd. (Mougin *et al.*, 2016). In addition, fat-soluble vitamins A, D, E, and K have been encapsulated in oil in water emulsions to develop their commercially available forms 'Vitalipid®' (Wang *et al.*, 2014). Microcapsules containing vitamin A, potassium iodate and ferric pyrophosphate were incorporated into consumable salts in Morocco. The triple fortified salt had an acceptable color change, 20 per cent iodine loss, and excellent vitamin A stability (~12 per cent loss) in comparison to the iodized salt after six months of storage period. Furthermore, the sensory properties of typical Moroccan dishes cooked with triple fortified salt or iodized salt were not significantly different and the salt showed a good overall acceptability. Such capsules can therefore be used for fortification of other dry matrices, such as flour and sugar (Wegmüller *et al.*, 2006).

Vitamin B group has also been encapsulated and incorporated into food products. Giroux *et al.* (2013) fortified cheese with encapsulated vitamin B_{12} as a marker for hydrophilic nutrients. Encapsulated vitamin B_{12} in double emulsions had reduced losses in whey and during *in vitro* gastric digestion and increased retention in cheese curd compared to non-encapsulated one. Encapsulation can therefore be used to protect vitamins during storage and passage through the digestive tract, increase bioactivity and intestinal uptake, control release and, finally, incorporate these bioactives in a stable form into different food products.

6.1.2 Phytochemicals

Phytochemicals are bioactive compounds with therapeutic and preventive potential for different diseases. However, they have low stability, solubility, bioavailability and target specificity in body. This is especially true of phenolic compounds, such as curcumin, resveratrol, quercetin, and (-)-epigalloatechingallate (EGCG) (Wang *et al.*, 2014). Encapsulation can protect these bioactives and extend their applications.

Catechin and curcumin have been encapsulated in double emulsions to fortify beverages. Encapsulation has led to increase in stability of catechin (~20 per cent) and curcumin (~50 per cent) in beverages. It was provided a co-delivery of nutraceuticals to elevate their health-enhancing and therapeutic potentials by acting synergically in the host body (Aditya *et al.*, 2015). In the study conducted by Fathi *et al.* (2013), milk fortified with hesperetin-loaded solid lipid nanoparticles (SLNs) and nanostructured lipid carriers (NLCs) had no significant differences compared to non-fortified milk and could remarkably cover the after-taste, bitter taste, and ameliorate poor solubility of hesperetin. Resveratrol is a phenolic compound found in peanuts and red wine; its addition to food products it might provide health benefits to consumers. However, it has a bitter taste and instability towards light which limits its applications. Resveratrol encapsulation in sodium caseinate matrix and subsequent addition of the obtained microcapsules to bars and gummies resulted in healthy products; also, bars containing encapsulated resveratrol had no significant difference in terms of overall acceptance as compared to the control sample, whereas the overall acceptance of gummies with resveratrol-loaded microcapsules was significantly lower than the control. These results suggest the potential application of resveratrol-loaded microcapsules into easy-to-consume foods to confer health-promoting effects on consumers (Koga *et al.*, 2016). Currently, there is a marketed encapsulated phytochemical in the food industry, i.e. Curcusome® which is a nanoliposome containing curcumin, quercetin and peperine for superior bioavailability and absorption (Mirafzali *et al.*, 2014).

6.1.3 Fatty Acids

Fatty acids, as the basic unit of lipids, are classified into three groups: omega-3, -6, and -9 fatty acids based on the distance between the double bond and methyl group in their structure. Among them, omega-3 fatty acids, such as α-linolenic acid, eicosapentaenoic acid (EPA) and docosahexaenoic acid (DHA), are essential for humans, due to the lack of key desaturase enzymes to synthesize these fatty acids in our body. Thus, consumption of foods containing high levels of these essential fatty acids is considered the best way to raise their plasma concentrations. However, these bioactive fatty acids and fish oil are prone to oxidation, which produces off-flavor and taste as well as harmful compounds. In this context, encapsulation is one of the effective methods to prevent oxidation reaction and protect sensitive fatty acids and oils from environmental stresses applied during food processing and storage. Encapsulated omega-3 fatty acids have been successfully added to various food products (Table 3).

The addition of fish oil-loaded nanocapsules into bread formulation resulted in a fortified bread with many advantages, including masking of unfavorable taste of fish oil (Siegrist *et al.*, 2007). Gökmen *et al.* (2011) have also used omega-3 fatty acids to produce functional bread. In their work, microparticles containing omega-3 fatty acids, which were obtained by spray drying method, were added to bread formulations at various levels. Lipid oxidation (hexanal and nonanal contents) decreased significantly in breads containing microparticles and higher levels of the particles in formulations led to a significant decrease in formation of hydroxymethyl furfural (HMF) and acrylamide in the obtained breads. In addition, it was observed by scanning electron microscopy (SEM) that the particles incorporated into dough remained unchanged in bread crumb, while they were partially destroyed in the crust. In comparison to non-encapsulated flaxseed oil, incorporation of its encapsulated form into dough resulted in improved quality and safety of products via reduce in lipid oxidation and production of hazardous compounds in bread during the baking process (Gökmen *et al.*, 2011). In another study, the addition of encapsulated omega-3 fatty acids (ROPUFA® '10' n-3 Food Powder, DSM Inc.) to cream-filled sandwich cookies showed that it is possible to provide a shelf-stable fortified food with high concentration of essential fatty acids without affecting sensory characteristics of the final product (Borneo *et al.*, 2007). The addition of omega-3 fatty acids to infant powder is quite challenging, mainly due to its long shelf-life period (~three years) and low number of ingredients permitted in formula. The microencapsulated omega-3 oils were first successfully added to the infant formula in Australia and New Zealand in 1998, without any negative impact on sensory features, shelf-life and ingredients of the original product. Afterwards, several commercial infant formulas have been incorporated with omega-3 fatty acids (Sanguansri and Augustin, 2006). Dairy products have also been fortified with encapsulated forms of essential fatty acids and oils (Tamjidi *et al.*, 2012; Ghorbanzade *et al.*, 2017). A low-calorie spread with long shelf-life has been produced by incorporating fish oil in vegetable fat blend, or butter-vegetable oil blends (Young *et al.*, 1993). In addition, it was found that the usage of encapsulated tuna oil (Gelphorm® and Driphorm®) in yogurt resulted in a product with superior sensory properties in comparison to yogurt with free tuna oil (Sharma *et al.*, 2003).

Microencapsulation techniques have been successfully applied to fortify meat and fish products with bioactive omega-3 fatty acids. For example, several Hans low-fat meat products, such as roast chicken, Strassburg, Vienna sausage and Bird's Eye crumbed fish fillets have been produced in Australia under the 'Smart Choice-Omega 3' brand (Sanguansri and Augustin,2006). Incorporation of nanoencapsulated omega-3 fatty acids into pork meat improved the fatty acid profile of the meat (Ojha *et al.*, 2017). It was reported that the addition of commercially available encapsulated fish oil (Vana-Sana NG Codliver; Friesland Campina kievit, Meppel, The Netherland) in Dutch-style fermented sausage is the best formulation in retaining overall quality of the product (Josquin *et al.*, 2012). Potential use of another commercial encapsulated fish oil 'Mega-3®' (Ocean Nutrition, Canada) has also been evaluated in production of beef burgers (Keenan *et al.*, 2015). Encapsulation has therefore the potential to protect sensitive oils and fatty acids from oxidation reaction, mask the unpleasant odor and taste of oils and increase shelf-life of food products containing these bioactive compounds during processing and storage.

6.1.4 *Essential Oils (EOs)*

Antimicrobial compounds are chemicals with a potential to suppress microbial growth and are commonly applied in combination with various preservation methods to preserve food products (Davidson *et al.*, 2013). Some essential oils produce antioxidant and antimicrobial effects and therefore many attempts have been made to fortify food products with these biologically active oils. Encapsulation could protect antimicrobial compounds against degradation, improve their bioavailability and targeted delivery and, finally, lower the required level of antimicrobial agents for successful food preservation (Table 3) (Ribeiro-Santos *et al.* 2017). For instance, nanoencapsulation of EOs in zein nanoparticles increased their water solubility without hindering their ability to control *Escherichia coli* growth or suppress free radicals; indeed, EOs-loaded zein nanoparticles facilitate EOs applications in food formulations (Wu *et al.* 2012). EOs have been added to edible films and coatings to improve quality and safety of fresh fruits and vegetables. Alginate films incorporated with cinnamon bark oil were used for cantaloupe coating. EOs-loaded coatings were able to increase quality and microbiological safety of the fruit (Zhang *et al.*, 2015). Similar results were obtained by the same authors by applying thyme oil emulsion for washing of cantaloupes (Zhang *et al.*, 2016). Edible coatings and films containing EOs have been also successfully used to enhance the quality and safety of meat products, such as Bologna and ham (Oussalah *et al.*, 2007), fresh beef (Zinoviadou *et al.*, 2009), fresh ground beef (Emiroğlu *et al.*, 2010) and fish (Gómez-Estaca *et al.*, 2010). Encapsulation also enables the application of Eos, as natural antioxidants, in dairy-based products. It was indicated that co-encapsulation of thyme oil and nisin in zein provided better inhibition towards microbial growth in comparison to non-encapsulated antimicrobial agents in low-fat milk (Xiao *et al.*, 2011). Successful incorporation of encapsulated EOs into cheese has been reported in the literature (Fernandes *et al.*, 2017; Bedoya-Serna *et al.*, 2018). Meat and bakery products containing encapsulated EOs often show improved sensory properties and storage stability (Ghaderi-Ghahfarokhi *et al.*, 2017; Gonçalves *et al.*, 2017). Eos' incorporation into food products and packaging materials can therefore inhibit microbial growth and improve oxidative stability. It is important to mention that EOs have health benefits, which may motivate consumers to use EOs-fortified food products.

6.1.5 Herbs and Spices

Herbs and spices play a crucial role in everyday life as key sources of natural flavoring and colorant agents in foods, beverages and pharmaceuticals. Spices have also antimicrobial, antioxidant, nutritional and medicinal properties, leading to increase in their applications. Encapsulation enables better protection as well as superior handling of herb extracts and active compounds. For example, liquid EOs or oleoresin extracts were encapsulated to free-flowing powders by spray drying of pre-made emulsions which enabled the flavor of encapsulated spices to be released in aqueous solutions (Peter, 2006).

Sawale *et al.* (2017) investigated the storage stability of chocolate vanilla dairy drink upon addition of free and encapsulated Arjuna herb. All dairy drinks showed a significant decrease in flavor, mouthfeel and overall acceptance. However, physicochemical properties of the drink containing encapsulated Arjuna ethanolic extract were changed at a slower rate as compared to the sample containing free extract and viscosity and sedimentation stability of the product containing encapsulated extract was also significantly higher during four months of storage time. The encapsulated form of Arjuna extract was therefore effective in improving shelf-life stability of dairy drinks. By designing a nanoencapsulation system with a stable and sustained-release potential for olive leaf phenolic extracts in whey protein concentrate/pectin-stabilized double emulsions, it was demonstrated that the antioxidative effect of the encapsulated bioactive compound as a natural antioxidant in soybean oil was greater when compared to free olive leaf extract (Mohammadi *et al.*, 2016).

6.2 Colors

Color is generally associated with food flavor and quality and therefore is considered as an indispensable parameter in identifying and acceptance of many food products (Underriner, 2012). In this way, natural colorants are receiving more attention among consumers due to their origin than artificial colorings. However, their industrial application is limited mainly because of low solubility and stability. Encapsulation techniques are one of the most frequently and successfully used strategies to improve functionality and application of natural colorants for production of functional foods with health-promoting effects.

Encapsulation of natural coloring agents is a promising way to amend their solubility, handling, shelf-life (six months to two years), antioxidant potential and antimicrobial activity, particularly phenolic-based ones, such as curcumin (Gibbs *et al.*, 1999; Rostamabadi *et al.*, 2019; Rafiee *et al.*, 2019). Thus, encapsulation techniques have the potential to increase the functionality and bioactivity of natural colorings without any remarkable changes in their quality and sensory characteristics (Table 3). Encapsulated curcumin has been used in many food products, such as semi-skimmed milk (Gomez-Estaca *et al.*, 2012), ice cream (Kumar *et al.*, 2016), yogurt (Almeida *et al.*, 2018), etc. Foods fortified with encapsulated curcumin showed acceptable sensory features and curcumin was mainly stable during harsh processing conditions of the products. Encapsulation has also been employed to boost chemical stability, bioavailability and water solubility of β-carotene for facilitating its utilization in water-based formulations. The addition of β-carotene–loaded liposomes into yogurt formulation did not affect the texture of the final product when compared to the samples containing synthetic colorant, suggesting that the artificial colorants can be substituted by the encapsulated β-carotene (Toniazzo *et al.*, 2014).

Encapsulating anthocyanins in maltodextrin, gum arabic and gelatin wall materials enhances their food applications. Anthocyanins in the encapsulated form were used as the artificial color substitute to fabricate coloring jelly powder. The obtained product with 7 per cent encapsulated color received better sensory scores in comparison to jelly with artificial colorant. Moreover, the encapsulated color did not affect the physicochemical properties of the product (i.e. texture, acidity, moisture content, ash content and hygroscopicity), while it did lead to a significant decrease in solubility and syneresis of the fortified sample (Mahdavi *et al.*, 2016). In another study regarding protective effect of encapsulation process on natural colorants, lycopene stability was improved by encapsulating in modified starch and subsequently spray drying to obtain powdered encapsulated lycopene for food application. Besides the obtained high stability of the encapsulated pigment, it was, also, released in a sustained manner in cake (Rocha *et al.*, 2012). Shrimp astaxanthin encapsulation in chitosan-alginate matrix through complex coacervation and its subsequent use of the encapsulated pigment in yogurt has been reported by Taksima *et al.* (2015). The antioxidant activity of astaxanthin has been maintained after encapsulation and the overall acceptance score of yogurt containing encapsulated pigment was over 6 (based on the 9-point hedonic scale test). In addition, the positive acceptance and purchase intent of the product were 86.2 per cent and 95.6 per cent, respectively. This technique was therefore suitable for sufficient protection of astaxanthin and production of the functional product for consumers.

6.3 Flavors

Food products usually undergo flavor and aroma losses upon exposure to environmental conditions during processing, transportation and storage. Encapsulation can protect flavoring agents against harsh conditions (light, oxygen and high temperatures) and unfavorable ingredient interactions. It can also prolong shelf-life (i.e. by converting liquid flavors to powdered forms), mask off-flavors and control release of flavors (Table 3) (Kohane *et al.*, 2015). Lemon-flavor loaded β-cyclodextrins were used in a fat-free yogurt system and it was indicated that the release of flavor was comparable to yogurt with regular fat content, suggesting the potential of β-cyclodextrin to tailor the flavor delivery in food systems (Kant *et al.*, 2004). Encapsulated clove and tea tree essential oils were successfully employed to suppress the growth of coliforms, mesophilic and psychotropic microorganisms on lettuce leaves (Ponce *et al.*, 2011). The encapsulation techniques can also improve solubility and distribution of flavors in food products; thymol encapsulation in sodium caseinate exhibited greater inhibitory effects against food pathogenic microorganisms in milk compared to the non-encapsulated counterpart and this was attributed to the improved solubility and distribution of flavor (Pan *et al.*, 2014). Moreover, encapsulated clove oil in liposomes had higher chemical stability and better antimicrobial effect towards *Staphylococcus aureus* growth in tofu (Cui *et al.*, 2016). Food flavors have been vastly used in food packaging materials to extend the storage stability of food products (Koontz and Marcy, 2007; Brasil *et al.*, 2012; Martiñon *et al.*, 2014; Cui *et al.*, 2018). For example, trans-cinnamaldehyde and β-cyclodextrin complex has been incorporated into multilayered chitosan/pectin-based edible coatings for fresh-cut papaya. The coated fruit had lower juice leakage, higher β-carotene content, better firmness and preserved its color (Brasil *et al.*, 2012). Similarly, carvacrol/methyl cinnamate-loaded strawberry puree-based edible films were successfully applied in clamshells to significantly delay and reduce strawberry fruit decay, and the berries also presented brighter color and firmer texture (Peretto *et al.*, 2014). In the interesting study by Cui *et al.* (2017), SiO_2-eugenol liposomes loaded electrospun nanofibrous membranes were developed and used as potent antioxidants for shelf-life improvement of beef. In general, encapsulation methods are introduced to improve stability and bioavailability of flavors, to widen their applications as natural antimicrobial, antioxidant and flavoring ingredients in various food products.

6.4 Sweeteners

Pharmaceutical and oral hygiene product manufactures are lowering sugar content and applying sweeteners as sugar substitutes in their products to inhibit glycemic effects and dental caries. Sweeteners are low-calorie substances that confer sweet taste to food products, such as soft drinks, yogurts, bakery products, ice creams, fruit juices and confectionery products. Nonetheless, certain sweeteners have intense sweet taste, after-taste, high hygroscopicity, reactivity and thermosensitivity, which may restrict their applications in food products. In this context, encapsulation process can be used to (i) lower after-taste, hygroscopicity (mainly for certain polyols), reactivity and bitterness; (ii) increase thermal stability (particularly for aspartame) and fluidity; (iii) adjust solubility of sweeteners in dispersions; (iv) protect sugar against moisture and other destabilization components; (v) inhibit denaturation of protein-based sweeteners, such as thaumatin; (vi) make stevioside more bioavailable and easily absorbable by intestine to exert its health-promoting effects; (vii) convert liquid sweeteners to powdered forms; and (viii) extend the sweetness sensation via controlled release (Favaro-Trindade *et al.*, 2015). Aspartame has been widely encapsulated in various wall materials in comparison to other sweeteners (Rocha-Selmi *et al.*, 2013; Savage *et al.*, 2004; Zyck and Yatka, 2007; Wetzel *et al.*, 1997; Wetzel and Bell, 1998). The recovery and degradation kinetics of encapsulated aspartame were evaluated in cake formulations after baking and during product storage. The encapsulated aspartame has greater recovery after baking and lower degradation rate in cakes than non-encapsulated aspartame. These results suggest that the encapsulated aspartame could be used in heat-treated food products with high moisture content and pH values greater than 6 (Wetzel and Bell, 1998). In another research, it was shown that encapsulated aspartame can be successfully employed to fabricate no-sugar-added cupcakes for home preparation, since it exhibited similar sweetness level and no significant difference in terms of consumer acceptability when compared to full-sugar cake (Wetzel *et al.*, 1997). Encapsulated aspartame has also been used in chewing gums to expand the release rate of its sweetness (Schobel and Yang, 1989; Savage *et al.*, 2004; Zyck and Yatka, 2007).

6.5   Salts and Minerals

The direct incorporation of mineral salts to food products is not simple; some micronutrients may lead to adverse effects on sensory characteristics of food, such as undesirable flavor, color changes and sandy textures, while others can induce chemical reactions, resulting in decreased nutritional quality and increased levels of toxins (Prichapan and Klinkesorn, 2014). Encapsulation technology provides solution to these problems. It was reported by Gharibzahedi and Jafari (2017) that encapsulation of mineral salts provides many advantages, such as masking off-flavor and tastes, avoiding discoloration, inhibiting unfavorable interactions with other ingredients, improving physical properties, controlling release rate of mineral salt and enhancing storage stability and microbiological safety of food products. In a successful study conducted by Wegmüller *et al.* (2006), iodine (potassium iodate), vitamin A (retinyl palmitate) and iron (ferric pyrophosphate) were simultaneously microencapsulated in hydrogenated palm fat to fortify salt and overcome their deficiencies in developing countries. About 80 and 88 per cent of iodine and retinyl palmitate were, respectively, retained in the triple fortified salt compared to iodized salt after six months of storage period. Typical Moroccan dishes were then cooked with triple fortified salt or iodized salt and sensory evaluation did not show any significant difference between dishes and the overall acceptability of the modified salt was superior. This implies that other dry matrices, such as flour and sugar, can be fortified with such capsules (Wegmüller *et al.*, 2006). Another study revealed that microencapsulation of iron (ferric ammonium sulfate) in polyglycerol monostearate and subsequent use of the encapsulated iron in milk reduced the common metallic and astringent tastes related to iron fortification (Kwak *et al.*, 2003). Encapsulated salt (NaCl) was also used instead of ordinary salt for preparation of bread. It was shown that about 50 per cent of the normal salt can be replaced with encapsulated salt crystals in bread, which led to lower sodium concentration and higher consumer acceptability (Noort *et al.*, 2012). Sodium chloride can accelerate and modify pathways of chemical reaction in food products; for instance, it leads to increased formation of hydroxymethyl furfural (HMF) and acrylamide through Maillard reaction routes (Claus *et al.*, 2008; Gökmen and Şenyuva, 2007; Troise and Fogliano, 2013). NaCl encapsulation can reduce its contribution to undesirable chemical reactions in food systems. Fiore *et al.* (2012) indicated that HMF content of cookies containing free NaCl increased up to 75 per cent as the salt level in the formulation increased up to 0.65 per cent; whilst, the addition of encapsulated salt resulted in a decrease in HMF concentration (18-61 per cent), and did not affect the color or sensorial attributes of the obtained product. HMF reduction was attributed to the ability of encapsulated salt to suppress pyrolytic degradation of sucrose and formation of fructofuranosyl cation. There are some commercially encapsulated salts available on the market – Supercoat™ (encapsulated ferrous fumarate from Wright Nutrition), Sunactive Fe® (micronized dispersible ferric pyrophosphate in liposomes, Tayio Kagaku, Japan), encapsulated dried ferrous sulfate (from Durkee Industrial Food Group, Cleveland, OH) and Biofer™ (encapsulated ferrous sulfate and ascorbic acid in liposome). These forms of salts usually have higher bioavailability than their non-encapsulated counterparts in food products (Moretti *et al.*, 2016). Shelf-life, nutritional properties and sensory quality of food products can be improved using salt encapsulation technology. However, it seems that some novel and stable delivery systems should be designed to provide better bioavailability rates of the mineral salts in fortified foods.

6.6 Probiotics

Probiotic microorganisms have health-promoting effects and play beneficial roles in gastrointestinal tract of the host (Heidebach *et al.*, 2012). Nonetheless, living probiotic cells, usually, undergo bioactivity loss during production, storage and digestion. Therefore, researchers are trying to design appropriate matrices to minimize probiotic loss (Siuta-Cruce and Goulet, 2001; Mattila-Sandholm *et al.*, 2002). Encapsulation is considered as the most promising approach for protection of probiotic bacteria against harsh food processing and/or gastrointestinal conditions (Nualkaekul *et al.*, 2013). Encapsulated probiotic microorganisms generally had higher survival rate during production and storage of various food products, such as yogurt (Picot and Lacroix, 2004; Kailasapathy, 2006; El Kadri *et al.*, 2018), dairy desserts (Shah and Ravula, 2000; Godward and Kailasapathy, 2003), cheese (van den Tempel *et al.*, 2002; Gardiner *et al.*, 2002; Özer *et al.*, 2009; Rodríguez-Huezo *et al.*, 2014), milk powder (Desmond *et al.*, 2002), meat products (Muthukumarasamy and Holley, 2006), mayonnaise (Khalil and Mansour, 1998), soy products (Wang *et al.*, 2002), fermented cereals (Laine *et al.*, 2003), fruit juice (Reid *et al.*, 2007; Ding and Shah, 2008; Tsen *et al.*, 2008) and biscuits (Reid *et al.*, 2007). In most cases, encapsulated probiotic showed not only the absence of adverse effects on sensory and textural properties of fortified products, but also resulted in functional food products. For example, functional cheese incorporated with encapsulated *Bifidobacterium* had similar appearance, flavor and texture when compared to the control sample (Dinakar and Mistry, 1994; Desmond *et al.*, 2002). Moreover, functional mayonnaise with encapsulated probiotics presented improved sensory properties (Khalil and Mansour, 1998). Additionally, probiotic-containing fermented tomato juice had better overall acceptance compared to the control sample along with higher viable probiotic cell counts during refrigeration storage (Tsen *et al.*, 2008).

Currently, probiotic encapsulation via Maillard reaction products (MRPs) has received much attention due to high solubility, better thermal, colloidal and oxidative stabilities and superb antioxidant activity of MRPs encapsulated material (Liu *et al.*, 2016; Nooshkam and Madadlou, 2016a, b; Liu *et al.*, 2017; Nooshkam *et al.*, 2018).

In a study conducted by Liu *et al.* (2017), viability of *Lactobacillus rhamnosus,* encapsulated via emulsification-cold gelation procedure in whey protein concentrate-isomaltooligosaccharide MRPs, has been evaluated in white-brined cheese during its storage period and under *in vitro* digestion conditions. Cheese sample with encapsulated probiotic had remarkably higher number of viable living cells during storage and after *in vitro* digestion (7.3 log cfu g^{-1}, which was even higher than the reported minimum therapeutic level). High viability of probiotic cells was due to the protective effect of carbohydrates against harsh gastrointestinal tract conditions. These studies proved that encapsulation processes can improve viability of probiotic microorganisms in food products and during gastric transition.

6.7   Proteins and Peptides

Food protein hydrolysates and peptides are receiving a great deal of attention as promising functional food ingredients. However, bitter taste, low bioavailability and hygroscopicity of protein hydrolysates and bioactive peptides may limit their commercial applications. Encapsulation can be applied to tackle these challenges and increase organoleptic properties and bioavailability of peptides (Mohan *et al.*, 2015; Hosseini *et al.*, 2017). Spray drying has been vastly used to reduce bitter taste of protein hydrolysates and peptides for their food applications (Fávaro-Trindade *et al.*, 2010; Yang *et al.*, 2012; Ma *et al.*, 2014). Gómez-Mascaraque *et al.* (2016) microencapsulated whey protein hydrolysate within micro-hydrogels to increase its stability towards lactic acid fermentation during yogurt manufacturing. It has been demonstrated that microencapsulation within chitosan could provide better protection of bioactive peptides against degradation induced by living starter cultures, in comparison to the peptides encapsulated in gelatin, suggesting that the encapsulation matrix has an important role in achieving the highest protection of biologically active peptides against harsh food processing conditions.

It is notably to mention that proteolytic degradation and interaction of a bioactive peptide (e.g. nisin) with other food components may lead to decrease in its bioactivity (e.g. antimicrobial activity). In this sense, encapsulation of antimicrobial peptides in liposomes can be important to overcome interaction with food components and stability issues. For example, the combination of nanovesicle-encapsulated nisin and low temperatures could be more effective in controlling *Listeria monocytogenes* growth in milk (da Silva Malheiros *et al.*, 2010). In addition, reverse micelle-loaded nisin had antimicrobial effect against *L. monocytogenes* and *Staphylococcus aureus* growth in lettuce leaves and minced meat (Chatzidaki *et al.*, 2018). Furthermore, nisin and bacteriocin-like substance P34 encapsulated in partially-purified soybean phosphatidylcholine liposomes presented the highest inhibitory effects towards *L. monocytogenes* after ten days of storage of Minas frescal cheese as compared to the free bacteriocins (da Silva Malheiros *et al.*, 2012). In another study, it was stated that Cheddar cheese fermentation was not severely disturbed by nisin-loaded liposomes and the encapsulated nisin had remarkable stability throughout the cheese temperature cycle (Laridi *et al.*, 2003). By coencapsulation of nisin and lysozyme in liposomes coated with pectin, Lopes *et al.* (2019) reported that the obtained liposomes were efficient in reducing the population of *L. monocytogenes* in whole milk (2 log cfu mL^{-1}) and skim milk (5 log cfu mL^{-1}) at 37°C. Interestingly, in skim milk containing nisin-lysozyme loaded liposomes *L. monocytogenes* population was reduced below detection limit after twenty-five days. These studies suggest that the incorporation of bacteriocins into vesicles can increase their stability and provide the appropriate controlled release of antimicrobial peptides in foodstuff.

Protein/peptide encapsulation has also been used for development of controlled release packaging. In these systems, bioactive compounds (e.g. antimicrobial and antioxidant proteins or peptides) are released at favorable rates from packaging materials to the surface of food products in order to render continuous replenishment of biologically-active protein/peptide, suppress microbial growth and expand shelf-life of food products. In this context, a novel controlled release system was developed based on the incorporation of the pH-responsive polyacrylic acid/lysozyme complex into a hydrophilic whey protein isolate film matrix. Lysozyme release from the packaging film was extended from less than 24 h up to 500 h. In addition, 50 per cent-free-lysozyme+50 per cent-polyarylic acid/lysozyme complex film presented a 5.7 log decrease in bacterial population within 72 h, while 100 per cent-free-lysozyme film was not able to inhibit the growth of *Listeria innocua* after 24 h. Therefore, lysozyme encapsulation in films could provide a long-lasting antimicrobial effect and the obtained edible films could be used for active food packaging applications to extend product shelf-life and safety (Ozer *et al.*, 2016).

Some food enzymes are incorporated as food additives to accelerate some chemical reaction. For instance, enzymes have an important role in cheese ripening, but their direct incorporation could lead to several problems: direct addition of free enzymes to milk before renneting results in significant enzyme loss in whey and the efficiency of the enzyme added at dry salting stage is dependent on its diffusion rate into the cheese. It was suggested that liposomes stabilized by phospholipids have more advantages over other encapsulation techniques to encapsulate enzymes for cheese making – since these carriers inherently exist in milk, they protect casein protein from proteolysis during the initial steps of cheese making, and are distributed more efficiently in the curd (Augustin and Sanguansri, 2012). Liposome-encapsulated enzymes have been applied to tune proteolysis rate and subsequently the formation of peptides with bitter taste during cheese manufacturing and ripening (Picon *et al.*, 1994, 1995).

The encapsulation of proteins, protein hydrolysates and bioactive peptides can therefore increase their stability towards undesirable food processing conditions and during passage through the digestion tract. In addition, encapsulated proteins/peptides, especially antimicrobial ones, can be used to extend shelf-life and safety of fortified food products.

7. Conclusion and Future Perspectives

Encapsulation process acts as a promising technology to preserve, deliver and benefit bioactive agents. There are many diverse encapsulation methods; some but not all are addressed and explained in this chapter. Diversity of encapsulation techniques, coating materials, particle size and protection of the target compounds should meet the essentials for their application in different food matrixes. Accordingly, some bioactive agents like probiotics encapsulation could play a key role in both protection and targeted delivery of probiotic to the gut. It is expected that attention to functional food with promoted nutraceutical values and desirable organoleptic features becomes a driving force for food scientists and industries to improve and benefit from encapsulation methods, targeting delivery and controlled release.

References

Aditya, N.P., Aditya, S., Yang, H.J., Kim, H.W., Park, S.O., Lee, J. and Ko, S. (2015). Curcumin and catechin co-loaded water-in-oil-in-water emulsion and its beverage application, *Journal of Functional Foods*, 15: 35-43.

Allahdad, Z., Varidi, M., Zadmard, R., Saboury, A.A. and Haertlé, T. (2018). Binding of β-carotene to whey proteins: Multi-spectroscopic techniques and docking studies, *Food Chemistry*, 277: 96-106.

Almeida, H.H., Barros, L., Barreira, J.C., Calhelha, R.C., Heleno, S.A., Sayer, C., Miranda, C.G., Leimann, F.V., Barreiro, M.F. and Ferreira, I.C. (2018). Bioactive evaluation and application of different formulations of the natural colorant curcumin (E100) in a hydrophilic matrix (yogurt), *Food Chemistry*, 261: 224-232.

Anandharamakrishnan, C. and Ishwarya S.P. (2015). *Spray Drying Techniques for Food Ingredient Encapsulation*, John Wiley & Sons Ltd. Chichester, West Sussex.

Augustin, M.A. and Sanguansri, L. (2012). Challenges in developing delivery systems for food additives, nutraceuticals and dietary supplements, pp. 19-48. *In*: N. Garti and D.J. McClements (Eds.), *Encapsulation Technologies and Delivery Systems for Food Ingredients and Nutraceuticals*, Woodhead Publishing. Sawston, United Kingdom.

Augustin, M.A. and Hemar, Y. (2009). Nano- and micro-structured assemblies for encapsulation of food ingredients, *Chemical Society Reviews*, 38: 902-912.

Azeredo, H.M., Santos, A.N., Souza, A.C., Mendes, K.C. and Andrade, M.I.R. (2007). Betacyanin stability during processing and storage of a microencapsulated red beetroot extract, *American Journal of Food Technology*, 2(4): 307-312.

Azevedo, M.A., Bourbon, A.I., Vicente, A.A. and Cerqueira, M.A. (2014). Alginate/chitosan nanoparticles for encapsulation and controlled release of vitamin B_2, *International Journal of Biological Macromolecules*, 71: 141-146.

Banville, C., Vuillemard, J.C. and Lacroix, C. (2000). Comparison of different methods for fortifying Cheddar cheese with vitamin D, *International Dairy Journal*, 10(5-6): 375-382.

Bedoya-Serna, C.M., Dacanal, G.C., Fernandes, A.M. and Pinho, S.C. (2018). Antifungal activity of nanoemulsions encapsulating oregano (*Origanum vulgare*) essential oil: *In vitro* study and application in Minas Padrão cheese, *Brazilian Journal of Microbiology*, 49(4): 929-935.

Borneo, R., Kocer, D., Ghai, G., Tepper, B.J. and Karwe, M.V. (2007). Stability and consumer acceptance of long-chain omega-3 fatty acids (eicosapentaenoic acid, 20: 5, n-3 and docosahexaenoic acid, 22: 6, n-3) in cream-filled sandwich cookies, *Journal of Food Science*, 72(1): S049-S054.

Brasil, I.M., Gomes, C., Puerta-Gomez, A., Castell-Perez, M.E. and Moreira, R.G. (2012). Polysaccharide-based multilayered antimicrobial edible coating enhances quality of fresh-cut papaya, *LWT – Food Science and Technology*, 47(1): 39-45.

Burgain, J., Gaiani, C., Linder, M. and Scher, J. (2011). Encapsulation of probiotic living cells: From laboratory scale to industrial applications, *Journal of Food Engineering*, 104: 467-483.

Champagne, C.P. and Fustier, P. (2007). Microencapsulation for the improved delivery of bioactive compounds into foods, *Current Opinion in Biotechnology*, 18: 184-190.

Chatterjee, D. and Bhattacharjee, P. (2013). Comparative evaluation of the antioxidant efficacy of encapsulated and un-encapsulated eugenol-rich clove extracts in soybean oil: Shelf-life and frying stability of soybean oil, *Journal of Food Engineering*, 117(4): 545-550.

Chatzidaki, M.D., Papadimitriou, K., Alexandraki, V., Balkiza, F., Georgalaki, M.,

Papadimitriou, V., Tsakalidou, E. and Xenakis, A. (2018). Reverse micelles as nanocarriers of nisin against food-borne pathogens, *Food Chemistry*, 255: 97-103.

Claus, A., Mongili, M., Weisz, G., Schieber, A. and Carle, R. (2008). Impact of formulation and technological factors on the acrylamide content of wheat bread and bread rolls, *Journal of Cereal Science*, 47(3): 546-554.

Comunian, T.A., Chaves, I.E., Thomazini, M., Moraes, I.C.F., Ferro-Furtado, R., de Castro, I.A. and Favaro-Trindade, C.S. (2017). Development of functional yogurt containing free and encapsulated echium oil, phytosterol and sinapic acid, *Food Chemistry*, 237: 948-956.

Coradini, K., Lima, F.O., Oliveira, C.M., Chaves, P.S., Athayde, M.L., Carvalho, L.M. and Beck, R.C.R. (2014). Co-encapsulation of resveratrol and curcumin in lipid-core nanocapsules improves their *in vitro* antioxidant effects, *European Journal of Pharmaceutics and Biopharmaceutics*, 88: 178-185.

Corcoran, B.M., Stanton, C., Fitzgerald, G.F. and Ross, R.P. (2005). Survival of probiotic *Lactobacilli* in acidic environments is enhanced in the presence of metabolizable sugars, *Applied and Environmental Microbiology*, 71: 3060-3067.

Cui, H., Ma, C. and Lin, L. (2016). Co-loaded proteinase K/thyme oil liposomes for inactivation of *Escherichia coli* O157: H7 biofilms on cucumber, *Food & Function*, 7(9): 4030-4040.

Cui, H., Yuan, L., Li, W. and Lin, L. (2017). Antioxidant property of SiO2-eugenol liposome loaded nanofibrous membranes on beef, *Food Packaging and Shelf-Life*, 11: 49-57.

Cui, H., Bai, M., Rashed, M.M. and Lin, L. (2018). The antibacterial activity of clove oil/chitosan nanoparticles embedded gelatin nanofibers against *Escherichia coli* O157: H7 biofilms on cucumber, *International Journal of Food Microbiology*, 266: 69-78.

da Silva Malheiros, P., Daroit, D.J., da Silveira, N.P. and Brandelli, A. (2010). Effect of nanovesicle-encapsulated nisin on growth of *Listeria monocytogenes* in milk, *Food Microbiology*, 27(1): 175-178.

da Silva Malheiros, P., Sant'Anna, V., de Souza Barbosa, M., Brandelli, A. and de Melo Franco, B.D.G. (2012). Effect of liposome-encapsulated nisin and bacteriocin-like substance P34 on *Listeria monocytogenes* growth in Minas frescal cheese, *International Journal of Food Microbiology*, 156(3): 272-277.

Davidson, P.M., Taylor, T.M. and Schmidt, S.E. (2013). Chemical preservatives and natural antimicrobial compounds, pp. 765-801. *In:* M.P. Doyle and R.L. Buchanan (Eds.), *Food Microbiology: Fundamentals and Frontiers*, American Society of Microbiology.

de Vos, P., Faas, M.M., Spasojevic, M. and Sikkema, J. (2010). Encapsulation for preservation of functionality and targeted delivery of bioactive food components, *International Dairy Journal*, 20: 292-302.

Desmond, C., Ross, R.P., O'Callaghan, E., Fitzgerald, G. and Stanton, C. (2002). Improved survival of *Lactobacillus paracasei* NFBC 338 in spray-dried powders containing gum acacia, *Journal of Applied Microbiology*, 93: 1003-1011.

Desobry, S.A., Netto, F.M. and Labuza, T.P. (1997). Comparison of spray-drying, drum-drying and freeze-drying for β-Carotene encapsulation and preservation, *Journal of Food Science*, 62: 1158-1162.

Dias, D.R., Botrel, D.A., Fernandes, R.V.D. and Borges, S.V. (2017). Encapsulation as a tool for bioprocessing of functional foods, *Current Opinion in Food Science*, 13: 31-37.

Dinakar, P. and Mistry, V.V. (1994). Growth and viability of *Bifidobacterium bifidum* in Cheddar cheese, *Journal of Dairy Science*, 77(10): 2854-2864.

Ding, W.K. and Shah, N.P. (2008). Survival of free and microencapsulated probiotic bacteria in orange and apple juices, *International Food Research Journal*, 15(2): 219-232.

El Kadri, H., Lalou, S., Mantzouridou, F. and Gkatzionis, K. (2018). Utilisation of water-in-oil-water (W1/O/W2) double emulsion in a set-type yogurt model for the delivery of probiotic *Lactobacillus paracasei*, *Food Research International*, 107: 325-336.

Emiroğlu, Z.K., Yemiş, G.P., Coşkun, B.K. and Candoğan, K. (2010). Antimicrobial activity of soy edible films incorporated with thyme and oregano essential oils on fresh ground beef patties, *Meat Science*, **86**(2): 283-288.

Farbod, F., Kalbasi, A., Moini, S., Emam-Djomeh, Z., Razavi, H. and Mortazavi, A. (2015). Effects of storage time on compositional, micro-structural, rheological and sensory properties of low-fat Iranian UF-Feta cheese fortified with fish oil or fish oil powder, *Journal of Food Science and Technology*, **52**(3): 1372-1382.

Fathi, M., Varshosaz, J., Mohebbi, M. and Shahidi, F. (2013). Hesperetin-loaded solid lipid nanoparticles and nanostructure lipid carriers for food fortification: Preparation, characterization, and modeling, *Food and Bioprocess Technology*, **6**(6): 1464-1475.

Fathi, M., Martín, A. and McClements, D.J. (2014). Nanoencapsulation of food ingredients using carbohydrate based delivery systems, *Trends in Food Science & Technology*, 39: 18-39.

Fávaro-Trindade, C.S., Santana, A.D.S., Monterrey-Quintero, E.S., Trindade, M.A. and Netto, F.M. (2010). The use of spray drying technology to reduce bitter taste of casein hydrolysate, *Food Hydrocolloids*, **24**(4): 336-340.

Favaro-Trindade, C.S., Rocha-Selmi, G.A. and dos Santos, M.G. (2015). Microencapsulation of Sweeteners, pp. 333-349. *In*: L.M. Sagis (Ed.), *Microencapsulation and Microspheres for Food Applications*, Academic Press. Massachusetts, United States.

Fernandes, R.V.D.B., Guimarães, I.C., Ferreira, C.L.R., Botrel, D.A., Borges, S.V. and de Souza, A.U. (2017). Microencapsulated rosemary (*Rosmarinus officinalis*) essential oil as a biopreservative in Minas Frescal cheese, *Journal of Food Processing and Preservation*, **41**(1): e12759.

Fernandez, A., Torres-Giner, S. and Lagaron, J.M. (2009). Novel route to stabilization of bioactive antioxidants by encapsulation in electrospun fibers of zein prolamine, *Food Hydrocolloids*, 23: 1427-1432.

Feyzi, S., Varidi, M., Housaindokht, M.R. and Es'haghi, Z. (2019). Binding of safranal to whey proteins in aqueous solution: Combination of headspace solid phase microextraction/gas chromatography with multi spectroscopic techniques and docking studies, *Food Chemistry*, 287: 313-323.

Fiore, A., Troise, A.D., Ataç Mogol, B., Roullier, V., Gourdon, A., El Mafadi Jian, S., Hamzalıoğlu, B.A., Gökmen, V. and Fogliano, V. (2012). Controlling the Maillard reaction by reactant encapsulation: Sodium chloride in cookies, *Journal of Agricultural and Food Chemistry*, **60**(43): 10808-10814.

Ganesan, B., Brothersen, C. and McMahon, D.J. (2011). Fortification of Cheddar cheese with vitamin D does not alter cheese flavor perception, *Journal of Dairy Science*, **94**(7): 3708-3714.

Gaonkar, A.G., Vasisht, N., Khare, A.R. and Sobel, R. (2014). *Microencapsulation in the Food Industry a Practical Implementation Guide*, Elsevier, San Diego, California.

Garavand, F. and Madadlou, A. (2014). Recovery of phenolic compounds from effluents by a microemulsion liquid membrane (MLM) extractor, *Colloids and Surfaces A: Physicochemical and Engineering Aspects*, 443: 303-310.

Garavand, F., Razavi, S.H. and Cacciotti, I. (2018). Synchronized extraction and purification of L-lactic acid from fermentation broth by emulsion liquid membrane technique, *Journal of Dispersion Science and Technology*, **39**(9): 1291-1299.

Gardiner, G.E., Bouchier, P., O'Sullivan, E., Kelly, J., Collins, J.K., Fitzgerald, G., Ross, R.P. and Stanton, C. (2002). A spray-dried culture for probiotic Cheddar cheese manufacture, *International Dairy Journal*, **12**(9): 749-756.

Ghaderi-Ghahfarokhi, M., Barzegar, M., Sahari, M.A., Gavlighi, H.A. and Gardini, F. (2017). Chitosan-cinnamon essential oil nano-formulation: Application as a novel additive for controlled release and shelf-life extension of beef patties, *International Journal of Biological Macromolecules*, 102: 19-28.

Gharibzahedi, S.M.T. and Jafari, S.M. (2017). The importance of minerals in human nutrition: Bioavailability, food fortification, processing effects and nanoencapsulation, *Trends in Food Science & Technology*, 62: 119-132.

Ghorbanzade, T., Jafari, S.M., Akhavan, S. and Hadavi, R. (2017). Nano-encapsulation of fish oil in nano-liposomes and its application in fortification of yogurt, *Food Chemistry*, 216: 146-152.

Gibbs, B.F., Kermasha, S., Alli, I. and Mulligan, C.N. (1999). Encapsulation in the food industry: A review, *International Journal of Food Sciences and Nutrition*, 50(3): 213-224.

Giroux, H.J., Constantineau, S., Fustier, P., Champagne, C.P., St-Gelais, D., Lacroix, M. and Britten, M. (2013). Cheese fortification using water-in-oil-in-water double emulsions as carrier for water soluble nutrients, *International Dairy Journal*, 29(2): 107-114.

Godward, G. and Kailasapathy, K. (2003). Viability and survival of free, encapsulated and co-encapsulated probiotic bacteria in yogurt, *Milchwissenschaft: Milk Science International*, 58: 396-399.

Gökmen, V. and Şenyuva, H.Z. (2007). Effects of some cations on the formation of acrylamide and furfurals in glucose–asparagine model system, *European Food Research and Technology*, 225(5-6): 815-820.

Gökmen, V., Mogol, B.A., Lumaga, R.B., Fogliano, V., Kaplun, Z. and Shimoni, E. (2011). Development of functional bread containing nanoencapsulated omega-3 fatty acids, *Journal of Food Engineering*, 105(4): 585-591.

Gómez-Estaca, J., De Lacey, A.L., López-Caballero, M.E., Gómez-Guillén, M.C. and Montero, P. (2010). Biodegradable gelatin–chitosan films incorporated with essential oils as antimicrobial agents for fish preservation, *Food Microbiology*, 27(7): 889-896.

Gomez-Estaca, J., Balaguer, M.P., Gavara, R. and Hernandez-Munoz, P. (2012). Formation of zein nanoparticles by electrohydrodynamic atomization: Effect of the main processing variables and suitability for encapsulating the food coloring and active ingredient curcumin, *Food Hydrocolloids*, 28(1): 82-91.

Gómez-Mascaraque, L.G., Miralles, B., Recio, I. and López-Rubio, A. (2016). Microencapsulation of a whey protein hydrolysate within micro-hydrogels: Impact on gastrointestinal stability and potential for functional yogurt development, *Journal of Functional Foods*, 26: 290-300.

Gonçalves, N.D., de Lima Pena, F., Sartoratto, A., Derlamelina, C., Duarte, M.C.T., Antunes, A.E.C. and Prata, A.S. (2017). Encapsulated thyme (*Thymus vulgaris*) essential oil used as a natural preservative in bakery product, *Food Research International*, 96: 154-160.

Güzey, D. and McClements, D.J. (2006). Influence of environmental stresses on O/W emulsions stabilized by β-lactoglobulin–pectin and β-lactoglobulin–pectin–chitosan membranes produced by the electrostatic layer-by-layer deposition technique, *Food Biophysics*, 1: 30-40.

Hasanvand, E., Fathi, M., Bassiri, A., Javanmard, M. and Abbaszadeh, R. (2015). Novel starch-based nanocarrier for vitamin D fortification of milk: Production and characterization, *Food and Bioproducts Processing*, 96: 264-277.

Heidebach, T., Först, P. and Kulozik, U. (2012). Microencapsulation of probiotic cells for food applications, *Critical Reviews in Food Science and Nutrition*, 52(4): 291-311.

Hosseini, S.F., Ramezanzade, L. and Nikkhah, M. (2017). Nano-liposomal entrapment of bioactive peptidic fraction from fish gelatin hydrolysate, *International Journal of Biological Macromolecules*, 105: 1455-1463.

Jemaa, M.B., Falleh, H., Neves, M.A., Isoda, H., Nakajima, M. and Ksouri, R. (2017). Quality preservation of deliberately contaminated milk using thyme free and nanoemulsified essential oils, *Food Chemistry*, 217: 726-734.

Jenjob, S., Sunintaboon, P., Inprakhon, P., Anantachoke, N. and Reutrakul, V. (2012). Chitosan-functionalized poly (methyl methacrylate) particles by spinning disk processing for lipase immobilization, *Carbohydrate Polymers*, 89: 842-848.

Josquin, N.M., Linssen, J.P. and Houben, J.H. (2012). Quality characteristics of Dutch-style fermented sausages manufactured with partial replacement of pork back-fat with pure, pre-emulsified or encapsulated fish oil, *Meat Science*, 90(1): 81-86.

Jun-xia, X., Hai-yan, Y. and Jian, Y. (2011). Microencapsulation of sweet orange oil by complex coacervation with soy bean protein isolate/gum arabic, *Food Chemistry*, 125: 1267-1272.

Kailasapathy, K. (2006). Survival of free and encapsulated probiotic bacteria and their effect on the sensory properties of yogurt, *LWT – Food Science and Technology*, 39(10): 1221-1227.

Kanakis, C.D., Petros, A.T., Tajmir-Riahi, H.A. and Polissiou, G. (2007). Crocetin, dimethylcrocetin, and safranal bind human seruma lbumin: Stability and antioxidative properties, *Journal of Agricultural and Food Chemistry*, 55: 970-977.

Kanakisa, C.D., Hasni, I., Bourassa, P., Tarantilis, P.A., Polissioua, M.G. and Tajmir-Riahi, H.A. (2011). Milk β-lactoglobulin complexes with tea polyphenols, *Food Chemistry*, 127: 1046-1055.

Kant, A., Linforth, R.S., Hort, J. and Taylor, A.J. (2004). Effect of β-cyclodextrin on aroma release and flavor perception, *Journal of Agricultural and Food Chemistry*, 52(7): 2028-2035.

Katouzian, I. and Jafari, S.M. (2017). Nanoencapsulation of vitamins, pp. 145-181. *In:* S.M. Jafari (Ed.), *Nanoencapsulation of Food Bioactive Ingredients: Principles and Applications*, Academic Press. Massachusetts, United States.

Katouzian, I. and Jafari, S.M. (2016). Nano-encapsulation as a promising approach for targeted delivery and controlled release of vitamins, *Trends in Food Science & Technology*, 53: 34-48.

Kaushik, R., Sachdeva, B. and Arora, S. (2017). Effect of calcium and vitamin D_2 fortification on quality characteristics of *dahi*, *International Journal of Dairy Technology*, 70(2): 269-276.

Keenan, D.F., Resconi, V.C., Smyth, T.J., Botinestean, C., Lefranc, C., Kerry, J.P. and Hamill, R.M. (2015). The effect of partial-fat substitutions with encapsulated and unencapsulated fish oils on the technological and eating quality of beef burgers over storage, *Meat Science*, 107: 75-85.

Khalil, A.H. and Mansour, E.H. (1998). Alginate encapsulated bifidobacteria survival in mayonnaise, *Journal of Food Science*, 63(4): 702-705.

Kheadr, E.E., Vuillemard, J.C. and El-Deeb, S.A. (2003). Impact of liposome-encapsulated enzyme cocktails on Cheddar cheese ripening, *Food Research International*, 36: 241-252.

Koga, C.C., Lee, S.Y. and Lee, Y. (2016). Consumer acceptance of bars and gummies with unencapsulated and encapsulated resveratrol, *Journal of Food Science*, 81(5): S1222-S1229.

Kohane, D.S., Yeo, Y., Given, P. and Langer, R.S. (2015). U.S. Patent No. 9,186,640, Washington, DC: U.S. Patent and Trademark Office.

Koontz, J.L. and Marcy, J.E. (2007). Controlled release of active ingredients from polymer food packaging by molecular encapsulation with cyclodextrins, *Polymer Preprints*, **48**(2): 742.

Kumar, D.D., Mann, B., Pothuraju, R., Sharma, R. and Bajaj, R. (2016). Formulation and characterization of nanoencapsulated curcumin using sodium caseinate and its incorporation in ice cream, *Food & Functions*, **7**(1): 417-424.

Kwak, H.S., Yang, K.M. and Ahn, J. (2003). Microencapsulated iron for milk fortification, *Journal of Agricultural and Food Chemistry*, **51**(26): 7770-7774.

Laine, R., Salminen, S., Benno, Y. and Ouwehand, A.C. (2003). Performance of bifidobacteria in oat-based media, *International Journal of Food Microbiology*, **83**(1): 105-109.

Lakkis, J.M. (2016). *Encapsulation and Controlled Release Technologies in Food Systems*, John Wiley & Sons Ltd. Chichester, West Sussex.

Laouini, A., Fessi, H. and Charcosset, C. (2012). Membrane emulsification: A promising alternative for vitamin encapsulation within nano-emulsion, *Journal of Membrane Science*, 423-424: 85-96.

Laridi, R., Kheadr, E.E., Benech, R.O., Vuillemard, J.C., Lacroix, C. and Fliss, I. (2003). Liposome encapsulated nisin Z: optimization, stability and release during milk fermentation, *International Dairy Journal*, **13**(4): 325-336.

Le Maux, S., Bouhallab, S., Giblin, L., Brodkorb A. and Croguennec, T. (2014). Bovine β-lactoglobulin/fatty acid complexes: Binding, structural, and biological properties, *Dairy Science & Technology*, 94: 409-426.

Leonard, S.W., Good, C.K., Gugger, E.T. and Traber, M.G. (2004). Vitamin E bioavailability from fortified breakfast cereal is greater than that from encapsulated supplements, *The American Journal of Clinical Nutrition*, **79**(1): 86-92.

Li, Z., Grün, I.U. and Fernando, L.N. (2000). Interaction of vanillin with soy and dairy proteins in aqueous model systems: A thermodynamic study, *Journal of Food Science*, 65: 997-1001.

Lin, H.S., Chean, C.S., Ng, Y.Y., Chan, S.Y. and Ho, P.C. (2000). 2-Hydroxypropyl-betacyclodextrin increases aqueous solubility and photostability of all-trans-retinoic acid, *Journal of Clinical Pharmacy and Therapeutics*, 25: 265-269.

Lin, L., Dai, Y. and Cui, H. (2017). Antibacterial poly (ethylene oxide) electrospun nanofibers containing cinnamon essential oil/beta-cyclodextrin proteoliposomes, *Carbohydrate Polymers*, 178: 131-140.

Liu, J. and Mori, A. (1993). Antioxidant and pro-oxidant activities of p-hydroxybenzyl alcohol and vanillin: Effect on free radicals, brain peroxidation and degradation of benzoate, deoxyribose, aminoacids and DNA, *Neurophnmiocotogy*, 32: 659-669.

Liu, L., Li, X., Zhu, Y., Bora, A.F.M., Zhao, Y., Du, L., Li, D. and Bi, W. (2016). Effect of microencapsulation with Maillard reaction products of whey proteins and isomaltooligosaccharide on the survival of *Lactobacillus rhamnosus*, *LWT – Food Science and Technology*, 73: 37-43.

Liu, L., Chen, P., Zhao, W., Li, X., Wang, H. and Qu, X. (2017). Effect of microencapsulation with the Maillard reaction products of whey proteins and isomaltooligosaccharide on the survival rate of *Lactobacillus rhamnosus* in white brined cheese, *Food Control*, 79: 44-49.

Lopes, N.A., Pinilla, C.M.B. and Brandelli, A. (2019). Antimicrobial activity of lysozyme-nisin co-encapsulated in liposomes coated with polysaccharides, *Food Hydrocolloids*, 93: 1-9.

Ma, J.J., Mao, X.Y., Wang, Q., Yang, S., Zhang, D., Chen, S.W. and Li, Y.H. (2014). Effect of spray drying and freeze drying on the immunomodulatory activity, bitter taste and hygroscopicity of hydrolysate derived from whey protein concentrate, *LWT – Food Science and Technology*, **56**(2): 296-302.

Madene, A., Jacquot, M., Scher, J. and Desobry, S. (2006). Flavor encapsulation and controlled release – A review, *International Journal of Food Science and Technology*, 41: 1-21.

Mahdavi, S.A., Jafari, S.M., Assadpour, E. and Ghorbani, M. (2016). Storage stability of encapsulated barberry's anthocyanin and its application in jelly formulation, *Journal of Food Engineering*, 181: 59-66.

Marcuzzo, E., Sensidoni, A., Debeaufort, D. and Voilley, A. (2010). Encapsulation of aroma compounds in biopolymeric emulsion based edible films to control flavor release, *Carbohydrate Polymers*, 80: 984-988.

Marsanasco, M., Márquez, A.L., Wagner, J.R., Alonso, S.D.V. and Chiaramoni, N.S. (2011). Liposomes as vehicles for vitamins E and C: An alternative to fortify orange juice and offer vitamin C protection after heat treatment, *Food Research International*, 44(9): 3039-3046.

Martiñon, M.E., Moreira, R.G., Castell-Perez, M.E. and Gomes, C. (2014). Development of a multilayered antimicrobial edible coating for shelf-life extension of fresh-cut cantaloupe (*Cucumis melo* L.) stored at 4 C, *LWT – Food Science and Technology*, 56(2): 341-350.

Mattila-Sandholm, T., Myllärinen, P., Crittenden, R., Mogensen, G., Fondén, R. and Saarela, M. (2002). Technological challenges for future probiotic foods, *International Dairy Journal*, 12(2-3): 173-182.

McClements, D.J. (2018). Recent developments in encapsulation and release of functional food ingredients: Delivery by design, *Current Opinion in Food Science*, 23: 80-84.

Mirafzali, Z., Thompson, C.S. and Tallua, K. (2014). Application of liposomes in the food industry, pp. 139-150. *In*: A.G. Gaonkar, N. Vasisht, A. Khare and R. Sobel (Eds.), *Microencapsulation in the Food Industry*, Academic Press. New Jersey, United States.

Mishra, M. (2016). *Handbook of Encapsulation and Controlled Release*, CRC Press, New York.

Mohammadi, A., Jafari, S.M., Assadpour, E. and Esfanjani, A.F. (2016). Nano-encapsulation of olive leaf phenolic compounds through WPC–pectin complexes and evaluating their release rate, *International Journal of Biological Macromolecules*, 82: 816-822.

Mohan, A., Rajendran, S.R., He, Q.S., Bazinet, L. and Udenigwe, C.C. (2015). Encapsulation of food protein hydrolysates and peptides: A review, *RSC Advances*, 5(97): 79270-79278.

Moretti, D., Zimmermann, M. and Lakkis, J.M. (2016). Assessing bioavailability and nutritional value of microencapsulated minerals, pp. 289-308. *In*: J.M. Lakkis (Ed.), *Encapsulation and Controlled Release Technologies in Food Systems*, John Wiley & Sons. New Jersey, United States.

Mougin, K., Bruntz, A., Severin, D. and Teleki, A. (2016). Morphological stability of microencapsulated vitamin formulations by AFM imaging, *Food Structure*, 9: 1-12.

Mozafari, M.R., Khosravi-Darani, K., Borazan, G.G., Cui, J., Pardakhty, A. and Yurdugul, S. (2008). Encapsulation of food ingredients using nanoliposome technology, *International Journal of Food Properties*, 11: 833-844.

Muthukumarasamy, P. and Holley, R.A. (2006). Microbiological and sensory quality of dry fermented sausages containing alginate-microencapsulated *Lactobacillus reuteri*, *International Journal of Food Microbiology*, 111(2): 164-169.

Nedovic, V., Kalusevica, A., Manojlovicb, V., Levica, S. and Bugarski, B. (2011). An overview of encapsulation technologies for food applications, *Procedia Food Science*, 1: 1806-1815.

Noort, M.W., Bult, J.H. and Stieger, M. (2012). Saltiness enhancement by taste contrast in bread prepared with encapsulated salt, *Journal of Cereal Science*, 55(2): 218-225.

Nooshkam, M. and Madadlou, A. (2016a). Maillard conjugation of lactulose with potentially bioactive peptides, *Food Chemistry*, 192: 831-836.

Nooshkam, M. and Madadlou, A. (2016b). Microwave-assisted isomerisation of lactose to lactulose and Maillard conjugation of lactulose and lactose with whey proteins and peptides, *Food Chemistry*, 200: 1-9.

Nooshkam, M., Babazadeh, A. and Jooyandeh, H. (2018). Lactulose: Properties, techno-functional food applications, and food grade delivery system, *Trends in Food Science & Technology*, 80: 23-34.

Nualkaekul, S., Cook, M.T., Khutoryanskiy, V.V. and Charalampopoulos, D. (2013). Influence of encapsulation and coating materials on the survival of *Lactobacillus plantarum* and *Bifidobacterium longum* in fruit juices, *Food Research International*, **53**(1): 304-311.

O'boyle, A.R., Aladin-Kassam, N.A.Z.N.I.N., Rubin, L.J. and Diosady, L.L. (1992). Encapsulated cured-meat pigment and its application in nitrite-free ham, *Journal of Food Science*, **57**(4): 807-812.

Ojha, K.S., Perussello, C.A., García, C.Á., Kerry, J.P., Pando, D. and Tiwari, B.K. (2017). Ultrasonic-assisted incorporation of nano-encapsulated omega-3 fatty acids to enhance the fatty acid profile of pork meat, *Meat Science*, 132: 99-106.

Oliveira, A.C., Moretti, T.S., Boschini, C., Baliero, J.C.C., Freitas, L.A.P., Freitas, O. and Favaro-Trindade, C.S. (2007). Microencapsulation of *B. lactis* (BI 01) and *L. acidophilus* (LAC4) by complex coacervation followed by spouted – bed drying, *Drying Technology*, 25: 1687-1693.

Oussalah, M., Caillet, S., Salmieri, S., Saucier, L. and Lacroix, M. (2007). Antimicrobial effects of alginate-based films containing essential oils on *Listeria monocytogenes* and *Salmonella typhimurium* present in bologna and ham, *Journal of Food Protection*, **70**(4): 901-908.

Özer, B., Kirmaci, H.A., Şenel, E., Atamer, M. and Hayaloğlu, A. (2009). Improving the viability of *Bifidobacterium bifidum* BB-12 and *Lactobacillus acidophilus* LA-5 in white-brined cheese by microencapsulation, *International Dairy Journal*, **19**(1): 22-29.

Ozer, B.B.P., Uz, M., Oymaci, P. and Altinkaya, S.A. (2016). Development of a novel strategy for controlled release of lysozyme from whey protein isolate-based active food packaging films, *Food Hydrocolloids*, 61: 877-886.

Pan, K., Chen, H., Davidson, P.M. and Zhong, Q. (2014). Thymol nanoencapsulated by sodium caseinate: Physical and antilisterial properties, *Journal of Agricultural and Food Chemistry*, **62**(7): 1649-1657.

Pando, D., Beltrán, M., Gerone, I., Matos, M. and Pazos, C. (2015). Resveratrol entrapped niosomes as yogurt additive, *Food Chemistry*, 170: 281-287.

Pegg, R.B. and Shahidi, F. (2007). Encapsulation, stabilization, and controlled release of food ingredients and bioactives, pp. 509-570. *In*: Rahman, M.S. (Ed.). *Handbook of Food Preservation*, Taylor & Francis Group, LLC, New York.

Peretto, G., Du, W.X., Avena-Bustillos, R.J., Sarreal, S.B.L., Hua, S.S.T., Sambo, P. and McHugh, T.H. (2014). Increasing strawberry shelf-life with carvacrol and methyl cinnamate antimicrobial vapors released from edible films, *Postharvest Biology and Technology*, 89: 11-18.

Peter, K.V. (2006). *Handbook of Herbs and Spices* (vol. 3), Woodhead Publishing.

Picon, A., Gaya, P., Medina, M. and Nunez, M. (1994). The effect of liposome encapsulation of chymosin derived by fermentation on Manchego cheese ripening, *Journal of Dairy Science*, **77**(1): 16-23.

Picon, A., Gaya, P., Medina, M. and Nunez, M. (1995). The effect of liposome-encapsulated *Bacillus subtilis* neutral proteinase on Manchego cheese ripening, *Journal of Dairy Science*, **78**(6): 1238-1247.

Picot, A. and Lacroix, C. (2004). Encapsulation of bifidobacteria in whey protein-based microcapsules and survival in simulated gastrointestinal conditions and in yogurt, *International Dairy Journal*, **14**(6): 505-515.

Ponce, A., Roura, S.I. and Moreira, M.D.R. (2011). Essential oils as biopreservatives: Different methods for the technological application in lettuce leaves, *Journal of Food Science*, **76**(1): 34-40.

Prichapan, N. and Klinkesorn, U. (2014). Factor affecting the properties of water-in-oil-in-water emulsions for encapsulation of minerals and vitamins. *Journal of Science & Technology*, **36**(6): 651-661.

Qi, Z.H. and Shieh, W.J. (2002). Aqueous media for effective delivery of tretinoin, *Journal of Inclusion Phenomena and Macrocyclic Chemistry*, 44: 133-136.

Qin, B.Y., Bewley, M.C., Creamer, L.K., Baker, H.M., Baker, E.N. and Jameson, G.B. (1998). Structural basis of the tanford transition of bovine β-lactoglobulin, *Biochemistry*, 37: 14014-14023.

Rafiee, Z., Nejatian, M., Daeihamed, M. and Jafari, S.M. (2019). Application of curcumin-loaded nanocarriers for food, drug and cosmetic purposes, *Trends in Food Science & Technology*, 88: 445-458.

Rao, J. and McClements, D.J. (2011). Food-grade microemulsions, nanoemulsions and emulsions: Fabrication from sucrose monopalmitate and lemon oil, *Food Hydrocolloids*, **25**(6): 1413-1423.

Rashidinejad, A., Birch, E.J., Sun-Waterhouse, D. and Everett, D.W. (2016). Effect of liposomal encapsulation on the recovery and antioxidant properties of green tea catechins incorporated into a hard low-fat cheese following in vitro simulated gastrointestinal digestion, *Food and Bioproducts Processing*, 100: 238-245.

Rashidinejad, A., Birch, E.J., Sun-Waterhouse, D. and Everett, D.W. (2014). Delivery of green tea catechin and epigallocatechin gallate in liposomes incorporated into low-fat hard cheese, *Food Chemistryi*, 156: 176-183.

Reid, A.A., Champagne, C.P., Gardner, N., Fustier, P. and Vuillemard, J.C. (2007). Survival in food systems of *Lactobacillus rhamnosus* R011 microentrapped in whey protein gel particles, *Journal of Food Science*, **72**(1): 31-37.

Reineccius, G.A. and Risch, S.J. (1986). Encapsulation of artificial flavors by β-cyclodextrin, *Perfumer & Flavorist*, **11**(4): 1-6.

Ribeiro-Santos, R., Andrade, M. and Sanches-Silva, A. (2017). Application of encapsulated essential oils as antimicrobial agents in food packaging, *Current Opinion in Food Science*, 14: 78-84.

Risch, S.J. and Reineccius, G.A. (1995). *Encapsulation and Controlled Release of Food Ingredients*, ACS Symposium Series; American Chemical Society, Washington, DC.

Rocha, G.A., Fávaro-Trindade, C.S. and Grosso, C.R.F. (2012). Microencapsulation of lycopene by spray drying: Characterization, stability and application of microcapsules, *Food and Bioproducts Processing*, **90**(1): 37-42.

Rocha-Selmi, G.A., Bozza, F.T., Thomazini, M., Bolini, H.M. and Fávaro-Trindade, C.S. (2013). Microencapsulation of aspartame by double emulsion followed by complex coacervation to provide protection and prolong sweetness, *Food Chemistry*, **139**(1-4): 72-78.

Rodríguez-Huezo, M.E., Estrada-Fernández, A.G., García-Almendárez, B.E., Ludena-Urquizo, F., Campos-Montiel, R.G. and Pimentel-González, D.J. (2014). Viability of *Lactobacillus plantarum* entrapped in double emulsion during Oaxaca cheese manufacture, melting and simulated intestinal conditions, *LWT – Food Science and Technology*, **59**(2): 768-773.

Rostamabadi, H., Falsafi, S.R. and Jafari, S.M. (2019). Nanoencapsulation of carotenoids within lipid-based nanocarriers, *Journal of Controlled Release*, 298: 38-67.

Ruiz Canizales, J., Velderrain Rodríguez, G.R., Domínguez Avila, J.A., Preciado Saldaña, A.M., Alvarez Parrilla, E., Villegas Ochoa, M.A. and González Aguilar, G.A. (2018). Encapsulation to protect different bioactives to be used as nutraceuticals and food ingredients, pp. 1-20. *In*: J.M. Mérillon and K. Ramawat (Eds.), *Bioactive Molecules in Food*, Reference Series in Phytochemistry. Springer, Cham.

Sanguansri, L. and Augustin, M.A. (2006). 12 microencapsulation and delivery of omega-3 fatty acids, pp. 297-327. *In*: J. Shi (Ed.), *Functional Food Ingredients and Nutraceuticals*, CRC Press. Florida, United States.

Sanguansri, L. and Augustin, M.A. (2010). Microencapsulation in functional food product development, pp. 3-23. *In*: J. Smith and E. Charter (Eds.), *Functional Food Product Development*, Blackwell Publishing Ltd., London.

Sauvant, P., Cansell, M., Sassi, A.H. and Atgié, C. (2012). Vitamin A enrichment: Caution with encapsulation strategies used for food applications, *Food Research International*, 46: 469-479.

Savage, W.D., Schnell, P.G., Aumann, R.A. and Yatka, R.J. (2004). U.S. Patent No. 6,759,066. Washington, DC: U.S. Patent and Trademark Office.

Sawale, P.D., Patil, G.R., Hussain, S.A., Singh, A.K. and Singh, R.R.B. (2017). Effect of incorporation of encapsulated and free Arjuna herb on storage stability of chocolate vanilla dairy drink, *Food Bioscience*, 19: 142-148.

Schobel, A.M. and Yang, R.K. (1989). U.S. Patent No. 4,824,681, Washington, DC: U.S. Patent and Trademark Office.

Shah, N.P. and Ravula, R.R. (2000). Microencapsulation of probiotic bacteria and their survival in frozen fermented dairy desserts, *Australian Journal of Dairy Technology*, **55**(3): 139.

Sharma, R., Sanguansri, P., Marsh, R., Sanguansri, L. and Augustin, M.A. (2003). Applications of microencapsulated omega-3 fatty acids in dairy products, *Australian Journal of Dairy Technology*, **58**(2): 211.

Shaw, L.A., McClements, D.J. and Decker, E.A. (2007). Spray-dried multilayered emulsions as a delivery method for ω-3 fatty acids into food systems, *Journal of Agricultural and Food Chemistry*, 55: 3112-3119.

Siegrist, M., Cousin, M.E., Kastenholz, H. and Wiek, A. (2007). Public acceptance of nanotechnology foods and food packaging: The influence of affect and trust, *Appetite*, **49**(2): 459-466.

Sipahi, R.E., Castell-Perez, M.E., Moreira, R.G., Gomes, C. and Castillo, A. (2013). Improved multilayered antimicrobial alginate-based edible coating extends the shelf-life of fresh-cut watermelon (*Citrullus lanatus*), *LWT – Food Science and Technology*, **51**(1): 9-15.

Siuta-Cruce, P. and Goulet, J. (2001). Improving probiotic survival rates, *Food Technology*, **55**(10): 36-42.

Stratulat, I., Britten, M., Salmieri, S., Fustier, P., St-Gelais, D., Champagne, C.P. and Lacroix, M. (2015). Enrichment of cheese with vitamin D3 and vegetable omega-3, *Journal of Functional Foods*, 13: 300-307.

Taguchi, K., Iwami, K., Ibuki, F. and Kawabata, M. (1992). Oxidative stability of sardine oil embedded in spray-dried egg white powder and its use for n-3 unsaturated fatty acid fortification of cookies, *Bioscience, Biotechnology, and Biochemistry*, **56**(4): 560-563.

Taksima, T., Limpawattana, M. and Klaypradit, W. (2015). Astaxanthin encapsulated in beads using ultrasonic atomizer and application in yogurt as evaluated by consumer sensory profile, *LWT – Food Science and Technology*, **62**(1): 431-437.

Tamjidi, F., Nasirpour, A. and Shahedi, M. (2012). Physicochemical and sensory properties of yogurt enriched with microencapsulated fish oil, *Food Science and Technology International*, **18**(4): 381-390.

Tavel, L., Moreau, C., Bouhallab, S., Li-Chan, E.C.Y., and Guichard, E. (2010). Interactions between aroma compounds and β-lactoglobulin in the heat-induced molten globule state, *Food Chemistry*, 119: 1550-1556.

Toniazzo, T., Berbel, I.F., Cho, S., Fávaro-Trindade, C.S., Moraes, I.C. and Pinho, S.C. (2014). β-carotene-loaded liposome dispersions stabilized with xanthan and guar gums: Physico-chemical stability and feasibility of application in yogurt, *LWT – Food Science and Technology*, **59**(2): 1265-1273.

Troise, A.D. and Fogliano, V. (2013). Reactants encapsulation and Maillard reaction, *Trends in Food Science & Technology*, **33**(1): 63-74.

Tsen, J.H., Lin, Y.P., Huang, H.Y. and King, V.A.E. (2008). Studies on the fermentation of tomato juice by using κ-carrageenan immobilized *Lactobacillus acidophilus*, *Journal of Food Processing and Preservation*, **32**(2): 178-189.

Ubbink, J. (2013). Flavor delivery systems, pp. 1-35. *In*: D. Othmer and R.E. Kirk (Eds.), *Kirk-Othmer Encyclopedia of Chemical Technology*, John Wiley & Sons, Inc. New Jersey, United States.

Underriner, E.W. (2012). *Handbook of Industrial Seasonings*, Springer Science & Business Media.

van den Tempel, T., Gundersen, J.K. and Nielsen, M.S. (2002). The microdistribution of oxygen in Danablu cheese measured by a microsensor during ripening, *International Journal of Food Microbiology*, **75**(1-2): 157-161.

Verardo, V., Ferioli, F., Riciputi, Y., Iafelice, G., Marconi, E. and Caboni, M.F. (2009). Evaluation of lipid oxidation in spaghetti pasta enriched with long chain n-3 polyunsaturated fatty acids under different storage conditions, *Food Chemistry*, **114**(2): 472-477.

Vu, K.D., Hollingsworth, R.G., Leroux, E., Salmieri, S. and Lacroix, M. (2011). Development of edible bioactive coating based on modified chitosan for increasing the shelf-life of strawberries, *Food Research International*, **44**(1): 198-203.

Wang, S., Su, R., Nie, S., Sun, M., Zhang, J., Wu, D. and Moustaid-Moussa, N. (2014). Application of nanotechnology in improving bioavailability and bioactivity of diet-derived phytochemicals, *The Journal of Nutritional Biochemistry*, **25**(4): 363-376.

Wang, Y.C., Yu, R.C. and Chou, C.C. (2002). Growth and survival of bifidobacteria and lactic acid bacteria during the fermentation and storage of cultured soymilk drinks, *Food Microbiology*, **19**(5): 501-508.

Weerawatanakorn, M., Wu, J., Pan, M. and Ho, C. (2015). Reactivity and stability of selected flavor compounds, *Journal of Food and Drug Analysis*, 23: 176-190.

Wegmüller, R., Zimmermann, M.B., Bühr, V.G., Windhab, E.J. and Hurrell, R.F. (2006). Development, stability, and sensory testing of microcapsules containing iron, iodine, and vitamin A for use in food fortification, *Journal of Food Science*, **71**(2): 181-187.

Wetzel, C.R. and Bell, L.N. (1998). Chemical stability of encapsulated aspartame in cakes without added sugar, *Food Chemistry*, **63**(1): 33-37.

Wetzel, C.R., Weese, J.O. and Bell, L.N. (1997). Sensory evaluation of no-sugar-added cakes containing encapsulated aspartame, *Food Research International*, **30**(6): 395-399.

Wu, Y., Luo, Y. and Wang, Q. (2012). Antioxidant and antimicrobial properties of essential oils encapsulated in zein nanoparticles prepared by liquid–liquid dispersion method, *LWT – Food Science and Technology*, **48**(2): 283-290.

Xiao, D., Davidson, P.M. and Zhong, Q. (2011). Spray-dried zein capsules with coencapsulated nisin and thymol as antimicrobial delivery system for enhanced antilisterial properties, *Journal of Agricultural and Food Chemistry*, **59**(13): 7393-7404.

Xue, J., Davidson, P.M. and Zhong, Q. (2015). Antimicrobial activity of thyme oil co-nanoemulsified with sodium caseinate and lecithin, *International Journal of Food Microbiology*, 210: 1-8.

Yang, S., Mao, X.Y., Li, F.F., Zhang, D., Leng, X.J., Ren, F.Z. and Teng, G.X. (2012). The improving effect of spray-drying encapsulation process on the bitter taste and stability of whey protein hydrolysate, *European Food Research and Technology*, **235**(1): 91-97.

Ying, D.Y., Phoon, M.C., Sanguansri, L., Weerakkody, R., Burgar, I. and Augustin, M.A. (2010). Microencapsulated *Lactobacillus rhamnosus* GG powders: Relationship of powder physical properties to probiotic survival during storage, *Journal of Food Science*, 75: 588-595.

Yokota, D., Moraes, M. and Pinho, S.C. (2012). Characterization of lyophilized liposomes produced with non-purified soy lecithin: A case study of casein hydrolysate microencapsulation, *Brazilian Journal of Chemical Engineering*, 29: 325-335.

Young, F.V.K., Barlow, S.M. and Madsen, J. (1993). Unhydrogenated fish oil in low-calorie spreads, *Inform.*, **4**(1140): 4.

Zawada, Z.H. (2004). Vesicles with a double bilayer, *Cellular and Molecular Biology Letters*, 9: 589-602.

Zhang, Y., Ma, Q., Critzer, F., Davidson, P.M. and Zhong, Q. (2015). Effect of alginate coatings with cinnamon bark oil and soybean oil on quality and microbiological safety of cantaloupe, *International Journal of Food Microbiology*, 215: 25-30.

Zhang, Y., Ma, Q., Critzer, F., Davidson, P.M. and Zhong, Q. (2016). Organic thyme oil emulsion as an alternative washing solution to enhance the microbial safety of organic cantaloupes, *Food Control*, 67: 31-38.

Zinoviadou, K.G., Koutsoumanis, K.P. and Biliaderis, C.G. (2009). Physico-chemical properties of whey protein isolate films containing oregano oil and their antimicrobial action against spoilage flora of fresh beef, *Meat Science*, **82**(3): 338-345.

Zuidam, N.J. and Heinrich, J. (2010). Encapsulation of aroma, pp. 127-160. *In*: Zuidam, N.J. and Nedovic, V.A. (Eds.), *Encapsulation Technologies for Food Active Ingredients and Food Processing*, Springer, Dordrecht, The Netherlands.

Zyck, D.J. and Yatka, R. (2007). U.S. Patent No. 7,244,454, Washington, DC: U.S. Patent and Trademark Office.

Methods for Analysis of Encapsulates

Vesna Tumbas Šaponjac*, Vanja Šeregelj, Jelena Vulić, Jasna
Čanadanović-Brunet and Gordana Ćetković

University of Novi Sad, Faculty of Technology, Bulevar cara Lazara 5,
21000 Novi Sad, Serbia

1. Introduction

Encapsulation relates to technologies which enable formulation of one or more active compounds inside individualized particles with a specific geometry and properties. This technology is developing rapidly and finds applications in a wide range of products and sectors, including pharmaceutical, agriculture, food, cosmetic, textile, electronic and environmental protection (Ponder *et al.*, 2001; Obare and Meyer, 2004). In the food sector, encapsulation is a useful tool to improve delivery of bioactive molecules (e.g. antioxidants, minerals, vitamins, carotenoids) and living cells (e.g. probiotics) into foods, keeping the bioactives fully functional (Nedović *et al.*, 2013).

Many encapsulation technologies are available (spray-drying, spray-cooling, spray-chilling, coacervation, extrusion, fluidized bed coating, molecular inclusion, and freeze-drying). Also, a range of edible materials can be used as wall materials for encapsulates (polysaccharides, proteins, fats and waxes). Generally, encapsulates can be classified according to their size: macrocapsules (>5.000 μm), microcapsules (0.2-5.000 μm) and nanocapsules (<0.2 μm). In terms of their shape and construction, encapsulates are basically particles with a core-wall structure (mononuclear), but some of them can have a more complex structure, e.g. in a form of multiple cores embedded in a matrix (polynuclear).

To be effective in providing protection against environmental conditions or target delivering, it is essential that encapsulates have a well-selected core and wall material with convenient physicochemical properties to enable quality control during formulation, storage and application. Also, the choice of encapsulation technique should be based on the final application of encapsulates, the production scale and the costs. Different process variables can impact the physical, mechanical and structural properties of encapsulates (Moran *et al.*, 2014). As a consequence of their size, encapsulates show different physicochemical properties and behaviors in

---

*Corresponding author: vesnat@uns.ac.rs

different food environments, which is also related to agglomeration, composition, particle size distribution, shape, solubility, dispersibility, morphology, surface charge, wall thickness, mechanical strength, glass transition temperature, degree of crystallinity, flowability, permeability, etc. A wide range of analytical methods for the characterization of encapsulates is available, including microscopy approaches, chromatography, spectroscopic and related techniques. This chapter briefly describes the principles and applications of existing methods which could provide necessary information about physicochemical characteristics of encapsulated active food ingredients.

2. Physicochemical Properties of Encapsulates

The physicochemical properties of food encapsulates include primary or fundamental properties (e.g. particle size, shape, density, etc.) and secondary or functional properties (e.g. settling velocity of particles, rehydration rate, etc.) (Barbosa-Canovas *et al.*, 2005). Since information obtained from fundamental characterization is not sufficient to predict the behavior of encapsulates during food processing, both groups of properties are needed for their application in functional foods and beverages (Cuq *et al.*, 2011). Celli *et al.* (2005) suggested that early investigations about encapsulates properties are very important to ensure their suitability for incorporation in particular food products, as well as to determine the best method of food production. Table 1 summarizes some studies that investigated encapsulates properties significant for the development of different food products.

Particle size and their distribution are the most important characteristics of encapsulates, affecting both the properties and performance of intermediate and final products. It may also greatly impact the nature and efficiency of the production process (Chan *et al.*, 2008). However, various studies have highlighted the effect of the particle size when encapsulates were used as bioactives in functional foods. Reducing the bioactives particle size may improve their availability, delivery properties, solubility and thus biological activity, due to the fact that biological activities mainly depend on the ability of bioactive molecules to cross the intestinal membranes and enter the blood (Shegokar and Muller, 2010). The average size and uniformity (polydispersity) of encapsulate particles also have a critical impact on release kinetics of encapsulated ingredients, as particles with the same size tend to have the same release rate (Ramos 2011; Rhine *et al.*, 1980; Wilkins, 1999).

Furthermore, particle size influences the bulk density, compressibility and flowability of the encapsulates (Zhao *et al.*, 2009). Knowledge of flowing properties is essential for the design of reliable storage systems, processing, packaging and distribution conditions. Apart from particle size, bulk density also depends on chemical and water composition. Moisture content represents the water composition in a food system and vital physical attribute of encapsulated food ingredients which is usually associated with its stability, quality and composition, that could affect its storage stability, packaging and processing (Yamashita *et al.*, 2017). Related characteristics are water activity which determines the availability of free water in a food system responsible for many biochemical reactions and hygroscopicity of encapsulates (Fitzpatrick and Ahrne, 2005).

Surface properties and morphology of particles are playing a major role in the manufacturing and use of food encapsulates. A specific particle morphology is often preferred for a specific encapsulation application (Mittal, 2013). Impacts of processing are also revealed by changes in microparticle morphology (Yang *et al.*, 2001). Among the different surface properties, shape, surface porosity, surface composition, surface energy, surface charge, etc. are commonly considered parameters. The shape of encapsulates can play an important role in bioactive delivery, target effect, degradation and transport. High-surface porosity can lead to initial burst release, which is undesirable for controlled delivery systems (Yoe and Park, 2004). Surface composition affects transport, delivery and biodistribution, especially when combining two or more types of materials. Surface energy is relevant to the aggregation of encapsulates. Aggregation causes increase in size and a decrease in the particle-number concentration of an encapsulate, which may result in changes in reactivity and functionality of these particles. Surface charge presents one of the most important characteristics in the determination of the coloidal stability. The colloidal stability is generally analyzed by the zeta potential of particles. Zeta potential is the value of the particle surface charge at specific environmental conditions of the medium, such as pH, temperature, ionic strength, etc. and can help to predict the stability of formulated microcapsule suspensions. Values of zeta potential are either positive or negative, and adequately controlled values may ensure the stability and avoid microparticles aggregation. If prolonged stability of a dispersed suspension of microcapsules is required, some experiments should be performed to maximize the positive or negative zeta potential away from the isoelectric point (zeta potential of zero) to prevent aggregation (Sharma *et al.*, 2019).

Particle size distribution and morphology also present an indication of stickiness and caking. These characteristics are related to the glass transition temperature (Tg) and can often be observed in encapsulates. Encapsulates produced by spray and freeze-drying techniques might contain components in amorphous glassy state, which are thermodynamically unstable and can become sticky when the temperature is raised above their Tg (by approximately 10-20°C). Fitzpatrick (2007) explained this phenomenon by the increase of surface energy of the particulates that will enable interactions with other molecules and adhesion to solid surfaces. Powders in the amorphous state are also prone to caking during storage, which will result in undesirable reactions and potential losses of the encapsulated bioactives and quality of encapsulate (Telis and Martínez-Navarrete, 2009). Caking can appear due to temperatures above the Tg, amount of amorphous material and absorption of moisture, which forms a liquid film on the surface of the particles (Goula and Adamopoulos, 2010) that increase their cohesiveness (Fitzpatrick *et al.*, 2007). Fitzpatrick (2007) also noted that cohesiveness could result from the surface composition.

Encapsulation efficiency, also known as loading efficiency, is a numerical measure of the amount of incorporated core material in microparticles, which is generally expressed as the percentage of core material encapsulated relative to the amount of initially added core material (Liu *et al.*, 2008). Encapsulation efficiency is one of the critical properties of microencapsulation that have important influence on subsequent applications. Microparticles with high encapsulation efficiencies are often desired as less carrier materials are required and less net amount of encapsulated products could deliver the required amount of core material (Lu *et al.*, 2011).

Encapsulates segregation effects can occur at any stage of industrial production, resulting in the particles with different dimensions and bulk densities to separate. Also, segregation can occur due to differences in particle shape, surface texture, angle of response, cohesiveness, etc. (Liu, 2008). This effect is very important when working with dry mixtures containing encapsulated bioactive compounds. Hence, the appropriate selection of ingredients that share similar physical characteristics as those of the encapsulated bioactive is recommended to avoid segregation of these particles and guarantee the homogeneity and concentration of the functional ingredient in the final product (Fitzpatrick and Ahrné, 2005).

Physical properties include the reconstitution properties as well, which involves wetting, dispersing and dissolution of the encapsulates. Wettability is the capacity of the encapsulate particles to absorb water on their surface and present the initial stage of the reconstitution process. Dispersibility is the distribution of particles in the liquid with little stirring and solubility refers to the extent and rate of dissolution of these particles. These mechanisms are closely related to the particle size, density, porosity, surface charge, surface area and morphology of the particles in the final powder product (Kwapińska and Zbiciński, 2005). Reconstitution properties are very important for products such as beverages when encapsulates are used for their preparation and present reconstituted drinks, or for chocolate as the encapsulate should be distributed throughout the chocolate mass and not form aggregates. On the other side, producing bakery items involves high shear and high temperatures which can have deleterious effects on the integrity of the encapsulates, stability and release of the bioactive ingredients. In this case, properties, such as mechanical strength, hardness and friability should be considered (Celli *et al.*, 2015).

The *in vitro* release test is one of the most important analyses to assure the functionality of an encapsulated ingredient (Wise, 2000). The release test provides an estimate of the behavior of microparticles in actual applications, by using similar environmental conditions (Rathbone and Butler, 2011). Another important goal of the release test is to evaluate the sensitivity of the designed release mechanism, i.e. release tests under different conditions are compared. Results of release tests are commonly known as release profiles, where cumulative concentration or percentage release of the core ingredient is plotted against time, and, based on such profiles, decisions are made on whether the release pattern meets the expectation or not (Zhang *et al.*, 2010).

3. Encapsulates Analysis by Microscopic Techniques

In recent years, microscopy analysis has been extensively used in studies of the food encapsulates characteristics that cannot be visualized by the naked eye. Imaging methods provide data on morphology, size and particle size distribution, agglomeration, shape and chemical composition. Also, these methods can be used to help to understand the fate of encapsulates in food products and determine the influence of their presence on sample structure (Castaneda *et al.*, 2008). The most common instruments for the visualization and characterization of micro- and nano-structured materials are optical microscopy, fluorescence and electron microscopy. These techniques differ in their mechanisms, which is very important to understand in order to choose the most appropriate method for a specific type of encapsulates. Table 2 summarizes the commonly used microscopy techniques for physicochemical characterization of encapsulates, with highlighted applications, advantages and disadvantages.

Table 1: Overview of Encapsulated Bioactive Compounds, Techniques, Wall Materials, Food Applications and Their Physicochemical Properties

Bioactive Compound	Encapsulation Technique	Wall Material	Food Product	Physicochemical Properties	Reference
Fish oil	Spray-drying	Barley protein	Milk and yogurt	Encapsulation efficiency Loading efficiency Moisture content Particle size Morphology	Wang et al., 2011
Hydroxycitric acid	Spray-drying	Whey protein	Pasta	Moisture content Encapsulation efficiency Morphology	Pillai et al., 2012
Carotenoids oil	Crosslinking followed by freeze-drying	Chitosan/sodium tripolyphosphate	Salad cream and commercial drink	Encapsulation efficiency Particle size Solubility Releasing profile	Thamaket and Raviyan, 2015
β-carotene	Crosslinking	Chitosan/sodium tripolyphosphate	Hamburger patties	Particle size Morphology	Ozvural and Huang, 2017
Phenolics and flavonoids	Liposome	Lecithin, chitosan, and maltodextrin	Cheddar cheese	Particle size distribution Zeta potential Micrography	El-Messery et al., 2019
Carotenoids	Complex coacervation	Chitosan, pectin, and xanthan gum	Yogurt and bread	Encapsulation efficiency Yield Morphology Thermal behavior Release profile	Rutz et al., 2017
Carotenoids and phenolics	Spray- and freeze-drying	Whey protein	Yogurt	Water activity Moisture content Higroscopicty	Šeregelj et al., 2019

Active compound	Encapsulation technique	Wall material	Food matrix	Parameters analyzed	Reference
Casein hydrolysate	Spray-drying	Maltodextrin	Protein bar	Solubility Flowabilty Encapsulation efficiency Particle size Morphology Color parameters Dissolution Higroscopicity Moisture content Morphology Particle size distribution Glass transition temperature Water activity	Rocha et al., 2009
Polyphenols and anthocyanins	Spray-drying	Maltodextrin, soybean protein isolates	Yogurt	Particle size Morphology	Robert et al., 2010
Anthocyanins	Liposomes followed by spray-drying	Chitosan, maltodextrin	Chocolate	Particle size distribution Zeta potential Aggregation Morphology	Gültekin-Özgüven et al., 2016
Polyphenols and anthocyanins	Freeze-drying	Whey and soy proteins	Cookies	Encapsulation efficiency Color parameters	Tumbas Šaponjac et al., 2016
Polyphenols	Spray-drying	Maltodextrin	Chocolate	Particle size distribution Color parameters	Lončarević et al., 2019
Polyphenols	Spray-drying	Maltodextrin	Chocolate	Particle size distribution Color parameters	Lončarević et al., 2018
Betacyanins and polyphenols	Freeze-drying	Soy protein	Water biscuits	Encapsulation efficiency Color parameters	Hidalgo et al., 2018

Encapsulation in Food Processing and Fermentation

Table 2: Overview of Commonly Used Microscopic Techniques for the Physiochemical Characterization of Encapsulates – Applications, Advantages and Disadvantages

Technique	Observed Properties	Advantages	Disadvantages	References
Optical or light microscopy	Size Shape Degree of aggregation Homogeneity	Optical imaging Technique simplicity	Limited resolution by light wavelengths Manual sample preparation Long scanning time (multiple measurements required)	Robson *et al.*, 2018
Near-field scanning optical microscopy	Size Shape Chemical bonding	Simultaneous fluorescence and spectroscopy measurement Surface analysis at ambient conditions Assessment of chemical information and interactions at a nano-scaled resolution	Long scanning time Small specimen area analyzed Incident light intensity insufficient to excite weak fluorescent molecules Difficulty in imaging soft materials Analysis limited to the nanomaterial surface	Maynard, 2000 Lin *et al.*, 2013
Fluorescence microscopy	Size Shape Hydrodynamic diameter Diffusion coefficient	High spatial and temporal resolution Low sample consumption Specificity for fluorescent probes Method for studying chemical kinetics and molecular diffusion	Limited for fluorescence samples Limited applications and inaccuracy due to lack of appropriate models Suffers from photobleaching and autofluorescence Fixation of samples may cause artifacts	Lin *et al.*, 2013 Sapsford *et al.*, 2011 Domingos *et al.*, 2009
Confocal laser scanning microscopy	Size Surface and internal properties	Superior image clarity over fluorescence microscopy Can provide a composite 3D image of the sample Capable of visualizing the internal structure	Limited for fluorescence samples	Domingos *et al.*, 2009 Kuyper *et al.*, 2006a Kuyper *et al.*, 2006b

Method	Properties	Advantages	Limitations	References
Scanning electron microscopy	Size and size distribution Shape Aggregation Dispersion	Direct measurement of the size/size distribution and shape of particles High resolution	Conducting sample or coating conductive materials required Dry samples required Sample analysis in non-physiological conditions Biased statistics of size distributions Expensive equipment	Bootz et al., 2004 Lin et al., 2013 Sapsford et al., 2011 Hall et al., 2007 Tiede et al., 2008
Environmental scanning electron microscope	Size and size distribution Shape Aggregation Dispersion	Direct measurement of the size/size distribution and shape of nanomaterials Images of biomolecules in natural state	Expensive equipment Cryogenic method required for most bioconjugates Reduced resolution	Bootz et al., 2004 Lin et al., 2013 Sapsford et al., 2011 Hall et al., 2007 Tiede et al., 2008
Transmission electron microscopy	Size and size distribution Shape Aggregation Dispersion	Direct measurement of the size/size distribution and shape of particles with higher spatial resolution than SEM Several analytical methods coupled with TEM for investigation of electronic structure and chemical composition of materials	Ultrathin samples required Samples in non-physiological condition Sample damage/alternation Poor sampling Expensive equipment	Domingos et al., 2009 Hall et al., 2007 Tiede et al., 2008 Lin et al., 2013 Mavrocordatos et al., 2004
Scanning tunneling microscopy	Size and size distribution Shape Aggregation Dispersion	Direct measurement High spatial resolution at atomic scale	Conductive surface required Surface electronic structure and surface topography unnecessarily having a simple connection	Binnig and Rohrer, 1983 Fleming et al., 2009 Lin et al., 2013 Wang and Chu, 2013
Atomic force microscopy	Size and size distribution Shape Structure Sorption Dispersion Aggregation Surface properties	3D sample surface mapping Sub-nanoscaled topographic resolution Direct measurement of samples in dry, aqueous or ambient environment	Overestimation of lateral dimensions Poor sampling and time consuming Analysis in general limited to the exterior of materials	Domingos et al., 2009 Tiede et al., 2008 Mavrocordatos et al., 2004 Lin et al., 2013 Sapsford et al., 2011

3.1   Optical or Light Microscopy Technique

Optical or light microscopy refers to microscopes that utilize visible light and an arrangement of lenses to magnify a field of view (Murphy and Davidson, 2012c). A conventional optical microscope (COM) has a limited resolution of 200 nm or greater, as restricted by the wavelength of light and, as such, are incapable of providing detailed information regarding the structures of small dimensions. For example, the typical dimensions of nanoparticles are below the diffraction limit of visible light so that they are outside of the range for optical microscopy. Generally, the size of single encapsulate particle can be determined. Digital and video images of a specimen are taken, followed by quantitative analysis using image software. In addition to particle size information, COM may also provide information on the tendency of microparticles toward agglomeration. A major drawback of optical microscopy is that it is time-consuming due to manual sample preparation (disposing samples on to slides) and low throughput. Multiple measurements are needed to examine a significant number of microparticles (typically 100-200) allowing calculation of the mean diameter (van Beers *et al.*, 2017).

3.2   Near-field Scanning Optical Microscopy Technique

Near-field scanning optical microscopy (NSOM), also known as scanning near-field optical microscopy (SNOM), is a scanning probe technique developed to surpass the spatial resolution constraints that traditionally limit optical microscopy. NSOM uses fiber optic probes to funnel light down to the nanometric dimension. By scanning these probes near sample surface, fluorescence measurements can be taken with a spatial resolution of tens of nanometers, which represent an order of magnitude improvement over conventional fluorescence microscopy. NSOM, therefore, helps bridge the gap in spatial resolution between far-field optical approaches and high-resolution techniques, such as electron microscopy (Huckabay *et al.*, 2013). Some disadvantages of implementing NSOM include lengthy scanning time for high-resolution images or large specimen area, low incident light intensity hindering excitation of weak fluorescent molecules, difficulty in imaging soft materials caused by the high spring constants of the optical fibers, particularly in shear-force mode and the ability to image only surface features.

3.3   Fluorescence Microscopy Technique

Fluorescence microscopy is a special form of light microscopy that exploits the ability of fluorochromes to emit light after being excited with light of a certain wavelength (Murphy and Davidson, 2012). A large range of fluorescent dyes with various chemical and photonic properties is commercially available. The choice of fluorescent dye is a critical step, as some dyes can induce large changes in the membrane and/ or cause experimental artifacts, resulting in inaccurate data interpretation (Bouvrais *et al.*, 2010). In addition, photo-induced lipid peroxidation can also lead to domain formation even in simple dye systems. Also, prolonged exposure to fluorescent light can also result in bleaching and loss of fluorescence intensity.

3.4   Confocal Laser Scanning Microscopy Technique

Confocal laser scanning microscopy (CLSM) presents an advancement in the area of fluorescence microscopy and can be applied as a non-destructive visualization technique for encapsulates characterization. CLSM minimizes scattered light from out-of-focus structures and permits the identification of several compounds through the use of different fluorescence labels. This technique allows characterization of the particle surface, but also particle internal structure if the material is sufficiently transparent and can be fluorescently labeled (Soottitantawat *et al.*, 2004). By collecting several coplanar cross sections, a three-dimensional reconstruction of the inspected objects can be performed.

3.5 Scanning Electron Microscopy Technique

Contrary to optical microscopy, which uses light sources and glass lenses to illuminate specimens to produce magnified images, electron microscopy uses beams of accelerated electrons and electrostatic or electromagnetic lenses to generate images of much higher resolution, based on the much shorter wavelengths of electrons than visible light photons (Lin *et al.*, 2013). Scanning electron microscopy (SEM) is a type of electron microscope that uses a focused beam of high-energy electrons to produce images of a sample by creating a variety of signals at the surface of the sample. The signals are generated by the electron-sample interaction providing the information about the sample such as external morphology, crystalline structure and chemical composition. The interaction of the sample and the electron beam generate secondary electrons, back-scattered electrons, X-rays, visible rays, absorbed and diffracted electrons. Among these emissions, detection of the secondary electrons is the most common mode in SEM and can capture details of size up to 5 nm (Sharma *et al.*, 2019). The high-resolution abilities of SEM provide a highly magnified three-dimensional image of surface structure of particles, information regarding the purity of the sample, degree of aggregation and also determine size, size distribution and shape of encapsulates. SEM application could be limited for the biomolecule samples that are non-conductive and therefore tend to acquire charge and insufficiently deflect the electron beam, leading to imaging faults or artifacts. Therefore, for this sample preparation procedure, coating an ultrathin layer of electrically conducting material onto the biomolecules is often required (Hall *et al.*, 2007; Suzuki, 2002).

3.6 Environmental Scanning Electron Microscope Technique

Preparative techniques, such as drying or contrasting samples can lead to shrinkage of the specimen and therefore to altering the characteristics of encapsulates. Environmental scanning electron microscope technique (ESEM) allows imaging of particles in their natural state without modification or preparation. The main feature of ESEM is the presence of water vapor in the microscope chamber. The ability to maintain a water-containing atmosphere around the sample that may be partially or even fully hydrated is made possible by the use of a multiple-aperture, graduated vacuum system that allows the imaging chamber to be sustained at pressures up to 55hPa (Bibi *et al.*, 2011; Ruozi *et al.*, 2011). The primary electron beam can generate secondary electrons that then encounter vapor molecules, leading to a cascade amplification of the signal before reaching the detector. Because of this, ESEM does not require sample preparation (Muscariello *et al.*, 2005). This technique allows variation in the sample environment through a series of pressure, temperature and gas compositions (Mohammed *et al.*, 2004), which is useful when determining how environmental changes affect the particles. A limitation of ESEM is that it cannot provide detailed information regarding internal architecture of the nanoscale structures, reduced resolution and expensive equipment (Ruozi *et al.*, 2011).

3.7   Transmission Electron Microscopy Technique

Transmission electron microscopy (TEM) is a high-resolution technique which is very important for structural and chemical characterization of samples by gathering information about the size, crystallinity, shape and the interaction of the particles. Compared with SEM, TEM provides better spatial resolution and capability for additional analytical measurements (Hall *et al.*, 2007). In TEM, the electron beam is accelerated with a high voltage (80-400kV), passed through an ultra-thin sample, during which the incident electrons interacting with it are transformed to unscattered electrons, elastically or inelastically scattered electrons (Williams and Carter, 2016). The main limitation of this technique is that a high vacuum and thin sample section are required for electron-beam penetration, which can cause losses in resolution. Specimens have to be thin enough to transmit sufficient electrons to produce images. For example, in cases, such as employed high-resolution TEM (HRTEM), the specimen thickness of less than 50 nm is required. Sample preparation for TEM analysis is a complex process which can lead to altering its structure that can be damaged by intense, high-voltage electron beams.

3.8   Scanning Tunneling Microscopy Technique

Scanning tunneling microscopy (STM) is another type of scanning probe microscopy for imaging surfaces at the atomic level, based on the quantum mechanical nature of electrons. STM uses quantum tunneling current to generate electron density images for conductive or semi-conductive surfaces and biomolecules attached on conductive substrates at the atomic scale. It can be performed in vacuum air, water, gas and various liquids at temperatures ranging from near zero Kelvin to over 1000°C, with very good resolution (0.1 nm lateral and 0.01 nm depth resolution). STM can directly observe the size, shape, structure and states of dispersion and aggregation of particles. The essential components of an STM include a sharp scanning tip, an xyz-piezo scanner controlling the lateral and vertical movement of the tip, a coarse control unit positioning the tip close to the sample within the tunneling range, a vibration isolation stage and feedback regulation electronics (Wiesendanger, 1994). As the tip–sample separation is maintained in the range of 4-7 A, a small voltage applied between the scanning tip and the surface causes tunneling of electrons by which variation of the responding current can be recorded while the tip moves across the sample in the x-y plane to generate a map of charge density (Bonnell, 2001). Therefore, the output (tunneling current) is a function of applied voltage, tip position and the local density of the sample. Information can be obtained by analyzing the current as the tip's position scans across the surface and is usually displayed in image form.

3.9   Atomic Force Microscopy Technique

Atomic force microscopy (AFM), also known as chemical force microscopy, belongs to the family of scanning probe microscopes. It works by running a sharp tip attached to a cantilever and sensor over the surface of a sample and measuring the surface forces between the probe and the sample (Sitterberg *et al.*, 2010). As the cantilever runs along the sample surface, it moves up and down due to the surface features and the cantilever deflects accordingly. This deflection is usually quantified using an optical sensor, with the laser beam being reflected on the back of the cantilever on to the light detector (Sitterberg *et al.*, 2010). AFM does not require a vacuum and can operate in ambient air or under fully liquid conditions (Liang *et al.*, 2004 a,b; Ruozi *et al.*, 2005, 2009). Although under liquid conditions, particles not fixed to a substrate will float around and eventually stick to the cantilever, which leads to imaging artifacts, both as smearing effects and changes in the cantilever oscillation properties, as the tip gains weight. This smearing effect could be minimized by using a non-contact scanning mode where the tip is not touching the particles but only feels its forces (Balnois *et al.*, 2007). AFM has an outstanding resolution in the order of fractions of a nanometer and can provide a 3D image with details on morphology, size distribution, homogeneity and stability.

4. Encapsulates Analysis by Spectroscopic Techniques

Given the quick and easy sample preparation, a wide range of spectroscopic methods are available and can provide information on the characteristics of encapsulates, including concentration, size and structure. Table 3 summarizes the commonly used spectroscopic techniques for the physicochemical characterization of encapsulates, with highlighted applications, advantages and disadvantages.

4.1 Dynamic Light Scattering

Scattering techniques useful for encapsulates characterization measure the amount of light that a substance scatters at certain wavelengths, incident angles and polarization angles, where the interference patterns can be detected to measure the molecular weight and particle size. Dynamic light scattering (DLS) is the main scattering technique which examines the Brownian motion of the particles in liquid and measures the rate at which the intensity of light scattered from the particles fluctuates as a function of time through a fast photon detector. Brownian motion is random and is related to the particle size. Small particles diffuse faster than larger ones, giving information on the size of particles. The angular scattering intensity data is used to identify the average size and size distribution of encapsulates, typically in the range of 10 nm to 3mm (Gray *et al.*, 2016). Some DLS instruments can also be used for zeta potential measurements by applying an electric field across the dispersion and measuring particle mobility. Sample characteristics, such as aggregation, folding or conformation, can be monitored as functions of various preparations, solvent conditions, temperature or time (Sharma *et al.*, 2019). The main advantages of DLS include its non-invasive manner, short experiment duration, accuracy in determining the hydrodynamic size of monodisperse samples and capabilities of measuring diluted samples, analyzing samples in a wide range of concentrations and detecting small amounts of higher molecular weight species, along with lower apparatus costs and more reproducible measurements than other methods (Brar and Verma, 2011; Filipe *et al.*, 2010; Lim *et al.*, 2013). However, the disadvantages of DLS are based on difficulty in correlating size fractions with a particular composition when certain amounts of aggregates are present (Bootz *et al.*, 2004; Brar and Verma, 2011; Filipe *et al.*, 2010). In addition, DLS has limited utility for analysis of samples with heterogeneous size distributions. DLS is also unsuited to accurately measuring the sizes of non-spherical materials (Bootz *et al.*, 2004; Brar and Verma, 2011; Filipe *et al.*, 2010; Uskokovic, 2012).

Table 3: Overview of commonly Used Spectroscopic Techniques for the Physicochemical Characterization of Encapsulates – Applications, Advantages and Disadvantages

Technique	Observed Properties	Advantages	Disadvantages	References
Dynamic light scattering	Hydrodynamic diameter or distribution	Non-destructive method Rapid and simple analysis Measures in any liquid media Hydrodynamic sizes accurately determined for monodisperse samples	Insensitive correlation of size fractions with a specific composition Influence of small number of large particles Limit in polydisperse sample measures Limited size resolution Assumption of spherical shape samples	Lin *et al.*, 2013 Sapsford *et al.*, 2011
Raman spectroscopy	Size and size distribution Structure Oxidation state	Non-destructive technique Minimal or no sample preparation requirement Widely applicable for various materials in all physical states	Expensive equipment Long processing time Interference of fluorescence	Beattie *et al.*, 2013
Ultraviolet-visible spectroscopy	Size and size distribution Aggregation Mass concentration	Small sample volume required Non-destructive technique	Lack of sensitivity and selectivity Limited to UV-Vis absorbing compounds	Pesika *et al.*, 2003
Fourier transform infrared spectroscopy	Structure and conformation of bioconjugate Surface properties	Non-destructive technique Rapid, reproducible and inexpensive analysis High resolution and high sensitivity Minimal or no sample preparation requirement	Low sensitivity in nanoscale analysis	Zhao *et al.*, 2008 Lin *et al.*, 2013

X-ray photoelectron spectroscopy	Size Shape Elemental composition Oxidation state	Non-destructive technique Surface sensitive technique Quantitative measurements are obtained Provides information about chemical bonding Elemental mapping	High vacuum is required Expensive equipment Long processing time Large area analysis required Low resolution (0.1-1.0 eV)	Schrick et al., 2004 Nurmi et al., 2005 Andrade et al., 1985

4.2   Raman Spectroscopy

Raman Spectroscopy (RS) is a widely used tool for structural characterization of encapsulates based on the phenomenon of light scattering. The principle of RS is to measure the inelastic scattering of photons possessing different frequencies from the incident light after interacting with electric dipoles of the molecule (Lin *et al.*, 2013). The process of RS results in frequency differences between the incident photons and the inelastically-scattered photons associated with the characteristics of the molecular vibrational states, during which the inelastically-scattered photons emitting frequencies lower than the incident photons refer to the Stokes lines in Raman spectrum and the inelastically-scattered photons emitting frequencies higher than the incident photons are named Anti-Stokes lines (Lin *et al.*, 2013). This information can be collected for molecular identification, oxidation state, structure and sizing analysis (Wong *et al.*, 2009).

4.3   Ultraviolet-visible Spectroscopy

Ultraviolet-visible spectroscopy (UV-Vis) is a widely used technique in chemistry for the quantitative determination of highly-conjugated organic compounds, biological macromolecules and transition metal ions. The spectroscopic analysis is mainly carried out in solutions, but solids and gases may be studied as well. UV-Vis spectroscopy refers to the absorption or reflectance spectroscopy in the UV-visible spectral region (Sharma *et al.*, 2019). Absorption of ultraviolet and visible radiation is associated with excitation of electrons, in both atoms and molecules, from lower to higher energy levels. This radiation has a spectral range of approximately around 190-800 nm, which also differs in terms of energy ranges and type of excitation from other related regions. Hence, the electron excitation tendency is directly proportional to the capacity to absorb the wavelength of the light. According to Beer's law, the absorbance is proportional to the concentration of the substance to determine and to the distance of the light when it passes through the sample during the irradiation. This dependency can be influenced by many parameters, such as the characteristics of the spectrophotometer, photo degradation of the molecules, presence of scattering or absorbing interferences in the sample, fluorescent compounds in the sample, interactions between the substance and the solvent and the pH (Passos and Saraiva, 2019).

4.4   Fourier Transform Infrared Spectroscopy

Fourier Transform Infrared Spectroscopy (FTIR) is a rapid, nondestructive and accurate technique for measuring many quality parameters of encapsulates. This technique can detect a range of functional groups and it is sensitive to changes in molecular structure, and therefore provides specific information based on chemical composition and physical state of the whole sample (Cocchi *et al.*, 2004). An infrared spectrum represents a fingerprint of a sample with absorption peaks which correspond to the frequencies of vibrations between the bonds of the atoms making up the material. Each different material represents a unique combination of atoms, and there are no two compounds that produce the exact same infrared spectrum. Therefore, infrared spectroscopy can result in a positive identification, i.e. qualitative analysis of every different kind of materials. In addition, the size of the peaks in the spectrum is a direct indication of the amount of material present (Murali Krisna *et al.*, 2013). FTIR is based on the measurement of the absorption of electromagnetic radiation with wavelengths within the mid-infrared region (4000-400 cm^{-1}). Infrared photons have enough energy to cause groups of atoms to vibrate with respect to the bonds that connect them. Like electronic transitions, these vibrational transitions correspond to distinct energies, and molecules absorb infrared radiation only at certain wavelengths and frequencies. In order to make identification, the measured interferogram signal cannot be interpreted directly. A means of 'decoding' the individual frequencies is required. This can be accomplished via a well-known mathematical technique, called the Fourier transformation. This transformation is performed by the computer which then presents the user with the desired spectral information for analysis.

4.5 X-ray Photoelectron Spectroscopy

X-ray spectroscopy comprises X-ray photoelectron (XPS), X-ray fluorescence (XRF), X-ray absorption spectroscopy (XAS) and X-ray diffraction (XRD). XPS is highly surface-specific quantitative spectroscopic method due to the short range of the photoelectrons that are excited from the solid sample and, therefore, XPS could be useful to characterize nanoparticle surfaces and coatings (Tiede *et al.*, 2008). XPS provides chemical information and elemental composition of the materials. Over the past years, XPS has been extensively applied to a wide variety of fields, expanding from chemistry and materials science, into many other areas associated with environmental and biological systems (Shchukarev and Ramstedt, 2017; Garcia-Bedoya *et al.*, 2017). The principle of XPS is based on irradiation of a material with a beam of X-rays, mostly Al and Mg sources, with simultaneously measuring the kinetic energy and number of electrons that escape from atoms on the surface of the analyzed material. XPS requires high vacuum or ultra-high vacuum conditions, in order to obtain the maximum electron count during the acquisition of spectra because the analyzer is usually one meter away from the X-ray irradiated surface. XPS spectrum and atomic surface composition are obtained according to the intensity of the electron escaping from the surface and the binding energy is calculated from the measured kinetic energy (Korin *et al.*, 2017). The surface composition of the sample can be estimated in terms of pure molecular components, rather than atomic surface composition, by use of different calculation methods (Kappel *et al.*, 2000).

5. Other Methods

Many other methods like zeta potential measurement, centrifugation, chromatography and electrophoresis have been used to analyze encapsulates.

The zeta potential is the electric potential on the shear surface, which is usually determined by measuring the velocity of the charged species towards the electrode in the presence of an external electric field across the sample solution (Pons *et al.*, 2006b; Sapsford *et al.*, 2011). It gives an indication of the potential stability of a colloidal system. The zeta potential with a value of ± 30 mv is generally chosen to infer particle stability, the absolute value greater than 30 mv indicates a stable condition, while a low zeta potential value of less than 30 mv indicates a condition towards instability, coagulation, aggregation, or flocculation (Sapsford *et al.*, 2011). If all the particles have a large negative or positive zeta potential, they will repel each other, leading to higher stability than if the particle charge is nearly neutral. The zeta potential is a measure of the net charge and there may be significant charge heterogeneities that can still lead to aggregation, even though the net zeta potential suggests otherwise. Information about the aggregation state of nanoparticle dispersions is highly valuable for nanoparticle fate and behavior studies (Tiede *et al.*, 2008). Addition of minerals present in food systems could also have a major effect on the zeta potential. For instance, minerals increase the ionic strength of the aqueous phase, which reduces the electrostatic repulsion between droplets through electrostatic screening. Some minerals bind to oppositely-charged groups on the surface of emulsion droplets, decreasing the magnitude of their zeta potential and thereby reducing the electrostatic repulsion between droplets (Hunter, 1986). Among the methods of evaluating zeta potential, the technique of electrophoretic light scattering (ELS), which can simultaneously measure the velocities of many charged particles in liquid, is most commonly used (Doane *et al.*, 2011; Xu, 2008).

Centrifugation is a well-known conventional methodology of separating and purifying mixed materials. Analytical ultracentrifugation (AUC) can be used to investigate the conformation, structure, stoichiometry and self-aggregation state, in addition to determining the size/size distribution, shape and molecular weight (Schaefer *et al.*, 2012).

Chromatographic techniques are rapid, sensitive and non-destructive. Usually chromatographic analyses allow usage of a range of solvents, but samples sometimes cannot be analyzed in their original media, which can cause sample alteration and sample solvent interaction. The best known technique for size separation is size exclusion chromatography (SEC). A size exclusion column, packed with porous beads as the stationary phase, retains in its pores the particles of the analyzed encapsulate, depending on their size and shape. SEC has good separation efficiency, but major disadvantages are possible interactions of the solute with the solid phase (Lead and Wilkinson, 2006) or the limited size separation range of the columns. Even if chromatographic methods, such as HPLC are used for analysis, the mono- and oligo-saccharides must usually be separated from other food components before chromatography.

In capillary electrophoresis (CE), unlike SEC, there are no solid-phase interactions. CE allows the separation of particles in different solutions based on the charge and size distribution of the components. However, mobile-phase interactions cannot be excluded. Electrophoresis (gel electrophoresis (GE) and capillary electrophoresis (CE)) are routinely used methods to partition and purify biomolecules (Sapsford *et al.*, 2011).

Hydrodynamic chromatography (HDC) separates particles based on their hydrodynamic radius. A HDC column is packed with non-porous beads building up flow channels in which particles are separated by flow velocity and the velocity gradient across the particle. HDC columns usually show size separation ranges from 5 to 1200 nm, depending on the column length, whereas the size separation range of a SEC column is dominated by its pore size distribution.

6. Conclusion and Future Perspectives

Encapsulation technology is of growing interest to provide many useful effects in the food sector. The main objectives of this technology are to improve the stability of bioactive compounds against adverse environmental conditions, their incorporation into food matrices conferring functional properties to food products and to enable their controlled release at a specific target of the gastrointestinal tract after food ingestion. Encapsulates must have particular physicochemical properties to ensure that they are suitable for incorporation into particular food products. Moreover, the encapsulates can also impact the end product, potentially influencing its sensory attributes.

As pointed out in this chapter, there is no single available technique able to detect and characterize all the important features of encapsulates used in food at once. In order to achieve quality control of encapsulates at all stages, from their production to the final application, all physicochemical, structural and mechanical properties discussed in this chapter are required to analyze. For the determination of each property multiple methods are available, giving the researcher a range of choices depending on the resolution required, the cost and ease of the experimental procedure. In future, more efforts should be made toward the development and validation of analytical techniques for encapsulates in the chase of a more complete understanding of their properties in order to facilitate progression in industrial applications. Also, the development of new techniques for analysis of encapsulates, that can extend the limits of current ones, be more accurate, more user-friendly, fully automated and of lower cost should be highly desirable.

References

Andrade J.D. (1985). X-ray Photoelectron Spectroscopy (XPS), pp. 105-195. *In:* J.D. Andrade (Ed.), *Surface and Interfacial Aspects of Biomedical Polymers*, Springer, Boston.

Balnois, E., Papastavrou, G. and Wilkinson, K.J. (2007). Force microscopy and force measurements of environmental colloids, pp. 405-468. *In:* K.J. Wilkinson and J.R. Lead (Eds.), *Environmental Colloids and Particles: Behavior, Structure and Characterization*, Wiley, Chichester.

Barbosa-Canovas, G.V., Ortega-Rivas, E., Juliano, P. and Yan, H. (200). *Food Powders: Physical Properties, Processing, and Functionality*, Springer Nature, Switzerland.

Beattie, J.R., McGarvey, J.J. and Stitt, A.W. (2013). Raman spectroscopy for the detection of AGEs/ALEs, *Methods in Molecular Biology*, 965: 297-312

Bibi, S., Kaur, R., Henriksen-Lacey, M., McNeil, S.E., Wilkhu, J., Lattmann E., Chhristensen, D., Mohammed, A.R. and Perrie, Y. (2011). Microscopy imaging of liposomes: From coverslips to environmental SEM, *International Journal of Pharmaceutics*, 417: 138-150.

Binnig, G. and Rohrer, H. (1983). Scanning tunneling microscopy, *Surface Science*, 126: 236-244.

Bonnell, D. (2001). *Scanning Probemicroscopy and Spectroscopy: Theory, Techniques, and Applications*, Wiley-Vch, New York.

Bootz, A., Vogel, V., Schubert, D. and Kreuter, J. (2004). Comparison of scanning electron microscopy, dynamic light scattering and analytical ultracentrifugation for the sizing of poly (butyl cyanoacrylate) nanoparticles, *European Journal of Pharmaceutics and Biopharmaceutics*, 57: 369-375.

Bouvrais, H., Pott, T., Bagatolli, L.A., Ipsen, J.H. and Meleard, P. (2010). Impact of membrane-anchored fluorescent probes on the mechanical properties of lipid bilayers, *Biochimimica and Biophysica Acta*, 1798: 1333-1337.

Brar, S.K. and Verma, M. (2011). Measurement of nanoparticles by light-scattering techniques, *TrAC Trends in Analytical Chemistry*, 30: 4-17.

Castaneda, L., Valle, J., Yang, N., Pluskat, S. and Slowinska, K. (2008). Collagen cross-linking with Au nanoparticles, *Biomacromolecules*, 9: 3383-3388.

Celli, G.B., Ghanem, A. and Brooks, M.S. (2015). Bioactive encapsulated powders for functional foods – A review of methods and current limitations, *Food Bioprocess Technology*, 8: 1825-1837.

Chan, L.W., Tan, L.H. and Heng, P.W.S. (2008). Process analytical technology: Application to particle sizing in spray drying, *AAPS PharmSciTech*, 9: 259-266.

Cocchi, M., Foca, G., Lucisano, M., Marchetti, A., Paean, M.A., Tassi, L. and Ulrici, A. (2004). Classification of cereal flours by chemometrie analysis of MIR spectra, *Journal of Agriculture and Food Chemistry*, 52: 1062-1067.

Cuq, B., Rondet, E. and Abecassis, J. (2011). Food powders engineering, between knowhow and science: Constraints, stakes and opportunities, *Powder Technology*, 208: 244-251.

Doane, T.L., Chuang, C.H., Hill, R.J. and Burda, C. (2011). Nanoparticle – Potentials, *Accounts of Chemical Research*, 45: 317-326.

Domingos, R.F., Baalousha, M.A., Ju-Nam, Y., Reid, M.M., Tufenkji, N., Lead, J.R., Leppard. G.G. and Wilkinson, K.J. (2009). Characterizing manufactured nanoparticles in the environment: Multimethod determination of particle sizes, *Environmental Science and Technology*, 43: 7277-7284.

El-Messery, T., El-Said, M.M. and Farahat, E.S.A. (2019). Production of functional processed cheese supplemented with nanoliposomes of mandarin peel extract, *Pakistan Journal of Biological Sciences*, 22: 247-256.

Filipe, V., Hawe, A. and Jiskoot, W. (2010) Critical evaluation of nanoparticle tracking analysis (NTA) by nanosight for the measurement of nanoparticles and protein aggregates, *Pharmaceutical Research*, 27: 796-810.

Fitzpatrick, J.J., Hodnett, M., Twomey, M., Cerqueira, P.S.M., O'Flynn, J. and Roos, Y.H. (2007). Glass transition and the flowability and caking of powders containing amorphous lactose, *Powder Technology*, 178: 119-128.

Fitzpatrick, J.J. and Ahrne, L. (2005). Food powder handling and processing: Industry problems, knowledge barriers and research opportunities, *Chemical Engineering and Processing*, 44: 209-214.

Fleming, C.J., Liu, Y.X., Deng, Z. and Liu, G.Y. (2009). Deformation and hyperfine structures of dendrimers investigated by scanning tunneling microscopy, *The Journal of Physical Chemistry, A*, 113: 4168-4174.

Garcia-Bedoya, D., Ramirez-Rodriguez, L.P., Mendivil-Reynoso, T., Quiroz-Castillo, J.M., De la Mora-Covarrubias, A. and Castillo, S.J. (2017). Direct XPS analysis of biological materials for environmental purposes, *Applied Ecology and Environmental Research*, 15: 501-509.

Goula, A.M. and Adamopoulos, K.G. (2010). A new technique for spray drying orange juice concentrate, *Innovative Food Science & Emerging Technologies*, 11: 342-351.

Gray, A., Egan, S., Bakalis, S. and Zhang, Z. (2016). Determination of microcapsule physicochemical, structural, and mechanical properties, Particuology, 24: 32-43.

Gultekin-Ozguven, M. Karadag, A., Duman, S., Ozkal, B. and Ozcelik, B. (2016). Fortification of dark chocolate with spray dried black mulberry (*Morus nigra*) waste extract encapsulated in chitosan-coated liposomes and bioaccessability studies, *Food Chemstry*, 201: 205-212.

Hall, J.B., Dobrovolskaia, M.A., Patri, A.K. and McNeil, S.E. (2007). Characterization of nanoparticles for therapeutics, *Nanomedicine*, 2: 789-803.

Hidalgo, A., Brandolini, A., Čanadanović-Brunet, J., Ćetković, G. and Tumbas Šaponjac, V. (2018). Microencapsulates and extracts from red beetroot pomace modifyantioxidant capacity, heat damage and color of pseudocereals-enriched einkorn water biscuits, *Food Chemistry*, 268: 40-48.

Huckabay, H.A., Armendariz, K.P., Newhart, W.H., Wildgen, S.M. and Dunn, R.C. (2013). Near-field scanning optical microscopy for high-resolution membrane studies, *Methods in Molecular Biology*, 950: 373-394.

Hunter, R.J. (1986). *Foundations of Colloid Science*, vol. 1. Oxford University Press, Oxford.

Kappel, G., Kapsammer, E., Rausch-Schott, S. and Retschitzegger, W. (2000). X-Ray – Towards integrating XML and relational database systems, pp. 339-353. *In:* A.H.F. Laender, S.W. Liddle, V.C. Storey (Eds.), *ER2000 Conference*, Springer-Verlag, Berlin, Heidelberg.

Korin, E., Froumin, N. and Cohen, S. (2017). Surface analysis of nanocomplexes by X-ray photoelectron spectroscopy (XPS), *ACS Biomaterials Science and Engineering*, 3: 6.

Kuyper, C.L., Budzinski, K.L., Lorenz, R.M. and Chiu, D.T. (2006a). Real-time sizing of nanoparticles in microfluidic channels using confocal correlation spectroscopy, *The Journal of the American Chemical Society*, 128: 730-731.

Kuyper, C.L., Fujimoto, B.S., Zhao, Y., Schiro, P.G. and Chiu, D.T. (2006b). Accurate sizing of nanoparticles using confocal correlation spectroscopy, *The Journal of Physical Chemistry, B*, 110: 24433-24441.

Kwapinska, M. and Zbicinski, I. (2005). Prediction of final product properties after cocurrent spray drying, *Drying Technology*, 23: 1653-1665. Lead, J.R. and Wilkinson, K.J. (2006). Aquatic colloids and nanoparticles: Current knowledge and future trends, *Environmental Chemistry*, 3: 159-171.

Li, Y., Lee, J., Lal, J., An, L. and Huang, Q. (2008). Effects of pH on the interactions and conformation of bovine serum albumin: Comparison between chemical force microscopy and small-angle neutron scattering, *Journal of Physical Chemistry, B*, 112: 3797-3806.

Liang, X., Mao, G. and Simon Ng, K.Y. (2004a). Mechanical properties and stability measurement of cholesterol-containing liposome on mica by atomic force microscopy, *Journal of Colloid and Interface Science*, 278: 53-62.

Liang, X., Mao, G. and Simon Ng, K.Y. (2004b). Probing small unilamellar EggPC vesicles on mica surface by atomic force microscopy, *Colloids and Surfaces B: Biointerfaces*, 34: 41-51.

Lim, J., Yeap, S.P., Che, H.X. and Low, S.C. (2013). Characterization of magnetic nanoparticle by dynamic light scattering, *Nanoscale Research Letters*, 8: 381.

Lin, P.C., Lin, S., Wang, P.C. and Sridhar, R. (2013). Techniques for physicochemical characterization of nanomaterials, *Biotechnology Advances*, 32: 711-726.

Liu, J., Chao, J., Liu, R., Tan, Z., Yin, Y., Wu, Y. and Jiang, G. (2009). Cloud point extraction as an advantageous preconcentration approach for analysis of trace silver nanoparticles in environmental waters, *Analytical Chemistry*, 81: 6496-6502.

Lončarević, I., Pajin, B., Fišteš, A., Tumbas Šaponjac, V., Petrović, J., Jovanović, P., Vulić, J. and Zarić, D. (2018). Enrichment of white chocolate with blackberry juice encapsulate: Impact on physical properties, sensory characteristics and polyphenol content, *LWT – Food Science and Technology*, 92: 458-464.

Lončarević, I., Pajin, B., Tumbas Šaponjac, V., Petrović, J., Vulić, J., Fišteš, A. and Jovanović, P. (2019). Physical, sensorial and bioactive characteristics of white chocolate with encapsulated green tea extract, *Journal of the Science of Food and Agriculture*, 99: 5834-5841.

Mavrocordatos, D., Pronk, W. and Boller, M. (2004). Analysis of environmental particles by atomic force microscopy, scanning and transmission electron microscopy, *Water Science and Technology*, 50: 9-18.

Maynard, A.D. (2000). Overview of methods for analyzing single ultrafine particles, *Philosophical Transactions of the Royal Society, A*, 358: 2593-2609.

Mittal, V. (2013). *Encapsulation Nanotechnologies*, Wiley Publishers, NY, USA.

Mohammed, A.R., Weston, N., Coombes, A.G., Fitzgerald, M. and Perrie, Y. (2004). Liposome formulation of poorly water soluble drugs: Optimization of drug loading and ESEM analysis of stability, *International Journal of Pharmaceutics*, 285: 23-34.

Moran, L.L., Yin, Y., Cadwallader, K.R. and Padua, G.W. (2014). Testing tools and physical, chemical, and microbiological characterization of microencapsulated systems, pp. 323-352. *In:* A.G. Gaonkar, N. Vasisht, A.R. Khare, R. Sobel (Eds.), *Microencapsulation in the Food Industry: A Practical Implementation Guide*, Elsevier, USA.

Murali Krisna, G., Muthukumaran, M., Krishnamoorthy, B. and Nishat, A. (2013). A critical review on fundamental and pharmaceutical analysis of FT-IR spectroscopy, *International Journal of Pharmacy*, 3: 396-402.

Murphy, D.B. and Davidson M.W. (2012). *Fundamentals of Light Microscopy and Electronic Imaging*, John Wiley & Sons, Hoboken, New York.

Muscariello, L., Rosso, F., Marino, G., Giordano, A., Barbarisi, M., Cafiero, G. and Barbarisi, A. (2005). A critical overview of ESEM applications in the biological field, *Journal of Cellular Physiology*, 205: 328-334.

Nedović, V., Kalušević, A., Manojlović, V., Petrović, T. and Bugarski, B. (2013). Encapsulation systems in the food industry, pp. 229-253. *In:* S. Yannotis, P. Taoukis, N.G. Stoforos and V. Karathanos (Eds.), *Advances in Food Process Engineering Research and Applications*, Springer, New York.

Nurmi, J.T., Tratnyek, P.G., Sarathy, V., Baer, D.R., Amonette, J.E., Pecher, K., Wang, C.M., Linehan, J.C., Matson, D.W., Penn, R.L. and Driessen, M.D. (2005). Characterization and properties of metallic iron nanoparticles: Spectroscopy, electrochemistry, and kinetics, *Environomental Science and Technology*, 39: 1221-1230.

Obare, S.O. and Meyer, G.J. (2004). Nanostructured materials for environmental remediation of organic contaminants in water, *Journal of Environmental Science and Health Part A – Toxic/Hazardous Substances & Environmental Engineering*, 39: 2549-2582.

Ozvural, E.B. and Huang, Q. (2017). Quality differences of hamburger patties incorporated with encapsulated β carotene both as an additive and edible coating, *Journal of Food Processing and Preservation*, 42: e13353.

Passos, M.L.C. and Saraiva, M.L. (2019). Detection in UV-visible spectrophotometry: Detectors, detection systems, and detection strategies, *Measurement*, 135: 896-904.

Pesika, N.S., Stebe, K.J. and Searson, P.C. (2003). Relationship between absorbance spectra and particle size distributions for quantum-sized nanocrystals, *Journal of Physical Chemistry, B*, 107: 10412-10415.

Pillai, D.S., Prabhasankar, P., Jena, B.S. and Anandharamakrishnan, C. (2012). Microencapsulation of Garcinia cowa fruit extract and effect of its use on pasta process and quality, *International Journal of Food Properties*, 15: 590-604.

Ponder, S.M., Darab, J.G., Bucher, J., Caulder, D., Craig, I., Davis, L., Edelstein, N., Lukens, W., Nitsche, H., Rao, L.F., Shuh, D.K. and Mallouk, T.E. (2001). Surface chemistry and electrochemistry of supported zerovalent iron nanoparticles in the remediation of aqueous metal contaminants, *Chemistry of Materials*, 13: 479-486.

Pons, T., Uyeda, H.T., Medintz, I.L. and Mattoussi, H. (2006b). Hydrodynamic dimensions, electrophoretic mobility, and stability of hydrophilic quantum dots, *Journal of Physical Chemistry B*, 110: 20308-20316.

Ramos, B.G.Z. (2011). Biopolymers Employed in Drug Delivery, pp. 559-573. *In: Biopolymers: Biomedical and Environmental Applications*, Scrivener Publishing, Salem.

Rathbone, M.J. and Butler, J.M. (2011). In vitro testing of controlled release dosage forms during development and manufacture, pp. 91-108. *In:* C.G. Wilson and P.J. Crowley (Eds.), *Controlled Release in Oral Drug Delivery*, Springer: New York.

Rhine, W.D., Hsieh, D.S. and Langer, R. (1980). Polymers for sustained macromolecule release: Procedures to fabricate reproducible delivery systems and control release kinetics, *Journal of Pharmaceutical Sciences*, 69: 265-270.

Robert, P., Gorena, T., Romero, N., Sepulveda, E., Chavez, J. and Saenz, C. (2010). Encapsulation of polyphenols and anthocyanins from pomegranate (*Punica granatum*) by spray drying, *Internatonal Journal of Food Science and Technology*, 45: 1386-1394.

Robson, A.L., Dastoor, P.C., Flynn, J., Palmer, W., Martin, A., Smith, D.W., Woldu, A. and Hua, S. (2018). Advantages and limitations of current imaging techniques for characterizing liposome morphology, *Frontiers in Pharmacology*, 9: 80.

Rocha, G.A., Trindade, M.A., Netto, F.M. and Favaro-Trindade, C.S. (2009). Microcapsules of a casein hydrolysate: Production, characterization, and application in protein bars, *Food Science and Technology*, 15: 407-413.

Ruozi, B., Belletti, D., Tombesi, A., Tosi, G., Bondioli, L., Forni F. and Vandelli, M.A. (2011). AFM, ESEM, TEM, and CLSM in liposomal characterization: A comparative study, *International Journal of Nanomedicine*, 6: 557-563.

Ruozi, B., Tosi, G., Forni, F., Fresta, M. and Vandelli, M.A. (2005). Atomic force microscopy and photon correlation spectroscopy: Two techniques for rapid characterization of liposomes, *European Journal of Pharmaceutical Science*, 25: 81-89.

Ruozi, B., Tosi, G., Tonelli, M., Bondioli, L., Mucci, A., Forni F. and Vandelli, M.A. (2009). AFM phase imaging of soft-hydrated samples: A versatile tool to complete the chemical-physical study of liposomes, *Journal of Liposome Research*, 19: 59-67.

Rutz, J.K., Borges, C.D., Zambazi, R.C., Crizel-Cardozo, M.M., Kuck, L.S. and Norena, C.P.Z. (2017). Microencapsulation of palm oil by complex coacervation for application in food systems, *Food Chemistry*, 220: 59-66.

Sapsford, K.E., Tyner, K.M., Dair, B.J., Deschamps, J.R. and Medintz, I.L. (2011). Analyzing nanomaterial bioconjugates: A review of current and emerging purification and characterization techniques, *Analytical Chemistry*, 83: 4453-4488.

Schaefer, J., Schulze, C., Marxer, E.E., Schaefer, U.F., Wohlleben, W., Bakowsky, U. and Lehr, C.M. (2012). Atomic force microscopy and analytical ultracentrifugation for probing nanomaterial protein interactions, *ACS Nano*, 6: 4603-4614.

Schrick, B., Hydutsky, B.W., Blough, J.L. and Mallouk, T.E. (2004). Delivery vehicles for zerovalent metal nanoparticles in soil and groundwater, *Chemistry of Materials*, 16: 2187-2193.

Šeregelj, V., Tumbas Šaponjac, V., Lević, S., Kalušević, A., Ćetković, G., Čanadanović-Brunet, J., Nedović, V., Stajčić, S., Vulić, J. and Vidaković, A. (2019). Application of encapsulated natural bioactive compounds from red pepper waste in yogurt, *Journal of Microencapsulaton*, DOI: 10.1080/02652048.2019.1668488

Sharma, S., Jaiswal, S., Duffy, B. and Jaiswal, A.K. (2019). Nanostructured materials for food applications: Spectroscopy, microscopy and physical properties, *Bioengineering*, 6: 26.

Shchukarev, A. and Ramstedt, M. (2016). Cryo-XPS: Probing intact interfaces in nature and life: Cryo-XPS allows studies of intact solid-aqueous solution interfaces, *Surface and Interface Analysis*, 49: 4.

Shegokar, R. and Muller, R.H. (2010). Nanocrystals – Industrially feasible multifunctional formulation technology for poorly soluble actives, *International Journal of Pharmaceutics*, 399: 129-139.

Sitterberg, J., Ozcetin, A., Ehrhardt, C. and Bakowsky, U. (2010). Utilising atomic force microscopy for the characterisation of nanoscale drug delivery systems, *European Journal of Pharmaceutics and Biopharmaceutics*, 74: 2-13.

Soottitantawat, A., Yoshii, H., Furuta, T., Ohkawara, M., Forssell, P., Partanen, R., Poutanen, K. and Linko, P. (2004). Effect of water activity on the release characteristics and oxidative stability of D-limonene encapsulated by spray drying, *Journal of Agricultural and Food Chemistry*, 52: 1269-1276.

Suzuki, E. (2002). High-resolution scanning electron microscopy of immunogold-labelled cells by the use of thin plasma coating of osmium, *Journal of Microscopy*, 208: 153-157.

Telis, V.R.N. and Martínez-Navarrete, N. (2009). Collapse and color changes in grapefruit juice powder as affected by water activity, glass transition, and addition of carbohydrate polymers, *Food Biophysics*, 4: 83-93.

Thamaket, P. and Raviyan, P. (2015). Preparation and physical properties of carotenoids encapsulated in chitosan cross-linked tripolyphosphate nanoparticles, *Food and Applied Bioscience Journal*, 3: 69-84.

Tiede, K., Boxall, A.B.A., Tear, S.P., Lewis, J., David, H. and Hassellöv, M. (2008). Detection and characterization of engineered nanoparticles in food and the environment, *Food Additives and Contaminants*, 25: 795-821.

Tumbas Šaponjac, V., Ćetković, G., Čanadanović-Brunet, J., Pajin, B., Đilas, S., Petrović, J., Lončarević, I., Stajčić, S. and Vulić, J. (2016). Sour cherry pomace extract encapsulated in whey and soy proteins: Incorporation in cookies, *Food Chemistry*, 207: 27-33.

Uskokovic, V. (2012). Dynamic light scattering based microelectrophoresis: Main prospects and limitations, *Journal of Dispersion Science and Technology*, 33: 1762-1786.

Van Beers, M.M.C., Slooten, C., Meulenaar, J., Sediq, A.S., Varrijk, R. and Jiskoot, W. (2017). Micro-flow imaging as a quantitative tool to assess size and agglomeration of PLGA microparticles, *European Journal of Pharmaceutics and Biopharmaceutics*, 117: 91-104.

Wang, H. and Chu, P.K. (2013). Surface characterization of biomaterials, pp. 105-174. *In:* B. Amit and B. Susmita (Eds.), *Characterization of Biomaterials*, Oxford: Academic Press.

Wang, R., Tian, Z. and Chen, L. (2011). A novel process for microencapsulation of fish oil with barley protein, *Food Research International*, 44: 2735-2741.

Wiesendanger, R. (1994). *Scanning Probe Microscopy and Spectroscopy: Methods and Applications*, Cambridge University Press, Cambridge.

Wilkins, R.M. (1999). Controlled-release granules, with emphasis on lignin-based methods. *In: Controlled Release Delivery Systems for Pesticides*, CRC Press: Boca Raton, FL.

Williams, D.B. and Carter, C.B. (2016). The transmission electron microscope, *Transmission Electron Microscopy*, Springer, Switzerland.

Wise, D.L. (2000). *Handbook of Pharmaceutical Controlled Release Technology*, CRC Press: Boca Raton, FL.

Wong, H.W., Choi, S.M., Phillips, D.L. and Ma, C.Y. (2009). RAMAN spectroscopic study of deamidated food proteins, *Food Chemistry*, 113: 363-370.

Xu, R. (2008). Progress in nanoparticles characterization: Sizing and zeta potential measurement, *Particuology*, 6: 112-115.

Yamashita, C., Chung, M.M.S., dos Santos, C., Mayer, C.R.M., Moraes, I.C.F. and Branco, I. (2017). Microencapsulation of an anthocyanin-rich blackberry (*Rubus* spp.) by-product extract by freeze-drying, *LWT – Food Science and Technology*, 84: 256-262.

Yang, Y.Y., Chung, T.S. and Ping, N.N. (2001). Morphology, drug distribution and *in vitro* release profiles of biodegradable polymeric microspheres containing protein fabricated by double-emulsion solvent extraction/evaporation method, *Biomaterial*, 22: 231-241.

Yoe, Y. and Park, K. (2004). A new microencapsulation method using an ultrasonic atomizer based on interfacial solvent exchange, *Journal of Controlled Release*, 100: 379-388.

Zhang, Z., Law, D. and Lian, G. (2010). Characterization methods of encapsulates, pp. 101-125. *In:* N.J. Zuidam and V. Nedović (Eds.), *Encapsulation Technologies for Active Food Ingredients and Food Processing*, Springer, New York.

Zhao, X., Yang, Z., Gai, G. and Yang, Y. (2009). Effect of superfine grinding on properties of ginger powder, *Journal of Food Engineering*, 91: 217-222.

Controlled Release Systems for Food Application

Verica Djordjević[1]*, Bojana Balanč[1], Ana Kalušević[2,3], Mina Volić[1], Nataša Obradović[1], Predrag Petrović[1] and Branko Bugarski[1]

[1] Faculty of Technology and Metallurgy, University of Belgrade, Karnegijeva 4, 11 060 Belgrade, Serbia

[2] Academy for Applied Studies, Zorana Đinđića 152a, 11070 Belgrade, Serbia

[3] Serbian Food Technology Council, Starci bb, Aleksandrovac, Serbia

1. Introduction

Micro- and nano-particulate encapsulating systems have the potential to play a key role in the future of delivery of nutraceuticals which is an essential characteristic of functional foods. For this purpose, different types of carriers are being constantly developed, such as core-shell structures or host-guest structures, nanospheres, capsules, hydrogels, micelles and nanoparticles. Such systems are designed with an aim of increasing flavor retention, stabilizing food ingredients and preserving functional properties, increasing nutrients solubility and/or bioavailability, masking strong flavors above certain concentrations of functional active ingredients (e.g. green tea) or unpleasant tastes (e.g. polyphenols), increasing viability (number of cells) and functionality ((glyco)proteins in the cell envelope have to be preserved) of probiotic cells during passage in the gastrointestinal tract). In particular, there is a strong demand for delivery systems that can target or control the release of bioactives at specific sites within the gastrointestinal tract (GIT) (such as mouth, stomach, small intestine or colon) because this may improve their efficacy, as well as reduce any undesirable side effects. Examples of those that should be freed in oral region are encapsulated oil-soluble flavors and nutraceuticals, while the majority of antioxidants (plant polyphenols, phytosterols, vitamins and carotenoids) are envisaged to be released in the intestinal region. In certain cases, bioactives (peptide, probiotic) would preferably be released in the stomach, for example, to fight *Helicobacter pylori* infections; but in most cases of getting probiotics from supplements and food, the aim is to preserve them upon reaching the final section of the digestive system, i.e. the colon.

---

*Corresponding author: vmanojlovic@tmf.bg.ac.rs

Controlled release systems have been growing also in food packaging due to the increase of the concept of 'active packaging'. The particular aim of controlled release systems intended for food packaging applications is to transfer the antimicrobial agent from the polymeric carrier to the food with the aim of maintaining a predetermined concentration of the antimicrobial agent in the packed food for a determined period of time. The final goal is to control or even prevent the growth of undesired bacteria responsible for the packed foodstuff degradation.

For the potential mentioned above to be realized, it is important to study the release characteristics of the encapsulated material from the particle matrix as a function of particle size, material properties and processing conditions. A huge number of scientific publications is dedicated on the studying and modeling of swelling and dissolution behavior of drug-delivery devices mostly for pharmaceutical applications. The results obtained are being transferred to the field of food technology, so that controlled released systems intended for delivery of active food compounds are currently one of the topics of nearly all journals on food science.

2. Factors Affecting Release Kinetics

The release profiles can be influenced by the properties of the polymer, i.e. molecular weight and concentration, the nature of the loaded active component, its concentration and distribution and the particle size distribution. These properties in turn may be, to a higher or less extent, controlled by the processing operating parameters during production of particles. Hydrogels, frequently used in drug delivery and controlled release, can be designed such to have desired porosity (through the combination of polymer volume fraction and cross-link density) and/or hydrogel-active affinity. In food, natural (and modified) biopolymers, such as polysaccharides (e.g. alginate, agar, agarose) and proteins (e.g. gelatin) are used to design the properties of the hydrogel network.

In general, smaller particles typically give rise to higher release rates due to the higher surface area to volume ratio and hence more rapid penetration by water. This rule was confirmed by Eltayeb *et al.* (2015) with the presented release of ethylvanillin from ethylcellulose nanoparticles with the mean particle size varied between 45 and 85 nm.

Nature of the loaded active component is another factor that influences release process. The solubility of the bioactive ingredients determines release mechanism from a matrix system. Release kinetics of hydrophilic compounds is determined by a combination of diffusion and erosion mechanisms while lipophilic compounds often have lower dissolution rates by an erosion mechanism (Kuang *et al.*, 2010; Ezhilarasi *et al.*, 2013). Moreover, lipophilic compounds are often incompletely released in food due to poor solubility. On the other hand, lipophilic compounds are favorable permeable through the intestine via active transport mechanism and accelerated diffusion, whereas hydrophilic compounds have much lower permeability through the intestine and become absorbed only by active transport mechanism.

3. Stimuli-controlled Release of Food Compounds

In the last decade, a special attention has been paid to the design of stimuli-responsive systems which are able to deliver an active molecule upon pH variation, light irradiation, temperature change, variation of the redox potential and magnitude of a magnetic field. These systems have been intended for pharmaceutical, agriculture, food, cosmetics, dyes, chemistry and related applications (Bruneau *et al.*, 2019).

3.1 Release Triggered by Temperature Change

Stimulating the release by inducing a temperature variation is possible to achieve by using a temperature-sensitive polymeric carrier. Such carriers have been made by mixing and modifying of some polyacrylamides and polyesters, among which poly(N-isopropylacrylamide) has been most frequently used. Zhang *et al.* (2009) encapsulated trehalose (bioprotectant) in a thermally responsive core-shell structured polymeric hydrogel nanocapsule made of Pluronic F127 and polyethylenimine. The nanocapsules were capable to withhold trehalose with negligible release in hours for cellular uptake at 37°C when its wall permeability is low; a quick release of the encapsulated sugar can be achieved by thermally cycling the nanocapsule between 37 and 22°C (or lower). Certain biopolymers undergo a conformational change, helix-coil or globular-coil, in response to an alteration in external temperature. For example, gelatin molecules tend to form helical structures at low temperatures that bond each other via hydrogen bondings and create hydrogel network, but when they are heated above their melting temperature (around 20-30°C), they undergo a helix-to-coil transition that causes the network to breakdown and release of embedded molecules (Zhang *et al.*, 2015a).

3.2 Release Triggered by pH Change

Delivery of one or several active molecule(s) through a change of the pH may occur due to (1) swelling of a carrier material induced by hydration phenomena and triggered by pH of the medium, (2) degradation or dissolving of a carrier material or due to (3) a change in solubility of the active compound. The drawback linked to this kind of systems is that they release the occluded molecules at higher or lower extent, whatever the pH is, because of the impossibility to totally prevent releasing. By applying pH-sensitive polymers that stay intact in the stomach (acidic environment) and then open and release bioactive molecules in specific parts of either the small intestine or large intestine, it is possible to control release in GI tract. If release in stomach is preferred, then usage of polyorthoesters can be proposed, since they are stable at higher pH (alkaline), while they disintegrate at acidic pH. Herein, it should be stressed that not only pH change, but digestive enzymes play important role in releasing process.

Different polymeric organic and inorganic materials expressed pH-responsive behavior. Barat *et al.* (2011) used mesoporous silica (MCM-41) as a carrier to perform the controlled release of folic acid. The release was low in acid environment (15-20 per cent) at pH=2, while at pH=7.5 more than 90 per cent was released. Moreover, the addition of 3-[2-(2-aminoethylamino)ethylamino] propyl-trimethoxysilane appeared to form a pH-sensitive barrier which further delayed the releasing process. Alginate and chitosan can also be accepted as carriers with pH-controlled released properties. Folic acid was releasing from nanolaminated alginate/chitosan films for 400 minutes at pH 7 reaching 100 per cent *versus* 20 per cent at pH 2 (Acevedo-Fani *et al.*, 2018). Hydrogel particles made of oppositely charged biopolymers (anionic polysaccharide and cationic protein) are held together by electrostatic attraction between them. As an example, Zhang *et al.* (2015b) have developed hydrogel particles containing an oil by electrostatic complexation of a protein (casein) and an anionic polysaccharide (alginate). Relatively small hydrogel particles were stable at pH values 4 to 5 (around isoelectric point of the protein) but aggregated or dissociated at lower or higher pH values, respectively; as a consequence, lipid droplets were released from hydrogel particles by a pH change. Protein nanohydrogels, produced by interaction between lactoferrin (multifunctional glycoprotein) and glycomacropeptide (acid glycosylated peptide naturally present in bovine milk) also demonstrated pH-dependent release of a lipophilic compound (curcumin): at pH 2 there was a clear release profile of curcumin, while at pH 7 curcumin was not released at all (Bourbon *et al.*, 2016). Manatunga *et al.*, (2017) designed iron oxide nanoparticles covered by a bi-layer of sodium alginate and hydroxyapatite for controlled delivery of food flavoring compounds which possess anti-cancer properties (curcumin and 6-gingerol). At pH=7.4 about 40-50 per cent of the encapsulated active compound was released after 180h, while at pH=5.3 the flavoring compounds released totally during the same time period.

3.3 Release Triggered by Multiple Stimuli-responsive

The multiple stimuli-responsive systems have promising potential for precisely controlled release of encapsulated molecules. Jiang *et al.* (2016) synthesized photo, temperature and pH responsive copolymer nanoparticles by quaternization between the spiropyran derivative (SPN-Cl) and dimethylaminoethyl units of poly(dimethylaminoethyl methacrylate); the release of coumarin 102 (a photosensitive molecule) was controlled by UV light irradiation (at 356 nm), pH stimulation (at pH 5 and pH 9), heating (at 60°C) separately, and synergistically. This superior behavior was explained by heterogenous morphology of the nanoparticles with the existence of spiropyran-quaternized segments and spiropyran-unquaternized segments.

4. Release in Gastrointestinal Tract

4.1 Models to Study Release in GIT

Release studies under simulated digestion typically include the oral, gastric and small intestinal phases, and occasionally large intestinal fermentation. Each phase is unique and mimics a region in GI tract and physiological conditions *in vivo*. This is mainly achieved by the addition of digestive enzymes into the medium where release form encapsulates occurs and by optimization of their concentrations, pH, the composition of surface-active components, concentration of salts, temperature, etc. Information about behavior of delivery systems in different GI regions and in different surrounding conditions is crucial for their application in food products. Understanding the factors that stimulate the release of actives is beneficial for the food industry in order to design foods that could increase, decrease, or control actives digestion within the GI tract. One can discover a vast amount of literature on delivery properties of various encapsulated systems; however, variations are present in literature when comparing results of release studies, i.e. release behavior and kinetics even for the same delivery system. This slows down realization of general outcomes. One of the reasons is that *in vitro* methods simulating digestion processes may differ a lot from one paper to another, in terms of source of enzymes (porcine, rabbit or human origin) used for the experiment, their activity and characterization, pH value, mineral type, ionic strength and digestion time, which alter enzyme activity and other phenomena. This fact inspired a large group of scientists who united together in the COST action INFOGEST22, an international network, with an aim to consolidate conditions for a digestion model, give 'recommendation and justification' for certain parameters, protocols and chemicals. The final outcome was the standardized model recommended to the scientific community (Minekus *et al.*, 2014) and validated by these scientists (Egger *et al.*, 2016). However, the majority of models reported in literature, including this one, are static ones, which means that they consider constant ratios of meal to enzymes, salt, bile acids, etc. at each step of digestion. Furthermore, the peristaltic contractions and antral grinding generated by the stomach induce mechanical forces which are important for digestion and could affect release kinetics, but they are neglected in the static digestion models currently used. As a consequence, there is complete lack of data on the impact of mechanical forces of gastro-intestinal motions on release behavior of the encapsulating systems. Only few computerized sophisticated models have been developed which allow the simulation of dynamic aspects of digestion, such as transport of digested food, variable enzyme concentrations and pH changes over time (Wickham *et al.*, 2012; Minekus *et al.*, 2014).

The mouth is the first region of the upper part of GIT and digestion model should include simulation of this part if food/delivery system of interest contains starch. To simulate this region, the food or delivery system has to be mixed with a complex fluid, simulated salivary fluid, which mimics the fluid present in the oral cavity – saliva. Usually saliva has neutral pH or it can vary between 5 and 7. It contains carbohydrate-degrading enzymes (α-amylase), salts (usually KCl, KH_2PO_4, $NaHCO_3$, $MgCl_2$, $(NH_4)_2CO_3$), buffers and polymers (mucin) (McClements, 2015; Zhang *et al.*, 2015b). The movements of food in the mouth can be mimicked by using a variety of commercial mincers which apply the mechanical forces typical for chewing (Corstens *et al.*, 2017). The retention time of food in mouth is between 5 s and 10 min, depending on food type or delivery system (solid or liquid) (Mandalari *et al.*, 2013; Zhang *et al.*, 2015b).

Further, the food goes in the gastric chamber where it is exposed to another set of complex environmental conditions. Gastric fluid pH may vary drastically between individuals and within a person over time from pH 1.3 to pH 4, which is regulated by secretion of hydrochloric acid. Minerals, surface active components (phospholipids, bile salts, proteins, surfactants) and enzymes (lipases and proteases) are also present there (McClements, 2015). The residence time of food in the gastric chamber is between 30 min and 4 h.

From gastric chamber food enters the small intestine mainly where the absorption of bioactive compounds occurs. Pancreatic enzymes (a complex mixture of proteases, amylases and lipases) and other digestive enzymes produced by the small intestine act together with bile salts, bicarbonate and phospholipids to degrade food. It is important to state that GIT ingredients are not only useful for breakdown of particles but also assist in solubilization of the fat in the water phase. Actually, bile salts are emulsifiers and triglycerides become hydrolyzed to free fatty acids by lipases; in turn, free fatty acids are water soluble at a pH of intestinal region (6 to 8.1) when the free carboxylic end of the fatty acid chain is ionized ($COO-$). The common period of time that particles spend in the small intestine is 1 to 2 h, sometimes longer (4h).

Undigested parts (such as indigestible dietary fibers or lipids) further go to the colon. A varied population of bacteria are present in the colon which produces digestive enzymes that can break down numerous food components. Hence, related *in vitro* models have to include microbes of the whole GIT, in addition to all other above-mentioned factors. The pH value is between 5 and 7, while the holding time is between

Table 1: Some of the Latest Studies which Use in vitro Digestion Models to Investigate Release of Encapsulated Active Compounds of Food Relevance

Type of Encapsulation	Matrix/Aactive Compound	GI Region	Conditions	Adjuncts	References
Electrostatic complexation	Casein+alginate/ fat (sunflower oil emulsion)	Mouth (M)	M: pH 6.8, 37°C, 10 min	+	Zhang et al., 2015b
Pickering emulsion Nanoemulsion	Oil phase Curcumin+ Corn oil or Medium chain triglyceride / water phase chitosan–tripolyphosphate nanoparticles	Mouth (M) Stomach (S) Intestine (I)	M: pH 6.8, 37°C, 10 min S: pH 3.0, 37°C, 2 min I: pH 7, 37°C, 2h	+	Shah et al., 2016
Nanoemulsion encapsulated in filled hydrogel complex	Methoxyl pectin and whey protein isolate/orange oil	Mouth	M: pH 7, 37°C, 2 min	-	Kwan and Davidov-Pardo, 2018
W/O/W emulsion	Soybin or orange oil/water phase apigenin in of ethanol	Mouth Stomach Intestine	M: pH 6.8, 37°C, 5 min S: pH 1.3, 37°C, 2 h I: pH 8.1, 37°C, 2 h	+	Kim et al., 2016
Nanoemulsions; Nanoemulsions mixed with caseinate; Hydrogel beads; Hydrogel beads containing caseinate	Oil phase flaxseed oil/ water phase Quillaja saponin and PBS Alginate, alginate/casein	Mouth Stomach Intestine	M: pH 6.8, 37°C, 2 min S: pH 2.5, 37°C, 2 h I: pH 7, 37°C, 2 h	+	Chen et al., 2017
Dropwise nanoparticles	Chitosan/Phenolic acids	Stomach Intestine	S: pH 1.5-2, 37°C, 1 h I: pH 7, 37°C, 2 h	-	Madureira et al., 2016
Emulsification and ionotropic gelation	Carboxymethyl sago cellulose /Red palm oil	Stomach Intestine	S: pH 1.5, 37°C, 2 h I: pH 6.9, 37°C, 1 h	+	Sathasivam et al., 2018
Emulsification	Alginate combined with starch, xanthan, chitosan, or whey protein isolate /caffeine	Mouth	M: pH 6.5 37°C, 30 min	+	Mohammadi et al., 2018

(Contd.)

Table 1: (*Contd.*)

Type of Encapsulation	Matrix /Aactive Compound	GI Region	Conditions	Adjuncts	References
Spray drying	Sodium caseinate or Whey protein concentrate and Anhydrous milk fat/ trans-resveratrol	Mouth Stomach Intestine	M: pH 6.5 37°C, 10 min S: pH 2.5, 37°C, 2 h I: pH 6.5, 37°C, 2 h	+	Koga et al., 2016
Spray drying	Whey Solution/L. Reuteri	Stomach Intestine	S: pH 1.9, 37°C, 30 min I: pH 7.5, 37°C, 60 min	+	Jantzen et al., 2013
Spray drying	Whey protein isolate or gum arabic/ bluberry extract	Mouth Stomach Intestine	M: pH 6.8 37°C, 5 min S: pH 1.3, 37°C, 2 h I: pH 8.1, 37°C, 2 h	+	Flores et al., 2014
Spray drying	Alginate/caffeine	Stomach	S: pH 2.0, 37°C, 1h	+	Bagheri et al., 2014
Spray drying	Glucan and cyclodextrin/ saffron	Mouth Stomach Intestine	M: pH 6.8 37°C, 5 min S: pH 3.0, 37°C, 1 h I: pH 8.0, 37°C, 4 h	+	Ahmad et al., 2018
Liposomes, spray drying	Lecithin and chitosan, maltodextrin/ black mulberry extract	Stomach Intestine	S: pH 2, 37°C, 2 h I: 37°C, 2 h	+	Gültekin-Özgüven et al., 2016
Liposomes, spray drying	Deoiled lecithin and whey protein isolate/ Quercetin	Stomach Intestine	S: pH 2.0, 37°C, 2 h I: pH 6.8, 37°C, 4 h	+	Frenzel and Steffen-Heins, 2015
Freeze drying	Milk fat sodium caseinate lecithin/lactase	Stomach Intestine	S: pH 1.8, 37°C, 2 h I: pH 7.0, 37 C, 4 h	+	Zhang and Zhong, 2018
Complex coacervation	Gelatin and gum arabic/structured lipids	Stomach Intestine	S: pH 1.8, 37°C, 2 h I: pH 7.8, 37°C, up to 3h	+	Yüksel-Bilsel and Şahin-Yeşilçubuk, 2019

Method	Material/Compound	Location	Conditions		Reference
Liposomes	Phospholipid and cholesterol/ Gallic acid	Mouth Stomach Intestine	M: pH 6.8, 37°C, 10 min S: pH 2.5, 37°C,2h I: pH 7.0, 37C, 2h	+	Zhang et al., 2019
Liposomes	L-α-phosphatidylcholine, L-α--phosphatidic acid and Chitosan/curcumin	Mouth Stomach Intestine	M: pH 6.8, 37°C, 10 min S: pH 2-3, 37°C,2h I: pH 7.0, 37°C, 2h	+	Cuomo et al. 2017
Liposomes	Pluronic and Lipoid S100/Curcumin	Mouth Stomach Intestine	M: pH 6.8, 37°C, 10 min S: pH 2.5, 37°C,2h I: pH 7.0, 37°C, 2h	+	Li et al., 2018
Liposomes	Soy lecithin/ theobromine, caffeine, catechin, epicatechin, and a cocoa extract	Mouth Stomach Intestine	M: pH 7,0, 37°C, 2 min S: pH 3,0, 37°C,2h I: pH 7.0, 37°C, 6h	+	Toro Uribe et al., 2018
Molecular inclusion complex	Spring Dextrin/Poly-unsaturated fatty acids	Stomach Intestine	S: pH 1.5, 37°C, 2h I: pH 7.4, 37°C, 24h	+	Xu et al., 2013
Layer-by-layer deposition	Lactoferrin, glycomacropeptide and chitosan / curcumin and caffeine	Stomach Intestine (dynamic)	S: pH 4.8 at t=0 to pH 1.7 at t=120 min I: pH 6.5, 6.8 and 7.2	+	Bourbon et al., 2018

12 and 24h. Digestion of microencapsulates for delivery of probiotics is expected to be examined using *in vitro* models which comprise the colon section.

Table 1 summarizes the latest studies which utilize *in vitro* GIT models to study release of encapsulated compounds of food relevance.

Apart from *in vitro* simulated GIT models, which were used in most of experiments with delivery systems for food, animal or human feeding studies have also been performed in some studies (Li *et al.*, 2012; Farfan *et al.*, 2015; Steingoetter *et al.*, 2015).

Apparatus of different design characteristics are in use for *in vitro* release, dissolution and digestion studies. In most cases, experiments are performed in a vessel in which a sample of encapsulates is placed in a basket or dialyzing bag, depending on the form of encapsulates (solid particles macro particles *versus* colloidal systems and powders) (Fig. 1 (1) and (2)). Modern concept of *in vitro* testing includes application of Franz diffusion cell, convenient for semisolid samples (Fig. 1 (3)). Release kinetics obtained thereof do not present a true picture of release rate due to existence of a synthetic membrane between the donor (containing the test product) and acceptor (filled with collection medium) compartments. The membrane is a barrier and creates resistance to drug diffusion in addition to the resistance originating from

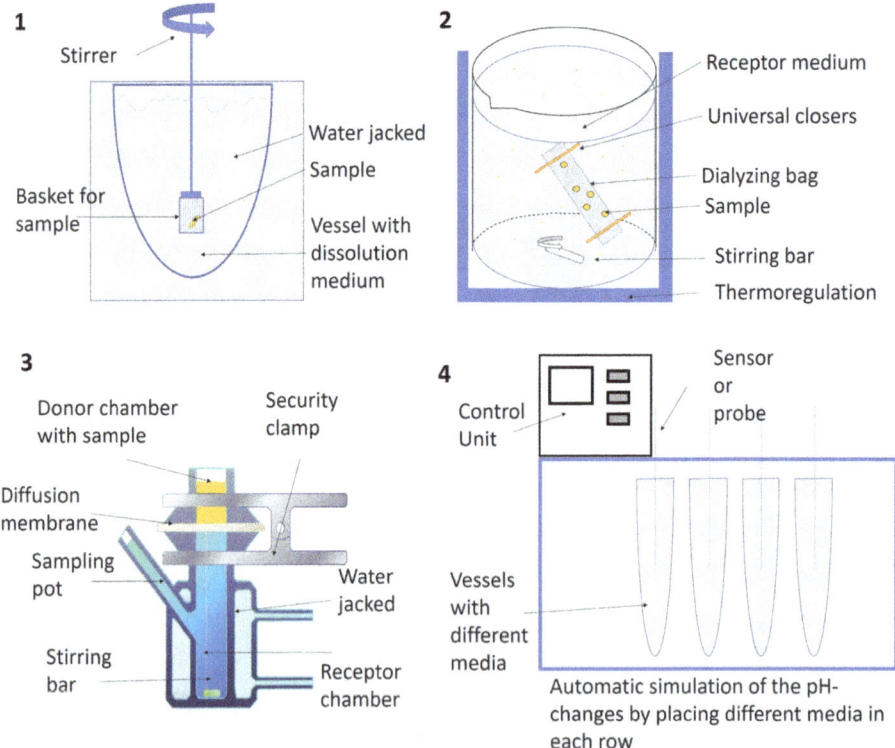

Fig. 1: Apparatus of different design characteristics mostly used for *in vitro* release, dissolution and digestion studies

matrix material, which actually does not exist in real. However, this system allows quantifying of mass transfer resistance generated solely by the carrier material. Novel devices commercially available are automatic dissolution testing machines convenient for experiments with diverse extended and sustained release dosage forms (Fig. 1 (4)). They are suited for simulating the pH-changes within the human body. One can find in literature other solutions for apparatus design where *in vitro* tests are performed and some of them are custom made for a purpose of testing probiotic cell viability (Chen *et al.*, 2005; Moumita *et al.*, 2017) and digestion of food constituents (Luo *et al.*, 2015). One of such models was designed as a series of bioreactors adapted from the simulated human intestinal microbial system, where each reactor corresponded to a specific stage of digestion (Chen *et al.*, 2005).

4.2   Gastrointestinal Fate of Nutraceutical-loaded Delivery Systems

The release kinetics of an encapsulated compound depends on the fate of the wall material in the upper and lower part of the GIT. Different wall materials express different extent of degradation. This section gives a summary on the fate of the most common encapsulating materials and systems when exposed to GIT. Herein, it should be stressed that not only the type of material used to design a delivery system, but the structure of the system itself have profound effect on behavior upon oral intake. For example, the pH variance within the gastrointestinal compartments can affect nanoparticles aggregation status and alter surface chemistry, particularly when zeta potential is highly dependent on pH, which at the end affects degradation rate and release kinetics. Another factor that could interfere is interaction between the carrier and core materials. Thus, polyphenols interact with the chitosan chains, acting as crosslinkers (Talón *et al.*, 2017) and also with proteins by creating protein-polyphenol complexes (Esfanjani and Jafari, 2016).

Starch is a convenient carrier since it is a major component of our diet and present in a number of food products; it is a polymeric carbohydrate consisting of glucose units joined by glycosidic bonds. Starches are easily digested and the conversion to oligosaccharides and glucose starts already in the mouth, by amylases activity. On the other hand, there is a variety of resistant starches. Resistant starch (RS) is generally defined as fraction of starch which is not hydrolyzed and absorbed in the small intestine in period of 120 min after consummation, but digested in the colon. RS is generally classified into three groups: RS1 is starch unreachable for digestive enzymes since it is locked in fiber material, RS2 is native starch granules protected from digestion by the granule conformation and RS3 is recrystallized amylose and amylopectin (Brouns *et al.*, 2007). Starch derivatives, prepared by physically, enzymatically, or chemically treating native starch, could enhance their delivery performance (Cortés *et al.*, 2014; Chranioti *et al.*, 2015; Hasanvand *et al.*, 2015; Zhu, 2017). For example, gels which were made of oxidized potato starch polymers, and chemically cross-linked by sodium trimetaphosphate (STMP) released no more than 16 per cent of encapsulated anthocyanins after incubation on gastric fluid, and 78 per cent in simulated intestinal fluid in the 15 min of incubation (Wang *et al.*, 2013). Cortés *et al.*, 2014 compared three different types of chemically modified starches (phosphorylated, acetylated and succinylated starch) from the aspect of probiotic survival in simulated gastrointestinal conditions; they established that OSA-starch provided the highest resistance (reduction was only 4.4 per cent of initial count) and associated this result to the more hydrophobic character than that of other two starches which resulted with retarded water penetration and enzyme access. Maltodextrins are obtained from edible starch by hydrolysis and they require the presence of pancreatic enzymes for initial digestion and mucosal oligosaccharidases for complete digestion. Pectins, polysaccharides rich in galacturonic acid, form hydrogels in the presence of divalent cations, which are non-digested in the upper intestinal tract but digested in the colon (Cabrera *et al.*, 2011).

Alginate, one of the most common polysaccharide materials used for manufacture of gel matrix delivery systems, expresses shrinkage in gastric medium which results from a decrease in the repulsive charge due to protonation of free carboxylate groups of alginate. Furthermore, due to dissociation of calcium ions at low pH and protonation of COO – groups, alginate chains come closer together and form hydrogen bonds; as a result, an acid gel can be formed in the stomach. In intestinal region, swelling of alginate gel occurs due to the electrostatic repulsive forces at a pH above the pKa of the uronic acid groups on the alginate and osmotic pressure originating from the negative charges of the ionized carboxyl groups. The presence of chelating agents (Gombotz and Wee, 1998) and electrostatic interactions induced by charged polymers (Bokkhim *et al.*, 2014) of the intestinal fluid also contributes to alginate denaturation and release of actives. *In vitro* studies have confirmed that alginate gel matrix is resistant in gastric medium and becomes digested in the intestinal region (Rayment *et al.*, 2009). Carrageenans are made up of repeating galactose units and 3,6 anhydrogalactose, both sulfated and nonsulfated. Carrageenan gels express similar behavior in GIT such alginate, but the pKa value of the sulfate groups is appreciably less (pKA≈2.0) than that of carboxyl groups (pKA≈3.5) (Stephen *et al.*, 2006). Carrageenan microgels are more susceptible to breakdown under simulated GIT conditions than alginate and release the actives more readily in the small intestine phase (Zhang *et al.*, 2016b; Chen *et al.*, 2018).

Chitosan, a linear polysaccharide composed of randomly distributed β-(1→4)-linked D-glucosamine units, swells in the acidic gastric juice and forms a hydrogel. Thus, it is used as an excipient compound in the floating pharmaceuticals that do not adhere to the gastric mucosa (Kumar *et al.*, 2016). When found in intestinal conditions, chitosan slows down the adsorption of bile salts and/or lipase by restricting the access of the enzyme to the active compound (Liu *et al.*, 2014). Digestion of chitosan-based encapsulates strongly depends on crosslinking agent. According to McConnell *et al.* (2008) non-crosslinked chitosan is digested by both human colonic bacteria and pancreatic enzymes, glutaraldehyde crosslinked chitosan is resistant to both pancreatic enzymatic digestion and colonic bacteria, while tripolyphosphate crosslinked chitosan is, to some extent, resistant to pancreatic enzymes but easily digested by human colonic bacteria.

Madureira *et al.* (2016) performed degradation and release studies of chitosan nanoparticles (made of low-molecular weight and high-molecular weight chitosan) ionically crosslinked with TPP and loaded with phenolic acids (protocatechuic and 2,5-dihydroxybenzoic acids). They revealed most relevant changes in particle size and release of phenolic compound in the intestinal stage (Fig. 2).

Gum arabic, a complex heteropolysaccharide with small protein content, with main and side chains linked by β-glycosidic bonds, is minimally digested in

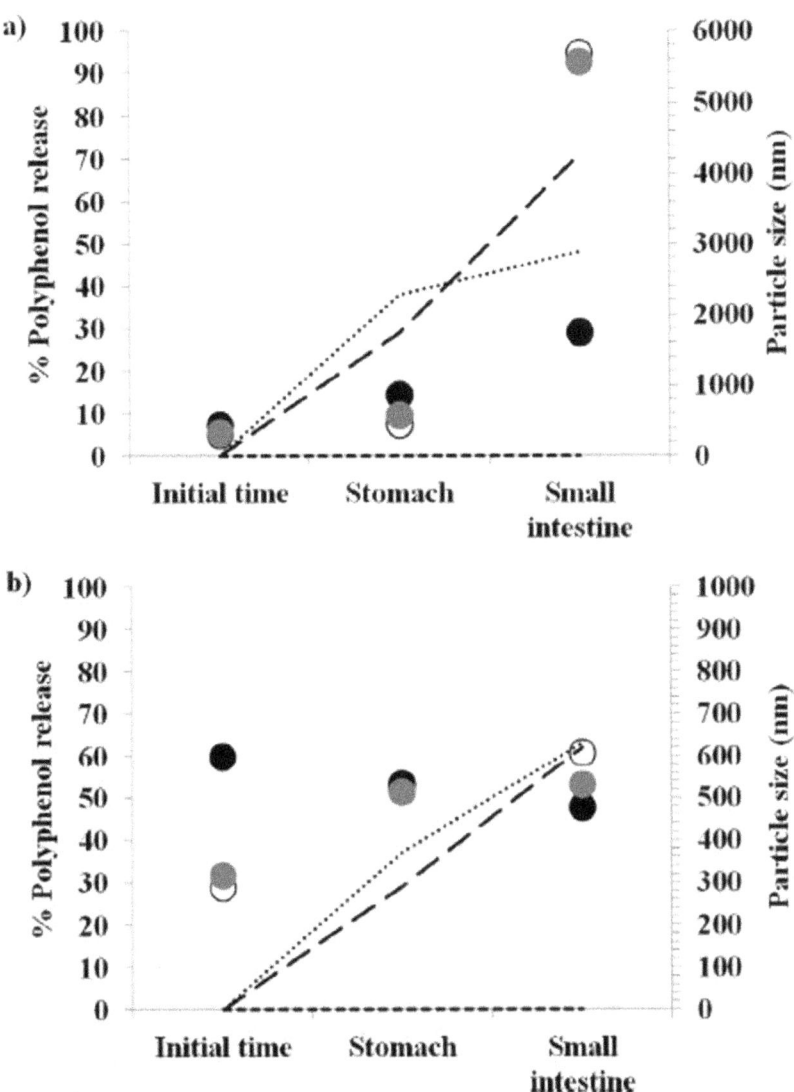

Fig. 2: Evolution of the nanoparticle sizes (y2 axis, round shapes) and polyphenol release (%) (y1 axis, lines) of the nanoparticles produced with a) low molecular weight chitosan and b) high molecular weight chitosan without polyphenol (○, - -), with protocatechuic acid (◐,- -) and with 2,5-dihydroxybenzoic acids (●, ·····) when exposed to the simulated GIT conditions (Reprinted from *Journal of Food Engineering*, vol. 174, Madureira, *et al.*, 2016 Copyright (2019), with permission from Elsevier.)

digestive fluids (containing mainly α-amylases and α-glucosidases) and may behave as a soluble fiber for the action of colonic bacteria. Flores *et al.* (2014) showed that during *in vitro* gastric digestion, 60-90 per cent of the total phenolics were released from gum arabic microcapsules, with the remaining fraction detected after simulated intestinal digestion. Guar gum is a polysaccharide composed of the sugars galactose and mannose. It has been considered a substitute for starch since it is undeniably resistant to human digestion and absorption. Upon oral administration, it can release an active compound in the colon, by delaying the release in mouth and gastrointestinal conditions (Yang *et al.*, 2019).

More recently, binary systems have attracted interest of scientists as they may combine advantages of two different compounds. For example, He *et al.* (2016) used a mixture of OSA-modified starch and xanthan gum to produce spray-dried microparticles encapsulating conjugated linoleic acid; the obtained microcapsules were not prone to gastric digestion, but digestible under intestinal conditions to release the active compound. Mixtures consisting of proteins and polysaccharides have been widely applied, since they exhibit high emulsifying capacity of proteins and may efficiently entrap even hydrophobic compounds within the polysaccharide gel network. The examples are re-constituted skim milk mixed with either starch or whey protein (Paéz *et al.*, 2012) or skim milk mixed with oligofructose (Fritzen-Freire *et al.*, 2012), alginate with whey, soy and hemp proteins (Harper *et al.*, 2013; Belščak-Cvitanović *et al.*, 2015; Liu *et al.*, 2018; Volić *et al.*, 2018). Properly selected binary systems exhibited higher survival rates, and superior *in vitro* gastrointestinal digestion and storage properties as compared to a single component system (Tanzina *et al.*, 2013). Protein materials (whey proteins, soybean proteins, caseinate) are being released in gastric fluid from binary composition capsules and digested due to the susceptibility to pepsin hydrolysis (Zhang *et al.*, 2016a, 2017). They rapidly swell in SIF (at pH values above their isoelectric point) which results in diffusion of pancreatic enzyme into the capsule matrix, degradation of protein-based capsules and release of bioactive molecules (Gunasekaran *et al.*, 2007; Hébrard *et al.*, 2010; Belščak-Cvitanović *et al.*, 2015; Samtlebe *et al.*, 2016; Zhang *et al.*, 2016a, 2017; Volić *et al.*, 2018). Polypeptides are then broken down into oligopeptides and are absorbed in the intestines to provide nutrition. In case of polyphenols (e.g. plant extracts) loaded gel systems based on polymer-polysaccharide matrix, the presence of polyphenols cause oligopeptides to bind with phenolics or complement their antioxidant activity (Buchweitz *et al.*, 2013; Flores *et al.*, 2014; Jakobek, 2015; Feng *et al.*, 2018; Volić *et al.*, 2018), which may contribute to the 'sustaining effect' of protein matrix.

Single protein gel system, if properly designed, can also provide sustained delivery in intestinal part of digestion system. For example, casein nanoparticles, fabricated by calcium-induced coacervation process were stable in simulated SGF while released folic acid in SIF within few hours (Penalva *et al.*, 2015). The authors' hypothesis is that the peptide bonds involving hydrophobic aromatic amino acids, which are susceptible to pepsin activity, become trapped and hidden from enzyme molecules once the network is formed. *In vitro* release results are in line with *in vivo* experiments on laboratory animals since those treated with folic acid-loaded casein nanoparticles displayed significantly higher serum levels than the animals which received an aqueous solution of the vitamin.

Coated microgel systems exhibited better protection properties in comparison to simple matrix system and the protective ability depends on coating layer material and its density. For example, the addition of a shellac coating layer diminished early release of anthocyanins from maltodextrin-pepsin capsules at some extent (Oidtmann *et al.*, 2012).

The structure of small droplets, implanted in larger ones, gives the opportunity to create food-grade microcompartments that will prevent early degradation of labile components during passage through GIT. Oil-soluble vitamins and functional oils (omega-3 and omega-6 fatty acids) are often encapsulated within emulsion-based delivery systems to facilitate their incorporation into aqueous-based products. The fate of single and multiple emulsions in GIT is rather difficult to explain with some general conclusions since it depends on many structural variables. So far it is clear that the emulsifier layers required for stabilizing the emulsion structure do have significant impact on the release of encapsulated ingredients. Long-chained emulsifiers, like proteins, are proved to reduce the release of entrapped ingredients since macromolecular proteins may form viscoelastic interfacial films around droplets which act as a barrier to diffusion of the embedded molecules. Frank *et al*. (2012) investigated stability of whey protein isolate-stabilized w/o/w-double emulsion (disperse phase content 30-80 per cent) with anthocyanin-rich bilberry extract encapsulated in the inner aqueous phase. The authors observed that the emulsion microstructure did not change during gastric incubation. After intestinal incubation the droplet size distributions were almost alike, independent of the original droplet size of the emulsions (related to the disperse phase content). The authors concluded that the droplets initially coalesced during digestion, after the emulsifier molecules were displaced from the interface by digestive enzymes and then the coalesced droplets were re-emulsified due to the action of bile acid, digestive enzymes and the mechanical energy input of the digestive system. In favor of protein protection ability is the fact that certain protein molecules are resistant toward pepsin digestion, such as β-lactoglobulin (whey protein). Thus, Sari *et al*. (2015) have demonstrated that the curcumin nanoemulsion stabilized by using whey protein concentrate was relatively resistant to pepsin digestion but pancreatin caused release of curcumin from the nanoemulsion. Recently, Chen *et al*. (2020) proved that the bioaccessibility of lipophilic nutrients (β-carotene) was highly related to the type of the protein used to stabilize the emulsions, such that compared with whey protein isolate and sodium casein, soy protein isolate-based emulsions possessed the highest bioaccessibility, probably due to the differences in the digestion behavior of these proteins; the micellization and lipolysis rate of whey protein isolate and sodium casein samples were lower than soy protein isolate. Maillard conjugate-stabilized emulsions are a new class of food-grade emulsions obtained through Maillard reaction. In these systems, the polysaccharide moiety reduces strong steric and sometimes electrostatic repulsion, and the protein in the conjugate can be attached to hydrophobic surfaces (Nooshkam and Varidi, 2020). However, a prerequisite is to control the Maillard reaction and stop it in its early stages to avoid the formation of brown melanoidin polymers and undesired advanced glycation products, such as acrylamide (Spivey, 2010). Emulsions stabilized with protein-polysaccharide conjugates in certain cases expressed better survival in GIT as compared to their analogues stabilized with unconjugated protein-polysaccharide. In specific, due to the Maillard conjugate-based cohesive and thick interfacial layer, these emulsions expressed higher stability to droplet flocculation and coalescence (smaller droplet size) during gastric digestion, inhibited pepsin interaction with emulsion droplets and, in turn, prevented proteolysis of interfacial protein layer, arduous accessibility to gastric lipases and improved stability towards enzymatic hydrolysis by trypsin and chymotrypsin (Lesmes and

McClements, 2012; Xu *et al.*, 2014; Davidov-Pardo *et al.*, 2015; Gumus *et al.*, 2016; Liu *et al.*, 2017; Zhong *et al.*, 2019).

Different hydrophilic and lipophilic bioactives (minerals, oils, probiotic bacteria, enzymes, vitamins and peptides, colorants) are encapsulated in fat matrices. Lipids offer advantage of providing barriers to oxygen, and moisture, as well as not dissolving in foods. Moreover, lipids which have melting point above 37°C exhibit limited dissolving properties in GIT. Specifically, in gastric solution no dissolution was observed for beeswax, carnauba wax and fractionated palm kernel oil particles while in duodenum solution fractionated palm kernel oil particles exhibited partial dissolution (the extent of erosion was a function of the particle size) unlike all waxes which remained unchanged (Raymond and Champagne, 2014). On the other hand, the encapsulated compound could simply slowly diffuse outside the wax or lipid particle without requiring its breakdown. Recently, Mazzocato *et al.* (2019) highlighted importance of the presence of soya lecithin (as a binding agent of the lipid crystals) on release properties of B_{12}-bearing solid lipid microparticles made from vegetable fats.

Liposomes are specific colloidal systems orally used in food, pharmaceutic and agricultural industries to entrap, protect and control the release of functional and unstable hydrophilic or lipophilic compounds, such as enzymes, antimicrobials, vaccines, and antioxidants (Emami *et al.*, 2016; Assadpour and Jafar, 2018). In the oral cavity, liposomes express swelling and aggregation (due to the binding of mucin and the bridging action of the salt and the polyelectrolyte in the saliva), but keep the compound entrapped (Zhang *et al.*, 2019; Mao and McClements, 2012). Liposomes are relatively stable in gastric environment and the compacted liposomal structures formed by assembly of phospholipids can inhibit the penetration of gastric enzyme and prevent release of the embedded molecules at some extent (Liu *et al.*, 2012; Tan *et al.*, 2014; Rodriguez *et al.*, 2018; Zhang *et al.*, 2019). For example, only about 18 per cent of gallic acid was released from gallic acid-loaded liposomes in SGF (Zhang *et al.*, 2019). In intestinal surrounding, swelling and disruption of the vesicles occurs caused by penetration of bile salts and pancreatic enzyme. According to degradation kinetics of gallic acid–loaded liposomes, liberation of free fatty acids increased rapidly within 40 min of SIF digestion, followed by a slow increase up to the final release rate stable around 40 per cent (Zhang *et al.*, 2019). It is known that the presence of cholesterol or its analogues from botanical sterols in the liposomal phospolipid bilayer reduces membrane permeability. Jovanović *et al.* (2019) showed that β-sitosterol reduced the membrane distortion under acidic conditions and retarded release of gentisic acid.

One novel way to improve delivery performance of lipid particulates is to either modify their membrane surface via forming the bioadhesive and polymeric layers or to embed them within a polymer network. The layer-by-layer deposition approach has become important for food manufacturers wishing to form thick polymer layers to stabilize dispersions such as emulsions or liposomes and enhance the tolerance of liposomes against gastrointestinal stresses (Gibis *et al.*, 2014). For instance, electrostatic adsorption of chitosan on liposomes has been used to design a novel delivery system (chitosomes) of carotenoids (lycopene, β-carotene, lutein and canthaxanthin) (Tan *et al.*, 2016) and a plant alkaloid, berberine hydrochloride (Nguyen *et al.*, 2014). The chitosan coating layer suppressed release of carotenoids in SGF and SIF in such manner that below 50 per cent of the encapsulated carotenoid was released in SIF after 10h of incubation time (Tan *et al*., 2016). Liu *et al* (2013) have proved that a polyelectrolyte delivery system based on sodium alginate and chitosan coated on the surface of nanoliposomes could better resist lipolytic degradation and facilitate a lower level of encapsulated component release in simulated gastrointestinal conditions. Different pluronics (triblock copolymers with a central hydrophobic poly(propylene oxide) chain with two hydrophilic poly (ethylene oxide) on each side) have been tested as modifiers of liposomal membranes which improve the stability and absorption of conventional liposomes and *in vitro* release in simulated gastrointestinal digestion (Hädicke and Blume, 2014; Li *et al*., 2018). Recently, Sun and Xia (2019) investigated the release of lipid nanoparticles immobilized in alginate hydrogel. Their dissolution study performed in SGF showed two kinds of release mechanism, lipid nanoparticles-mediated release and micelle-mediated release and the proportion of the two were related to the alginate concentration used to prepared alginate beads. Emulsion-filled gel beads were prepared by the combination of heat-denatured whey protein isolate emulsification and alginate gelation for delivery of α-tocopherol; the release of α-tocopherol from all emulsion beads was basically close to or < 10 per cent after gastric digestion for 2h and about 20 per cent or less after gastrointestinal digestion for 6h (Feng *et al*., 2018). Mun *et al*. (2015a, b) demonstrated that starch-based emulsion-filled gels gave a higher β-carotene bioaccessibility than starch gels and emulsions in simulated GIT; the hydrolysis pattern of lipids by lipase in the gels depended on the type of starch.

5. Mathematical Modeling of Release Kinetics

In order to find release kinetics of bioactive compounds from release curves in food, the data are fitted to different mathematical models, which allow researchers to determine the release rate constants and to explain the release mechanism. In addition, mathematical modeling of drug release can be very helpful to speed up product development. The diffusion of low molecular weight compounds in macromolecular polymeric systems is generally governed by two simultaneously occurring phenomena: (1) the penetrant flows exclusively driven by a concentration gradient (related to Brownian motion) which is a substantially stochastic phenomenon; (2) relaxation phenomenon driven by the distance of the local system from the equilibrium. There are several recent papers which give an overview on the mathematical models which can be used to predict the release of bioactives from biopolymer particles differently structured, either as homogenous spheres or porous spheres (e.g. polysaccharide/ protein particles) or coated spheres; the models refer to release predominantly controlled by diffusional mass transport (Siepmann and Siepmann, 2012; McClements, 2015). Table 2 summarizes mathematical models applied in recent literature on release kinetics of encapsulates which are of food significance.

Two steps of release kinetics have been recognized in most cases of matrix-type delivery system: 'burst' release of the encapsulated compound near the surface of the particles due to hydration and slower release from the polymer matrix as it degrades. The specific phenomenon involved in controlled release systems in food packaging differ from other controlled release systems – the polymeric matrix does not dissolve during the release of the antimicrobial agent.

Table 2: Mathematical Models Used to Describe the Controlled Release of Bioactives from Encapsulated Systems which are Applicable in Food Sector

Model	System	Characteristics	References
$$\frac{d}{dt}\int_{V(t)}(\rho \cdot C_w)\,dV = \int_{S(t)}\left[D_F^w \cdot \frac{\partial}{\partial x}(\rho \cdot C_w)\cdot x \cdot \vec{n}\right]\cdot dS$$ $$\frac{d}{dt}\int_{V(t)}(\rho \cdot C_L)\,dV = \int_{S(t)}\left[D_F^L \cdot \frac{\partial}{\partial x}(\rho \cdot C_w)\cdot x \cdot \vec{n}\right]\cdot dS$$	Lysozyme (antimicrobial agent) release kinetics from PVA films	water diffusion, macromolecular matrix relaxation kinetic, and active agent diffusion are taken into account	Buonocore *et al.*, 2003

V - an arbitrary volume fixed in space

S - the surface of the volume V

x - the axial coordinate

V(t) - the volume that contains an amount of macromolecular matrix equal to the initial one

S(t) - the surface of the volume V(t)

ρ - the density of the polymeric matrix defined as the ratio between the weight of the matrix and the volume of the mixture (that is, water plus matrix) expressed as g matrix/cm^3 hydrated matrix

C_w - the local water concentration expressed as g water/g dry polymer

C_L - the lysozyme concentration expressed as g lysozyme/g dry polymer

D_F^w - the water diffusion coefficient

D_F^L -the lysozyme diffusion coefficient

n – number of elements

$\dot{x}$ - the axial versor

$\vec{n}$ - the versor normal to the surface

Model	Application	Result / Kinetics	Reference
Higuchi model $Q = KH \cdot t^{(1/2)}$ (Higuchi 1963) Q – the amount of active component released over time t K_H – Higuchi dissolution constant $Q = Q_0 - K_0 t$ Q_0 – the initial amount of active component in the particles K_0 – the zero order release constant	Ethylvanillin encapsulated in ethylcellulose nanoparticles	Zero order kinetics	Eltayeb et al., 2015
	Gallic acid encapsulated with native starch, native inulin, acetylated starch or acetylated inulin	pH 5.5, at 25 °C	Robert et al., 2012
Hixson-Crowell model $Q^{1/3} = Q_0^{1/3} - K_C t$ Kc – the rate constant $\ln Q = \ln Q_0 K_1 t$	Ethylvanillin encapsulated in ethylcellulose nanoparticles	First order $R^2 = 0.906$	Eltayeb et al., 2015
	Vitamins B12 and C encapsulated using chitosan, modified chitosan and sodium alginate by spray-drying		Estevinho and Rocha, 2017
Korsmeyer-Peppas model $\dfrac{M_t}{M_\infty} = kt^n$ k – kinetic constant incorporating geometric and structural characteristics of the particles n=0.45 for Fickian diffusion mechanism n=0.89 for Case II release mechanism 0.45<n<0.89 for non-Fickian diffusion mechanism n>0.89 for Super Case II release mechanism (Peppas 1985, Peppas and Ritger 1987)	Ethylvanillin encapsulated in ethylcellulose nanoparticles	$R^2 = 0.985$	Eltayeb et al., 2015
	Quercetin (dietary flavonoid) encapsulated in lipid nanoparticles immobilized in alginate beads	$R^2 = 0.99$	Sun and Xia, 2019
	Gallic acid encapsulated with native starch, native inulin, acetylated starch or acetylated inulin	pH 5.5, at 25 °C $R^2 = 0.93 - 0.99$ non-Fickian diffusion, (anomalous diffusion)	Robert et al., 2012

(Contd.)

Table 2: (*Contd.*)

Model	System	Characteristics	References
	Vitamin D encapsulated in high amylose corn starch nanoparticles	SGF followed by SIF Fickian diffusion	Hasanvand et al., 2015
	Gallic acid encapsulated in liposomes	Case II transport	Zhang et al., 2019
First-order model $\frac{M_t}{M_\infty} = 1 - \exp(-kt)$	Curcumin encapsulated in pluronics modified liposomes	$R^2=0.99$ at pH 7.4 in 37 °C	Li et al., 2018
Diffusion-relaxation model $\frac{M_t}{M_\infty} = k_d t^m + k_r t^{2m}$ k_d - the diffusion rate constant k_r - the relaxation rate constant m - constant	Gallic acid encapsulated with native starch, native inulin, acetylated starch or acetylated inulin	$R^2=0.96-0.98$	Robert et al., 2012
Linear superposition model (Berens and Hopfenberg 1978) $\frac{M_t}{M_\infty} = X\left[1 - \frac{6}{\pi^2}\exp(-k_F t)\right] + (1-X)\left[1 - \exp(k_R t)\right]$ X - the fraction of compound released by Fickian transport k_F - the Fickian diffusion rate constant k_R - the relaxation rate constants	Resveratrol encapsulated in Ca-alginate submicron particles	Accounts for both Fickian and Case II transport effects in hydrophilic matrices pH=1.2 $R^2=0.985$ pH=7.2 $R^2=0.993$	Istenič et al., 2015
	Vitamin B2 encapsulated in alginate/chitosan nanoparticles	pH=2, pH=7, at 25 °C and 37 °C $R^2>0.90$	Azevedo et al., 2014

Model/Equation	Application	Conditions	Reference
Logistic equation (Cambel 1993) $\dfrac{\partial v(t)}{\partial t} = \mu v(t) - b v(t)^2$ $v(t) = \dfrac{v_{eq} v_0 e^{\mu t}}{v_{eq} - v_0 + v_0 e^{\mu t}}$ v – volume of released oil $\mu = 1/\tau_R \mu$ – kinetic constant (specific rate of oil release) τ_R relaxation time $b = \mu/v_{eq}$ $v_{eq} = v(t_{eq})$	Curcumin/caffeine encapsulated in lactoferrin-glycomacropeptide nanogels	pH=2, pH=7, at 37 °C	Bourbon *et al.*, 2016
	Thyme oil encapsulated in alginate/soy protein beads	SGF+SIF	Volić *et al.*, 2018
Baker–Lonsdale mode (1974) $\dfrac{3}{2}\left[1 - \left(1 - \dfrac{M_t}{M_\infty}\right)^{2/3}\right] - \dfrac{M_t}{M_\infty} = kt$ k - the release rate constant	Resveratrol encapsulated using Arabic gum by spray-drying		Cardoso *et al.*, 2019
Weibull model $M_t = M_\infty\left[1 - e^{-\left(\frac{t - t_0}{\tau_d}\right)^\beta}\right]$ t_0 – lag time of release (min) β – parameter which represent the shape of release curve τ_d – the time when 63.2% of Mt has been released	Vitamins B12 and C encapsulated using chitosan, modified chitosan and sodium alginate by spray-drying	in coconut oil, at 37 °C R^2=0.81-0.83	Estevinho and Rocha, 2017
	Resveratrol encapsulated using Arabic gum by spray-drying	in SIF	Cardoso *et al.*, 2019

(Contd.)

Table 2: (*Contd.*)

Model	System	Characteristics	Reference
	Morus nigra extract encapsulated in Eudragit L-100 Electrospun Fibers		Soares *et al.*, 2017
	Thymol and carvacrol encapsulated in maltodextrin and soy proteins	SGF followed by SIF	Ulloa *et al.*, 2017
	Vitamin D encapsulated in high amylose corn starch nanoparticles		Hasanvand *et al.*, 2015

M_t – the amount of active compound released at time t
M_∞ - the amount of active compound released at infinite time

6. Conclusion and Future Perspectives

Bioactive compounds are highly prone to decomposition during the production process, storage and passage through GIT. Therefore, a new field in food engineering has been opened focused on designing of novel food-grade delivery systems for potential application as functional ingredients in food products. One of the demands of these systems is a possibility for controllable release, which means release at targeted place and at desired rate. Many types of materials (biopolymers, silica and oxides) can be used as carrier in relation to their compatibility with the active molecule and the type of releasing medium, but the current focus in on cheap and nutritious materials, such as plant proteins, phospholipids and carbohydrate gels. Single and multiple stimuli (pH, temperature, light irradiation) responsive polymeric systems show promising potential as new carriers for precisely-controlled release of encapsulated molecules. Nanotechnology based colloidal systems (including oil-in water emulsions, nanoparticles, complex coacervates, nanoparticles and nanogels) and complex systems (emulsion filled microgels, liposomes-in-gel microparticles, coated microgels) conceived from a combination of the polymer and the lipid-based delivery systems provide sustained release of encapsulated drug and superior resistance to severe GIT conditions, which make them useful for the intestinal and colon delivery nutraceuticals. The Maillard protein-polysaccharide conjugate based delivery systems are the most protective since in these systems, the conjugate provides a thick and continuous layer around the encapsulated compound, which lowers its degradation rate in GIT. A large panel of theoretical and empirical mathematical models able to fit release kinetics at various conditions has been identified, some of them able to clarify mechanism of release. More *in vitro* and *in vivo* studies are needed in the future to reliably test the performance during digestion. Serious investigation is needed on molecular level on the relationship between the effects of digestion on the interfacial layer of colloidal systems in order to design safe structures able to deliver a wide range of bioactive compounds in food.

References

Acevedo-Fani, A., Soliva-Fortuny, R. and Martín-Belloso, O. (2018). Photo-protection and controlled release of folic acid using edible alginate/chitosan nanolaminates, *Journal of Food Engineering*, 229: 72-82.

Ahmad, M., Ashraf, B., Gani, A. and Gani, A. (2018). Microencapsulation of saffron anthocyanins using β glucan and β cyclodextrin: Microcapsule characterization, release behaviour and antioxidant potential during *in-vitro* digestion, *International Journal of Biological Macromolecules*, 109: 435-442.

Assadpour, E. and Jafar, S.M. (2018). A systematic review on nanoencapsulation of food bioactive ingredients and nutraceuticals by various nanocarriers, *Critical Reviews in Food Science and Nutrition*, 1-47.

Azevedo, M.A., Bourbon, A.I., Vicente, A.A. and Cerqueira, M.A. (2014). Alginate/chitosan nanoparticles for encapsulation and controlled release of vitamin B_2, *International Journal of Biological Macromolecules*, 71: 141-146.

Bagheri, L., Madadlou, A., Yarmand, M. and Mousavi, M.E. (2014). Spray-dried alginate microparticles carrying caffeine-loaded and potentially bioactive nanoparticles, *Food Research International*, 62: 1113-1119.

Baker, R.W. and Lonsdale, H.K. (1974). Controlled release: Mechanisms and rates, pp. 15-71. *In*: A.C. Tanquary and ILE. Lacey (Eds.), *Controlled Release of Biologically Active Agents*, Plenum, New York.

Barat, J., Pérez-Esteve, É., Bernardos, A. and Martínez-Mañez, R. (2011). Nutritional effects of folic acid controlled release from mesoporous materials, *Procedia Food Science*, 1: 1828-1832.

Belščak-Cvitanović, A., Đorđević, V., Karlović, S., Pavlović, V., Komes, D., Ježek, D., Bugarski, B. and Nedović, V. (2015). Protein-reinforced and chitosan-pectin coated alginate microparticles for delivery of flavan-3-ol antioxidants and caffeine from green tea extract, *Food Hydrocolloids*, 51: 361-374.

Berens, A.R. and Hopfenberg, H.B. (1978). Diffusion and relaxation in glassy polymer powders: 2. Separation of diffusion and relaxation parameters, *Polymers*, 19: 489-496.

Bokkhim, H., Bansal, N., Grøndahl, L. and Bhandari, B. (2014). Characterization of alginate-lactoferrin beads prepared by extrusion gelation method, *Food Hydrocolloids*, 53: 270-276.

Bourbon, A.I., Pinheiro, A.C., Cerqueira, M.A. and Vicente, A.A. (2018). *In vitro* digestion of lactoferrin-glycomacropeptide nanohydrogels incorporating bioactive compounds: Effect of a chitosan coating, *Food Hydrocolloids*, 84: 267-275.

Bourbon, A.I., Cerqueira, M.A. and Vicente, A.A. (2016). Encapsulation and controlled release of bioactive compounds in lactoferrin-glycomacropeptide nanohydrogels: Curcumin and caffeine as model compounds, *Journal of Food Engineering*, 180: 110-119.

Brouns, F., Arrigoni, E., Langkilde, A.M., Verkooijen, I., Fässler, C., Andersson, H., Kettlitz, B., van Nieuwenhoven, M., Philipsson, H. and Amadò, R. (2007). Physiological and metabolic properties of a digestion-resistant maltodextrin, classified as type 3 retrograded resistant starch, *Journal of Agricultural Food Chemistry*, 55: 1574-1581.

Bruneau, M., Bennicia, S., Brendlea, J., Dutourniea, P., Limousya, L. and Pluchon, S. (2019). Systems for stimuli-controlled release: Materials and applications, *Journal of Controlled Release*, 294: 355-371.

Buchweitz, M., Speth, M., Kammerer, D.R. and Carle, R. (2013). Impact of pectin type on the storage stability of black currant (*Ribes nigrum* L.) anthocyanins in pectic model solutions, *Food Chemistry*, 139: 1168-1178.

Buonocore, G.G., Del Nobile, M., Panizza, A., Battaglia, G. and Nicolais, L. (2003). Modeling the lysozyme release kinetics from antimicrobial films intended for food packaging applications, *Journal of Food Science*, 68: 1365-1370.

Cabrera, J.C., Cambier, P. and Cutsem, P.V. (2011). Drug encapsulation in pectin hydrogel beads – A systematic study of simulated digestion media, *International Journal of Pharmacy and Pharmaceutical Sciences*, 3: 292-299.

Cambel, A.B. (1993). *Applied Chaos Theory: A Paradigm for Complexity*, Academic Press, Inc., Boston/London, New York.

Cardoso, T., Gonçalves, A., Estevinho, B.N. and Rocha, F. (2019). Potential food application of resveratrol microparticles: Characterization and controlled release studies, *Powder Technology*, 355: 593-601.

Chen, F., Deng, Z., Zhang, Z., Zhang, R., Xu, Q., Fan, G., Luo, T. and McClements, D.J. (2018). Controlling lipid digestion profiles using mixtures of different types of microgel: Alginate beads and carrageenan beads, *Journal of Food Engineering*, 238: 156-163.

Chen, F., Fan, G.-Q., Zhang, Z., Zhang, R., Deng, Z.-Y. and McClements, D.J. (2017). Encapsulation of omega-3 fatty acids in nanoemulsions and microgels: Impact of delivery system type and protein addition on gastrointestinal fate, *Food Research International*, 100: 387-395.

Chen, H., Ouyang, W., Jones, M., Haque, T., Lawuyi, B. and Prakash, S. (2005). *In-vitro* analysis of APA microcapsules for oral delivery of live bacterial cells, *Journal of Microencapsulation*, 22: 539-547.

Chen, L., Yokoyama, W., Liang, R. and Zhong F. (2020). Enzymatic degradation and bioaccessibility of protein encapsulated β carotene nano-emulsions during *in vitro* gastro-intestinal digestion, *Food Hydrocolloids*, 100: 105-177.

Chranioti, C., Nikoloudaki, A. and Tzia, C. (2015). Saffron and beet root extracts encapsulated in maltodextrin, gum arabic, modified starch and chitosan: Incorporation in a chewing gum system, *Carbohydrate Polymers*, 127: 252-263.

Corstens, M.N., Berton-Carabin, C.C., de Vries, R., Troost, F.J., Masclee, A.A.M. and Schroën, K. (2017). Food-grade micro-encapsulation systems that may induce satiety via delayed lipolysis: A review, *Critical Reviews in Food Science and Nutrition*, 57: 2218-2244.

Cortés, R.N.F., Martínez, M.G., Guzmán, I.V., Llano, S.L.A., Grosso, C.R.F. and Bustos, F.M. (2014). Evaluation of modified amaranth starch as shell material for encapsulation of probiotics, *Cereal Chemistry*, 91: 300-308.

Cuomo, F., Cofelice, M., Venditti, F., Ceglie, A., Miguel, M., Lindman, B. and Lopez, F. (2017). *In-vitro* digestion of curcumin-loaded chitosan-coated liposomes, *Colloids and Surfaces B: Biointerfaces*, 168: 29-34.

Davidov-Pardo, G., Pérez-Ciordia, S., Marın-Arroyo, M.R. and McClements, D.J. (2015). Improving resveratrol bioaccessibility using biopolymer nanoparticles and complexes: Impact of protein-carbohydrate Maillard conjugation, *Journal of Agricultural and Food Chemistry*, 63: 3915-3923.

Egger, L., Ménard, O., Delgado-Andrade, C., Alvito, P., Assunçao, R., Balance, S., Barberá, R., Brodkorb, A., Cattenoz, T., Clemente, A., Comi, I., Dupont, D., Garcia-Llatas, G., Lagarda, M.J., Le Feunteun, S., Janssen Duijghuijsen, L., Karakaya, S., Lesmes, U., Mackie, A.R., Martins, C., Meyneir, A., Miralles, B., Murray, B.S., Pihlanto, A., Picariello, G., Santos, C.N., Simsek, S., Recio, I., Rigby, N., Rioux, L.-E., Stoffers, H., Tavares, A., Tavares, L., Turgeon, S., Ulleberg, E.K., Vegarud, G.E., Vergères, G. and Portmann, R. (2016). The harmonized INFOGEST *in vitro* digestion method: From knowledge to action, *Food Research International*, 88: 217-225.

Eltayeb, M., Stride, E. and Edirisinghe, M. (2015). Preparation, characterization and release kinetics of ethylcellulose nanoparticles encapsulating ethylvanillin as a model functional component, *Journal of Functional Foods*, 14: 726-735.

Emami, S., Azadmard-Damirchi, S., Peighambardoust, S.H., Valizadeh, H. and Hesari J. (2016) Liposomes as carrier vehicles for functional compounds in food sector, *Journal of Experimental Nanoscience*, 11: 737-759.

Esfanjani, A.F. and Jafari, S.M. (2016). Biopolymer nano-particles and natural nano-carriers for nano-encapsulation of phenolic compounds, *Colloids and Surfaces B: Biointerfaces*, 146: 532-543.

Estevinho, B.N. and Rocha, F. (2017). Kinetic models applied to soluble vitamins delivery systems prepared by spray drying, *Drying Technology*, 35: 1249-1257.

Ezhilarasi, P.N., Karthik, P, Chhanwal, N. and Anandharamakrishnan, C. (2013). Nanoencapsulation Techniques for Food Bioactive Components: A Review, *Food and Bioprocess Technology*, 6: 628-647.

Farfan, M., Villalon, M.J., Ortiz, M.E., Nieto, S. and Bouchon, P. (2015). *In vivo* postprandial bioavailability of interesterified-lipids in sodium-caseinate or chitosan based O/W emulsions, *Food Chemistry*, 171: 266-271.

Feng, W., Yue, C., Wusigale, Ni, Y. and Liang, L. (2018). Preparation and characterization of emulsion-filled gel beads for the encapsulation and protection of resveratrol and α-tocopherol, *Food Research International*, 108: 161-171.

Flores, F.P., Singh, R.K., Kerr, W.L., Pegg, R.B. and Kong, F. (2014). Total phenolics content and antioxidant capacities of microencapsulated blueberry anthocyanins during *in vitro* digestion, *Food Chemistry*, 153: 272-278.

Frank, K., Walz, E., Graf, V., Greiner, R., Köhler, K. and Schuchmann, P.H. (2012). Stability of anthocyanin-rich W/O/W-emulsions designed for intestinal release in gastrointestinal environment, *Journal of Food Science*, 12: 50-57.

Frenzel, M. and Steffen-Heins, A. (2015). Whey protein coating increases bilayer rigidity and stability of liposomes in food-like matrices, *Food Chemistry*, 173: 1090-1099.

Fritzen-Freire, C.B., Prudêncio, E.S., Amboni, R.D.M.C., Pinto, S.S., Negrão-Murakami, A.N. and Murakami, F.S. (2012). Microencapsulation of bifidobacteria by spray drying in the presence of prebiotics, *Food Research International*, 45: 306-312.

Gibis, M., Zeeb, B. and Weiss, J. (2014). Formation, characterization, and stability of encapsulated hibiscus extract in multilayered liposomes, *Food Hydrocolloids*, 38: 28-39.

Gombotz, W.R. and Wee, S. (1998). Protein release from alginate matrices, *Advanced Drug Delivery Reviews*, 31: 267-285.

Gültekin-Özgüven, M., Karadag, A., Duman, S., Özkal, D. and Özçelik, B. (2016). Fortification of dark chocolate with spray dried black mulberry (*Morus nigra*) waste extract encapsulated in chitosan-coated liposomes and bioaccessibility studies, *Food Chemistry*, 201: 205-212.

Gumus, C.E., Davidov-Pardo, G. and McClements, D.J. (2016). Lutein-enriched emulsion-based delivery systems: Impact of Maillard conjugation on physicochemical stability and gastrointestinal fate, *Food Hydrocolloids*, 60: 38-49.

Gunasekaran, S., Ko, S. and Xiao, L. (2007). Use of whey proteins for encapsulation and controlled delivery applications, *Journal of Food Engineering*, 83: 31-40.

Hädicke, A. and Blume, A. (2014). Interactions of Pluronic block copolymers with lipid vesicles depend on lipid phase and Pluronic aggregation state, *Colloid and Polymer Science*, 293: 267-276.

Harper, B.A., Barbut, S., Lim, L.-T. and Marcone, M.F. (2013). Characterization of 'wet' alginate and composite films containing gelatin, whey or soy protein, *Food Research International*, 52: 452-459.

Hasanvand, E., Fathi, M., Bassiri, A., Javanmard, M. and Abbaszadeh, R. (2015). Novel starch-based nanocarrier for vitamin D fortification of milk: Production and characterization, *Food and Bioproducts Processing*, 96: 264-277.

He, H., Hong, Y., Gu, Z., Liu, G., Cheng, L. and Li, Z. (2016). Improved stability and controlled release of CLA with spray-dried microcapsules of OSA-modified starch and xanthan gum, *Carbohydrate Polymers*, 147: 243-250.

Hébrard, G., Hoffart, V., Beyssac, E., Cardot, J.-M., Alric, M. and Subirade, M. (2010). Coated whey protein/alginate microparticles as oral controlled delivery systems for probiotic yeast, *Journal of Microencapsulation*, 27: 292-302.

Higuchi, T. (1963). Mechanism of sustained-action medication. Theoretical analysis of rate of release of solid drugs dispersed in solid matrices, *Journal of Pharmaceutical Sciences*, 52: 1145-1149.

Istenič, K., Balanč, B.D., Djordjević, V.B., Bele, M., Nedović, V.A., Bugarski, B.M. and Poklar Ulrih, N. (2015). Encapsulation of resveratrol into Ca-alginate submicron particles, *Journal of Food Engineering*, 167: 196-203.

Jakobek, L. (2015). Interactions of polyphenols with carbohydrates, lipids and proteins, *Food Chemistry*, 175: 556-567.

Jantzen, M., Göpel, A. and Beermann, C. (2013). Direct spray drying and microencapsulation of probiotic *Lactobacillus reuteri* from slurry fermentation with whey, *Journal of Applied Microbiology*, 115: 1029-1036.

Jiang, F., Chen, S., Cao, Z. and Wang, G. (2016). A photo, temperature, and pH responsive spiropyran-functionalized polymer: Synthesis, self-assembly and controlled release, *Polymers*, 83: 85-91.

Jovanović, A.A., Balanč, B.D., Djordjević, V.B., Ota, A., Skrt, M., Šavikin, K.P., Bugarski, B.M., Nedović, V.A. and Poklar Ulrih, N. (2019). Effect of gentisic acid on the structural-functional properties of liposomes incorporating β-sitosterol, *Colloids and Surfaces B: Biointerfaces*, 183: 110422.

Kim, B.-K., Cho, A.-R. and Park, D.J. (2016). Enhancing oral bioavailability using preparations of apigenin-loaded W/O/W emulsions: *In vitro* and *in vivo* evaluations, *Food Chemistry*, 206: 85-91.

Koga, C.C., Ferruzzi, M.G., Andrade, J.E. and Lee, Y. (2016). Stability of trans-resveratrol encapsulated in a protein matrix produced using spray drying to UV light stress and simulated gastro-intestinal digestion, *Journal of Food Science*, 81: 292-300.

Kuang, S.S., Oliveira, J.C. and Crean, A.M. (2010). Microencapsulation as a tool for incorporating bioactive ingredients into food, *Critical Reviews in Food Science and Nutrition*, 50: 951-968.

Kumar, A., Vimal, A. and Kumar, A. (2016). Why Chitosan? From properties to perspective of mucosal drug delivery, *International Journal of Biological Macromolecules*, 91: 615-622.

Kwan, A. and Davidov-Pardo, G. (2018). Controlled release of flavor oil nanoemulsions encapsulated in filled soluble hydrogels, *Food Chemistry*, 250: 46-53.

Lesmes, U. and McClements, D.J. (2012). Controlling lipid digestibility: Response of lipid droplets coated by β-lactoglobulin-dextran Maillard conjugates to simulated gastrointestinal conditions, *Food Hydrocolloids*, 26: 221-230.

Li, Y., Kim, J., Park, Y. and McClements, D.J. (2012). Modulation of lipid digestibility using structured emulsion-based delivery systems: Comparison of *in vivo* and *in vitro* measurements, *Food & Function*, 3: 528-536.

Li, Z., Peng, S., Chen, X., Zhu, Y., Zou, L., Liu, W. and Liu, C. (2018). Pluronics modified liposomes for curcumin encapsulation: Sustained release, stability and bioaccessibility, *Food Research International*, 108: 246-253.

Liu, F., Gao, Y. and Yuan, F. (2014). Effects of chitosan addition on *in vitro* digestibility of protein-coated lipid droplets, *Journal of Dispersion Science and Technology*, 36: 1556-1563.

Liu, F., Ma, C., Zhang, R., Gao, Y. and McClements, D.J. (2017). Controlling the potential gastrointestinal fate of β-carotene emulsions using interfacial engineering: Impact of coating lipid droplets with polyphenol-protein-carbohydrate conjugate, *Food Chemistry*, 221: 395-403.

Liu, H., Gong, J., Chabot, D., Miller, S.S., Cui, S.W., Zhong, F. and Wang, Q. (2018). Improved survival of *Lactobacillus zeae* LB1 in a spray dried alginate-protein matrix, *Food Hydrocolloids*, 78: 100-108.

Liu, W., Ye, A., Liu, C., Liu, W. and Singh H. (2012). Structure and integrity of liposomes prepared from milk- or soybean-derived phospholipids during *in vitro* digestion, *Food Research International*, 48: 499-506.

Liu, W., Liu, J., Liu, W., Li, T. and Liu, C. (2013). Improved physical and *in vitro* digestion stability of a polyelectrolyte delivery system based on layer-by-layer self-assembly alginate-chitosan-coated nanoliposomes, *Journal of Agricultural and Food Chemistry*, 61: 4133-4144.

Luo, Q., Boom, R.M. and Janssen, A.E.M. (2015). Digestion of protein and protein gels in simulated gastric environment, *LWT – Food Science and Technology*, 63: 161-168.

Madureira, A.R., Pereira A. and Pintado, M. (2016). Chitosan nanoparticles loaded with 2,5-dihydroxybenzoic acid and protocatechuic acid: Properties and digestion, *Journal of Food Engineering*, 174: 8-14.

Manatunga, D.C., de Silva, R.M., de Silva, K.M.N., de Silva, N., Bhandari, S., Yap, Y.K. and Costha, N.P. (2017). pH responsive controlled release of anti-cancer hydrophobic drugs from sodium alginate and hydroxyapatite bi-coated iron oxide nanoparticles, *European Journal of Pharmaceutics and Biopharmaceutics*, 117: 29-38.

Mandalari, G., Bisignano, C., Filocamo, A., Chessa, S., Sarò, M., Torre, G., Faulks, R.M. and Dugo, P. (2013). Bioaccessibility of pistachio polyphenols, xanthophylls, and tocopherols during simulated human digestion, *Nutrition*, 29: 338-344.

Mao, Y. and McClements, D.J. (2012). Influence of electrostatic hetero aggregation of lipid droplets on their stability and digestibility under simulated gastrointestinal conditions, *Food and Function*, 3: 1025-1034.

Mazzocato, M.C., Thomazini, M. and Favaro-Trindade, C.S. (2019). Improving stability of vitamin B_{12} (Cyanocobalamin) using microencapsulation by spray chilling technique, *Food Research International*, 126: 108663.

McClements, D.J. (2015). Encapsulation, protection, and release of hydrophilic active components: Potential and limitations of colloidal delivery systems, *Advances in Colloid and Interface Science*, 219: 27-53.

McConnell, E.L., Murdan, S. and Basit, A.W. (2008). An investigation into the digestion of chitosan (noncrosslinked and crosslinked) by human colonic bacteria, *Journal of Pharmaceutical Sciences*, 97: 3820-3829.

Minekus, M., Alminger, M., Alvito, P., Ballance, S., Bohn, T., Bourlieu, C., Carriere, F., Boutrou, R., Corredig, M., Dupont, D., Dufour, C., Egger, L., Golding, M., Karakaya, S., Kirkhus, B., Le Feunteun, S., Lesmes, U., Macierzanka, A., Mackie, A., Marze, S., McClements, D.J., Menard, O., Recio, I., Santos, C.N., Singh, R.P., Vegarud, G.E., Wickham, M.S.J., Weitschies, W. and Brodkorb, A. (2014). A standardized static *in vitro* digestion method suitable for food – An international consensus, *Food and Function*, 5: 1113-1124.

Mohammadi, N., Ehsani, M.R. and Bakhoda, H. (2018). Development of caffeine-encapsulated alginate based matrix combined with different natural biopolymers, and evaluation of release in simulated mouth conditions, *Flavor and Fragrance Journal*, 33: 357-366.

Moumita, S., Goderska, K., Johnson, E.M., Das, B., Indira, D., Yadav, R., Kumari, S. and Jayabalan, R. (2017). Evaluation of the viability of free and encapsulated lactic acid bacteria using *in-vitro* gastrointestinal model and survivability studies of synbiotic microcapsules in dry food matrix during storage, *LWT*, 77: 460-467.

Mun, S., Kim, Y. R. and McClements, D.J. (2015a). Control of β-carotene bioaccessibility using starch-based filled hydrogels, *Food Chemistry*, 173: 454-461.

Mun, S., Kim, Y.R., Shin, M. and McClements, D.J. (2015b). Control of lipid digestion and nutraceutical bioaccessibility using starch-based filled hydrogels: Influence of starch and surfactant type, *Food Hydrocolloids*, 44: 380-389.

Nguyen, T.X., Huang, L., Liu, L., Abdalla, A.M.E., Gauthier, M. and Yang, G. (2014). Chitosan-coated nano-liposomes for the oral delivery of berberine hydrochloride, *Journal of Materials Chemistry B*, 2: 7149-7159.

Nooshkam, M. and Varidi M. (2020). Maillard conjugate-based delivery systems for the encapsulation, protection, and controlled release of nutraceuticals and food bioactive ingredients: A review, *Food Hydrocolloids*, 100: 105389.

Oidtmann, J., Schantz, M., Mäder, K., Baum, M., Berg, S., Betz, M., Kulozik, U., Leick, S., Rehage, H., Schwarz, K. and Richling, E. (2012). Preparation and comparative release characteristics of three anthocyanin encapsulation systems, *Journal of Agricultural and Food Chemistry*, 60: 844-851.

Paéz, R., Lavari, L., Vinderola, G., Audero, G., Cuatrin, A., Zaritzky, N. and Reinheimer, J. (2012). Effect of heat treatment and spray drying on lactobacilli viability and resistance to simulated gastrointestinal digestion, *Food Research International*, 48: 748-754.

Penalva, R., Esparza, I., Agüeros, M., Gonzalez-Navarro, C.J., Gonzalez-Ferrero, C. and Irache, J.M. (2015). Casein nanoparticles as carriers for the oral delivery of folic acid, *Food Hydrocolloids*, 44: 399-406.

Peppas, N.A. (1985). Analysis of Fickian and non-Fickian drug release from polymers, *Pharmaceutica Acta Helvetiae*, 60: 110-111.

Peppas, N.A. and Ritger, P. (1987). A simple equation for description of solute release I. Fickian and non-Fickian release from non-swellable devices in the form of slabs, spheres, cylinders or discs, *Journal of Controlled Release*, 5: 23-36.

Rayment, P., Wright, P., Hoad, C., Ciampi, E., Haydock, D. and Gowland, P. (2009). Investigation of alginate beads for gastrointestinal functionality, part 1: *In vitro* characterization, *Food Hydrocolloids*, 23: 816-822.

Raymond, Y. and Champagne, C.P. (2014). Dissolution of lipid-based matrices in simulated gastrointestinal solutions to evaluate their potential for the encapsulation of bioactive ingredients for foods, *International Journal of Food Science*, 2014: 749630.

Robert, P., García, P., Reyes, N., Chávez, J. and Santos J. (2012). Acetylated starch and inulin as encapsulating agents of gallic acid and their release behavior in a hydrophilic system, *Food Chemistry*, 134: 1-8.

Rodriguez, E.B., Almeda, R.A., Vidallon, M.L.P. and Reyes, C.T. (2018). Enhanced bioactivity and efficient delivery of quercetin through nanoliposomal encapsulation using rice bran phospholipids, *Journal of the Science of Food and Agriculture*, 99: 1980-1989.

Samtlebe, M., Ergin, F., Wagner, N., Neve, H., Küçükçetin, A., Franz, CMAP., Heller, K.J., Hinrichs, J. and Atamer, Z. (2016). Carrier systems for bacteriophages to supplement food systems: Encapsulation and controlled release to modulate the human gut microbiota, *LWT – Food Science and Technology*, 68: 334-340.

Sari, T.P., Mann, B., Kumar, R., Singh, R.R.B., Sharma, R., Bhardwaj, M. and Athira S. (2015). Preparation and characterization of nanoemulsion encapsulating curcumin, *Food Hydrocolloids*, 43: 540-546.

Sathasivam, T., Muniyandy, S., Chuah, L.H. and Janarthanan, P. (2018). Encapsulation of red palm oil in carboxymethyl sago cellulose beads by emulsification and vibration technology: Physicochemical characterization and *in vitro* digestion, *Journal of Food Engineering*, 231: 10-21.

Shah, B.R., Zhang, C., Li, Y. and Li, B. (2016). Bioaccessibility and antioxidant activity of curcumin after encapsulated by nano and Pickering emulsion based on chitosan-tripolyphosphate nanoparticles, *Food Research International*, 89: 399-407.

Siepmann, J. and Siepmann, F. (2012). Modeling of diffusion controlled drug delivery, *Journal of Controlled Release*, 161: 351-362.

Soares, J.M.D., Almeida, J.R.G.S. and de Oliveira, H.P. (2017). Controlled release of extract of *Morus nigra* from Eudragit L-100 electrospun fibers: Toxicity and *in vitro* release evaluation, *Current Traditional Medicine*, 3: 146-154.

Spivey, A. (2010). A matter of degrees: advancing our understanding of acrylamide, *Environmental Health Perspectives*, 118: 161-167.

Steingoetter, A., Radovic, T., Buetikofer, S., Curcic, J., Menne, D., Fried, M., Schwizer, W. and Wooster, T.J. (2015). Imaging gastric structuring of lipid emulsions and its effect on gastrointestinal function: A randomized trial in healthy subjects, *The American Journal of Clinical Nutrition*, 101: 714-724.

Stephen, A.J., Phillips, G.O. and Williams, P.A. (2006). *Food Polysaccharides and Their Applications* (second edition), Boca Raton, FL.: CRC Press.

Sun, R. and Xia, Q. (2019). Release mechanism of lipid nanoparticles immobilized within alginate beads influenced by nanoparticle size and alginate concentration, *Colloid and Polymer Science*, 297: 1183-1198.

Talón, E., Trifkovic, K.T., Nedovic, V.A., Bugarski, B., Vargas, M., Chiralt, A. and González-Martínez C. (2017). Antioxidant edible films based on chitosan and starch containing polyphenols from thyme extracts, *Carbohydrate Polymers*, 157: 1153-1161.

Tan, C., Feng, B., Zhang, X., Xia, W. and Xia, S. (2016). Biopolymer-coated liposomes by electrostatic adsorption of chitosan (chitosomes) as novel delivery systems for carotenoids, *Food Hydrocolloids*, 52: 774-784.

Tan, C., Zhang, Y., Abbas, S., Feng, B., Zhang, X. and Xia, S. (2014). Modulation of the carotenoid bioaccessibility through liposomal encapsulation, *Colloids and Surfaces B: Biointerface*, 123: 692-700.

Tanzina, H., Avik, K., Ruhul, A.K., Riedl, B. and Lacroix, M. (2013). Encapsulation of probiotic bacteria in biopolymeric system, *Critical Reviews in Food Science and Nutrition*, 53: 909-916.

Toro Uribe, S., Ibanez, E., Decker, E., McClements, D.J., Zhang, R., Lopez-Giraldo, L.J. and Herrero, M. (2018). Design, fabrication, characterization and *in vitro* digestion of alkaloid-, catechin-, and cocoa extract-loaded liposomes, *Journal of Agricultural and Food Chemistry*, 66: 12051-12065.

Ulloa, P.A., Guarda, A., Valenzuela, X., Rubilar, J.F. and Galotto, M.J. (2017). Modeling the release of antimicrobial agents (thymol and carvacrol) from two different encapsulation material, *Food Science and Biotechnology*, 26: 1763-1772.

Volić, M., Pajić-Lijaković, I., Djordjević, V., Knežević-Jugović, Z., Pećinar, Z., Dajić-Stevanović, Z., Veljović, Dj., Hadnadjev, M. and Bugarski B. (2018). Alginate/soy protein system for essential oil encapsulation with intestinal delivery, *Carbohydrate Polymers*, 200: 15-24.

Wang, Z., Li, Y., Chen, L., Xin, X. and Yuan, Q. (2013). A study of controlled uptake and release of anthocyanins by oxidized starch microgels, *Journal of Agricultural and Food Chemistry*, 61: 5880-5887.

Wickham, M.J.S., Faulks, R.M., Mann, J. and Mandalari, G. (2012). The design, operation, and application of a dynamic gastric model, *Dissolution Technologies*, 19: 15-22.

Xu, D., Yuan, F., Gao, Y., Panya, A., McClements, D.J. and Decker, E.A. (2014). Influence of whey protein–beet pectin conjugate on the properties and digestibility of β-carotene emulsion during *in vitro* digestion, *Food Chemistry*, 156: 374-379.

Xu, J., Zhao, W., Ning, Y., Bashari, M., Wu, F., Chen, H., Yang, N., Jin, Z., Xu, B., Zhang, L. and Xu, X. (2013). Improved stability and controlled release of ω3/ω6 polyunsaturated fatty acids by spring dextrin encapsulation, *Carbohydrate Polymers*, 92: 1633-1640.

Yang, W., Zhang, M., Li, X., Jiang, J., Sousa, A.M., Zhao, Q., Pontious, S. and Liu, L. (2019). Incorporation of tannic acid in foodgrade guar gum fibrous mats by electrospinning technique, *Polymers*, 11: 141.

Yüksel-Bilsel, A. and Şahin-Yeşilçubuk N. (2019). Production of probiotic kefir fortified with encapsulated structured lipids and investigation of matrix effects by means of oxidation and *in vitro* digestion studies, *Food Chemistry*, 296: 17-22.

Zhang, W., Rong, J., Wang, Q. and He, X. (2009). The encapsulation and intracellular delivery of trehalose using a thermally responsive nanocapsule, *Nanotechnology*, 20: 275101.

Zhang, Y. and Zhong, Q. (2018). Freeze-dried capsules prepared from emulsions with encapsulated lactase as a potential delivery system to control lactose hydrolysis in milk, *Food Chemistry*, 241: 397-402.

Zhang, Y., Pu, C., Tang, W., Wang, S. and Sun Q. (2019). Gallic acid liposomes decorated with lactoferrin: Characterization, *in vitro* digestion and antibacterial activity, *Food Chemistry*, 293: 315-322.

Zhang, Z., Zhang, R., Chen, L., Tong, Q. and McClements D.J. (2015a). Designing hydrogel particles for controlled or targeted release of lipophilic bioactive agents in the gastrointestinal tract, *European Polymer Journal*, 72: 698-716.

Zhang, Z., Zhang, R., Decker, E.A. and McClements D.J. (2015b). Development of food-grade filled hydrogels for oral delivery of lipophilic active ingredients: pH-triggered release, *Food Hydrocolloids*, 44: 345-352.

Zhang, Z., Zhang, R., Zou, L. and McClements, D. J. (2016a). Protein encapsulation in alginate hydrogel beads: Effect of pH on microgel stability, protein retention and protein release, *Food Hydrocolloids*, 58: 308-315.

Zhang, Z.P., Zhang, R.J., Zou, L.Q., Chen, L., Ahmed, Y., Al Bishri, W., Balamash, K. and McClements, D.J. (2016b). Encapsulation of curcumin in polysaccharide-based hydrogel beads: Impact of bead type on lipid digestion and curcumin bioaccessibility, *Food Hydrocolloids*, 58: 160-170.

Zhang, Z., Zhang, R. and McClements, D.J. (2017). Control of protein digestion under simulated gastrointestinal conditions using biopolymer microgels, *Food Research International*, 100: 86-94.

Zhong, L., Ma, N., Wu, Y., Zhao, L., Ma, G., Pei, F. and Hu, Q. (2019). Gastrointestinal fate and antioxidation of β-carotene emulsion prepared by oat protein isolate-*Pleurotus ostreatus* β-glucan conjugate, *Carbohydrate Polymers*, 221: 10-20.

Zhu, F. (2017). Encapsulation and delivery of food ingredients using starch-based systems, *Food Chemistry*, 229: 542-552.

Encapsulation of Plant Extracts

**Ana Kalušević[1,2]*, Ana Salević[3], Aleksandra Jovanović[4], Kata Trifković[5],
Mile Veljović[2,3], Radoslava Pravilović[6] and Viktor Nedović[3]**

[1] Academy of Applied Studies, Zorana Đinđića 152a, 11070 Belgrade, Serbia
[2] Serbian Food Technology Council, Starci bb, Aleksandrovac, Serbia
[3] Faculty of Agriculture, University of Belgrade, Nemanjina 6, 11080 Zemun, Serbia
[4] Innovation Center of Faculty of Technology and Metallurgy, University of Belgrade,
  Karnegijeva 4, 11060 Belgrade, Serbia
[5] Teagasc Food Research Centre, Moorepark, Fermoy, Co. Cork, Ireland
[6] Faculty of Technology and Metallurgy, University of Belgrade, Karnegijeva 4, 11060
  Belgrade, Serbia

1. Introduction

The plant kingdom represents an excellent source of various actives: minerals, vitamins, amino acids, lipids, fibers, polyphenols, essential oils, flavors, fragrances and pigments. Their usage for medical purposes has a long history. The health-beneficial effects from the perspective of antimicrobial, anticancerogenic, antioxidant and anti-inflammatory properties have been extensively studied and mainly related to phenolic compounds. Based on these findings, bioactive compounds derived from plant materials, especially herbs and spices, are of great interest for the development of novel dietary supplements and functional food products (Rubió *et al.* 2013).

Critical factors for the implementation of bioactive compounds extracted from the natural sources as functional food ingredients are: unpleasant taste, susceptibility to various factors during food processing and storage (temperature, oxygen, light) or in the gastrointestinal tract (pH, enzymes, presence of other nutrients), rapid intestinal metabolism, low permeability and/or solubility. In this term, the use of encapsulated bioactive compounds instead of free ones, is a promising approach to preserve their stability, bioactivity and bioavailability which is the main challenge to achieve their effectiveness (Fang and Bhandari, 2010; Đorđević *et al.*, 2015). The other possible benefits of encapsulated bioactive compounds in food applications are superior handling, taste masking and creation of visual and textural effects (Zuidam and Shimoni, 2010). Generally, encapsulation may be defined as a process

*Corresponding author: ana.kalusevic@vhs.edu.rs

to entrap bioactive compounds within a carrier material(s) (Nedovic *et al.*, 2011). Various techniques and materials have been proposed for encapsulation of bioactive compounds, while selection of the optimal ones depends on stability and release type requirements, potential restrictions, concentration and aimed functionality of the encapsulate (Nedović *et al.*, 2013). Currently, many encapsulation techniques, such as spray-drying, spray-cooling, extrusion, coacervation, and emulsification have been used in turn to protect and enhance the stability of bioactive compounds. This area is of practical interest in food industry, but pharmaceutical, medicinal and cosmetic as well.

Among various encapsulation techniques, spray-drying is most commonly used for plant extracts stabilization, and generally, in the food industry. The main reasons are: low operating costs, high production rates, easy handling and available equipment (Gharsallaoui *et al.*, 2007; Đorđević *et al.*, 2015). The process of encapsulation by spray-drying involves the following steps: preparation and homogenization of a dispersion or emulsion of carrier material(s) and plant extract(s), atomization of the dispersion or emulsion in a heated air stream and dehydration of the atomized particles (Gharsallaoui *et al.*, 2007). However, prior to these steps optimized and adequate extraction of bioactive compounds from plant materials is required.

2. Extraction Techniques

The majority of the above-mentioned phytochemicals are very expensive or even more impossible to synthesize in the laboratory conditions; thus the best option of the manufacture of phytochemicals-rich products is the extraction from plant matrix (Pompeu *et al.*, 2009; Bucić-Kojić *et al.*, 2011). Therefore, the extraction procedures have been widely established in order to obtain different natural compounds for food, pharmaceutical and cosmetics industry (Wang and Weller, 2006; Pompeu *et al.*, 2009; Mustafa and Turner, 2011; Franco-Vega *et al.*, 2016). Since polyphenols, essential oils constituents and other bioactive compounds of various plants origin differ structurally, it is very difficult to establish a standardized extraction technique that would simultaneously extract all target compounds from each natural source. The extraction efficiency of the active compounds depends on the characteristics of plant source (conditions of cultivation, collection period, part of the plant, etc.), physicochemical properties of target components, extraction conditions (type of the solvent, solid:solvent ratio, extraction time, temperature, pressure, pH, the selected extraction technique, as well as on the presence of interfering substances) (Mustafa and Turner, 2011, Dent *et al.*, 2013; Jovanović *et al.*, 2017). For this reason, it is very important to optimize extraction conditions for each target component from each plant material (e.g. particle size of plant matrix, solid:solvent ratio, extraction time and particularly type of extraction medium and technique).

Traditional extraction procedures include maceration, heat-assisted extraction, percolation, digestion, hydrodistillation, Soxhlet extraction (Jovanović *et al.*, 2017; Vuleta *et al.*, 2012; Wang and Weller, 2006). Maceration represents solid-liquid extraction technique that is performed at room temperature by using a properly fragmented plant material and an appropriate solvent. Diffusion lasts until the concentrations of the active substances are equalized intracellular and extracellular. The resulting balance is disturbed by intense mixing during the entire extraction process, which shortens the extraction time to 10-30 min. The intensive agitation results in the growth of the concentration gradient in the solvent and the diffusion acceleration, with a consequent increase in the extraction yield. In order to achieve a higher degree of herbal material utilization, double maceration is applied to the larger particles of herbal matrix and the already extracted matrix is exposed to a fresh solvent, thus the concentration gradient is re-established (Vuleta *et al.*, 2012). In heat-assisted extraction, the application of high temperature increases the extraction efficiency as follows (1) the damage to the plant cell leads to an increase in the permeability of the membrane, and (2) the breaking of intermolecular connections (such as polyphenols-lipoproteins) increases the solubility of the active substances and accelerates the thermal movement of the molecules in the liquid phase and the extraction kinetics (Vuleta *et al.*, 2012; Jovanović *et al.*, 2017). Soxhlet extraction is a well-established technique as it outperforms other traditional extraction procedures, but with the limited field of application (unsuitable for thermosensitive components). During Soxhlet procedure, the surrounding solvent is usually recovered by evaporation, thus the evaporation temperature has a significant influence on the extract quality (Wang and Weller, 2006).

In recent studies, several modern extraction techniques have been established, including ultrasound- and microwave-assisted extraction, supercritical fluid extraction, pressurized liquid extraction (Wang and Weller, 2006; Mustafa and Turner, 2011; Milutinović *et al.*, 2015; Rai *et al.*, 2016; Jovanović *et al.*, 2017). In ultrasound-assisted extraction, due to the ultrasound waves, mechanical and thermal effects induce destruction of cell walls, release of cell contents and a better penetration of solvent into plant material, which lead to intensive mass transfer and higher extraction yield (Jovanović *et al.*, 2016). In microwave-assisted extraction, intracellular water heats and evaporates, resulting in a significant increase of the pressure, which causes damage to the plant cell wall and thus the intensive release of the active substances into the extraction medium (Milutinović *et al.*, 2015; Franco-Vega *et al.*, 2016). Supercritical fluid extraction is an extraction strategy which utilizes supercritical fluids as extraction solvents. The most commonly used supercritical solvent is carbon dioxide (CO_2). It is considered as green and non-toxic solvent, which has similar density as liquids, low viscosity and high diffusivity (Garavand *et al.*, 2019). Pressurized liquid extraction represents a new technique based on the use of liquid solvents at elevated temperature and pressure high enough to maintain the liquid state. The mentioned conditions enhance the extraction efficiency, in comparison to the procedures carried out at lower temperature and atmospheric pressure (Eikani *et al.*, 2007; Mustafa and Turner, 2011). The advantages and disadvantages of different conventional and modern extraction techniques are represented in Table 1.

Plant products, such as polyphenols-rich extracts and essential oils, obtained by using different plant sources, species, extraction techniques and solvents are represented in Table 2. The number of studies dealing with extraction of bioactive compounds from plants is constantly increasing. However, Table 2 gives a short review of studies wherein active components were extracted from different plant materials in order to be encapsulated.

The naturally derived extracts are commonly unstable, bitter, inconvenient to use, storage, or distribute, etc. Thus, the main aim is to transform the extracts into forms that allow easier handling, reduced stickiness, masked the unpleasant smell, with improved water dissolution rate. Additionally, the resulting forms have to be stable under different storage conditions, without affecting bioactivity and

Table 1: The Advantages and Disadvantages of Traditional and Novel Extraction Techniques

Extraction Technique	Advantages	Disadvantages	References
Maceration	(1) Simplicity (2) Suitable for extraction of various thermosensitive compounds	(1) Long extraction time (2) Large amount of solvent (3) Not possible to achieve the complete extraction	Vuleta *et al.*, 2012; Jovanović *et al.*, 2017; Garavand *et al.*, 2019; Bilušić *et al.*, 2020
Heat-assisted Extraction	(1) Faster kinetics (2) The efficient extraction and higher yield (3) Simple apparatus and technique	(1) Thermal degradation of active components (2) Large amount of solvent (3) Limited choice of plant material and medium	Bucić-Kojić *et al.*, 2011; , Mustafa and Turner, 2011 ; Jovanović *et al.*, 2017.
Soxhlet Extraction	(1) Intensive mass transfer by bringing fresh solvent into contact with the plant matrix (2) A relatively high extraction efficiency (3) No filtration requirement	(1) Long extraction time (2) High solvent consumption (3) The agitation cannot be provided (4) Thermal decomposition of the target Compounds	Wang and Weller, 2006; Eikani *et al.*, 2007; Andrade *et al.*, 2017; Garavand *et al.*, 2019
Ultrasound-assisted Extraction	(1) Better extraction yield (2) Fast kinetics (3) Reduction the operating temperature (4) Simple equipment and procedure (compared to other novel extraction techniques) (5) Numerous of potential extraction media	(1) The effects of ultrasound depend on the nature of the plant material (2) The active part of ultrasound is restricted to a zone located in the vicinity of the ultrasound probe (3) Potential degradation of antioxidants	Li *et al.*, 2005; Wang and Weller, 2006; Assami *et al.* 2012; Vajić *et al.*, 2015; Jovanović *et al.*, 2017; Garavand *et al.*, 2019

(Contd.)

Table 1: *(Contd.)*

Extraction Technique	Advantages	Disadvantages	References
Microwave-assisted Extraction	(1) Enhancement of the extraction efficiency (2) Significant reduction of extraction time and solvent consumption (3) Higher yields of antioxidant compounds (4) Simplicity and economy (5) Protection from oxygen	(1) Thermal decomposition of sensitive substances (2) Lower yield of non-polar and volatile constituents (3) The effects of microwaves depend on the polarity of surrounding solvent (4) A potential extraction of ballast substances	Wang and Weller, 2006; Amirah *et al.*, 2012; Cui *et al.*, 2015; Milutinović *et al.* 2015; Franco-Vega *et al.* 2016; Garavand *et al.* 2019.
Supercritical fluid Extraction	(1) Selective extractions and fractionations (2) No concentration process requirement (3) Suitable for volatiles (4) Rapid mass transfer and larger extraction rate (5) More environmentally friendly extraction procedure	(1) Uneconomic process (2) Complicated experimental conditions	Wang and Weller, 2006; Chatterjee and Bhattacharjee 2013; Giacometti *et al.*, 2018; Gallego *et al.*, 2019; Garavand *et al.*, 2019
Pressurized liquid Extraction	(1) Enhanced solubility and mass transfer (2) Better extraction yield and selectivity (3) 'Green' technology (4) Suitable for extraction of polar and non-polar lipids (5) The decrease of time and solvent consumption	(1) A risk of thermolabile compounds degradation (2) The influence of pressure on release of most compounds is usually negligible (3) The expensive equipment	Eikani *et al.*, 2007; Mustafa and Turner, 2011; Giacometti *et al.*, 2018; Gallego *et al.*; 2019; Garavand *et al.*, 2019.

Table 2: List of the Extracted Active Components for Further Encapsulation by Using Different Plant Sources, Techniques and Solvents

Plant Species	Extraction Techniques	Extraction Medium	Active Components	References
Bark				
Hardy rubber tree (*E. ulmoides*), Mango (*M. indica*), Oak (*Q. robur*), Physic nut (*J. curcas*), Spruce (*P. abies*)	Heat-, ultrasound- and microwave-assisted extraction	Water, methanol/water mixture, ethanol/water mixture, aqueous β-cyclodextrin solution	Polyphenolic compunds (phenolic acids – Syringic, sinapic, *p*-coumaric acid, Flavonoids – naringenin, mangiferin)	Li *et al.*, 2005 ; Amirah *et al.*, 2012 ; Bouras *et al.*, 2015 ; Ghitescu *et al.*, 2015 ; Mura *et al.*, 2015
Flowers				
Butterfjee pea (*C. ternatea*), Clove (*S. aromaticum*), Hibiscus (*H. sabdariffa*), Saffron (*C. sativus*), Sea lavender (*L. sinuatum*)	Maceration, Soxhlet, heat, ultrasound, microwave, supercritical fluid and high pressure extraction	Water, methanol, ethanol, alcohol/water mixture, petroleum ether, diethyl ether, methanol/acetonitrile, CO_2	Polyphenolic compounds (flavonoids), essential oil, vitamins, pigments (β-carotene)	Chatterjee and Bhattacharjee, 2013; Gibis *et al.*, 2014; Pasukamonset *et al.*, 2016, Xu *et al.*, 2017; Gallego *et al.*, 2019; Garavand *et al.*, 2019
Fruits				
Bilberry (*V. myrtillus*), Chokeberry (*A. melanocarpa*), Hibiscus (*H. sabdariffa*), Noni (*M. citrifolia*)	Maceration, extraction at 30-35°C, supercritical fluid extraction	Ethanol/water mixture, ethyl acetate, ethanol/potassium chlorate buffer, CO_2	Polyphenols (flavonoids - Quercetin, anthocyanins - Cyanidin)	Krishnaiah *et al.*, 2012; Ćujić *et al.*, 2016; Nguyen *et al.*, 2018; Ćujić Nikolić *et al.*, 2018, 2019; Gallego *et al.*, 2019
Herbs				
Oregano (*O. vulgare*), Rosemary (*R. officinalis*), Sage (*S. officinalis*), St John's wort (*H. perforatum*), Wild thyme (*T. serpyllum*)	Maceration, heat, ultrasound, microwave, supercritical fluid and high pressure extraction	Water, methanol, ethanol, methanol/water mixture, ethanol/water mixture	Polyphenols (phenolic acids – rosmarinic, salvianolic acid, Flavonoids - luteolin, apigenin)	Kalogeropoulos *et al.*, 2010; Stojanović *et al.*, 2011; Dent *et al.*, 2013; Trifković *et al.*, 2014; Belščak-Cvitanović *et al.*, 2015b; Giacometti *et al.*, 2018; Gallego *et al.*, 2019; Bilušić *et al.*, 2020

(Contd.)

Table 2: (*Contd.*)

Plant Species	Extraction Techniques	Extraction Medium	Active Components	References
		Leaves		
Green tea (*C. sinensis*), Hawthorn (*C. laevigata*), Nettle (*U. dioica*), Olive (*O. europea*), Raspberry (*R. idaeus*), Yarrow (*A. millefolium*)	Maceration, heat- and ultrasound-assisted extraction	Water, methanol/water mixture, ethanol/water mixture	Polyphenols (phenolic acids – caffeic, rosmarinic acid, flavonoids), minerals	Belščak-Cvitanović et al., 2011; Belščak-Cvitanović et al., 2015a,b; Vajić et al., 2015; Giacometti et al., 2018
		Marc/pomace		
Apple (*M. domestica*), Blackberry (*R. fruticosus* L.), Cranberry (*V. macrocarpon*), Grape (*V. vinifera*), Olive (*O. europea*), Raspberry (*R. idaeus*), Sour cherry (*P. cerasus*)	Maceration, heat, ultrasound, supercritical fluid and high pressure extraction	Alcohol/water mixture, aqueous β-cyclodextrin solution, CO$_2$	Polyphenols (phenolic acids, flavonoids, stilbens), oil, lipids, proteins, sugars	Cilek et al., 2012; Spigno et al., 2013; Kalusevic et al., 2016; Cai et al., 2018; Gallego et al., 2019
		Peel/Skin/Coat		
Gac (*M. cochinchinensis*), Mandarin (*C. reticulata*), Mango (*M. indica*), Onion (*Allium cepa*), Pomegranate (*P. granatum*), Grape (*V. vinifera*), Black soybean (*Glycine max* (L.), Beetroot (*B. vulgaris*), Red pepper (*C. annuum* L)	Maceration followed by heat, ultrasound- and microwave-assisted extraction	Water, methanol, ethanol, acidified ethanol, alcohol/water mixture, ethyl acetate, diethyl ether, acetone, acetone/ethanol mixture	Polyphenolic compounds (phenolic acids, flavonoids, anthocyanins, tannins), carotenoids, betanin, isobetanin	Kaderides et al., 2015; Kalušević et al., 2017a,b,c; Hidalgo et al., 2018; Velderrain-Rodríguez et al., 2018; Chuyen et al., 2019; Ghatak and Iyyaswami, 2019; Hu et al., 2019; Vulić et al., 2019

	Extraction methods	Solvents	Compounds	References
Root				
Angelica (*A. archangelica*), Ginger (*Z. officinale*), Knotweed (*P. cuspidatum*), Licorice (*G. glabra*), Pigeon pea (*C. cajan*)	hydrodistillation, Soxhlet, heat, ultrasound, microwave and supercritical fluid extraction	water, methanol, ethanol, methanol/water mixture, ethanol/water mixture, CO_2	phenolic stilbenes (resveratrol, polydatin), fatty acids, essential oil, inositol, polysaccharides	Doneanu and Anitescu, 1998; Mantegna *et al.*, 2012; Cui *et al.*, 2015; Karami *et al.*, 2015; Akhtar and Shahzad, 2017; Tavares da Silva *et al.*, 2018
Seeds				
Caraway (*C. carvi*), Cardamom (*E. cardamomum*), Coriander (*C. sativum*), Passion fruit (*P. edulis*), Pink pepper (*S. terebinthifolius*)	hydrodistillation, Soxhlet, ultrasound, supercritical fluid and high pressure extraction	water, ethanol, ethyl acetate, hexane, CO_2	oil, essential oil, polyphenolic compounds (phenolic acids, flavonoids), monoterpenoid (linalool)	Eikani *et al.*, 2007; Assami *et al.*, 2012; Sereshti *et al.*, 2012; Andrade *et al.*, 2017; Oliveira *et al.*, 2017

antioxidant activity of plant extracts. Encapsulation is a technology that allows all above-mentioned issues. To be precise, the formation of a physical barrier between a bioactive compound extracted from plant material and its surrounding (e.g. food matrix) is an effective way of protection.

3. Encapsulation Techniques

As could be concluded from the numerous research papers cited in Table 2, the most commonly used techniques for encapsulation of plant extracts are spray-drying, freeze-drying, extrusion and encapsulation in cyclodextrins. Thus, this section gives short review of these techniques and their (potential) application.

3.1 Spray-drying

Spray-drying is a widely used and relatively low-cost technique that provides very short contact time between material and drying medium. Although, a too high operation temperature may lead to some quality losses, so parameters strongly depend on composition and sensitivity of liquid component that should be dried. Some of the main advantages of the spray dried powder forms of plant extracts designed for food applications are reduced bulk size, easier transportation and handling and long-term stability (Sansone *et al.*, 2011). On the other hand, there are several disadvantages, such as non-uniform size and shape of the particles and their tendency towards aggregation (Đorđević *et al.*, 2015).

Thus, spray-drying technique was employed to develop a novel form of functional, natural extracts from Makoni tea (*Fadogia ancylantha*), green tea (*Camellia sinensis*), oregano (*Origanum vulgare*), thyme *(Thymus vulgaris),* lemonbalm (*Melissa officinalis*) and coltsfoot (*Tussilago farfara*) that are very attractive to food industry due to their high content of phenolic compounds, antioxidant and antimicrobial activity (Sansone *et al.*, 2011; Belščak-Cvitanović *et al.*, 2015b; Bilušić *et al.*, 2020). Various carriers have been used, like pectin, alginate, inulin, protein isolates and gums. The obtained powders showed improved organoleptic, physicochemical, technological and functional properties that are desired for food application. Generally, the microencapsulated powder forms allowed easier handling, reduced stickiness, masked the unpleasant smell and improved water dissolution rate compared to the free extracts. Storage monitorings in different conditions have showed that encapsulated formulations protected sensitive bioactives from degradation even upon heating or photodegradation (Nedović *et al.*, 2017). Moreover, the resulting powders were stable under harsh storage conditions, since neither the bioactive compounds nor antioxidant activity was significantly affected. Above-mentioned adventages have been the reasons to use spray-drying in order to encapsulate and produce natural pigments extracted from plant materials (anthocyanins, carotenoids, betalains, betaxanthin, betacyanins) in fine powdered and stabilized forms (Kalušević *et al.*, 2017b, c; Ćujić Nikolić *et al.*, 2018, 2019; Vulić *et al.*, 2019; Šeregelj *et al.*, 2019).

The superiority of this technique could be also seen through examples given in Table 3. Most of the described encapsulates that have found application in final food products have been obtained by spray-drying. As carrier materials maltodextrin, gums, protein isolates were most frequently used.

3.2 Freeze-drying

Freeze-drying is a low temperature dehydration process, carried out at lower temperatures compared to spray-drying. This technique has been commonly offered as an alternative for spray-drying of heat sensitive plant bioactives (e.g. polyphenols). Furthermore, high percent of encapsulation efficiency is possible to achieve by this drying technique. On the other hand, the long processing time and lot of energy (comparing to spray-drying) make it costly and critical in large-scale applications. However, it is economically sustainable technology only for high-price products, which is more typical for pharmaceutical than food products (Nedović *et al.*, 2011, 2013).

There are many authors that compared encapsulates of plant extracts obtained by spray- and freeze-drying. For example, freeze-drying was proved to be better technique of red pepper bioactives encapsulation compared to the spray-drying, in terms of technological properties, such as water activity, moisture content, solubility, flowing and color properties (Šeregelj *et al.*, 2019). On the other hand, a number of research papers comparing spray- and freeze-dried encapsulates showed that spray-dried particles were superior to freeze-dried (Nedović *et al.*, 2017). However, for numerous properties of encapsulates, the crucial effect had carrier and technique influence has been insignificant.

Frequently, freeze-drying has been employed as additional technique after encapsulation of plant extracts or essential oils in different systems, e.g. hydrogel beads (Balanč *et al.*, 2016; Ćujić *et al.*, 2016; Kokina *et al.*, 2019). Plant processing byproducts and waste are also common source of bioactives for this technique of encapsulation. Different plant sources can be found in literature, such as sour cherry pomace (Šaponjac *et al.*, 2016), beetroot pomace (Hidalgo *et al.*, 2018), red pepper waste (Šeregelj *et al.*, 2019), grapeskin or soybean coat (Kalušević, 2017a). Freeze-dried encapsulates of waste-derived extracts, rich in polyphenols and based on polysaccharide and protein carriers, have been incorporated in several food products, such as cake, cookies, bisquits, yogurt (Popović *et al.*, 2019).

3.3 Electrostatic Extrusion

Extrusion methods as techniques for encapsulation of plant extracts or essential oils are mainly based on the dispersing of aqueous solutions of the polymer (e.g. alginate, pectin) and plant bioactives into droplets dropping into a gelling bath. The dripping tool could be a syringe, a pipette, vibrating or spraying nozzle, jet cutter, or an atomizing disk. The alginate is most often used as a carrier material for beads formation, where sodium alginate solution upon ion exchange with calcium cations from the gelling solution creates a dense matrix network (Nedović *et al.*, 2011, 2013).

Encapsulates of plant extracts in the form of hydrogel (micro)beads are also an interesting candidate as delivery systems of bioactive compounds and for the development of functional food products. In this regard, the use of electrostatic extrusion is highlighted since it allows production of very small (down to 50 µm) and uniform beads which is favorable from the aspect of textural and sensorial properties of food products (Manojlovic *et al.*, 2008; Đorđević *et al.*, 2015). The process is simple, precise, efficient and economical without need for severe conditions in terms of temperature and solvent (Đorđević *et al.*, 2015; Ćujić *et al.*, 2016). It is based on extrusion of a solution of carrier material and active agent though a needle and an action of an electrostatic force that disrupts the solution filament at the needle tip forming a charged stream of droplets. Further, physical (e.g. cooling or heating) or chemical (e.g. gelation) process is employed for solidification of the formed droplets into beads. A major limitation for a wide application of this technique represents a low production rate (Manojlovic *et al.*, 2008; Đorđević *et al.*, 2015).

For example, electrostatic extrusion was applied to encapsulate carqueja (*Pterospartum tridentatum* L.) extract that has been identified as a source of various bioactive compounds, such as alkaloids and flavonoids (Balanč *et al.*, 2016). In this way, Ca-alginate and Ca-alginate-inulin hydrogel microbeads loaded with carqueja extract were produced. In addition, dry microbeads were also produced, as being more appropriate for long-term storage, by subjecting the hydrogel beads to freeze-drying. All examined formulations showed a significant content of phenolic compounds and antioxidant activity. The presence of inulin protected the bead structure during freeze-drying, delayed disintegration and reduced swelling rate of the beads in simulated intestinal fluid. Based on the studied properties, the authors proposed the Ca-alginate-inulin beads as a delivery system of carqueja extract in functional food products. In addition, very similar results have been confirmed with encapsulates of anthocyanin-rich extracts of chokeberry (Ćujić *et al.*, 2016). Also, this technique has been successfully applied in order to encapsulate herbs extracts with chitosan (e.g. thyme, *Thymus serpyllum* L.) (Stojanović *et al.*, 2011; Trifković *et al.*, 2014).

It should be added that this technique showed up equally thriving in encapsulation of various types of oils. Among numerous published papers dealing with this topic, some studies confirmed successful application of such systems in final food products as well. Some good examples are beef emulsions with encapsulated pumpkin seed oil and fermented sausages with encapsulated grapeseed oil (Stajić *et al.*, 2014, 2020). The main results showed that backfat substitution with the encapsulated pumpkin seed oil could be to the level of 25 per cent and that encapsulating agents (alginate, pectin) altered mostly color parameters, especially these indicating yellow tones (Stajić *et al.*, 2020).

3.4   Encapsulation in Cyclodextrins

Molecular encapsulation in cyclodextrins represents an interesting way to increase the stability of herbal bioactives (Kalogeropoulos *et al.*, 2010). Cyclodextrins are cyclic oligosaccharides, shaped as a truncated cone, with a hydrophilic outer surface and an internal hydrophobic cavity. This three-dimensional structure of cyclodextrins allows encapsulation of active agents within their cavity, resulting in the inclusion complex formation and improved physicochemical properties of the entrapped bioactives (Mura, 2014). The inclusion complexation is based on co-precipitation of an aqueous solution of cyclodextrins and active agent subjected to stirring/sonicating and/or heating. In addition, spray-drying and freeze-drying are usually employed to prepare dried complexes, that are mainly amorphous and fast-dissolving (Đorđević *et al.*, 2015). Limited loading capacity and high costs are the main issues for wide exploitation of this technique (Nedovic *et al.*, 2011).

In this regard, a flavonoid-rich St John's wort (*Hypericum perforatum*) extract with potent antioxidant activity was encapsulated in an inclusion complex with β-cyclodextrin and subsequently freeze-dried (Kalogeropoulos *et al.*, 2010). The encapsulation protected the active compounds contained in the extract against thermal degradation. Namely, at temperature where thermal oxidation of the free extract takes place, the inclusion complex remaining intact. Therefore, the authors proposed a potential use of the solid inclusion complex of β-cyclodextrin and St John's wort extract as a novel supplement to enhance flavonoid content and antioxidant activity of food products, especially those expected to be thermally processed.

Encapsulated formulations obtained in the above-described approaches have been tested in different food products: (white) chocolate (Lončarević *et al.*, 2018, 2019), hazelnut paste (Kaderides *et al.*, 2015), ice cream (Çam *et al.*, 2014; Yilmaztekin *et al.*, 2019), biscuits and cookies (Saponjac *et al.*, 2016; Hidalgo *et al.*, 2018), yogurt (Azeredo *et al.*, 2007; Kalušević 2017a; Šeregelj *et al.*, 2019) and soft drinks (Burin *et al.*, 2011) (Table 3). In relation to the sensory assessments reported in the literature, majority of the panelists accepted the food products with encapsulates. This provides support for industrial application and great potential as functional foods.

Generally, encapsulated delivery systems of bioactive compounds have an increasing interest in the development of health-promoting food products. In this regard, many encapsulation processes have been extensively studied for the delivery of various bioactive compounds to food. Table 3 gives an overview of features and advantages achieved by the application of encapsulated bioactive compounds from natural sources into real food products.

4. Trends in Development of Innovative Functional Food Products

Experts and scientists of different fields (nutrition and food, ecology, chemistry, engeenering, economy etc) are interested in developing of innovative functional foods in accordance with circular economy and/or zero waste concept. Namely, food lifecycle creates enormous amounts of plant processing byproducts and waste that can be used for production of valuable bioactives and potential food additives. However, most of the world, especially underdeveloped and developing countries still think and work linear (the traditional production and consumption). Currently, only around 9 per cent of material waste streams in the global economy are recycled (PACE, 2019).

In addition, consumers' tendency to a healthy lifestyle has initiated the development of various novel and functional food products. High amounts of by-products, such as peels, seeds, and stones, are discarded during plant, especially fruit and vegetables, processing. It represents a problem from the environmental and the economic point of view. Then again, the resulting waste and byproducts are potential sources of numerous bioactives. Consequently, plant processing by-products such as substrates for the extraction of polyphenolic compounds, natural colorants, dietary fibers, protein isolates and oils attract great interest of industry as well (Salević *et al.*, 2018).

For instance, food with encapsulates containing bioactives and materials from plant processing byproducts and waste, as active component and carrier could be only one of directions. These bioactives still posses health-promoting effects, colour, flavor and may be stabilized by (nano/micro)encapsulation technology. Some promising approaches are: encapsulation of pomegranate peel polyphenols in orange juice industry by-product (Kaderides *et al.*, 2020), encapsulation of carotenoids from red pepper waste in whey proteins (Vulić *et al.*, 2019), utilization of grape pomaces and brewery waste for the production of microencapsulated pigments (Rubio *et al.*, 2020).

Table 3: Application of Encapsulated Plant Extracts for Development of Functional Food Products

Active Agent	Type of Encapsulate	Foodstuff	Main Results	References
		Herbs		
Rosemary (*R. officinalis* L.) lyophilized aqueous extract	Ca-alginate microbeads obtained by atomization/coagulation	Cottage cheese	The addition of free and encapsulated rosemary extract provided antioxidant activity that was more efficiently preserved during storage with the encapsulated extract without affecting the nutritional value of the cheese.	Ribeiro *et al.*, 2016
Clove buds (*S. aromaticum* L.) supercritical carbon dioxide extract	Spray dried maltodextrin/gum arabic encapsulates	Soybean oil	Although no significant difference was found between antioxidant activity of un-encapsulated and encapsulated clove extracts in soybean oil, the use of the encapsulated form is preferred due to controlled release of active compound and its no pro-oxidative activity at the initial stage of storage.	Chatterjee and Bhattacharjee, 2013
Ginger rhizome (*Z. officinale Roscoe*) ethanolic extract	Nanoliposomal structures	Sunflower oil	Nanoliposomal form of the ginger extract showed a higher performance in delaying the oil oxidation compared to the unencapsulated extract and synthetic antioxidant – butylated hydroxytoluene.	Ganji and Sayyed-Alangi, 2017
Arjuna (*T. arjuna*) ethanolic extract	Maltodextrin/gum arabic microcapsules obtained by mechanical homogenization and tray drying	Vanilla chocolate dairy drink	Effectiveness of the encapsulates in enhancing the storage stability of flavored dairy drinks.	Sawale *et al.*, 2017
Sage (*S. officinalis* L.) and oregano (*O. vulgaris* L.) extracts prepared with acetone and ethanol, respectively	Liposomes prepared by ultrasonication and microfluidization	Salad dressings	Strong antioxidative effects of the sage liposomes prepared by microfluidization, similar to the activity of butylated hydroxytoluene liposomes, in salad dressings during storage in the dark at ambient temperature and at 40°C, while all liposomal formulations were non-effective at 60°C.	Abdalla and Roozen, 2001

Plant source	Encapsulation system	Food product	Findings	Reference
Yerba mate (*I. paraguariensis*) lyophilized aqueous extract	Dried Ca-alginate and Ca-alginate-chitosan beads loaded with yerba mate extract	Instant vegetable soups	The addition of the encapsulates increased content of phenolic compounds without modifying flowability and sensory properties of the soups.	Deladino et al., 2013
Green tea (*C. sinensis* L.) aqueous extract	Alginate, alginate/chitosan, pectin/chitosan hydrogel microbeads	Pilsner beer and radler	Radler enriched with spray dried form was preferred in terms of content of total phenolic compounds and sensory profile.	Belščak-Cvitanović et al., 2017
	Freeze and spray dried green tea extract	Chocolate	The significant impact of the encapsulates on product viscosity, polyphenol content and sensorial profile of the chocolate.	Lončarević et al., 2019
Fennel (*F. vulgare* Mill.) and chamomile (*M. recutita* L.) lyophilized aqueous extracts	Lyophilized Ca-alginate microbeads obtained by atomization/coagulation	Cottage cheese	Increased antioxidant activity of the cheese functionalized with the encapsulated extract during storage.	Caleja et al., 2016
Olive leaf (*Olea europaea* L.) lyophilized methanol/water extract	Maltodextrin-based freeze dried microencapsulates	Tomato paste	Accelerated shelf-life testing predicted an ability of the encapsulated extract to maintain the original quality of the paste very well according to changes in color and pH indices.	Ganje et al., 2016
Peppermint (*M. piperita*) essential oil	Ca-alginate based beads Carnauba wax microparticles Gelatine/alginate coacervates	Ice cream	Incorporation of encapsulated peppermint oil into ice cream up to 0.3 per cent (w/w) might be a suitable option to add functional properties of peppermint into ice cream without damaging textural properties.	Yilmaztekin et al., 2019
Saffron (*C. sativus* L.) and beetroot aqueous extracts	Freeze dried microencapsulates based on various carrier materials: maltodextrin, gum arabic, gum arabic/modified starch, modified starch/chitosan and modified starch/ maltodextrin/chitosan	Chewing gum	Color parameters of the chewing gum incorporated with encapsulated saffron pigments were depended on the type of carrier material and storage conditions; gum arabic/modified starch provided the highest color stability.	Chranioti et al., 2015

(Contd.)

Table 3: (*Contd.*)

Active Agent	Type of Encapsulate	Foodstuff	Main Results	References
		Fruit and vegetables		
Blackberry juice	Maltodextrin-based spray dried microencapsulates	White chocolate	Viscosity of the chocolate mass and chocolate hardness increased. The sweetness of white chocolate was reduced and has pleasant fruity flavor. The addition of 10 per cent of encapsulated blackberry juice increased total polyphenol content of chocolate 3.8-fold.	Lončarević *et al.,* 2018
Beetroot extract	Maltodextrin-based spray dried microencapsulates	Yogurt	Betacyanin degradation rates during storage were increased by light exposure which decreased with maltodextrin/beetroot extract ratio. The addition of the microcapsules was not perceived in terms of flavour by panelists.	Azeredo *et al.,* 2007
Garlic oil	β-cyclodextrin capsules	Fresh-cut tomatoes	The panelists evaluated the smell of tomatoes with free garlic oil as 'unacceptable", while there was no significant difference between the control and the tomatoes with microencapsulated garlic oil. Microencapsulated garlic oil showed the lowest microbial growth.	Ayala-Zavala and Gonzalez-Aguilar, 2010
Malabar spinach (*Basella rubra* L.) fruits juice	Lecithin nanoliposomes	Vegan gummy candies	Sensory score was acceptable for betalains nanoliposomes fortified vegan gummy candies. Betalains nanoliposomes vegan gummy candies exhibited superior antioxidant activity.	Kumar *et al.,* 2020
Cactus pear (*Opuntia ficus-indica*) fruits juice	Ca-alginate hydrogel beads	Gummy candies	The gummy candies with encapsulated juice showed appropriate gelling and morphological properties to be used in confectionery industry. Encapsulated betalains have a potential to be successfully used as colouring additives in the food industry.	Otálora *et al.,* 2019

Fruit/Vegetable/Crop Processing Byproducts

Sour cherry pomace extract	Whey and soy protein-based freeze dried microparticles	Cookies	Encapsulated sour cherry pomace bioactives have positively influenced functional characteristics of fortified cookies and their preservation.	Šaponjac et al., 2016
Beetroot pomace extracts	Soy protein-based freeze dried microparticles	Pseudocereal-enriched einkorn water biscuits	Improvement of some nutritional characteristics of baked products.	Hidalgo et al., 2018
Red pepper waste	Whey protein-based spray and freeze dried microparticles	Yogurt	Freeze-drying was more efficient for the development of functional food products with improved nutritional, color and bioactive properties.	Šeregelj et al. 2019
Grapeskin extract	Maltodextrin-, maltodextrin/γ-cyclodextrin-, maltodextrin/gum arabic-based spray dried microcapsules	Soft drinks	The combination of maltodextrin/arabic gum led to the longest anthocyanin half-lifetime and lowest degradation constant under all conditions evaluated, and thus provided better protection of the anthocyanin pigments.	Burin et al., 2011
Grape marc extract	An oil/water nanoemulsion; a powder obtained by maltodextrin-assisted spray-drying of the previous and an ethanol/solid–lipid nanoemulsion	Hazelnut paste	Encapsulation improved phenolic efficiency against lipid oxidation, by increasing extract dispersability in the paste and preserving the antioxidant activity, with the oil/water nanoemulsion resulting as the best system.	Spigno et al., 2013
Grapeskin and Soybean coat Extracts	Maltodextrin, gum arabic and skim milk-based spray and freeze dried microencapsulates	Yogurt	Encapsulation techniques did not affect sensorial characteristics, but plant source and carriers showed significant influence. The highest acceptance was for yogurts with grapeskin extract/maltodextrin microencapsulates.	Kalušević, 2017a

(Contd.)

Table 3: (*Contd.*)

Active Agent	Type of Encapsulate	Foodstuff	Main Results	References
Grape seed oil	Alginate gels	Fermented sausages	Treatment was similar to control and did not differ significantly in terms of internal product colour and surface color, either before or after storage.	Stajić *et al.*, 2014
Pumpkin seed oil	Alginate and pectin beads	Beef emulsions	Backfat substitution with the encapsulated pumpkin seed oil to the level of 25 per cent as well as the encapsulating agents altered mostly colour parameters, especially these indicating yellow tones.	Stajić *et al.*, 2020
Pomegranate peel	Skim milk powder Maltodextrin Gum arabic Whey protein isolate Orange waste powders	Ice cream Hazelnut paste Cookies	Sensorial analyses showed no significant difference between the mean scores of free and icecream with microencapsulates. About 70-90 per cent of the panelists preferred cookies with the encapsulated extract for their color and odor compared to cookies with the uncoated extract.	Çam *et al.*, 2014 Kaderides *et al.*, 2015, 2020

Above mentioned and similar examples seem to be in accordance with zero waste model. Zero waste is a core principle of the circular economy. Simplified, zero waste is a design principle that minimizes waste, reduces consumption, maximizes recycling, and ensures that products are made in that way to be reused, repaired or recycled back to the nature or the marketplace. In this case, market of food products is the goal point where the reused and utilized byproducts or waste should finish its life.

The challenge to utilize plant waste and transform it into innovative functional food products has attracted the interest of academia and food industry and urged the EU towards a zero-waste economy by 2025. The issues that have to be overcome are scale up, technology transfer and successful commercialization.

Regarding to this and having in mind amounts of food packaging thrown away daily, one of the most important issues is the replacement of traditional packaging materials by eco-friendly ones. In that term, next segment of this chapter is focused on active food packaging.

4.1 Active Food Packaging

Apart from the medicinal purposes, the use of plants, specially spices and herbs in food preservation has also been known and respected since ancient times, but recently it has sparked a great interest in the design of a novel concept in food packaging, called 'active food packaging' (Silva-Weiss *et al.*, 2013; Zhang *et al.*, 2017). This concept is based on the incorporation of various active agents within biodegradable polymer matrices with the aim to improve the packaging's functionality, maintain or enhance product quality, safety and shelf-life, as well as reduce environmental pollution generated by the packaging industry (Valdés *et al.*, 2014; Vilela *et al.*, 2018). In this sense, the use of natural extracts from herbs and spices in active food packaging formulations is highlighted due to their antioxidant (Embuscado, 2015) and antimicrobial properties (Shan *et al.*, 2007) that are essential in preventing the food deterioration caused by oxidative reactions and microbial growth (Valdés *et al.*, 2014; Souza *et al.*, 2017). However, direct application of natural extracts on to the food surface is limited by their instability and rapid diffusion within the food bulk. For this reason, the incorporation of active agents into packaging materials is a promising alternative to maintain their activity and achieve an optimal effect during the total storage period (Lagarón *et al.*, 2011; Valdés *et al.*, 2014). The technique of the incorporation and the nature of polymer are critical factors in the formulation of the active systems that should meet the following challenges: the controlled release rate of the active compounds, minor changes to the physical and chemical properties of the polymeric matrix and compliance with the food packaging materials requirements (Mellinas *et al.*, 2016; Khaneghah *et al.*, 2018).

Solution-casting is a widely used technique for active film preparation at a laboratory scale. The process does not require specific equipment and includes three steps: preparation of film-forming solution by dissolving a polymer along with plasticizers and active agents in a suitable solvent, casting of the solution onto a flat surface or a heated drum drier and finally, drying which leads to solvent evaporation and the film formation. A wide variety of biopolymer-based active films has been developed by solution casting. However, manufacturing capacity and time-consuming are limiting factors for the application of this technique at commercial scale (Rhim and Ng, 2007; Nur Hanani *et al.*, 2012; Etxabide *et al.*, 2018).

Green tea (*Camellia sinensis* L.) and basil (*Ocimum basilicum* L.) extracts, as sources of compounds with antioxidant activity and ability to change color in a pH-dependent manner, were studied as components for development of starch-based active and intelligent biodegradable packaging by casting (Medina-Jaramillo *et al.*, 2017). The films loaded with these natural extracts exhibited high thermal stability and lower water vapor permeability, retaining their flexibility compared to typical thermoplastic starch-based materials. The presence of phenolic compounds in the extracts led to significant antioxidant activity of the developed films, while chlorophyll and carotenoids induced color changes in the films when being immersed in acid and basic media. Therefore, the starch-based films loaded with the natural extracts can be used to delay oxidation processes of food products, but also as indicators of food quality. In addition, the developed films can be considered as environmentally friendly materials, due to their very fast degradation in soil (less than two weeks).

As an example of pH indicator films, chitosan-based smart films were developed using blueberry and blackberry pomace extracts as active agents (Kurek *et al.*, 2018). The concept of the film production could be considered as eco-friendly contributing to the reduction of fruit pomace usually wasted material. Visible and significant colour changes of the dry pH indicator films occurred with changing pH values. Furthermore, pH indicator films could be produced with plant extracts previously encapsulated in micro- or nano-systems. For example, agar-based films were produced with microencapsulates of anthocyanin-rich extracts, obtained from waste materials (grapeskin and soybean coat) employing different carriers and encapsulation techniques (Fig. 1) (Kalušević, 2017a).

From an industrial point of view, the use of classical polymer technological processes, such as extrusion, has a great potential in the production of active films (Del Nobile *et al.*, 2009). Extrusion is a thermo-mechanical processing technique that uses one or two rotating screws extruder that induces high temperature and pressure to disrupt the polymer granules, mix the film components and push the mixture forward through a slit or flat film die where expansion may take place which is followed by a take-off device for orientation and collection (Liu *et al.*, 2009; Nur Hanani *et al.*, 2012; Mellinas *et al.*, 2016). This technique is interesting for film production due to

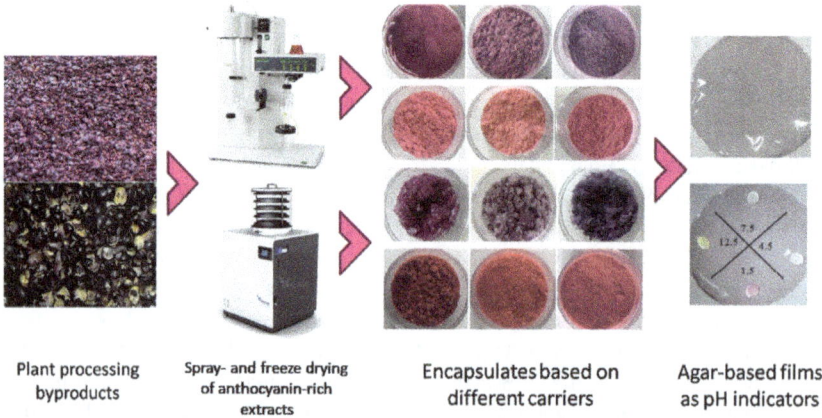

| Plant processing byproducts | Spray- and freeze drying of anthocyanin-rich extracts | Encapsulates based on different carriers | Agar-based films as pH indicators |

Fig. 1: Anthocyanin-rich microencapsulates in agar-based films as pH indicator

possibility of multiple-injection and processing of high-viscosity polymers without the use of solvents, large operational flexibility and control of the residence time and the degree of mixing (Liu *et al.*, 2009). However, the high temperature and pressure during extrusion processing can negatively affect the efficiency of active compounds due to their limited thermal stability (Del Nobile *et al.*, 2009).

In this regard, green tea (*Camellia sinensis* L.) extract, a source of compounds with potent antioxidant activity, as previously mentioned, was also incorporated in a hydrophilic plastic layer, ethylene vinyl alcohol copolymer, by flat extrusion (López de Dicastillo *et al.*, 2011). The extruded films were translucent and the extract addition improved their thermal resistance, as well as the water and oxygen barrier performance at low and medium relative humidity, but also increased water sensitivity. Although a partial degradation of the extract active compounds during extrusion was reported, the films exhibited efficiency against DPPH$^{\cdot}$ and ABTS$^{\cdot+}$free radicals. The release studies of green tea extract components into various food simulants indicated the potential of the developed films to be used as antioxidant active packaging for both aqueous and fatty food products.

One of the innovative solutions for the design of active food packaging relies on electrohydrodynamic processing. In this regard, electrospinning technique has gained a lot of interest in the incorporation of natural extracts within ultrathin solid fibers ranging from micrometers to nanometers, with a high surface to volume ratio (Zhang *et al.*, 2017; Etxabide *et al.*, 2018). This technique employs a high-voltage power supply, a syringe pump, a syringe connected to a metal needle and a grounded collector (Etxabide *et al.*, 2018). The process is based on the action of an electrostatic force imposed on a polymer solution producing continuous polymer fibers (Echegoyen *et al.*, 2017). In this way, a voltage applied between the needle and the collector induces an electric charge and a cone shape (called Taylor cone) of a polymer solution droplet held by its surface tension at the needle tip. When electrostatic force overcomes the surface tension, the polymeric filament is extruded from the Taylor cone and attracted by the collector, leading to the solvent evaporation and resulting in mats consisting of discharged fibers (Zhang *et al.*, 2017; Etxabide *et al.*, 2018). The characteristics, such as continuous fabricating capability, facile operating process, no need for high temperatures, high trapping efficiency and cost-effectiveness, as well as possibility to scale up the process, make electrospinning technique a promising candidate for active packaging fabrication (Echegoyen *et al.*, 2017; Etxabide *et al.*, 2018).

Thus, a solid dispersion of sage (*Salvia officinalis* L.) extract with a potent antioxidant and antimicrobial activity was incorporated within ultrathin electrospun poly(ε-caprolactone) fibers that were subsequently coalesced by an annealing step using a hydraulic press (Salević *et al.*, 2019). In this way, thin, continuous and hydrophobic films with good contact transparency were produced. The sage extract incorporation did not affect thermal stability and tensile properties of the poly(ε-caprolactone) films, while the water vapor and aroma permeability were increased. The analysis of the functional properties revealed a remarkable, sage extract content-dependent potential to scavenge free radicals and inhibit bacterial growth.

Generally, active food packaging is a very dynamic research area with continuous advances. Extensive studies on this matter are being carried out exploring a wide range of natural extracts, polymers, and techniques. Table 4 shows the potential of packaging systems based on biodegradable materials and natural extracts for application in food products preservation.

Table 4: Application of Films and Coatings Loaded with Natural Extracts for Food Products Preservation

Active Agent	Type of Film or Coating	Foodstuff	Main Results	References
Stinging nettle (*U. dioica* L.) lyophilized aqueous extract	Whey protein isolate coatings containing electrospun poly(ε-caprolactone) nanofibers loaded with the extract	Rainbow trout fillets	Potent antioxidant and antimicrobial activity of the coatings reflected in extended quality and shelf-life of the fresh fish fillets up to 15 days at 4°C.	Erbay *et al.*, 2017
Green tea (*C. sinensis* L.) aqueous extract	Casted films based on chitosan	Pork sausages	The ability of the films to maintain physical and sensory qualities and prolong shelf-life of the refrigerated sausages by inhibition of lipid oxidation and microbial growth.	Siripatrawan and Noipha, 2012
Olive leaf (*O. europaea* L.) aqueous extract	Casted films based on methylcellulose	Kasar cheese slices	Antimicrobial effect of the films against *S. aureus* inoculated on to the cheese.	Ayana and Turhan, 2009
Yerba mate (*I. paraguariensis*, St. Hil.) aqueous extract	Casted films based on cassava starch	Palm oil	The efficiency of the film containing yerba mate extract to inhibit lipid oxidation of palm oil.	Reis *et al.*, 2015
Green tea (*C. sinensis* L.) commercial extract	Blown extruded films based on polylactic acid	Smoked salmon slices	Effective protection of the model fatty food from lipid oxidation.	Martins *et al.*, 2018
Sage (*S. officinalis* L.) and laurel (*L. nobilis* L.) ethanol/water extracts	Casted films based on whey protein isolate	Meatballs	Effective antioxidant strategy to control lipid oxidation and lipid hydrolysis of the cooked and subsequently frozen comminuted meat product.	Akcan *et al.*, 2017
Cocoa (*T. cacao* L.) commercial extract	Casted films based on ethylene-vinyl alcohol copolymer	Infant milk formula	Growth inhibition of *Listeria monocytogenes* previously inoculated into the infant milk formula.	Calatayud *et al.*, 2013
Boldo (*P. boldus Molina*) ethanolic extract	Casted films based on plain chitosan and blend of chitosan and gelatin loaded with the extract	Beef hamburger	A protective effect against lipid oxidation and microbial growth.	Bonilla Lagos and Sobral, 2019

Mango leaf (*M. indica* L., *Neelam* variety) ethanolic extract	Casted films based on chitosan	Cashew nuts	More effective preservation of the cashew nuts against oxidation was achieved by the chitosan-based films containing mango leaf extract than by a commercial polyamide/polyethylene film during 28 days storage.	Rambabu *et al.*, 2019
Moringa (*M. oleifera* L.) lyophilized ethanolic extract	Casted films based on gelatin	Cube-type Gouda cheese	Inhibition of microbial growth and lipid oxidation during storage.	Lee *et al.*, 2016
Pineapple (*A. comosus*) peel extract (encapsulated, spray-dried peel juice, milled freeze-dried pineapple peel)	Casted, crosslinked alginate films	Beef meat products	Microbial spoilage control, color preservation, and barrier to lipid oxidation of beef steaks.	Lourenço *et al.*, 2020

5. Conclusion and Future Perspectives

Plants represent a natural source of various bioactives with diverse biological activities offering health-beneficial effects and/or even treatments for various diseases. Plant extracts are complex mixtures of bioactive compounds, that have antioxidant, antiviral, antibiotic, antifungal, anticancer, hypoglycemic and anti-hypertensive characteristics. The extraction procedures require the use of organic solvents, that complicates the formulations but direct usage of the plant extracts as well. Ecoinovative extraction procedures like supercritical carbon dioxide extraction, enzyme-, ultrasound-, microwave- and pressure-assisted extractions expanded popularity in the last decade for production of plant extracts and certain bioactive compounds. The main difficulty associated with the usage of plant extracts for different applications includes their complex composition, instability, optimized release, even toxicity risks.

On the other hand, encapsulation has been widely proven as useful in overcoming these issues and in improvement of the physico-chemical stability of plant extracts, as well as their bioavailability and efficacy. Also, encapsulated forms of plant extracts enabled and simplified the incorporation into a broad assortment of food products. Encapsulates of plant extracts (microparticles, beads, powders) can be used in food industry for enhancement of food stability, texture, aroma profile, nutrient and antimicrobial properties. Furthermore, their application is possible as active component of packaging materials with the aim to achieve antimicrobial and antioxidant activity of the materials, thus providing longer shelf-life of packed products.

This chapter focuses on the researches and achievements done in the development of encapsulates of plant extracts and their addition in food systems. Authors tried to give an overview of commonly used techniques and materials for extraction and encapsulation of plant extracts, as well as their applications in the food industry as food or packanging component.

References

Abdalla, A.E. and Roozen, J.P. (2001). The effects of stabilised extracts of sage and oregano on the oxidation of salad dressings, *European Food Research and Technology*, **212**(5): 551-560.

Akcan, T., Estévez, M. and Serdaroğlu, M. (2017). Antioxidant protection of cooked meatballs during frozen storage by whey protein edible films with phytochemicals from *Laurus nobilis* L. and *Salvia officinalis, LWT – Food Science and Technology*, 77: 323-331.

Akhtar, R. and Shahzad, A. (2017). Alginate encapsulation in *Glycyrrhiza glabra* L. with phyto-chemical profiling of root extracts of *in vitro* converted plants using GC-MS analysis, *Asian Pacific Journal of Tropical Biomedicine*, 7: 855-861.

Amirah, D.M., Reddy, P. and Maksudur, R.K. (2012). Comparison of extraction techniques on extraction of gallic acid from stem bark of *Jatropha curcas*, *Journal of Applied Science*, 12: 1106-1111.

Andrade, K., Poncelet, D. and Ferreira, S. (2017). Sustainable extraction and encapsulation of pink pepper oil, *Journal of Food Engineering*, 204: 38-45.

Assami, K., Pingret, D., Chemat, S., Meklati, B. and Chemat, F. (2012). Ultrasound induced intensification and selective extraction of essential oil from *Carum carvi* L. seeds, *Chemical Engineering and Processing*, 62: 99-105.

Ayala-Zavala, J.F. and González-Aguilar, G.A. (2010). Optimizing the use of garlic oil as antimicrobial agent on fresh-cut tomato through a controlled release system, *Journal of Food Science*, **75**(7): M398-M405.

Ayana, B. and Turhan, K.N. (2009). Use of antimicrobial methylcellulose films to control *Staphylococcus aureus* during storage of Kasar cheese, *Packaging Technology and Science: An International Journal*, **22**(8): 461-469.

Azeredo, H.M.C., Santos, A.N., Souza, A.C.R., Mendes, K.C. and Andrade, M.I.R. (2007). Betacyanin stability during processing and storage of a microencapsulated red beetroot extract, *American Journal of Food Technology*, **2**(4): 307-312.

Balanč, B., Kalušević, A., Drvenica, I., Coelho, M.T., Djordjević, V., Alves, V.D., Sousa, I., Moldao-Martins, M., Rakić, V., Nedović, V. and Bugarski, B. (2016). Calcium–alginate–inulin microbeads as carriers for aqueous carqueja extract, *Journal of Food Science*, **81**(1): E65-E75.

Belščak-Cvitanović, A., Stojanović, R., Manojlović, V., Komes, D., Juranović Cindrić, I., Nedović, V. and Bugarski B. (2011). Encapsulation of polyphenolic antioxidants from medicinal plant extracts in alginate-chitosan system enhanced with ascorbic acid by electrostatic extrusion, *Food Research International*, 44: 1094-1101.

Belščak-Cvitanović, A., Komes, D., Durgo, K., Vojvodić, A. and Bušić, A. (2015a). Nettle (*Urtica dioica* L.) extracts as functional ingredients for production of chocolates with improved bioactive composition and sensory properties, *Journal of Food Science and Technology*, 52: 7723-7734.

Belščak-Cvitanović, A., Lević, S., Kalušević, A., Špoljarić, I., Đorđević, V., Komes, D., Mršić, G. and Nedović, V. (2015b). Efficiency assessment of natural biopolymers as encapsulants of green tea (*Camellia sinensis* L.) bioactive compounds by spray-drying, *Food and Bioprocess Technology*, 8: 2444-2460.

Belščak-Cvitanović, A., Nedović, V., Salević, A., Despotović, S., Komes, D., Nikšić, M., Bugarski, B. and Čukalović, I.L. (2017). Modification of functional quality of beer by using microencapsulated green tea (*Camellia sinensis* L.) and Ganoderma mushroom (*Ganoderma lucidum* L.) bioactive compounds, *Chemical Industry & Chemical Engineering Quarterly*, **23**(4): 457-471.

Bilušić, T., Drvenica, I., Kalušević, A., Marijanović, Z., Jerković, I., Mužek, M.N. and Nedović, V. (2020). Influences of freeze- and spray-drying vs. encapsulation with soy and whey proteins on gastrointestinal stability and antioxidant activity of Mediterranean aromatic herbs, *International Journal of Food Science & Technology*, 1-15.

Bonilla Lagos, M.J. and Sobral, P.J.D.A. (2019). Application of active films with natural extract for beef hamburger preservation, *Ciência Rural*, **49**(1).

Bouras, M., Chadni, M., Barba, F., Grimi, N., Bals, O. and Vorobiev, E. (2015). Optimization of microwave-assisted extraction of polyphenols from *Quercus* bark, *Industrial Crops and Products*, 77: 590-601.

Bucić-Kojić, A., Planinić, M., Tomas, S., Mujić, I., Bilić, M. and Velić, D. (2011). Effect of extraction conditions on the extractability of phenolic compounds from lyophilised fig fruits (*Ficus Carica* L.), *Polish Journal of Food and Nutrition Sciences*, 61: 195-199.

Burin, V.M., Rossa, P.N., Ferreira-Lima, N.E., Hillmann, M.C. and Boirdignon-Luiz, M.T. (2011). Anthocyanins: Optimization of extraction from Cabernet Sauvignon grapes, microcapsulation and stability in soft drink, *International Journal of Food Science & Technology*, **46**(1): 186-193.

Cai, R., Yuan, Y., Cui, L., Wang, Z. and Yu, T. (2018). Cyclodextrin-assisted extraction of phenolic compounds: Current research and future prospects, *Trends in Food Science & Technology*, 79: 19-27.

Calatayud, M., López-de-Dicastillo, C., López-Carballo, G., Vélez, D., Muñoz, P.H. and Gavara, R. (2013). Active films based on cocoa extract with antioxidant, antimicrobial and biological applications, *Food Chemistry*, **139**(1-4): 51-58.

Caleja, C., Ribeiro, A., Barros, L., Barreira, J.C.M., Antonio, A.L., Oliveira, M.B.P.P., Barreiro, M.F. and Ferreira, I.C.F.R. (2016). Cottage cheeses functionalized with fennel and chamomile extracts: Comparative performance between free and microencapsulated forms, *Food Chemistry*, 199: 720-726.

Çam, M., İçyer, N.C. and Erdoğan, F. (2014). Pomegranate peel phenolics: Microencapsulation, storage stability and potential ingredient for functional food development, *LWT – Food Science and Technology*, **55**(1): 117-123.

Chatterjee, D. and Bhattacharjee, P. (2013). Comparative evaluation of the antioxidant efficacy of encapsulated and un-encapsulated eugenol-rich clove extracts in soybean oil: Shelf-life and frying stability of soybean oil, *Journal of Food Engineering*, 117(4): 545-550.

Chranioti, C., Nikoloudaki, A. and Tzia, C. (2015). Saffron and beetroot extracts encapsulated in maltodextrin, gum arabic, modified starch and chitosan: Incorporation in a chewing gum system, *Carbohydrate Polymers*, 127: 252-263.

Chuyen, H., Roach, P., Golding, J., Parks, S. and Nguyen, M. (2019). Encapsulation of carotenoid-rich oil from Gac peel: Optimization of the encapsulating process using a spray drier and the storage stability of encapsulated powder, *Powder Technology*, 344: 373-379.

Cilek, B., Luca, A., Hasirci, V., Sahin, S. and Sumnu, G. (2012). Microencapsulation of phenolic compounds extracted from sour cherry pomace: Effect of formulation, ultrasonication time and core to coating ratio, *European Food Research and Technology*, 235: 587-596.

Cui, Q., Peng, X., Yao, X.-H., Wei, Z.-F., Luo, M., Wang, W., Zhao, C.-J., Fu, Y.-J. and Zu, Y.-G. (2015). Deep eutectic solvent-based microwave-assisted extraction of genistin, genistein and apigenin from pigeon pea roots, *Separation and Purification Technology*, 150: 63-72.

Ćujić, N., Trifković, K., Bugarski, B., Ibrić, S., Pljevljakušić, D. and Šavikin, K. (2016). Chokeberry (*Aronia melanocarpa* L.) extract loaded in alginate and alginate/inulin system, *Industrial Crops and Products*, 86: 120-131.

Ćujić Nikolić, N., Stanisavljević, N., Šavikin, K., Kalušević, A., Nedović, V., Samardžić, J. and Janković, T. (2018). Application of gum arabic in the production of spray-dried chokeberry polyphenols, microparticles characterization and *in vitro* digestion method, *Lekovite Sirovine*, 38: 10-18.

Ćujić-Nikolić, N., Stanisavljević, N., Šavikin, K., Kalušević, A., Nedović, V., Samardžić, J. and Janković, T. (2019). Chokeberry polyphenols preservation using spray-drying: Effect of encapsulation using maltodextrin and skimmed milk on their recovery following *in vitro* digestion, *Journal of Microencapsulation*, **36**(8): 693-703.

Del Nobile, M.A., Conte, A., Buonocore, G.G., Incoronato, A.L., Massaro, A. and Panza, O. (2009). Active packaging by extrusion processing of recyclable and biodegradable polymers, *Journal of Food Engineering*, **93**(1): 1-6.

Deladino, L., Navarro, A.S. and Martino, M.N. (2013). Carrier systems for yerba mate extract (*Ilex paraguariensis*) to enrich instant soups. Release mechanisms under different pH conditions, *LWT – Food Science and Technology*, **53**(1): 163-169.

Dent, M., Dragović-Uzelac, V., Penić, M., Brnčić, M., Bosiljkov, T. and Levaj, B. (2013). The effect of extraction solvents, temperature and time on the composition and mass fraction of polyphenols in Dalmatian wild sage (*Salvia officinalis* L.) extracts, *Food Technology and Biotechnology*, 51: 84-91.

Doneanu, C. and Anitescu, G. (1998). Supercritical carbon dioxide extraction of *Angelica archangelica* L. root oil, *Journal of Supercritical Fluids*, 12: 59-67.

Đorđević, V., Balanč, B., Belščak-Cvitanović, A., Lević, S., Trifković, K., Kalušević, A., Kostić, I., Komes, D., Bugarski, B. and Nedović, V. (2015). Trends in encapsulation technologies for delivery of food bioactive compounds, *Food Engineering Reviews*, 7(4): 452-490.

Echegoyen, Y., Fabra, M.J., Castro-Mayorga, J.L., Cherpinski, A. and Lagaron, J.M. (2017). High throughput electro-hydrodynamic processing in food encapsulation and food packaging applications: Viewpoint, *Trends in Food Science & Technology*, 60: 71-79.

Eikani, M., Golmohammad, F. and Rowshanzamir, S. (2007). Subcritical water extraction of essential oils from coriander seeds (*Coriandrum sativum* L.), *Journal of Food Engineering*, 80: 735-740.

Embuscado, M.E. (2015). Spices and herbs: Natural sources of antioxidants – A mini review, *Journal of Functional Foods*, 18: 811-819.

Erbay, E.A., Dağtekin, B.B.G., Türe, M., Yeşilsu, A.F. and Torres-Giner, S. (2017). Quality improvement of rainbow trout fillets by whey protein isolate coatings containing electrospun poly (ε-caprolactone) nanofibers with *Urticadioica* L. extract during storage, *LWT*, 78: 340-351.

Etxabide, A., Garrido, T., Uranga, J., Guerrero, P. and de la Caba, K. (2018). Extraction and incorporation of bioactives into protein formulations for food and biomedical applications, *International Journal of Biological Macromolecules*, 120: 2094-2105.

Fang, Z. and Bhandari, B. (2010). Encapsulation of polyphenols – A review, *Trends in Food Science & Technology*, **21**(10): 510-523.

Franco-Vega, A., Ramírez-Corona, N., Palou, E. and Lopez-Malo, A. (2016). Estimation of mass transfer coefficients of the extraction process of essential oil from orange peel using microwave assisted extraction, *Journal of Food Engineering*, 170: 136-143.

Gallego, R., Bueno, M. and Herrero, M. (2019). Sub- and super-critical fluid extraction of bioactive compounds from plants, food-by-products, seaweeds and microalgae – An update, *Trends in Analytical Chemistry*, 116: 198-213.

Ganje, M., Jafari, S.M., Dusti, A., Dehnad, D., Amanjani, M. and Ghanbari, V. (2016). Modeling quality changes in tomato paste containing microencapsulated olive leaf extract by accelerated shelf-life testing, *Food and Bioproducts Processing*, 97: 12-19.

Ganji, S. and Sayyed-Alangi, S.Z. (2017). Encapsulation of ginger ethanolic extract in nanoliposome and evaluation of its antioxidant activity on sunflower oil, *Chemical Papers*, **71**(9): 1781-1789.

Garavand, F., Rahaee, S., Vahedikia, N. and Jafarie, S.M. (2019). Different techniques for extraction and micro/nanoencapsulation of saffron bioactive ingredients, *Trends in Food Science & Technology*, 89: 26-44.

Gharsallaoui, A., Roudaut, G., Chambin, O., Voilley, A. and Saurel, R. (2007). Applications of spray-drying in microencapsulation of food ingredients: An overview, *Food Research International*, **40**(9): 1107-1121.

Ghatak, D. and Iyyaswami, R. (2019). Selective encapsulation of quercetin from dry onion peel crude extract in reassembled casein particles, *Food and Bioproducts Processing*, 115: 100-109.

Ghitescu, R.-E., Volf, I., Carausu, C., Bühlmann, A.-M., Gilca, I.A. and Popa, V. (2015). Optimization of ultrasound-assisted extraction of polyphenols from spruce wood bark, *Ultrasonics Sonochemistry*, 22: 535-541.

Giacometti, J., Bursać Kovačević, D., Putnik, P., Gabrić, D., Bilušić, T., Krešić, G., Stulić, V., Barba, F., Chemat, F., Barbosa-Cánovas, G. and Režek Jambrak, A. (2018). Extraction of bioactive compounds and essential oils from mediterranean herbs by conventional and green innovative techniques: A review, *Food Research International*, 113: 245-262.

Gibis, M., Zeeb, B. and Weiss, J. (2014). Formation, characterization, and stability of encapsulated hibiscus extract in multilayered liposomes, *Food Hydrocolloids*, 38: 28-39.

Hidalgo, A., Brandolini, A., Čanadanović-Brunet, J., Ćetković, G. and Šaponjac, V.T. (2018). Microencapsulates and extracts from red beetroot pomace modify antioxidant capacity, heat damage and color of pseudocereals-enriched einkorn water biscuits, *Food Chemistry*, 268: 40-48.

Hu, Y., Kou, G., Chen, Q., Li, Y. and Zhou, Z. (2019). Protection and delivery of mandarin (*Citrus reticulata* Blanco) peel extracts by encapsulation of whey protein concentrate nanoparticles, *LWT*, 99: 24-33.

Jovanović, A., Đorđević, V., Zdunić, G., Šavikin, K., Pljevljakušić, D. and Bugarski, B. (2016). Ultrasound-assisted extraction of polyphenols from *Thymus serpyllum* and its antioxidant activity, *Hemijska Industrija*, 70: 391-398.

Jovanović, A., Đorđević, V., Zdunić, G., Pljevljakušić, D., Šavikin, K., Gođevac, D. and Bugarski, B. (2017). Optimization of the extraction process of polyphenols from *Thymus serpyllum* L. herb using maceration, heat- and ultrasound-assisted techniques, *Separation and Purification Technology*, 179: 369-380.

Kaderides, K., Goula, A. and Adamopoulos, K. (2015). A process for turning pomegranate peels into a valuable food ingredient using ultrasound-assisted extraction and encapsulation, *Innovative Food Science and Emerging Technologies*, 31: 204-215.

Kaderides, K., Mourtzinos, I. and Goula, A.M. (2020). Stability of pomegranate peel polyphenols encapsulated in orange juice industry by-product and their incorporation in cookies, *Food Chemistry*, 310: 125849.

Kalogeropoulos, N., Yannakopoulou, K., Gioxari, A., Chiou, A. and Makris, D. (2010). Polyphenol characterization and encapsulation in β-cyclodextrin of a flavonoid-rich *Hypericum perforatum* (St John's wort) extract, *LWT – Food Science and Technology*, 43: 882-889.

Kalusevic, A., Salevic, A., Dordevic, R., Veljovic, M. and Nedovic, V. (2016). Raspberry and blackberry pomaces as potential sources of bioactive compounds, *Ukrainian Food Journal*, **5**(3): 485-491.

Kalušević, A. (2017a). Microencapsulation of Bioactive Compounds of Food Industry Byproducts, doctoral thesis, University of Belgrade, Belgrade, Serbia (in Serbian).

Kalušević, A.M., Lević, S.M., Čalija, B.R., Milić, J.R., Pavlović, V.B., Bugarski, B.M. and Nedović, V.A. (2017b). Effects of different carrier materials on physicochemical properties of microencapsulated grape skin extract, *Journal of Food Science and Technology*, **54**(11): 3411-3420.

Kalušević, A., Lević, S., Čalija, B., Pantić, M., Belović, M., Pavlović, V., Milić, J., Bugarski, B., Žilić, S. and Nedović, V. (2017c). Microencapsulation of anthocyanin-rich black soybean coat extract by spray-drying using maltodextrin, gum arabic and skimmed milk powder, *Journal of Microencapsulation*, **34**(5): 475-487.

Karami, Z., Emam-Djomeh, Z., Mirzaee, H.A., Khomeiri, M., Mahoonak, A.S. and Aydani, E. (2015). Optimization of microwave assisted extraction (MAE) and soxhlet extraction of phenolic compound from licorice root, *Journal of Food Science and Technology*, 52: 3242-3253.

Khaneghah, A.M., Hashemi, S.M.B. and Limbo, S. (2018). Antimicrobial agents and packaging systems in antimicrobial active food packaging: An overview of approaches and interactions, *Food and Bioproducts Processing*, 111: 1-19.

Kokina, M., Salević, A., Kalušević, A., Lević, S., Pantić, M., Pljevljakušić, D., Šavikin, K., Shamtsyan, M., Nikšić, M. and Nedović, V. (2019). Characterization, antioxidant and antibacterial activity of essential oils and their encapsulation into biodegradable material followed by freeze-drying, *Food Technology and Biotechnology*, **57**(2): 282-289.

Krishnaiah, D., Sarbatly, R. and Nithyanandam, R. (2012). Microencapsulation of *Morinda citrifolia* L. extract by spray-drying, *Chemical Engineering Research and Design*, 90: 622-632.

Kumar, S.S., Chauhan, A.S. and Giridhar, P. (2020). Nanoliposomal encapsulation mediated enhancement of betalain stability: Characterization, storage stability and antioxidant activity of *Basella rubra* L. fruits for its applications in vegan gummy candies, *Food Chemistry*, 333: 127442.

Kurek, M., Garofulić, I.E., Bakić, M.T., Ščetar, M., Uzelac, V.D. and Galić, K. (2018). Development and evaluation of a novel antioxidant and pH indicator film-based on chitosan and food waste sources of antioxidants, *Food Hydrocolloids*, 84: 238-246.

Lagarón, J.M., Ocio, M.J. and López-Rubio, A. (2011). Antimicrobial packaging polymers. A general introduction, pp. 1-22. *In:* J.M. Lagarón, M.J. Ocio and A. López-Rubio (Eds.), *Antimicrobial Polymers*, John Wiley & Sons, Inc., Hoboken.

Lee, K.Y., Yang, H.J. and Song, K.B. (2016). Application of a puffer fish skin gelatin film containing *Moringa oleifera* Lam. leaf extract to the packaging of Gouda cheese, *Journal of Food Science and Technology*, **53**(11): 3876-3883.

Li, H., Chen, B. and Yao, S. (2005). Application of ultrasonic technique for extracting chlorogenic acid from *Eucommia ulmodies* Oliv. (*E. ulmodies*), *Ultrasonics Sonochemistry*, 12: 295-300.

Liu, H., Xie, F., Yu, L., Chen, L. and Li, L. (2009). Thermal processing of starch-based polymers, *Progress in Polymer Science*, **34**(12): 1348-1368.

Lončarević, I., Pajin, B., Fišteš, A., Šaponjac, V.T., Petrović, J., Jovanović, P., Vulić, J. and Zarić, D. (2018). Enrichment of white chocolate with blackberry juice encapsulate: Impact on physical properties, sensory characteristics and polyphenol content, *LWT*, 92: 458-464.

Lončarević, I., Pajin, B., Tumbas Šaponjac, V., Petrović, J., Vulić, J., Fišteš, A. and Jovanović, P. (2019). Physical, sensorial and bioactive characteristics of white chocolate with encapsulated green tea extract, *Journal of the Science of Food and Agriculture*, **99**(13): 5834-5841.

López de Dicastillo, C., Nerin, C., Alfaro, P., Catalá, R., Gavara, R. and Hernández-Munoz, P. (2011). Development of new antioxidant active packaging films based on ethylene vinyl alcohol copolymer (EVOH) and green tea extract, *Journal of Agricultural and Food Chemistry*, **59**(14): 7832-7840.

Lourenço, S.C., Fraqueza, M.J., Fernandes, M.H., Moldão-Martins, M. and Alves, V.D. (2020). Application of edible alginate films with pineapple peel active compounds on beef meat preservation, *Antioxidants*, **9**(8): 667.

Manojlovic, V., Rajic, N., Djonlagic, J., Obradovic, B., Nedovic, V. and Bugarski, B. (2008). Application of electrostatic extrusion – Flavour encapsulation and controlled release, *Sensors*, **8**(3): 1488-1496.

Mantegna, S., Binello, A., Boffa, L., Giorgis, M., Cena, C. and Cravotto, G. (2012). A one-pot ultrasound-assisted water extraction/cyclodextrin encapsulation of resveratrol from *Polygonum cuspidatum*, *Food Chemistry*, 130: 746-750.

Martins, C., Vilarinho, F., Silva, A.S., Andrade, M., Machado, A.V., Castilho, M.C., Sá, A., Cunha, A., Vaz, M.F. and Ramos, F. (2018). Active polylactic acid film incorporated with green tea extract: Development, characterization and effectiveness, *Industrial Crops and Products*, 123: 100-110.

Medina-Jaramillo, C., Ochoa-Yepes, O., Bernal, C. and Famá, L. (2017). Active and smart biodegradable packaging based on starch and natural extracts, *Carbohydrate Polymers*, 176: 187-194.

Mellinas, C., Valdés, A., Ramos, M., Burgos, N., Garrigos, M.D.C. and Jiménez, A. (2016). Active edible films: Current state and future trends, *Journal of Applied Polymer Science*, **133**(2).

Milutinović, M., Radovanović, N., Ćorović, M., Šiler-Marinković, S., Rajilić-Stojanović, M. and Dimitrijević-Branković, S. (2015). Optimisation of microwave-assisted extraction parameters for antioxidants from waste *Achillea millefolium* dust, *Industrial Crops and Products*, 77: 333-341.

Mura, P. (2014). Analytical techniques for characterization of cyclodextrin complexes in aqueous solution: A review, *Journal of Pharmaceutical and Biomedical Analysis*, 101: 238-250.

Mura, M., Palmieri, D., Garella, D., Di Stilo, A., Perego, P., Cravotto, G. and Palombo, D. (2015). Simultaneous ultrasound-assisted water extraction and β-cyclodextrin encapsulation of polyphenols from *Mangifera indica* stem bark in counteracting TNFα-induced endothelial dysfunction, *Natural Product Research*, **29**(17): 1657-1663.

Mustafa, A. and Turner, C. (2011). Pressurized liquid extraction as a green approach in food and herbal plants extraction: A review, *Analytica Chimica Acta*, 703: 8-18.

Nedovic, V., Kalusevic, A., Manojlovic, V., Levic, S. and Bugarski, B. (2011). An overview of encapsulation technologies for food applications, *Procedia Food Science*, 1: 1806-1815.

Nedović, V., Kalušević, A., Manojlović, V., Petrović, T. and Bugarski, B. (2013). Encapsulation systems in the food industry, pp. 229-253. *In:* S. Yanniotis, P. Taoukis, N.G. Stoforos and V.T. Karathanos (Eds.), *Advances in Food Process Engineering Research and Applications*, Springer, New York.

Nedović, V.A., Mantzouridou, F.T., Đorđević, V.B., Kaluševič, A.M., Nenadis, N. and Bugarski, B. (2017). Isolation, purification and encapsulation techniques for bioactive compounds from agricultural and food production waste, pp. 159-194. *In:* Q.V. Voung (Ed.), *Utilization of Bioactive Compounds from Agricultural and Food Production Waste*, CRC Press, Boca Raton, Florida.

Nguyen, T., Phan-Thi, H., Pham-Hoang, B.-N., Ho, P.-T., Tran, T. and Waché, Y. (2018). Encapsulation of *Hibiscus sabdariffa* L. anthocyanins as natural colours in yeast, *Food Research International*, 107: 275-280.

Nur Hanani, Z.A., Beatty, E., Roos, Y.H., Morris, M.A. and Kerry, J.P. (2012). Manufacture and characterization of gelatin films derived from beef, pork and fish sources using twin screw extrusion, *Journal of Food Engineering*, **113**(4): 606-614.

Oliveira, D.A., Mezzomo, N., Gomes, C. and Ferreira, S.R.S. (2017). Encapsulation of passion fruit seed oil by means of supercritical antisolvent process, *The Journal of Supercritical Fluids*, 129: 96-105.

Otálora, M.C., de Jesús Barbosa, H., Perilla, J.E., Osorio, C. and Nazareno, M.A. (2019). Encapsulated betalains (*Opuntia ficus-indica*) as natural colorants. Case study: Gummy candies, *LWT*, 103: 222-227.

PACE (The Platform for Accelerating the Circular Economy) (2019). The Circularity Gap Report.

Pasukamonset, P., Kwon, O. and Adisakwattana, S. (2016). Alginate-based encapsulation of polyphenols from *Clitoria ternatea* petal flower extract enhances stability and biological activity under simulated gastrointestinal conditions, *Food Hydrocolloids*, 61: 772-779.

Pompeu, D.R., Silva, E.M. and Rogez, H. (2009). Optimization of the solvent extraction of phenolic antioxidants from fruits of *Euterpe oleracea* using response surface methodology, *Bioresource Technology*, 100: 6076-6082.

Popović, D.A., Milinčić, D.D., Pešić, M.B., Kalušević, A.M., Tešić, Ž.Lj. and Nedović, V.A. (2019). Encapsulation technologies for polyphenol-loaded microparticles in food industry, pp. 335-367. *In*: F. Chemat and E. Vorobiev (Eds.). *Green Food Processing Techniques*, Elsevier Academic Press.

Rai, A., Mohanty, B. and Bhargava, R. (2016). Supercritical extraction of sunflower oil: A central composite design for extraction variables, *Food Chemistry*, 192: 647-659.

Rambabu, K., Bharath, G., Banat, F., Show, P.L. and Cocoletzi, H.H. (2019). Mango leaf extract incorporated chitosan antioxidant film for active food packaging, *International Journal of Biological Macromolecules*, 126: 1234-1243.

Reis, L.C.B., de Souza, C.O., da Silva, J.B.A., Martins, A.C., Nunes, I.L. and Druzian, J.I. (2015). Active biocomposites of cassava starch: The effect of yerba mate extract and mango pulp as antioxidant additives on the properties and the stability of a packaged product, *Food and Bioproducts Processing*, 94: 382-391.

Rhim, J.W. and Ng, P.K.W. (2007). Natural biopolymer-based nanocomposite films for packaging applications, *Critical Reviews in Food Science and Nutrition*, **47**(4): 411-433.

Ribeiro, A., Caleja, C., Barros, L., Santos-Buelga, C., Barreiro, M.F. and Ferreira, I.C.F.R. (2016). Rosemary extracts in functional foods: Extraction, chemical characterization and incorporation of free and microencapsulated forms in cottage cheese, *Food & Function*, 7(5): 2185-2196.

Rubió, L., Motilva, M.J. and Romero, M.P. (2013). Recent advances in biologically active compounds in herbs and spices: A review of the most effective antioxidant and anti-inflammatory active principles, *Critical Reviews in Food Science and Nutrition*, **53**(9): 943-953.

Rubio, F.T.V., Haminiuk, C.W.I., Martelli-Tosi, M., da Silva, M.P., Makimori, G.Y.F. and Favaro-Trindade, C.S. (2020). Utilization of grape pomaces and brewery waste *Saccharomyces cerevisiae* for the production of bio-based microencapsulated pigments, *Food Research International*, 136: 109470.

Salević, A.S., Kalušević, A.M., Lević, S.M. and Nedović, V.A. (2018). Encapsulation of bioactive compounds derived from fruit processing by-products, *Journal of Agricultural Sciences*, **63**(2): 113-137.

Salević, A., Prieto, C., Cabedo, L., Nedović, V. and Lagaron, J.M. (2019). Physicochemical, antioxidant and antimicrobial properties of electrospun poly (ε-caprolactone) films containing a solid dispersion of sage (*Salvia officinalis* L.) extract, *Nanomaterials*, **9**(2): 270.

Sansone, F., Mencherini, T., Picerno, P., d'Amore, M., Aquino, R.P. and Lauro, M.R. (2011). Maltodextrin/pectin microparticles by spray-drying as carrier for nutraceutical extracts, *Journal of Food Engineering*, **105**(3): 468-476.

Šaponjac, V.T., Ćetković, G., Čanadanović-Brunet, J., Pajin, B., Djilas, S., Petrović, J., Lončarević, I. and Vulić, J. (2016). Sour cherry pomace extract encapsulated in whey and soy proteins: Incorporation in cookies, *Food Chemistry*, 207: 27-33.

Sawale, P.D., Patil, G.R., Hussain, S.A., Singh, A.K. and Singh, R.R.B. (2017). Effect of incorporation of encapsulated and free Arjuna herb on storage stability of chocolate vanilla dairy drink, *Food Bioscience*, 19: 142-148.

Šeregelj, V., Tumbas Šaponjac, V., Levic, S., Kalušević, A., Ćetković, G., Čanadanović-Brunet, J., Nedović, V., Stajčić, S., Vulić, J. and Vidaković, A. (2019). Application of encapsulated natural bioactive compounds from red pepper waste in yogurt, *Journal of Microencapsulation*, **36**(8): 704-714.

Sereshti, H., Rohanifar, A., Bakhtiari, S. and Samadi, S. (2012). Bifunctional ultrasound assisted extraction and determination of *Elettaria cardamomum* Maton essential oil, *Journal of Chromatography, A*, 1238: 46-53.

Shan, B., Cai, Y.Z., Brooks, J.D. and Corke, H. (2007). The *in vitro* antibacterial activity of dietary spice and medicinal herb extracts, *International Journal of Food Microbiology*, **117**(1): 112-119.

Silva-Weiss, A., Ihl, M., Sobral, P.J.A., Gómez-Guillén, M.C. and Bifani, V. (2013). Natural additives in bioactive edible films and coatings: Functionality and applications in foods, *Food Engineering Reviews*, **5**(4): 200-216.

Siripatrawan, U. and Noipha, S. (2012). Active film from chitosan incorporating green tea extract for shelf-life extension of pork sausages, *Food Hydrocolloids*, **27**(1): 102-108.

Souza, V.G.L., Fernando, A.L., Pires, J.R.A., Rodrigues, P.F., Lopes, A.A. and Fernandes, F.M.B. (2017). Physical properties of chitosan films incorporated with natural antioxidants, *Industrial Crops and Products*, 107: 565-572.

Spigno, G., Donsì, F., Amendola, D., Sessa, M., Ferrari, G. and De Faveri, M. (2013). Nanoencapsulation systems to improve solubility and antioxidant efficiency of a grape marc extract into hazelnut paste, *Journal of Food Engineering*, 114: 207-214.

Stajić, S., Kalušević, A., Tomasevic, I., Rabrenović, B., Božić, A., Radović, P., Nedović, V. and Živković, D. (2020). Technological properties of model system beef emulsions with encapsulated pumpkin seed oil and shell powder, *Polish Journal of Food and Nutrition Sciences*, **70**(2): 159-168.

Stajić, S., Živković, D., Tomović, V., Nedović, V., Perunović, M., Kovjanić, N., Lević, S. and Stanišić, N. (2014). The utilization of grapeseed oil in improving the quality of dry fermented sausages, *International Journal of Food Science & Technology*, **49**(11): 2356-2363.

Stojanović, R., Belščak-Cvitanović, A., Manojlović, V., Komes, D., Nedović, V. and Bugarski, B. (2011). Encapsulation of thyme (*Thymus serpyllum* L.) aqueous extract in calcium alginate beads. *Journal of the Science of Food and Agriculture*, 92: 685-696.

Tavares da Silva, F., Furtado da Cunha, K., Fonseca, L., Antunes, M., Mello El Halal, S.L., Fiorentini, Â.M., Zavareze, E. and Guerra Dias, A.R. (2018). Action of ginger essential oil (*Zingiber officinale*) encapsulated in proteins ultrafine fibers on the antimicrobial control *in situ*, *International Journal of Biological Macromolecules*, 118: 107-115.

Trifković, K., Milašinović, N., Djordjević, V., Krušić Kalagasidis, M., Knežević-Jugović, Z., Nedović, V. and Bugarski, B. (2014). Chitosan microbeads for encapsulation of thyme (*Thymus serpyllum* L.) polyphenols, *Carbohydrate Polymers*, 111: 901-907.

Vajić, U.-J., Grujić-Milanović, J., Živković, J., Šavikin, K., Gođevac, D., Miloradović, Z., Bugarski, B. and Mihailović-Stanojević, N. (2015). Optimization of extraction of stinging nettle leaf phenolic compounds using response surface methodology, *Industrial Crops and Products*, 74: 912-917.

Valdés, A., Mellinas, A.C., Ramos, M., Garrigós, M.C. and Jiménez, A. (2014). Natural additives and agricultural wastes in biopolymer formulations for food packaging, *Frontiers in Chemistry*, 2: 6.

Velderrain-Rodríguez, G., Torres-Moreno, H., Villegas-Ochoa, M., Ayala-Zavala, F., Robles-Zepeda, R., Wall-Medrano, A. and González-Aguilar, G. (2018). Gallic acid content and an antioxidant mechanism are responsible for the antiproliferative activity of 'Ataulfo' mango peel on LS180 cells, *Molecules*, 23: 1-15.

Vilela, C., Kurek, M., Hayouka, Z., Röcker, B., Yildirim, S., Antunes, M.D.C., Nilsen-Nygaard, J., Kvalvåg Pettersen, M. and Freire, C.S.R. (2018). A concise guide to active agents for active food packaging, *Trends in Food Science & Technology*, 80: 212-222.

Vuleta, G., Milić, J. and Savić, S. (2012). *Farmaceutska tehnologija*, Faculty of Pharmacy, University of Belgrade, Belgrade, Serbia (in Serbian).

Vulić, J., Šeregelj, V., Kalušević, A., Lević, S., Nedović, V., Tumbas Šaponjac, V., Čanadanović-Brunet, J. and Ćetković, G. (2019). Bioavailability and bioactivity of encapsulated phenolics and carotenoids isolated from red pepper waste, *Molecules*, **24**(15): 2837.

Wang, L. and Weller, C. (2006). Recent advances in extraction of nutraceuticals from plants, *Trends in Food Science & Technology*, 17: 300-312.

Xu, D.-P., Zheng, J., Zhou, Y., Li, Y., Li, S. and Li, H.-B. (2017). Ultrasound-assisted extraction of natural antioxidants from the flower of *Limonium sinuatum*: Optimization and comparison with conventional methods, *Food Chemistry*, 217: 552-559.

Yilmaztekin, M., Lević, S., Kalušević, A., Cam, M., Bugarski, B., Rakić, V., Pavlović, V. and Nedović, V. (2019). Characterization of peppermint (*Mentha piperita* L.) essential oil encapsulates, *Journal of Microencapsulation*, **36**(2): 109-119.

Zhang, W., Ronca, S. and Mele, E. (2017). Electrospun nanofibres containing antimicrobial plant extracts, *Nanomaterials*, **7**(2): 42.

Zuidam, N.J. and Shimoni, E. (2010). Overview of microencapsulates for use in food products or processes and methods to make them, pp. 3-29 *In:* N.J. Zuidam and V. Nedović (Eds.), *Encapsulation Technologies for Active Food Ingredients and Food Processing*, Springer, New York.

Alcoholic Beverages Produced by Immobilised Microorganisms

Ronnie G. Willaert[1]* and Viktor Nedovic[2]

[1] Department of Bioengineering Sciences, Vrije Universiteit Brussel, Pleinlaan 2, 1050 Brussel, Belgium
[2] Department of Food Technology and Biochemistry, Faculty of Agriculture, University of Belgrade, Serbia

1. Introduction

Historical sources indicate that alcoholic beverages have been produced by mankind since many centuries. Wine may have an archaeological record going back to the early Neolithic period (ca. 6000-5800 BC), to the territory of Georgia, South Caucasus region, where the earliest biomolecular archaeological evidence for grape wine and viniculture has been discovered (McGovern et al., 1996; McGovern et al., 2017). Clear evidence of intentional winemaking first appears in the representations of wine presses that date back to the reign of Udimu in Egypt, some 5,000 years ago. The direct fermentation of fruit juices, such as that of grape, had doubtlessly taken place for many thousands of years before early thinking man developed beer brewing and, probably coincidentally, bread baking (Hardwick et al., 1995). One of the oldest historical evidence of formal brewing dates back to about 6000 BC in ancient Babylonia. Chemical analyses of ancient organics revealed that a mixed fermented beverage of rice beer, honey mead, and hawthorn fruit wine was also being produced as early as the 7th millennium BC in Jiahu, China (McGovern et al., 2004).

The metabolic activity of microbial cells has been exploited for the production of a variety of fermented alcoholic beverages. Especially the yeast *Saccharomyces cerevisiae* (Brewers's or Baker's yeast) has been used at an industrial scale to produce alcoholic beverages due to the many advantages such as its availability, safety for food production (it is considered as a GRAS (Generally Recognized as Safe) microorganism), ease of handling and disposal, low cost, and high catalytic activity for a variety of substrates.

*Corresponding author: Ronnie.Willaert@vub.be

Whole-cell immobilisation can be defined as the physical confinement or localisation of intact cells to a certain defined region of space with the preservation of some desired activity (Karel *et al.*, 1985). Immobilized cell technology (ICT) that is based on the immobilisation of cells, has been developed as a tool to obtain high volumetric productivities during fermentation. Despite the technological and economic advantages of ICT, the commercialisation of this technology for alcoholic beverage production experienced only limited success during the last 30 years, mainly due to unpredictable effect of the cell confinement on yeast physiology (Djordjevic *et al.*, 2016). In this chapter, the knowledge of ICT for the production of alcoholic beverages, i.e. beer, wine, and cider, is reviewed.

2. Cell Immobilisation Methods

Immobilization methods can be classified based on natural or artificially induced immobilization and on the physical location and the nature of the microenvironment (Fig. 1). Natural immobilisation can be further subdivided in self-aggregation (such as the formation of yeast flocs and fungal pellets), and natural adhesion of cells on a surface and producing a biofilm or adhesion in a porous matrix. Artificially induced immobilization can be subdivided into surface attachment by using linking agents, entrapment within porous matrices (in hydrogels or preformed supports), containment

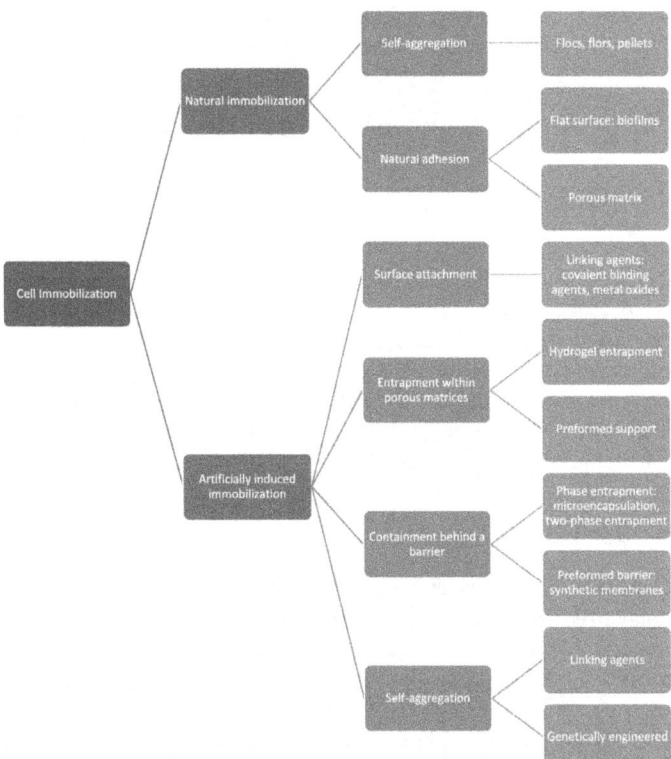

Fig. 1: Classification of immobilization method according to natural or artificially induced immobilization

behind a barrier (which can be by phase entrapment or by preformed synthetic membranes), and self-aggregation by using linking agents or genetically engineered cells (Karel *et al.*, 1985; Willaert and Baron, 1996).

Since *S. cerevisiae* and *S. pastorianus* are the main yeast species that are used in the production of alcoholic beverages (beer and wine) and many strains form aggregates during fermentation, thus the natural immobilisation method, i.e. 'self-aggregation' deserves special attention (Fig. 1). The aggregation or flocculation is caused by cell-cell interaction by yeast adhesins, i.e. the Flo proteins (Goossens and Willaert, 2010; Willaert, 2018) (Fig. 2). The Flo adhesin protein family of *S. cerevisiae* contains a group of proteins that is encoded by genes including *FLO1, FLO5, FLO9* and *FLO10*. The gene products of *FLO1, FLO5, FLO9* and – to a lesser extent – *FLO10*, promote cell-cell adhesion and form multicellular clumps (flocs), which sediment out of solution and therefore, they are called flocculins. The flocculation mechanism is based on the self-interaction of Flo proteins and this interaction is established in two stages, involving both glycan-glycan and protein-glycan interactions (Goossens *et al.*, 2015).

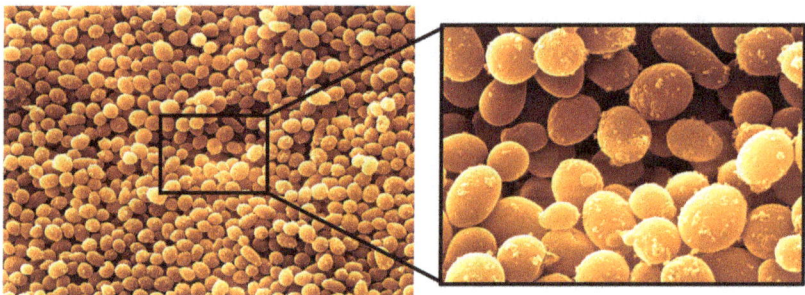

Fig. 2: *S. cerevisiae* cells interacting to produce a floc (Scanning Electron Microscopy, provided by Ronnie G. Willaert)

The flocculation phenomenon is exploited in the brewery industry as an easy, convenient and cost-effective way to separate the aggregated yeast cells from the beer at the end of the primary fermentation. The timing of flocculation is crucial for brewers as the quality of beer highly depends on it. When cells start to flocculate too early, the fermentation will be incomplete with undesirable aromas and too many residual sugars. On the other hand, when the flocculation is delayed, problems can arise during beer filtration (Willaert, 2007).

3. Beer Production

Traditional beer fermentation technology uses freely suspended yeast cells to ferment wort in a non-stirred batch reactor (currently, mostly a cylindrically-*conical* tank (CCT) reactor is used). The fermentation is time consuming. The traditional primary fermentation for lager beer takes approximately 7 days with a subsequent secondary fermentation (maturation) of several weeks. However, the resulting beer has a well-balanced flavor profile, which is well accepted by the consumer. Nowadays, large breweries use an accelerated fermentation scheme, which is based on using a higher fermentation temperature and a selected specific yeast strain. This allows the production of finished lager beer in 12-15 days.

Since the 1950s and 1960s, various continuous fermentation systems have been developed. These systems can be classified as: (1) stirred *versus* unstirred tank reactors, (2) single-vessel systems *versus* systems consisting of a number of vessels connected in series, (3) vessels which allow yeast to overflow freely with the beer *versus* vessels which have abnormally high yeast concentrations (Wellhoener, 1954; Coutts, 1957; Bishop, 1970). These continuous beer fermentation processes were not commercially successful due to many practical problems, such as contamination problems and changes in beer flavor (Thorne, 1968). One of the well-known exceptions is the successful implementation of a continuous beer production process in New Zealand by Morton Coutts, which is still in use today (DB Breweries, Heineken). In the 1970s, there was a revival in developing continuous beer fermentation systems due to the progress in research on immobilization bioprocesses using living cells. Important advantages of immobilized cell processes are the increase of the volumetric productivity due to intense fermentation caused by high cell densities per reactor volume that can be obtained, resulting in a drastic increase in fermentation productivities compared to the traditional time-consuming batch fermentation processes; and that the process can be performed continuously due to the retainment of the cells in the immobilization matrix. The last 30 years, immobilized cell technology has been extensively examined and some designs have reached commercial exploitation.

Beer fermentation processes using immobilized cells have been designed for different stages in the beer fermentation process such as natural wort acidification (brewing according to the German 'Reinheitsgebot'), the primary (main) fermentation, the secondary fermentation, the combined main and secondary fermentation and Fermentation for the production of alcohol-free (AFB) or low-alcohol beers (Table 1).

The objective of ICT application in the production of acidified wort is to reduce the pH of the wort according to the 'Reinheitsgebot' using a 'natural' method by using lactic acid bacteria, before the start of the boiling process. Acidification of wort could be increased by an ICT process where *Lactobacillus amylovorus* was immobilized on DEAE-cellulose beads (Table 1) (Pittner *et al.*, 1993). The produced acidified wort was stored in a holding tank and used during wort production to acidify the wort.

During the first (1914-1918) and second (1939-1945) World War, beers with a low alcohol content were produced since there was a shortage of raw materials (Brányik *et al.*, 2012). Also, low alcohol beers were produced during the prohibition period (1919-1933) in the USA (Meussdoerffer, 2009). Today, low-alcohol or alcohol-free beer is no longer just a niche product but an increasing segment of the beer market (Liguori *et al.*, 2018; Mangindaan *et al.*, 2018). Brewers can increase the overall production by bringing out new products in countries with highly competitive markets and for beverage markets in countries where alcohol consumption if forbidden for religious reasons (Brányik *et al.*, 2012). This product category offers economic benefits in the form of a steadily growing market and often a lower tax burden (Bellut *et al.*, 2018). These beers provide consumers a beverage that is appropriate for driving motor vehicles and will not negatively influence the workplace productivity or increase accidents. Recently, consumers interest increased also due to the associated health aspects such as the presence of healthy beer components (antioxidants, bioactive phenolic substances, soluble fiber, vitamins, and minerals) (Bamforth, 2002; Sohrabvandi *et al.*, 2012); lower energy intake (reduce overweight) and the absence of negative aspects of alcohol consumption (Bellut *et al.*, 2018).

Table 1: Selected Applications of Cell Immobilization Technology for the Production of Beer

Immobilization Method	Carrier Material	Reactor Type	Reference
Wort acidification			
Artificially induced	DEAE-cellulose beads	Fixed bed (continuous)	Pittner *et al.*, 1993
Production of alcohol-free beer			
Artificially induced	DEAE-cellulose beads	Fixed bed (continuous)	Lommi *et al.*, 1990, 1997; van Iersel *et al.*, 1995
Artificially induced	Porous glass beads	Fixed bed (continuous)	Aivasidis *et al.*, 1991
Artificially induced	Porous silicon carbide rods	Monolith (continuous)	Van de Winkel *et al.*, 1991
Main fermentation			
Natural immobilization	Self-aggregation	Tank (batch)	Holle *et al.*, 2012
Natural immobilization	Self-aggregation	Tower (continuous)	Ault *et al.*, 1969
Natural immobilization	Self-aggregation	Gas lift (continuous)	Pires *et al.*, 2014
Natural immobilization	Gluten pellets	Fixed bed (continuous)	Bardi *et al.*, 1997
Natural immobilization	Spent grains/ corncobs	Gas lift (continuous)	Brányik *et al.*, 2002, 2004, 2006
Natural immobilization	Wood chips	Fixed bed (continuous)	Linko *et al.*, 1997; Kronlöf and Virkajärvi, 1999
Artificially induced	Alginate hydrogel beads	Gas lift (batch and continuous)	Nedovic *et al.*, Nedovic *et al.*, 1997
Artificially induced	Alginate/chitosan hydrogel beads with liquid core	Bottle (batch)	Naydenova *et al.*, 2013, 2014
Artificially induced	κ-Carrageenan hydrogel Beads	Gas lift (continuous)	Mensour *et al.*, 1996
Artificially induced	Pectate hydrogel beads	Gas lift	Smogrovicová *et al.*, 1997
Artificially induced	Polyvinyl alcohol beads	Gas lift (continuous)	Smogrovicová *et al.*, 2001
Artificially induced	Porous chitosan beads	Fluidized bed (continuous)	Unemoto *et al.*, 1998
Artificially induced	Ceramic beads	Fluidized bed (continuous)	Inoue *et al.*, 1995

Artificially induced	Porous glass	Fluidized bed (continuous)	Virkajärvi and Krönlof, 1998; Tata *et al.*, 1999
Artificially induced	Porous silicon carbide rods	Monolith (continuous)	Andries *et al.*, 1996; Tata *et al.*, 1999
Flavor maturation			
Artificially induced	Alginate beads	Fixed bed (continuous)	Shindo *et al.*, 1994
Artificially induced	DEAE-cellulose beads	Fixed bed (continuous)	Pajunen and Grönqvist, 1994
Artificially induced	Polyvinyl alcohol beads	Fixed bed (continuous)	Smogrovicová *et al.*, 2001
Artificially induced	Porous glass beads	Fixed bed (continuous)	Linko *et al.*, 1993; Aivasidis, 1996

The main production methods to produce low alcohol or alcohol-free beverages can be classified based on the type of method such as biological or pre-production methods and physical or post-production methods (Brányik *et al.*, 2012; Jackowski and Trusek, 2018; Mangindaan *et al.*, 2018). The physical methods use specific equipment, which remove ethanol from regular beer. Thermal processes such as vacuum distillation (Narziss *et al.*, 1993; Andrés-Iglesias *et al.*, 2016) or evaporation (Eckert *et al.*, 1990; Zufall and Wackerbauer, 2000), occur at low pressure to reduce temperature damage to the beer since some volatile components are thermally degraded, which impact the taste and flavor of the beer (Blanco *et al.*, 2016). The other physical methods to remove ethanol make use of semipermeable membranes such as dialysis, reverse osmosis, nanofiltration, membrane contactor, and pervaporation (Brányik et al., 2012; Mangindaan *et al.*, 2018). At an industrial scale, dialysis and reverse osmosis are applied. Membrane processes have less thermal impact on the produced beer.

Biological methods to produce alcohol-free or low alcohol beers are based on changes in the mashing regime or in the fermentation process (Brányik *et al.*, 2012; Jackowski and Trusek, 2018; Mangindaan *et al.*, 2018). By adapting the mashing process, the fermentable carbohydrate content is changed in such a way that the fermentability is reduced, which results in beers with a low ethanol content (Muller, 1990, 2000; Ivanov *et al.*, 2016). Several strategies can be applied to reduce the fermentable carbohydrate content of wort. Firstly, the saccharifying β-amylase (which produces most of the fermentable maltose) is inactivated by high temperature mashing (75-80°C). Increasing the mashing temperature from 65°C to 80°C had only a slight effect on extract but reduced wort fermentability from over 70 per cent to less than 30 per cent (Muller, 1991). Spent grains can be remashed to produce a second extract with a low content of fermentable carbohydrates (Zurcher and Gruss, 1991). Methods based on modified mashing schemes are seldom successful for the production of AFB beers and they have to be combined with further actions such as vigorous wort boiling to lower the concentration of aldehydes, wort acidification, limited fermentation, color and bitterness adjustment (Brányik *et al.*, 2012). The classical technology to produce alcohol-free or low alcohol beer is based on the suppression of alcohol formation by arrested or limited fermentation (Narziss *et al.*, 1992). Beers produced in this way are characterized by an undesirable wort aroma since the wort aldehydes have only been reduced to a limited degree (Collin *et al.*, 1991; Debourg *et al.*, 1994; van Iersel *et al.*, 1998). The reduction of these wort aldehydes can be quickly achieved by a short contact time with immobilized yeast cells at a low temperature without undesirable cell growth and ethanol production. A disadvantage of this short contact process is the production of only a low concentration of desirable esters. Controlled ethanol production for low-alcohol and alcohol-free beers have been successfully achieved by partial fermentation using DEAE-cellulose as carrier material, which was packed in a column reactor (Collin *et al.*, 1991; Van Dieren, 1995) (Table 1). This technology has been successfully implemented by Bavaria Brewery (The Netherlands) to produce malt beer on an industrial scale (150000 hl/year) (Pittner *et al.*, 1993).

4. Wine Production

Requirements that ICT must meet to be used in wine production methods are: the method should be cheap, easily performed in an industrial situation, not liable to cause oxidation of the wine, robust, not susceptible to contamination, able to impart correct flavor changes to the wine and should use commercially acceptable supports and organisms (Janssen, 1993). Various methods have been evaluated that were focussed on the alcoholic fermentation, the malolactic fermentation or a combination of both (Table 2). Additionally, immobilization systems have been developed to perform bottle fermentation for the production of sparkling wine.

Various yeast cell immobilization systems have been evaluated to perform the alcoholic or primary wine fermentation (Table 2). Initially, yeast cells were mostly immobilized in a hydrogel such as Ca^{++} alginate. Recently, immobilization methods based on the natural adsorption onto fruit pieces, such as grape skins, guava pieces, or watermelon rinds have been successfully used to accelerate the fermentation. *Saccharomyces cerevisiae* and a filamentous fungus *Penicillium chrysogenum* spontaneous co-aggregate and upon shaking spherical beads that were called 'yeast biocapsules' since they are hollow, are formed (Peinado *et al.*, 2005, 2006). These yeast biocapsules are smooth, elastic, strong, creamy-colored; the diameter depends on the particular shaking rate and time of residence in the formation medium (García-Martínez *et al.*, 2011). Placing the biocapsules in fermentation medium caused yeast cells to colonize and invade all hyphae, thereby causing the fungus to die and remain as a porous support for the yeast.

After the primary fermentation, a secondary fermentation known as malolactic fermentation (MLF) is often undertaken, depending on the style of wine that the winemaker seeks to achieve (Maicas *et al.*, 2001a; Sumby *et al.*, 2019). Species of lactic acid bacteria, i.e. *Lactobacillus, Pediococcus, Leuconostoc* and *Oenococcus*, are involved this fermentation. The first reason for inducing MLF is deacidification of the wine. The degradation of malic acid (dicarboxylic) to lactic acid (monocarboxylic) results in a drop in titratable acidity and a small increase in pH. MLF also influences the microbiological stability and organoleptic quality of the wine (Davis *et al.*, 1988; Kunkee, 1967). *Oenococcus oeni* (Dicks *et al.*, 1995) is the major bacterial species found in wines during MLF since it is well adapted to the low pH and high ethanol concentration of wine (Wibowo *et al.*, 1985).

In traditional sparkling wine production, lees removal is a labor-intensive and time-consuming process. Industrial use of immobilized yeasts in the classic technology of wine champagnization makes it possible to reduce and simplify the riddling (*remuage*) and disgorging procedures (Torresi *et al.*, 2011). Cell immobilization systems for bottle fermentation of the wine has the advantages that the area of the production facilities can be half of the area needed by using the classical method (Lallement, 1991) and the net cost can be reduced by 80 per cent (Loureiro, 1990) (Table 2).

Table 2: Selected Applications of Cell Immobilization Technology for the Production of Wine

Immobilization Method	Carrier Material	Organism/ Operation Mode	Reference
Alcoholic Fermentation			
Artificially induced	Ca alginate beads	*S. cerevisiae*, batch	Bakoyianis *et al.*, 1997; Yajima and Yokotsuka, 2001
Artificially induced	Ca alginate beads	*S. cerevisiae*, continuous	Bakoyianis *et al.*, 1997
Artificially induced	κ-Carrageenan beads	*S. cerevisiae*, continuous	Uematsu *et al.*, 1988
Artificially induced	Gluten pellets	*S. cerevisiae*, batch, continuous	Bardi *et al.*, 1996; Sipsas *et al.*, 2009
Natural immobilization	Grape skins	*S. cerevisiae*, batch	Mallouchos *et al.*, 2002, 2003
Natural immobilization	Guava pieces	*S. cerevisiae*, batch	Reddy *et al.*, 2006
Natural immobilization	Raisins (dried)	*S. cerevisiae*, batch	Tsakiris *et al.*, 2004, 2006
Natural immobilization	Watermelon rind pieces	*S. cerevisiae*, batch	Reddy *et al.*, 2008
Natural immobilization	"Yeast biocapsules"	Spontaneous co-aggregation of *S. cerevisiae* and *Penicillium chrysogenum*	Peinado *et al.*, 2005, 2006; García-Martínez *et al.*, 2013, 2015
Malolactic fermentation			
Artificially induced	Ca alginate beads	*O. oeni*, batch	Spettoli *et al.*, 1982
Artificially induced	Ca alginate – oak charcoal powder beads	*Issatchenkia orientalis*, batch	Hong *et al.*, 2010
Artificially induced	κ-Carrageenan beads	*O. oeni*, batch	McCord and Ryu, 1985
Artificially induced	κ-Carrageenan beads	*Lactobacillus* sp., continuous	Crapisi *et al.*, 1987
Artificially induced	Ca pectate and modified chitosan beads	*L. casei*, batch	Kosseva *et al.*, 1998, 2004
Artificially induced	Cellulose sponge	*O. oeni*, batch	Maicas *et al.*, 2001b

(Contd.)

Table 2: *(Contd.)*

Immobilization Method	Carrier Material	Organism/ Operation Mode	Reference
Artificially induced	Delignified cellulose	*L. casei, O. oeni,* batch	Agouridis *et al.,* 2005; 2008
Natural immobilization	Corn cobs, grape skins, grape stems	*O. oeni,* batch	Genisheva *et al.,* 2013
Natural immobilization	grape skins	continuous	Genisheva *et al.,* 2014
Combined Alcoholic and malolactic fermentation			
Artificially induced	Two-layer delignified cellulose and starch gel	Co-entrapment of *S. cerevisiae* and *O. oeni*; batch	Servetas *et al.,* 2013
	Ca alginate beads	Co-entrapment of *S. cerevisiae* and *O. oeni*; batch	Bleve *et al.,* 2016
Natural immobilization	Grape skins	*S. cerevisiae, O. oeni,* Continuous: series of packed bed reactors	Genisheva *et al.,* 2014
Bottle-fermented Wine			
Artificially induced	Ca alginate beads	Immobilization of *S. cerevisiae* or *S. bayanus*	Fumi *et al.,* 1987
Artificially induced	Ca alginate beads	*S. cerevisiae* entrapped in double-layer beads	Yokotsuka *et al.,* 1997
Natural immobilization	'Yeast biocapsules'	Spontaneous co-aggregation of *S. cerevisiae* and *Penicillium chrysogenum*	Puig-Pujol *et al.,* 2013

5. Cider Production

Cider is a fermented alcoholic beverage that is made from apple juice. It is one of the oldest fermented beverages. The transformation of apple juice into cider requires – in a first step – the activity of yeast for the alcoholic fermentation, and – in a second step – lactic acid bacteria to accomplish the malolactic fermentation. The primary yeast fermentation is traditionally performed at a temperature of 4-16°C and can take 5 weeks to 3 months (Nedović *et al.,* 2015). The malolactic fermentation is performed at 10-30°C. Two procedures to carry out this fermentation are used: the malolactic fermentation proceeds after the alcoholic fermentation reaching attenuation, or both fermentation processes occur simultaneous. When the processes are performed separately, the alcoholic and the malolactic fermentation can be performed in a cascade of 2 reactors, of which one fermenter vessel contains immobilised yeast and the other immobilised bacterium (Simon *et al.*, 1996) (Table 3). The process can also start with free (Cabranes *et al.*, 1998) or immobilized (Scott & O'Reilly, 1996) yeast cells and adding an immobilised bacterium. A continuous malolactic fermentation was performed efficiently using a tubular reactor with *O. oeni* immobilised in polyvinyl alcohol (PVA) hydrogel LentiKats (Durieux *et al.*, 2000). Processes with co-immobilisation of the yeast and the bacterium in the same matrix allows complete fermentation of the apple juice in one fermentor (Nedovic *et al.*, 2000) as was demonstrated for the production of wine (see before) (Bleve *et al.*, 2016).

Table 3: Selected Applications of Cell Immobilization Technology for the Production of Cider

Immobilization Method	Carrier Material	Organism/Operation Mode	Reference
Artificially induced	Ca alginate beads	*S. bayanus* and *O. oeni*; continuous: cascade of 2 APV tower fermenters	Simon *et al.*, 1996
	Cellulose sponge	*S. cerevisiae* and *L. plantarum* separately immobilised	Scott and O'Reilly, 1996
Artificially induced	Ca alginate beads	*O. oeni* immobilised, *S. cerevisiae* as free cells	Cabranes *et al.*, 1998
Artificially induced	Ca alginate beads	*S. bayanus* and *O. oeni* co-immobilised; continuous	Nedovic *et al.*, 2000
Artificially induced	PVA LentiKats	*O. oeni* immobilised (only malolactic fermentation)	Durieux *et al.*, 2000
Artificially induced	Ca alginate beads	*O. oeni* immobilised (only malolactic fermentation)	Herrero *et al.*, 2001

6. Conclusion and Future Perspective

The chapter summarizes major cell immobilization methods, as well as applications of immobilized cell technology in beer, wine and cider production. In spite of the fact that it was recognized by many studies at laboratory and pilot scale as a very promising technology in the sense of high cell densities, high fermentation rates, need for smaller reactor sizes, possibility for conducting of continuous process without the risk of cell washout from the reactor, and so on, there are still not so many commercial applications at the industrial level. The main reason lies in altered cell physiology and morphology during the long running continuous operations that causes unbalanced formation of volatile compounds and unbalanced flavor profile of the final product. However, based on recent progresses in investigation and design of new materials for cell immobilization, new methods of cell immobilization as well as new reactor concepts, there are still a lot of potential of this technology not only for the bioprocesses that are less complex, like alcohol-free beer production and secondary beer fermentation are, but also for the main beer fermentation, wine and cider alcoholic and malolactic fermentation.

Acknowledgments

The Belgian Federal Science Policy Office (Belspo) and the European Space Agency (ESA) PRODEX program supported this work. The Research Council of the *Vrije Universiteit Brussel* (Belgium) and the University of Ghent (Belgium) are acknowledged to support the Alliance Research Group VUB-UGent NanoMicrobiology (NAMI), and the International Joint Research Group (IJRG) VUB-EPFL BioNanotechnology & NanoMedicine (NANO).

References

Agouridis, N., Bekatorou, A., Nigam, P. and Kanellaki, M. (2005). Malolactic fermentation in wine with *Lactobacillus casei* cells immobilized on delignified cellulosic material, *Journal of Agricultural and Food Chemistry*, 53: 2546-2551.

Agouridis, N., Kopsahelis, N., Plessas, S., Koutinas, A. and Kanellaki, M. (2008). *Oenococcus oeni* cells immobilized on delignified cellulosic material for malolactic fermentation of wine, *Bioresource Technology*, 99: 9017-9020.

Aivasidis, A., Wandrey, C., Eils, H.-G. and Katzke, M. (1991). Continuous fermentation of alcohol-free beer with immobilized yeast cells in fluidized bed reactors, *Proceedings European Brewery Convention Congress*, 569-576.

Aivasidis, A. (1996). Another look at immobilized yeast systems, *Cerevisia*, 21: 27-32.

Andrés-Iglesias, C., Blanco, C.A., García-Serna, J., Pando, V. and Montero, O. (2016). Volatile compound profiling in commercial lager regular beers and derived alcohol-free beers after dealcoholization by vacuum distillation, *Food Analytical Methods*, 9: 3230-3241.

Andries, M., Van Beveren, P.C., Goffin, O. and Masschelein, C.A. (1996). Design and application of an immobilized loop bioreactor for continuous beer fermentation, pp. 672-678. *In*: R.H. Wijffels, R.M. Buitelaar, C. Bucke and J. Tramper (Eds.), *Immobilized Cells: Basics and Applications*, Elsevier, Amsterdam.

Ault, R.G., Hampton, A.N., Newton, R. and Roberts, R.H. (1969). Biological and biochemical aspects of tower fermentation, *Journal of the Institute of Brewing*, 75: 260-277.

Bakoyianis, V., Koutinas, A.A., Agelopoulos, K. and Kanellaki, M. (1997). Comparative study of kissiris, g-alumina, and calcium alginate as supports of cells for batch and continuous winemaking at low temperatures, *Journal of Agricultural and Food Chemistry*, 45: 4884-4888.

Bamforth, C.W. (2002). Nutritional aspects of beer – A review, *Nutrition Research*, 22: 227-237.

Bardi, E.P., Bakoyianis, V., Koutinas, A.A. and Kanellaki, M. (1996). Room temperature and low temperature wine making using yeast immobilized on gluten pellets, *Process Biochemistry*, 31: 425-430.

Bardi, E., Koutinas, A.A. and Kanellaki, M. (1997). Room and low temperature brewing with yeast immobilized on gluten pellets, *Process Biochemistry*, 32: 691-696.

Bellut, K., Michel, M., Zarnkow, M., Hutzler, M., Jacob, F., De Schutter, D.P., Daenen, L., Lynch, K.M., Zannini, E. and Arendt, E.K. (2018). Application of non-*Saccharomyces* yeasts isolated from kombucha in the production of alcohol-free beer, *Fermentation*, 4: 66.

Bishop, L.R. (1970). A system of continuous Fermentation, *Journal of the Institute of Brewing*, 76: 172-181.

Blanco, C.A., Andrés-Iglesias, C. and Montero, O. (2016). Low-alcohol beers: Flavor compounds, defects, and improvement strategies, *Critical Reviews in Food Science and Nutrition*, 56: 1379-1388.

Bleve, G., Tufariello, M., Vetrano, C., Mita, G. and Grieco, F. (2016). Simultaneous alcoholic and malolactic Fermentation by *Saccharomyces cerevisiae* and *Oenococcus oeni* cells co-immobilized in alginate beads, *Frontiers in Microbiology*, 7: 943.

Brányik, T., Vicente, A., Cruz, J.M. and Teixeira, J. (2002). Continuous primary beer fermentation with brewing yeast immobilized on spent grains, *Journal of the Institute of Brewing*, 108: 410-415.

Brányik, T., Vicente, A.A., Cruz, J.M.M. and Texeira, J.A. (2004). Continuous primary fermentation of beer with yeast immobilized on spent grains – The effect of operational conditions, *J. Am. Soc. Brew. Chem.*, 62: 29-34.

Brányik, T., Silva, D.P., Vicente, A.A., Lehnert, R., Almeida E Silva, J.B., Dostálek, P. and Teixeira, J.A. (2006). Continuous immobilized yeast reactor system for complete beer fermentation using spent grains and corncobs as carrier materials, *Journal of Industrial Microbiology and Biotechnology*, 33: 1010-1018.

Brányik, T., Silva, D.P., Baszczynski, M., Lehnert, R. and Almeida E Silva, J.B. (2012). A review of methods of low alcohol and alcohol-free beer production, *Journal of Food Engineering*, 108: 493-506.

Cabranes, C., Moreno, J. and Mangas, J.J. (1998). Cider production with immobilized *Leuconostoc oenos*, *Journal of the Institute of Brewing*, 104: 127-130.

Collin, S., Montesinos, M., Meersman, E., Swinkels, W. and Dufour, J.P. (1991). Yeast dehydrogenase activities in relation to carbonyl compounds removal from wort and beer, *Proceedings European Brewery Convention Congress*, 409-416.

Coutts, M.W. (1957). A continuous process for the production of beer, UK Patents 872,391-400.

Crapisi, A., Nuti, M.P., Zamorani, A. and Spettoli, P. (1987). Improved stability of immobilized *Lactobacillus* sp. cells for the control of malolactic fermentation in wine, *American Journal of Enology and Viticulture*, 38: 310-312.

Davis, C.R., Wibowo, D., Fleet, G.H. and Lee, T.H. (1988). Properties of wine lactic acid bacteria: Their potential enological significance, *Am. J. Enol. Vitic.*, 39: 290-301.

Debourg, A., Laurent, M., Goossens, E., Van De Winkel, L. and Masschelein, C.A. (1994). Wort aldehyde reduction potential in free and immobilized yeast systems, *J. Am. Soc. Brew. Chem.*, 52: 100-106.

Dicks, L.M.T., Dellaglio, F. and Collins, M.D. (1995). Proposal to reclassify *Leuconostoc oenos* as *Oenococcus oeni* (corig.) gen. nov., comb. Nov, *Int. J. System. Bacteriol.*, 45: 395-397.

Djordjevic, V., Willaert, R., Gibson, B. and Nedovic, V. (2016).Immobilized yeast cells and secondary metabolites. *In*: J.-M. Mérillon and K.G. Ramawat (Eds.), *Fungal Metabolites*, pp. 599-639. Springer International Publishing, Switzerland.

Durieux, A., Nicolay, X. and Simon, J.P. (2000). Continuous malolactic fermentation by *Oenococcus Oeni* entrapped in LentiKats, *Biotechnology Letters*, 22: 1679-1684.

Eckert, M., Baumann, G. and Gierschner, K. (1990). Dealcoholization of beer by evaporation, *Food Biotechnology*, 4(1): 278. 1990 Proceedings of the International Conference on Biotechnology and Food, Stuttgart, West Ger, 20 February 1989-24 February 1989, 13349; retrieved from: https://www.scopus.com/record/display. uri?eid=2-s2.0-0025209094&origin=inward&txGid=d96c2513de1a6045daa127324b 843b0e

Fumi, M., Trioli, G. and Colagrande, O. (1987). Preliminary assessment on the use of immobilized yeast cells in sodium alginate for sparkling wine process, *Biotechnology Letters*, 9: 339-342.

García-Martínez, T., López de Lerma, N., Moreno, J., Peinado, R.A., Millán, M.C. and Mauricio, J.C. (2013). Sweet wine production by two osmotolerant *Saccharomyces cerevisiae* strains, *Journal of Food Science*, 78: M874-M879.

García-Martínez, T., Moreno, J., Mauricio, J.C. and Peinado, R. (2015). Natural sweet wine production by repeated use of yeast cells immobilized on *Penicillium chrysogenum*, *LWT – Food Science and Technology*, 61: 503-509.

García-Martínez, T., Peinado, R.A., Moreno, J., García-García, I. and Mauricio, J.C. (2011). Co-culture of *Penicillium chrysogenum* and *Saccharomyces cerevisiae* leading to the immobilization of yeast, *Journal of Chemical Technology and Biotechnology*, 86: 812-817.

Genisheva, Z., Mota, A., Mussatto, S., Oliveira, J.M. and Teixeira, J.A. (2014). Integrated continuous winemaking process involving sequential alcoholic and malolactic Fermentation with immobilized cells, *Process Biochemistry*, 49: 1-9.

Genisheva, Z., Mussatto, S., Oliveira, J.M. and Teixeira, J.A. (2013). Malolactic fermentation of wines with immobilized lactic acid bacteria – Influence of concentration, type of support material and storage conditions, *Food Chemistry*, 138: 1510-1514.

Goossens, K.V.Y. and Willaert, R.G. (2010). Flocculation protein structure and cell-cell adhesion mechanism in *Saccharomyces cerevisiae*, *Biotechnology Letters*, 32: 1571-1585.

Goossens, K.V., Ielasi, F.S., Nookaew, I., Stals, I., Alonso-Sarduy, L., Daenen L., Van Mulders, S.E., Stassen, C., van Eijsden, R.G., Siewers, V., Delvaux, F.R., Kasas, S., Nielsen, J., Devreese, B. and Willaert, R.G. (2015). Molecular mechanism of flocculation self-recognition in yeast and its role in mating and survival, *MBio*, **6**(2).

Hardwick, W.A., van Oevelen, D.E.J., Novellie, L. and Yoshizawa, K. (1995). Kinds of beer and beerlike beverages, pp. 53-85. *In*: W.A. Hardwick (Ed.), *Handbook of Brewing*, Marcel Dekker, New York.

Herrero, M., Laca, A., Garcia, L.A. and Díaz, M. (2001). Controlled malolactic fermentation in cider using *Oenococcus oeni* immobilized in alginate beads and comparison with free cell fermentation, *Enzyme and Microbial Technology*, 28: 35-41.

Holle, A.V., Machado, M.D. and Soares, E.V. (2012). Flocculation in ale brewing strains of *Saccharomyces cerevisiae*: Re-evaluation of the role of cell surface charge and hydrophobicity, *Applied Microbiology and Biotechnology*, 93: 1221-1229.

Hong, S.K., Lee, H.J., Park, H.J., Hong, Y.A., Rhee, I.K., Lee, W.H., Choi, S.W., Lee, O.S. and Park, H.D. (2010). Degradation of malic acid in wine by immobilized *Issatchenkia orientalis* cells with oriental oak charcoal and alginate, *Letters in Applied Microbiology*, 50: 522-529.

Inoue, T. (1995). Development of a two-stage immobilized yeast fermentation system for continuous beer brewing, *Proceedings of European Brewery Convention Congress*, 25-36.

Ivanov, K., Petelkov, I., Shopska, V., Denkova, R., Gochev, V. and Kostov, G. (2016). Investigation of mashing regimes for low-alcohol beer production, *Journal of the Institute of Brewing*, 122: 508-516.

Jackowski, M. and Trusek, A. (2018). Non-alcoholic beer production – An overview, *Polish Journal of Chemical Technology* 20: 32-38.

Janssen, D. (1993). Immobilized-cell malolactic fermentation, *Viticulture and Enology Science*, 48: 219.

Karel, S.F., Libicki, S.B. and Robertson, C.R. (1985). The immobilization of whole cells: Engineering principles, *Chem. Eng. Sci.*, **40**(8): 1321-1354.

Kosseva, M., Beschkov, V., Kennedy, J.F. and Lloyd, L.L. (1998). Malolactic fermentation in Chardonnay wine by immobilized *Lactobacillus casei* cells, *Process Biochemistry*, 33: 793-798.

Kosseva, M. and Kennedy, J.F. (2004). Encapsulated lactic acid bacteria for control of malolactic fermentation in wine, *Artificial Cells Blood Substitutes and Biotechnology*, 32: 55-65.

Kronlöf, J. and Virkajärvi, I. (1999). Primary fermentation with immobilized yeast, *Proceedings of European Brewery Convention Congress*, 761-770.

Kunkee, R.E. (1967). Malo-lactic fermentation, *Advances in Applied Microbiology*, 9: 235-279.

Lallement, A. (1991). Les levures incluses pour la prise de mousse, *Revue Oenologues*, 58: 29-31.

Liguori, L., Russo, P., Albanese, D. and Di Matteo, M. (2018). Production of low-alcohol beverages: Current status and perspectives, pp. 347-382. *In*: A.M. Grumezescu and A.M. Holban (Eds.), *Food Processing for Increased Quality and Consumption*, Elsevier.

Linko, M., Suihko, M.-L., Kronlöf, J. and Home, S. (1993). Use of brewer's yeast expressing a-acetolactate decarboxylase in conventional and immobilized Fermentation, *MBAA Tech. Q.*, 30: 93-97.

Linko, M., Virkajärvi, I., Pohjala, N., Lindborg, K., Kronlöf, J. and Pajunen, E. (1997). Main fermentation with immobilized yeast – A breakthrough? *Proceedings of European Brewery Convention Congress*, 385-394.

Lommi, H., Grönqvist, A. and Pajunen, E. (1990). Immobilized yeast reactor speeds beer production, *Food Techn.*, 5: 128-133.

Lommi, H., Swinkels, W. and Van Dieren, B. (1997). Process for the production of non-alcoholic or low alcohol malt beverage, U.S. Patent, 5,612,072.

Loureiro, V. (1990). Contributo portoghese sull'immobilizzazione dei lieviti per la produzione dei vini spumanti, *Ind. Bevande*, 19: 501-506.

Maicas, S. (2001a). The use of alternative technologies to develop malolactic fermentation in wine, *Applied Microbiology and Biotechnology*, 56: 35-39.

Maicas, S., Pardo, I. and Ferrer, S. (2001b). The potential of positively-charged cellulose sponge for malolactic fermentation of wine, using *Oenococcus oeni*, *Enzyme and Microbial Technology*, 28: 415-419.

Mallouchos, A., Reppa, P., Aggelis, G., Kanellaki, M., Koutinas, A.A. and Komaitis, M. (2002). Grape skins as a natural support for yeast immobilization, *Biotechnology Letters*, 24: 1331-1335.

Mallouchos, A., Skandamis, P., Loukatos, P., Komaitis, M., Koutinas, A. and Kanellaki, M. (2003). Volatile compounds of wines produced by cells immobilized on grape skins, *Journal of Agricultural and Food Chemistry*, 51: 3060-3066.

Mangindaan, D., Khoiruddin, K. and Wenten, I.G. (2018). Beverage dealcoholization processes: Past, present and future, *Trends in Food Science and Technology*, 71: 36-45.

McCord, J.D. and Ryu, D.D.Y. (1985). Development of malolactic fermentation process using immobilized whole cells and enzymes, *American Journal of Enology and Viticulture*, 36: 214-218.

McGovern, P.E., Glusker, D.L., Exner, L.J. and Voigt M.M. (1996). Neolithic resinated wine, *Nature*, 381: 480-481.

McGovern, P.E., Zhang, J., Tang, J., Zhang, Z., Hall, G.R., Moreau, R.A., Nuñez, A., Butrym, E.D., Richards, M.P., Wang, C.S., Cheng, G., Zhao, Z. and Wang, C. (2004). Fermented beverages of pre- and proto-historic China, *Proceedings of the National Academy of Science USA*, 101: 17593-17598.

McGovern, P., Jalabadze, M., Batiuk, S., Callahan, M.P., Smith, K.E., Hall, G.R., Kvavadze, E., Maghradze, D., Rusishvili, N., Bouby, L., Failla, O., Cola, G., Mariani, L., Boaretto, E., Bacilieri, R., This, P., Wales, N. and Lordkipanidze, D. (2017).

Early neolithic wine of Georgia in the South Caucasus, *Proceedings of the National Academy of Science USA*, **114**(48): E10309-E10318.

Mensour, N., Margaritis, A., Briens, C.L., Pilkington, H. and Russell, I. (1996). Applications of immobilized yeast cells in the brewing industry, pp. 661-671. *In*: R.H. Wijffels, R.M. Buitelaar, C. Bucke and J. Tramper (Eds.), *Immobilized Cells: Basics and Applications*, Elsevier, Amsterdam.

Meussdoerffer, F.G. (2009). A comprehensive history of beer brewing, pp. 1-42. *In*: H.M. Esslinger (Ed.), *Handbook of Brewing*, Wiley-VCH Verlag GmbH & Co., Weinheim.

Muller, M. (1991). The effects of mashing temperature and mash thickness on wort carbohydrate composition, *Journal of the Institute of Brewing*, 97: 85-92.

Muller, R. (1990). The production of low-alcohol and alcohol-free beers by limited Fermentation, *Ferment*, 3: 224-230.

Muller, R. (2000). A mathematical model of the formation of fermentable sugars from starch hydrolysis during high-temperature mashing, *Enzyme and Microbial Technology*, 27: 337-344.

Narziss, L., Miedaner, H., Kern, E. and Leibhard, M. (1992). Technology and composition of non-alcoholic beers, *Brauwelt International*, 4: 396.

Narziss, L., Back, W. and Stich, S. (1993). Alcohol removal from beer by countercurrent distillation in combination with rectification, *Brauwelt*, 133: 1806-1820.

Naydenova, V., Vassilev, S., Kaneva, M. and Kostov, G. (2013). Encapsulation of brewing yeast in alginate/chitosan matrix: Comparative study of beer fermentation with immobilized and free cells, *Bulgarian Journal of Agricultural Science*, 19: 123-127.

Naydenova, V., Badova, M., Vassilev, S., Iliev, V., Kaneva, M. and Kostov, G. (2014). Encapsulation of brewing yeast in alginate/chitosan matrix: Lab-scale optimization of lager beer fermentation, *Biotechnology and Biotechnological Equipment*, 28: 277-284.

Nedovic, V., Obradovic, B., Vunjak-Novakovic, G. and Leskosek-Cukalovic, I. (1993). Kinetics of beer fermentation with immobilized yeast (in Serbian), *Hem. Ind.*, **47**(11-12): 168-172.

Nedovic, V.A., Pesic, R., Leskosek-Cukalovic, I., Laketic, D. and Vunjal-Novakovic, G. (1997). Analysis of liquid axial dispersion in an internal loop gas-lift bioreactor for beer fermentation with immobilized yeast cells, *Proceedings of Second European Conference on Fluidization*, Bilbao, 627-635.

Nedovic, V.A., Durieux, A., Van Nedervelde, L., Rosseels, P., Vandegans, J., Plaisant, A. and Simon, J. (2000). Continuous cider fermentation with co-immobilized yeast and *Leuconostoc oenos* cells, *Enzyme and Microbial Technology*, 26: 834-839.

Nedović, V., Gibson, B., Mantzouridou, T.F., Bugarski, B., Djordjević, V., Kalušević, A., Paraskevopoulou, A., Sandell, M., Šmogrovičová, D. and Yilmaztekin, M. (2015). Aroma formation by immobilized yeast cells in fermentation processes, *Yeast*, 32: 173-216.

Pajunen, E. and Grönqvist, A. (1994). Immobilized yeast fermenters for continuous lager beer maturation, pp. 101-103. *Proceedings 23rd Convention Inst. of Brew*, Australia and New Zealand Section, Sydney.

Peinado, R.A., Moreno, J.J., Maestre, O. and Mauricio, J.C. (2005). Use of a novel immobilization yeast system for winemaking, *Biotechnology Letters*, 27: 1421-1424.

Peinado, R.A., Moreno, J.J., Villalba, J.M., González-Reyes, J.A., Ortega, J.M. and Mauricio, J.C. (2006). Yeast biocapsules: A new immobilization method and their applications, *Enzyme and Microbial Technology*, 40: 79-84.

Pires, E.J., Teixeira, J.A., Tomás Brányik, T. and Vicente, A.A. (2014). Carrier-free, continuous primary beer fermentation, *Journal of the Institute of Brewing*, 120: 500-506.

Pittner, H., Back, W., Swinkels, W., Meersman, E., Van Dieren, B. and Lomni, H. (1993). Continuous production of acidified wort for alcohol-free-beer with immobilized lactic acid bacteria, *Proceedings European Brewery Convention Congress*, 323-329.

Puig-Pujol, A., Bertran, E., García-Martínez, T., Capdevila, F., Mínguez, S. and Mauricio, J.C. (2013). Application of a new organic yeast immobilization method for sparkling wine production, *American Journal of Enology and Viticulture*, 64: 386-394.

Reddy, L.V.A., Reddy, Y.H.K. and Reddy, O.V.S. (2006). Wine production by guava piece immobilized yeast from Indian cultivar grapes and its volatile composition, *Biotechnology*, 5: 449-454.

Reddy, L.V., Reddy, Y.H.K., Reddy, L.P.A. and Reddy, O.V.S. (2008). Wine production by novel yeast biocatalyst prepared by immobilization on watermelon (*Citrullus vulgaris*) rind pieces and characterization of volatile compounds, *Process Biochemistry*, 43: 748e-752.

Scott, J.A. and O'Reilly, A.M. (1996). Co-immobilization of selected yeast and bacteria for controlled flavor development in an alcoholic cider beverage, *Process Biochemistry*, 31: 111-117.

Servetas, I., Berbegal, C., Camacho, N., Bekatorou, A., Ferrer, S., Nigam, P., Drouza, C. and Koutinas, A.A. (2013). *Saccharomyces cerevisiae* and *Oenococcus oeni* immobilized in different layers of a cellulose/starch gel composite for simultaneous alcoholic and malolactic wine Fermentation, *Process Biochemistry*, 48: 1279-1284.

Shindo, S., Sahara, H. and Koshino, S. (1994). Suppression of a-acetolactate formation in brewing with immobilized yeast, *Journal of the Institute of Brewing*, 100: 69-72.

Simon, J.P., Durieux, A., Pinnel, V., Garré, V., Vandegans, J., Rosseels, P., Godan, N., Plaisant, A.M., Defroyennes, J.-P. and Foroni, G. (1996). Organoleptic profiles of different ciders after continuous fermentation (encapsulated living cells) versus batch fermentation (free cells), *Progress in Biotechnology*, 11: 615-621.

Sipsas, V., Kolokythas, J., Kourkoutas, Y., Plessas, S., Nedovic, V.A. and Kanellaki, M. (2009). Comparative study of batch and continuous multi-stage fixed-bed tower (MFBT) bioreactor during winemaking using freeze-dried immobilized cells, *Journal of Food Engineering*, 90: 495-503.

Sohrabvandi, S., Mortazavian, A.M. and Rezaei, K. (2012). Health-related aspects of beer: A review, *International Journal of Food Properties*, 15: 350-373.

Spettoli, P., Bottacin, A., Nuti, M.P. and Zamorani, A. (1982). Immobilization of *Leuconostoc oenos* ML 34 in calcium alginate gels and its application to wine technology, *American Journal of Enology and Viticulture*, 33: 1-5.

Smogrovicová, D., Dömény, Z., Gemeiner, P., Malovíková, A. and Sturdík, E. (1997). Reactors for the continuous primary beer fermentation using immobilised yeast, *Biotechnology Techniques*, 11: 261-264.

Smogrovicová, D., Dömény, Navrátil, M. and Dvorák, P. (2001). Continuous beer fermentation using polyvinyl alcohol entrapped yeast, *Proceedings European Brewery Convention Congress*, 50: 1-9.

Sumby, K.M., Bartle, L., Grbin, P.R. and Jiranek, V. (2019). Measures to improve wine malolactic fermentation, *Applied Microbiology and Biotechnology*, 103: 2033-2051.

Tata, M., Bower, P., Bromberg, S., Duncombe, D., Fehring, J., Lau, V.V., Ryder, D. and Stassi, P. (1999). Immobilized yeast bioreactor systems for continuous beer fermentation, *Biotechnology Progress*, 15: 105-113.

Thorne, R.S.W. (1968). Continuous fermentation in retrospect, *Brewing Digest*, 43: 50-55.

Torresi, S., Frangipane, M.T. and Anelli, G. (2011). Biotechnologies in sparkling wine production. Interesting approaches for quality improvement: A review, *Food Chemistry*, 129: 1232-1241.

Tsakiris, A., Sipsas, V., Bekatorou, A., Mallouchos, A. and Koutinas, A.A. (2004). Red wine making by immobilized cells and influence on volatile composition, *Journal of Science of Food and Agriculture*, 82: 1357-1363.

Tsakiris, A., Kourkoutas, Y., Dourtoglou, V.G., Koutinas, A.A., Psarianos, C. and Kanellaki, M. (2006). Wine produced by immobilized cells on dried raisin berries in sensory evaluation comparison with commercial products, *Journal of Science of Food and Agriculture*, 86: 539-543.

Uematsu, K., Fong, D. and Ryu, D.D.Y. (1988). Development of continuous fermentation using immobilized yeast cells, *Applied Biochemistry and Biotechnology*, 19: 177-178.

Unemoto, S., Mitani, Y. and Shinotsuka, K. (1998). Primary fermentation with immobilized yeast in a fluidized bed reactor, *MBAA Technical Quarterly*, 35: 58-61.

Van De Winkel, L., Van Beveren, P.C. and Masschelein, C.A. (1991). The application of an immobilized yeast loop reactor to the continuous production of alcohol-free beer, pp. 307-314. *Proceedings of European Brewery Convention Congress.*

Van Dieren, D. (1995). Yeast metabolism and the production of alcohol-free beer, pp. 66-76. *In:* EBC Monograph XXIV, EBC Symposium on Immobilized Yeast Applications in the Brewing Industry.

van Iersel, M.F.M., Meersman, E., Swinkels, W., Abee, T. and Rombouts, F.M. (1995) Continuous production of non-alcohol beer by immobilized yeast at low temperature, *Journal of Industrial Microbiology*, 14: 495-501.

van Iersel, M.F.M., Meersman, E., Arntz, M., Ronbouts, F.M. and Abee, T. (1998). Effect of environmental conditions on flocculation and immobilization of brewer's yeast during production of alcohol-free beer, *Journal of the Institute of Brewing*, 104: 131-136.

Virkajärvi, I. and Krönlof, J. (1998). Long-term stability of immobilized yeast columns in primary fermentation, *Journal of the American Society of Brewing Chemists*, 56: 70-75.

Wellhoener, H.J. (1954). Ein Kontinuierliches Gär – und Reifungsverfahren für Bier, *Brauwelt*, 94: 624-626.

Wibowo, D., Eschenbruch, R., Davis, C.R., Fleet, G.H. and Lee, T.H. (1985). Occurrence and growth of lactic acid bacteria in wine: A review, *American Journal of Enology and Viticulture*, 36: 302-313.

Willaert, R. (2007). The beer brewing process: Wort production and beer fermentation, pp. 443-506. *In:* Y.N. Hui (Ed.), *Handbook of Food Products Manufacturing*, John Wiley & Sons, Hoboken, New Jersey.

Willaert, R.G. (2018). Adhesins of yeasts: Protein structure and interactions, *Journal of Fungi (Basel)*, **4**(4).

Willaert, R. and Baron, G.V. (1996). Gel entrapment and micro-encapsulation: Methods, applications and engineering principles, *Reviews in Chemical Engineering*, 12: 1-205.

Yajima, M. and Yokotsuka, K. (2001). Volatile compound formation in white wines fermented using immobilized and free yeast, *American Journal of Enology and Viticulture*, 52: 210-218.

Yokotsuka, K., Yajima, M. and Matsudo, T. (1997). Production of bottle-fermented sparkling wine using yeast immobilized in double-layer gel beads or strands, *American Journal of Enology and Viticulture*, 48: 471-481.

Zurcher, C. and Gruss, R. (1991). Method of making alcohol-free or nearly alcohol-free beer, US Patent 5077,061.

Zufall, C. and Wackerbauer, K. (2000). Process engineering parameters for the dealcoholization of beer by means of falling film evaporation and its influence on beer quality, *Monatsschrift für Brauwissenschaft*, 53: 124-137.

Encapsulated/Immobilized Biocatalysts for Production of Dairy Products

Dimitra Dimitrellou[1,2], Valentini Santarmaki[1], Anastasios Nikolaou[1], Gregoria Mitropoulou[1], Panagiotis Kandylis[3] and Yiannis Kourkoutas[1]*

[1] Laboratory of Applied Microbiology and Biotechnology, Department of Molecular Biology and Genetics, Democritus University of Thrace, Alexandroupolis, 68100, Greece

[2] Department of Food Science and Technology, Ionian University, 28100 Argostoli, Kefalonia, Greece

[3] Department of Food Science and Technology, School of Agriculture, Aristotle University of Thessaloniki, P.O. Box 235, 54124 Thessaloniki, Greece

1. Introduction

Encapsulation/immobilization is a process according to which one substance is entrapped into a wall material, producing particles of nanometer, micrometer or millimeter scale (Ray *et al.*, 2016). Although the encapsulation technology is already successfully applied in the agricultural, biotechnological and the pharmaceutical sector, only recently it has attracted the interest of the food industry for protection of sensitive ingredients during processing and storage (Ray *et al.*, 2016; Shori, 2017).

The term 'biocatalyst' refers to living (biological) systems or their parts that speed up (catalyze) biochemical reactions. In such processes, natural catalysts, such as enzymes, whole microbial cells or bioactive compounds are used for conducting/ blocking biochemical reactions.

Although the consumers' demand for health-promoting foods is rapidly increasing nowadays, the dairy industry faces specific technical challenges for insertion of functional biocatalysts (enzymes, probiotic cultures, bioactive compounds, etc.) into food matrices, associated mainly with stability and survival issues during processing, storage and commercialization, in addition to passage through the gastro-intestinal (GI) tract. It is also possible that the incorporated components are not compatible with

*Corresponding author: ikourkou@mbg.duth.gr

the food matrix, resulting in a significant reduction of their efficacy (Mitropoulou *et al.*, 2013). Regarding the sensory characteristics of the final products, the modification of food texture, production of undesirable flavors or strange odors that might affect the consumer's acceptability are considered as major difficulties (Iravani *et al.*, 2015).

One of the most popular applications of encapsulation technology in food sector is definitely encapsulation/immobilization of probiotic cells aimed at controlled and continuous delivery of health-promoting microbes in the gut. The potential benefit of such a strategy is to maintain high cell viability, as probiotics are defined as 'live microorganisms which, administered in adequate amounts, confer a beneficial physiological effect on the host', according to the Food and Agriculture Organization and World Health Organization (FAO/WHO, 2002).

In this context, microencapsulation/immobilization methods have been proposed as an efficient tool to improve the survival of probiotic cells and stability of functional ingredients in foods and through the GI tract passage (Dimitrellou *et al.*, 2016a; Prasanna and Charalampopoulos, 2018; Rajam *et al.*, 2012; Shori, 2017; Wu and Zhang, 2018), providing a shelter against adverse conditions during food processing and storage, such as acidic environments, concentration of hydrogen peroxide and dissolved oxygen in dairy products, cold shock (in the case of ice-cream or frozen yogurt), high concentration of sodium chloride, homogenization and packaging materials, etc. (Levinson *et al.*, 2016).

Considering that the probiotic market is expanding rapidly and is estimated to reach a value of USD 69.3 billion worldwide by 2023, and that dairy products are the most popular source of functional foods (Oster, 2017), the present chapter focuses on application of encapsulation/immobilization of probiotic microorganisms in functional dairy products. However, applications of encapsulated enzymes and bioactive ingredients are also included.

2. Encapsulation/Immobilization Methods

The ideal strategy to deliver microbial cells, enzymes or an ingredient in a food matrix is a matter of high importance. Various encapsulation/immobilization methods allow the attachment/entrapment of different 'biocatalyst' types (including cells and their organelles, enzymes, biological active compounds, etc.) (Martins *et al.*, 2013), covering a wide range of applications in the biotechnological, pharmaceutical, agricultural and food sector (Peinado *et al.*, 2005; Champagne *et al.*, 2010).

2.1 Encapsulation/Immobilization Prerequisites

Carrier selection is an important factor depending mostly on the application (Zacheus *et al.*, 2000). The main prerequisites a carrier has to meet for adequate encapsulation/ immobilization are:

(1) Metabolic stability against enzymes, temperature and pressure fluctuations, solvents, shear forces, etc.
(2) Large surface area, such as porous-like or other structures (e.g. natural cavities), able to facilitate cell adhesion.
(3) Abundance, easy handling of the encapsulation/immobilization protocols, suitability for scale-up and process cost-effectiveness.

(4) Protection, enhancement of viability (for cells) and activity, and avoidance of potentially negative effects on cell physiology and metabolism.
(5) Easy-to-handle biocatalysts after encapsulation/immobilization process, easy-to-carry and regeneration ability (Freeman and Lilly, 1998; Kourkoutas *et al.*, 2004; Bekatorou *et al.*, 2016).

For applications and products intended for human consumption, the carriers should also:

(1) Be of food-grade purity and accepted by consumers (Mitropoulou *et al.*, 2013).
(2) Allow regeneration after long-term storage, be resistant and stable and have no effect on the quality of the final products (Sipsas *et al.*, 2009; Genisheva *et al.*, 2012; Genisheva *et al.*, 2014).

Physical characteristics of the carrier (porosity, swelling, compression, solubility etc.), are usually application-oriented (Górecka and Jastrzębska, 2011) and may further require the carrier to be insoluble, non-biodegradable, non-toxic, non-polluting, of light weight, flexible, chemically stable, present selectivity and/or high biomass retention, etc. (Zacheus *et al.*, 2000; Martins *et al.*, 2013).

Concerning the mechanisms employed (Fig. 1), encapsulation/immobilization methods can be divided into the following categories (Nedovic *et al.*, 2001; Kourkoutas *et al.*, 2004): a) Attachment/adsorption on carrier surfaces, b) Entrapment, c) Co-aggregation, d) Mechanical containment.

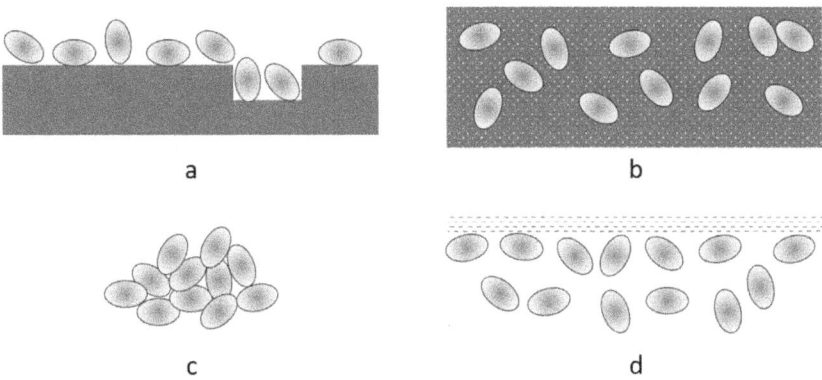

a   b

c   d

Fig. 1: Encapsulation/immobilization techniques: a) Attachment/adsorption on carrier surfaces, b) Entrapment, c) Co-aggregation, d) Mechanical containment

2.2   Encapsulation/Immobilization Methods

2.2.1 Attachment/Adsorption on Suitable Carrier Surfaces

Cells, enzymes or functional ingredients may be attached on a solid carrier surface either by physical adsorption or due to electrostatic and covalent binding forces developed between the cell membrane, enzyme or the functional ingredient and the carrier. Alternatively, cells may grow on physical cavities on the carrier surface until spacial containment and/or in addition with other forces (Bekatorou *et al.*, 2016). After immobilization, a biofilm is formed on the carrier, ranging from 1 cell to

1 mm in thickness, depending on the microorganism used and the nature of the carrier surface (Kourkoutas *et al.*, 2004).

2.2.2 Entrapment

This immobilization method relies on the entrapment of cells, enzymes or functional ingredients in a porous matrix (in high densities) or the formation of a porous material carrier *in situ* into a cell culture or an enzyme/functional ingredient solution. In some cases though, immobilized cell proliferation on the outer layers of a gel carrier may result in a situation where immobilized and free cells, enzymes or ingredients co-exist (Freeman and Lilly, 1998). In order to avoid such implications, a second layer can block the cell, enzyme or ingredient leakage from the porous matrix (Taillandier *et al.*, 1994; Ramon-Portugal *et al.*, 2003; Kourkoutas *et al.*, 2004; Petrov *et al.*, 2007).

2.2.3 Cell Co-aggregation

Free cells in liquid cultures tend to form aggregates in a natural way (flocculation), adhere in clumps and sediment rapidly (Jin and Speers, 1998; Kourkoutas *et al.*, 2004; Zhao and Bai, 2009). Flocculation can be considered as an immobilization method, since large-size aggregates may be exploited in bioreactor systems (packed-bed, fluidized-bed and continuous stirred-tank bioreactors). Although this technique is mostly effective among yeast and plant cells, the use of special artificial agents can enhance cross-linking in cell cultures that do not naturally flocculate (Bekatorou *et al.*, 2016).

2.2.4 Mechanical Containment

Cells, enzymes or functional ingredients may be physically contained behind a barrier which could be either a microporous membrane, a microcapsule or even the area between two liquid substances that do not mix (Park and Chang, 2000). Although this method is widely used in cell recycling and continuous processes (Bekatorou *et al.*, 2016), in some cases, the excessive biomass build-up could cause contaminations, membrane blockage or even membrane rupture (Lebeau *et al.*, 1998; Gryta, 2002; Kourkoutas *et al.*, 2004).

2.2.5 Encapsulation

Encapsulation methods, on the other hand, rely on the formation of semi-permeable, spherical, thin and strong coatings around a microorganism, enzyme or a functional ingredient fully contained in a capsule sized a few μm to 1 mm, thus, achieving high concentrations, cell viability and stability of the enzymes/functional ingredients (Anal and Singh, 2007; Mitropoulou *et al.*, 2013; Rathore *et al.*, 2013). However, the reduction of particle size in many cases (e.g. during probiotic cell encapsulation) is a matter of high interest, due to consumer and marketing reasons (Mitropoulou *et al.*, 2013). The reservoir and the matrix type are the most common encapsulate types available (Fig. 2). The reservoir type has a shell around the active agent (encapsulates with several reservoir chambers are also available), while the matrix type facilitates dispersion of the active agent, even on the carrier surface (Zuidam and Shimoni, 2010).

Various techniques offer different production rates, volumes and yields, as well as diverse initial capital and operating costs (Ré *et al.*, 2016). Nevertheless, each methodology exploits different physical, physicochemical, or chemical procedures

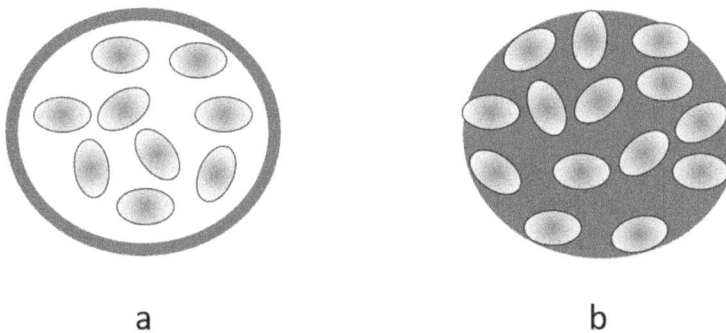

a   b

Fig. 2: Encapsulate types: a) Reservoir type, b) Matrix type

(Shahidi, and Han, 1993; Desai and Park, 2005; Champagne and Fustier, 2007) and results in varying particle size and biocatalyst payload. These techniques involve freeze- and vacuum-drying, spray-drying, spray-cooling, melt injection or extrusion, emulsification (and multilayer emulsification), encapsulation by rapid expansion of supercritical solutions (RESS), fluid bed coating, coacervation, liposome entrapment, etc. (Zuidam and Shimoni, 2010; Chávarri *et al.*, 2012). In each case, criteria like morphology, core and shell material, pre- and post-processing procedures, equipment needed, material availability and scaling-up are process-dependent and should be also considered (Oxley, 2016).

Regarding the materials used, carbohydrates, like starch (Dimantov *et al.*, 2004) and cellulose, gums (due to their gelling properties and heat resistance) and lipids (based on their hydrophobicity) are commonly used (Ré *et al.*, 2016). Similarly, alginates (Shah and Ravula, 2000; Truelstrup-Hansen, 2002; Chandramouli *et al.*, 2004; Krasaekoopt *et al.*, 2004; Kim *et al.*, 2008), chitosan (Hyndman *et al.*, 1993), xanthan (Chen and Chen, 2007), and carrageenan (Adhikari *et al.*, 2000; Mangione *et al.*, 2005) are also very popular materials that can be used alone or in combination (Sun and Griffiths, 2000; Sultana *et al.*, 2000; Chávarri *et al.*, 2012). Another approach involves the use of food proteins, in respect to their high nutritional value, abundance, range of matrices they can form, wide acceptability as naturally occurring food components and potential application in the development of innovative functional food products (Chen *et al.*, 2006; Ré *et al.*, 2016). Gelatin, for example, is a non-toxic, inexpensive,and commercially available protein, while milk proteins (casein and whey proteins) are natural, biocompatible and demonstrate excellent gelling properties for probiotic cell encapsulation (Picot and Lacroix, 2004; Heidebach *et al.*, 2009a,b; Livney, 2010).

Encapsulation technology has found application in numerous processes, such as functional food production, encapsulated biocatalysts in fermentation processes, environmental bioremediation, etc.

3. Application of Encapsulation/Immobilization Technologies in the Dairy Industry

3.1 Applications in Cheeses

The characteristic composition and formulation of cheese matrix makes it an ideal carrier for the survival and delivery of beneficial microorganisms, including probiotics, which may be incorporated either as starter or as adjunct cultures (Shori, 2015). Most cheeses are usually characterized by high fat and low water content and the low temperature ripening conditions can retain sufficient cell viability for up to several months (Fenster *et al.*, 2019). Table 1 summarizes the main applications of encapsulated/immobilized probiotic cells in cheese products.

Cheddar is cheese with the largest number of encapsulated probiotic applications. Characteristics mentioned previously combined with the relatively high pH (pH 5.5) and fat content compared to other cheese products, makes Cheddar a perfect carrier for probiotic microorganisms (Burgain *et al.*, 2011). Dinakar and Mistry (1994) encapsulated *Bifidobacterium bifidum* with an emulsification technique, using κ-carrageenan, and the obtained beads were freeze-dried. The encapsulated probiotic cells were incorporated in the cheese matrix after curd formation, milling and salting, resulting in low metabolic activity of bifidobacteria, which remained viable and increased in numbers during the 24-week study, but without affecting the flavor intensity, texture, appearance and chemical composition. An observed slight increase in the ash content was attributed to the inorganic salts derived by κ-carrageenan. However, the rate of addition of the bifidobacteria preparation was too low for a fundamental change in the composition. The absence of glycolysis substrates may also explain the lack of normal bifidobacteria metabolism. An additional reason would be the fact that the cheeses were ripened at 6-7°C, which is below the optimal growth temperature for bifidobacteria. The deficient vigorous metabolic activity was also confirmed by the sensory evaluation, which showed no differences among cheeses at any age.

Amine *et al.* (2014) produced probiotic Cheddar cheese by incorporating encapsulated *Bifidobacterium longum* cells in milk prior to cheese-making. Two microencapsulation methods were examined: i) droplet extrusion, and ii) emulsion, using two polymers (native and palmitoylated alginate). After 21 days of storage, Cheddar cheese containing encapsulated *B. longum* in native and palmitoylated alginate polymers produced with the emulsion process showed satisfactory survival rate, as only 2 log CFU/mL reduction was noticed compared to droplet extrusion encapsulated *B. longum* and free cells, recording 3 and 4 log CFU/mL reductions, respectively. The immobilized bacteria in both polymers proved to be more resistant to simulated gastric and intestinal environments by a factor of 30 than free cells, while droplet extrusion method maintained higher cell viability after 24 h of freezing at −80°C with no viability loss compared to the emulsion process and free cells, for which a 0.8 and 1.5 log CFU/mL loss was noted, respectively.

Another method to produce encapsulated probiotic formulations is through spray-drying. Gardiner *et al.* (2002) produced spray-dried probiotic milk powder using reconstituted skim milk containing *Lactobacillus paracasei* that was used for Cheddar cheese production. The probiotic culture in spray-dried form remained stable during refrigerated or room temperature storage for at least seven weeks. Cell addition into Cheddar cheese had no direct effect on cheese composition and on sensory characteristics. The advantages of the proposed method include cost-effectiveness of the process and the applicability to large-scale product development in comparison to the cultures preserved by more traditional methods, such as freezing or freeze-drying.

Accelerating Cheddar cheese-ripening period is a desirable concept and faster degradation of small peptides has been investigated by free and encapsulated recombinant aminopeptidase enzymes derived from lactic acid bacteria (Azarnia

Table 1: Applications of Encapsulated/Immobilized Probiotic Cells in Cheeses

Encapsulation Materials & Methods Size of Capsules	Probiotic Cells	Product	Positive Effect	Reference
Alginate microgel particles, Spray aerosol method (0.1-1000 μm)	*Lactobacillus rhamnosus*	Reduced-fat cream cheese	Less reduction in the viable counts of encapsulated cells after 35 days of refrigerated storage (final counts > 6.63 logCFU/g). No negative effect at pH, moisture, protein or fat content.	Ningtyas *et al.*, 2019
Na-alginate solution; Emulsification; ionic gelation (60-500 μm)	*Lactobacillus acidophilus*	Manaba fresh white cheese	Higher viability of encapsulated cells incorporated to films after 20 days of storage at 25 °C (final counts > 9.79 logCFU/g). Improved inhibition activity of mesophilic bacteria growth due to bacteriocins production.	Santacruz and Castro, 2018
Microencapsulation whey proteins and isomaltooligosaccharides emulsification (183 μm)	*Lactobacillus rhamnosus* 6134	White brined cheese	Effective protection of bacteria. Cells remained > 9.0 logCFU/g after 90 days and exhibited good survival under simulated gastro-intestinal tract conditions.	Liu *et al.*, 2017
Ca-alginate Emulsification canola oil	*Lactobacillus acidophilus* LA-5	UF cheese	Increase of probiotic bacteria survival rate due to encapsulation in *in vitro* simulated gastro-intestinal conditions. Viability remained > 6.5 logCFU/g after 60 days storage.	Nejati *et al.*, 2017
Alginate O-palmitoylation Extrusion, Emulsion canola oil (200-250 μm)	*Bifidobacterium longum* ATCC 15708	Cheddar cheese	No loss of viability during a 4 weeks storage period at -80°C.	Amine *et al.*, 2014
Na-alginate, Extrusion	*Lactobacillus acidophilus, Bifidobacterium longum, Bifidobacterium lactis*	Pecorino cheese	After 60 days, cell numbers were > 6.84 log CFU/g.	Santillo *et al.*, 2012

(Contd.)

Table 1: (*Contd.*)

Encapsulation Materials & Methods Size of Capsules	Probiotic Cells	Product	Positive Effect	Reference
Ca-alginate Emulsification canola oil	Bifidobacterium lactis BB-12	Ultra Filterated (UF) cheese	Satisfactory resistance of encapsulated probiotic bacteria at low pH values and at high concentrations of bile salts.	Nejati et al., 2011
Microencapsulation Extrusion Na-alginate (0.2-0.3 mm) or Emulsion κ-carragenan (0.3-0.4 mm)	Bifidobacterium bifidum BB-12, Lactobacillus acidophilus LA-5	White-brined cheese	Effective techniques in keeping the numbers of probiotic bacteria $>10^7$ CFU/g. Microencapsulation did not adversely affect sensory characteristics.	Özer et al., 2009
Microencapsulation Extrusion Na-alginate or Emulsion κ-carragenan (0.5-1 mm)	Lactobacillus acidophilus LA-5, Bifidobacterium bifidum BB-12	Kasar cheese	No difference in bacterial counts, proteolysis and sensory properties of the final products between techniques. Satisfactory viability during storage for 90 days.	Özer et al., 2008
Ca-alginate Hi Maize™ starch	Lactobacillus acidophilus DD 910, Bifidobacterium lactis DD 920	Feta cheese	No significant affection on textural parameters.	Kailasapathy and Masondole, 2005
Reconstituted skim milk Spray-dried	Lactobacillus paracasei NFBC 338	Cheddar cheese	High survival rates (84.5%). No adversely effects in cheese quality.	Gardiner et al., 2002
Gelled beads with κ-carragenan, Freeze-drying	Bifidobacterium bifidum	Cheddar cheese	No change in cheese composition. Increased cell numbers in cheese. Bifidobacteria did not affect flavor intensity, texture and appearance.	Dinakar and Mistry, 1994
Ca alginate, Freeze-drying	Bifidobacterium bifidum, Bifidobacterium infantis, Bifidobacterium longum	Crescenza cheese	Cells showed great adaptability to cheese environment. Similar sensory characteristics to conventional cheese.	Gobbetti et al., 1998

et al., 2011). In this case, application of encapsulation technology provides protection to water-soluble enzymes and ensures that the enzymes are released into the cheese matrix. Azarnia *et al.* (2011) suggested that higher concentrations of encapsulated recombinant peptidase led to shorter cheese-ripening period and the products received higher score at the sensory analysis compared to cheeses with no added biocatalysts. A one-month earlier ripening period was also documented when encapsulated protease enzymes into gum capsules were added during cheese-making (Kailasapathy and Lam, 2005). The encapsulated enzymes resulted in high β-casein hydrolysis rate and high release of free amino acids, but at the same time, the new cheese products had in general lower score for textural parameters, compared to control products. The study concluded that the type of encapsulating materials had an important impact on proteolysis rate, moisture and overall sensory quality of the new food products (Kailasapathy and Lam, 2005).

Prior to cheese ripening, immobilized milk-clotting enzymes on alginate-pectate gels have been investigated for potentially high operational stability (Kumari Narwal *et al.*, 2016). The immobilized biocatalysts were reused for up to ten cycles and for the first five cycles no loss in activity was observed. Although it was of great interest that pH and temperature values of reactions were broad in the case of immobilized enzymes, no significant impact on activity was detected as compared to soluble enzymes (Kumari Narwal *et al.*, 2016).

In the same context, addition of a mixture of different bifidobacteria species (*Bifidobacterium bifidum, Bifidobacterium infantis,* and *Bifidobacterium longum)* encapsulated in calcium alginate beads in Crescenza, a soft, rindless, Italian cheese with a short ripening time, has been evaluated (Gobbetti *et al.*, 1998). The products containing the bifidobacteria mixture were similar to the conventional Crescenza cheese regarding flavor, appearance, as well as microbial and physicochemical features, but pronounced enzyme activity was detected. The presence of bifidobacteria did not influence the aerobic microflora, the growth of *Streptococcus thermophilus* used as starter, or the gross composition of the cheese.

Encapsulated probiotic bacteria in alginates have been also incorporated in cream cheese (Ningtyas *et al.*, 2019). Although the addition of probiotics did not significantly alter the pH, moisture, protein and fat content, it resulted in a firmer and thicker cream cheese compared to the non-encapsulated product. Furthermore, encapsulation provides protection to probiotic cells, which maintain their viability during 35 days of storage.

Liu *et al.* (2017) microencapsulated *Lactobacillus rhamnosus* cells using Maillard reaction products of whey protein and isomaltooligosaccharide through cold gelation within an emulsification system and evaluated its application in white brined cheese production. The results showed that microencapsulation was able to increase the viability of the probiotic cells in white brined cheese after 90 days of storage. In addition, probiotic bacteria exhibited a satisfactory survival rate under simulated GI tract conditions. Interestingly, the sensory properties of the cheeses were not affected by the addition of the microencapsulated probiotic bacteria.

In another study with white brined cheese, two probiotic strains, namely *Bifidobacterium bifidum* BB-12 and *Lactobacillus acidophilus* LA-5, were used after microencapsulation, applying extrusion (alginates) or emulsion (κ-carragenan and corn oil as emulsifying agent) techniques (Özer *et al.*, 2009). Both processes were effective in preserving the numbers of the probiotic bacteria higher than the required level for conferring a health benefit ($>10^7$ CFU/g) and retained their viability in higher numbers (approximately only 1 log CFU reduction) than free cells (approximately 3 log CFU reduction) during 90 days of ripening. Although cheeses containing microencapsulated probiotics did not differ from the control cheese regarding the sensory properties, microencapsulation induced the formation of acetaldehyde and diacetyl and increased the concentration of long-chain free fatty acids (FFAs).

The same probiotic microorganisms and the same encapsulation techniques were also applied in Kasar cheese (Özer *et al.*, 2008), a dry-salted semi-hard Turkish cheese, scalded at 50-52°C for one to two min. The addition of microencapsulated cells did not affect the chemical composition and the effect on proteolysis as well as on sensory properties was limited. Although scalding caused a drastic decline in the counts of probiotic bacteria, the numbers of the encapsulated bacteria remained well above the threshold for a probiotic effect (10^7 CFU/g), due to the protective environment created by microencapsulation.

Incorporation of encapsulated cells in cheese, usually, occurs in the first part of cheese-making, in milk. Santillo *et al.* (2012) proposed the incorporation of encapsulated probiotics (*L. acidophilus*, *B. longum*, and *B. lactis*) in lamb rennet paste and then their use in Pecorino cheese production. The results showed that Ca-alginate encapsulation of probiotic bacteria and their addition to rennet could be a successful way for protecting cells during cheese-making.

Encapsulation has been also proposed for continuous acidification of milk, aimed at the production of fresh cheese. Mesophilic streptococci (*Streptococcus lactis*, *Streptococcus lactis* subsp. *diacetylactis* and *Streptococcus cremoris*) were entrapped in Ca-alginate and used for continuous acidification of milk (Prevost and Divies, 1987). The process applied resulted in a product with constant acidity, diacetyl and acetoine concentration. In addition, it allowed incubation- time reduction of 50% in comparison to a starter culture used in industry. In a second study, four strains of mesophilic lactic acid bacteria (*Lactococcus lactis* subsp. *lactis*, *Lactococcus lactis* subsp. *cremoris*, *Leuconostoc mesenteroides*, *Lactococcus lactis* subsp. *lactis* biovar *diacetylactis*) were entrapped separately in κ-carrageenan/locust bean gum gel beads and used in a stirred bioreactor operated during 8 h daily cycles for up to 7 weeks with different milks (Sodini *et al.*, 1997). The results were promising and confirmed the industrial feasibility of a continuous milk pre-fermentation process for fresh cheese manufacture, using immobilized cells.

Enhancing the viability of probiotic cells is often linked with an extended metabolic activity and shelf-life of the product due to the continuous production of antimicrobial substances (Santacruz and Castro, 2018). For example, Manaba is a fresh artisanal white cheese from Ecuador with many microbiological issues, since the presence of pathogen microorganisms, like *Salmonella* spp., is usual (Santacruz and Castro, 2018). Indeed, encapsulated *L. acidophilus* cells incorporated into edible films retards the growth of *Salmonella* spp. in white fresh cheese due to production of bacteriocins (Santacruz and Castro, 2018).

The use of natural supports (of food-grade purity) for immobilizing probiotic cells to be employed in cheese production as starters or adjunct cultures is another interesting approach. Such supports not only provide protection to probiotic cells during processing and storage, but they also contribute positively to organoleptic properties. Cheese is a product with a high protein content and milk proteins are characterized as natural vehicles for delivery of beneficial ingredients, due to their structural and physicochemical properties (Livney, 2010), since they are the most contractual agents for the cheese structure and thus do not affect negatively its textural characteristics. Moreover, the use of milk proteins as immobilization supports leads to a significant increase in the protein content, improving the nutritional value of the final products. Therefore, they have been proposed for microencapsulation and delivery of probiotics and bioactive compounds (Anal and Singh, 2007; Heidebach *et al.*, 2009b). For example, casein and whey proteins have been successfully tested as immobilization supports of probiotic cells for the production of several dairy products (Dimitrellou *et al.*, 2008; Dimitrellou *et al.*, 2009; Dimitrellou *et al.*, 2014a; Dimitrellou *et al.*, 2014b; Dimitrellou *et al.*, 2015; Dimitrellou *et al.*, 2016b; Dimitrellou *et al.*, 2017; Katechaki *et al.*, 2009; Sidira *et al.*, 2017). Hence, Feta-type, hard-type and whey cheeses were produced using lactobacilli and kefir cultures immobilized on milk proteins, resulting in improved quality and profile of aroma-related compounds. In whey cheeses produced by immobilized cells, an extension of the preservation time and protection from spoilage and pathogens was also reported (Dimitrellou *et al.*, 2009). Similarly, novel cheese products were also developed using probiotic cells immobilized on fruit (apple and pear) pieces (Kourkoutas *et al.*, 2006) and the preliminary sensory evaluation revealed their fruity flavor.

Although the majority of studies concluded that encapsulation improves the viability of probiotic cells during cheese production and ripening, Kailasapathy and Masondole (2005), working with microencapsulated *Lactobacillus acidophilus* and *Bifidobacterium lactis*, found that calcium-induced alginate starch microcapsules were not capable in protecting probiotic cells during Feta cheese production and ripening.

Another important application is the use of appropriate microencapsulation materials that can mask flavors and odors derived from metabolic products, like organic acids and volatile compounds (lactic, acetic, succinic and citric acid, acetaldehyde, diacetyl and acetoin) (Levinson *et al.*, 2016). Positive effects on the sensorial characteristics of the products can be attributed to the metabolic volatile compounds produced by probiotic cells over the starter cultures – an aspect that is strain-specific (Dimitrellou *et al.*, 2019a).

The general trend of functional foods has raised great concerns about the nutritional value and composition of milk fat, which mainly consists of saturated fatty acids. Hence, insertion of encapsulated plant oils, rich in omega fatty acids, is a very promising approach that might improve the nutritional value of dairy products and reinforce the health benefits (Table 2). Muñoz-Tébar *et al.* (2019) showed that enrichment of sheep cheese with chia seeds oil emulsions neither inhibited the normal cheese microbial flora, nor affected the production and the ripening process, even at a concentration equal to 5 g/L. In contrast, a positive effect on the physico-chemicals parameters was noted and considering the α-linoleic acid content, the levels reached values of 1.6 g/100 g of cheese, in agreement with the daily intake recommendations (Baró *et al.*, 2017). The human body and generally the animal cells cannot synthesize the essential linoleic and α-linoleic fatty acids, which exist in large quantities in plants extracts, such as flaxseed, walnut, sunflower or corn oil, and is claimed to prevent cardiovascular diseases (Muñoz-Tébar *et al.*, 2019). Studies on incorporation of plant oils as a source of essential fatty acids into soft or Cheddar cheese have been already published (Khalifa *et al.*, 2017; Ullah *et al.*, 2018). Moreover, fortification of dairy foods with animal origin omega-3 fatty acids (mainly from fish oil) has been investigated, but problems with an unpleasant "fishy taste" have been raised (Bermúdez-Aguirre and Barbosa-Cánovas, 2011).

Table 2: Examples of Applicability of Encapsulation Technology into Dairy Food Matrices

Encapsulation Materials & Methods	Size of Capsules	Dairy Product	Application	Positive Effect	Reference
Calcium caseinate (emulsification method)		Sheep cheese	Supplementation with chia (*Salvia hispanica* L.) oil	Positive impact on cheese yield, fat and dry matter. No interference with production and ripening process. Higher levels of α-linoleic acid (1.6/100 g of cheese). Pleasant odor and taste.	Muñoz-Tébar *et al.*, 2019
Gelatin and gum arabic (complex coacervation or freeze drying method)		Kefir beverage	Addition of encapsulated structured lipids produced from tricaprylin and Echium oil fatty acids	Freeze-dried capsules showed higher oxidative stability after 10 days of storage at 4°C. Higher consumer's acceptance of kefir products with freeze-dried capsules. Appropriate oil release during digestion simulations. Production of fortified kefir products with medium and long chain FAAs.	Yüksel Bilsel, and Şahin-Yeşilçubuk, 2019
Liposomes (lecithin powder) were covered with chitosan (secondary liposomes) Chitosan-coated liposomes were also covered with maltodextrin (tertiary liposomes)	250–450 nm	Processed cheese consisting of Cheddar cheese, Ras cheese, butter and skim milk powder	Encapsulation of dried mandarin peel extract as a source of phenolic compounds	Supplementation of processed cheese samples with nanoliposomes had no impact on chemical composition and physical properties of the new products after 3 months of storage. Gallic acid and catechin were highly concentrated into processed cheese.	El-Messery *et al.*, 2019

Material/Method	Size	Product	Application	Findings	Reference
Phosphatidylcholine fraction from soy beans (thin layer hydration method)	115-161 nm	Yogurt	Encapsulation of garlic essential oil	Antibacterial activity against *E. coli* and *S. aureus*. Stable antioxidant activity during storage at 4°C. Better sensory properties over non-encapsulated garlic essential oil.	Nazari *et al.*, 2019
Solid-in-oil-in-water (S/O/W) emulsions	292 nm	Milk products	Encapsulated lactase	Gradually release of encapsulated lactase resulted in increased lactose hydrolysis after 3 weeks of storage.	Zhang and Zhong, 2018
Glyceryl palmitostearate, medium chain triglycerides and polyethylene glycol (hot melt homogenization method)	140-183 nm	Butter	Encapsulation of beta-sitosterol	Fortification with beta-sitosterol had no effect on acid and peroxide values of butter after 3 months of storage at 4°C. The encapsulated beta-sitosterol had increased antioxidant capacity during storage compared to the free form.	Bagherpour *et al.*, 2017
Re-assembled casein micelles or polysorbate-80 (emulsification method)	89 nm (casein micelles) 7 nm (polysorbate-80)	Fat-free yogurt	Enrichment with vitamin D_3	Increase of the vitamin D_3 levels serum bioavailability by 8 ng/mL. High protection during shelf life. Improved rheological characteristics. Pleasant taste and texture.	Levinson *et al.*, 2016

In the same manner, nanoemulsions, encapsulating oregano essential oil, have been examined in semi-hard cured Brazilian cheese, and it was concluded that the inhibitory effect on yeast growth could allow application of the technology in cheese maturation chambers (Bedoya-Serna *et al.*, 2018).

3.2   Applications in Fermented Milk and Yogurt

According to Codex Alimentarius (Section 3.3), fermented milk is a milk product obtained by fermentation of milk, which may have been manufactured from milk with or without compositional modification by the action of suitable microorganisms, resulting in reduction of pH with or without coagulation (iso-electric precipitation). The starter microorganisms shall be viable, active and abundant to the date of minimum durability (shelf-life). If the product is heat treated after fermentation, the requirement for viable microorganisms is not applied (FAO/WHO, 2011). Several fermented milk products have been developed through centuries around the world with many different local names, but with few actual variations, such as yogurt, buttermilk, acidophilus milk, kefir, koumiss, etc (Iravani *et al.*, 2015).

Table 3 summarizes the main applications of encapsulated/immobilized probiotic cells in fermented milk products. Yogurt is the fermented milk product that has been tested in the majority of studies. Microencapsulation and alginates are the key technique and carrier material usually applied for encapsulation of probiotic cells in fermented milk products.

The use of alginates as a sole material for microencapsulation of probiotics provided satisfactory results in terms of cell survival; specifically, alginates capsules protected *Lactobacillus acidophilus* ATCC 4356 (Ortakci and Sert, 2012) and *Lactobacillus casei* ATCC 393 cells (Dimitrellou *et al.*, 2019a) during the tests in artificial gastric and bile salts solutions, as well as in refrigerated storage in yogurt and fermented milk. A negative effect on sensory characteristics and especially on body texture was detected and was correlated to the large size (2 mm diameter) of beads (Ortakci and Sert, 2012). However, the use of smaller probiotic capsules (0.587 mm diameter) eliminated such issues and the fermented milk with encapsulated probiotic cells received similar scores with the commercial products (Dimitrellou *et al.*, 2019a). An extrusion spraying technique may be used in order to produce even smaller probiotic capsules with a diameter ranging 0.082-0.149 mm. As the concentration of Na-alginate increased the microcapsule size, encapsulation efficiency and survival rate of bacteria also increased. The microencapsulated yogurt's starter culture was used for the production of a spray-dried sweetened yogurt powder and increased survival rates were recorded (Seth *et al.*, 2017).

Even the yogurt starter culture bacteria, *Streptococcus thermophilus* and *Lactobacillus delbrueckii* subsp. *bulgaricus*, have been microencapsulated in chitosan-alginate capsules (De Prisco *et al.*, 2017). Encapsulation influences the metabolic activity of the starter cultures, resulting in various concentrations of several volatile compounds, but acetaldehyde, acetoin and diacetyl are not affected. The use of encapsulated starter cultures has led to a final product where both encapsulated and free cells are present, but encapsulation improves cell viability during yogurt storage and simulates GI passage, improving yogurt's quality and functionality.

Another interesting approach is the use of herbal based polymers like psyllium and fenugreek in a blend with alginate polymer as a matrix for probiotic formulation (Haghshenas *et al.*, 2015; Nami *et al.*, 2017). These new preparations add prebiotic

Table 3: Application of Encapsulated/Immobilized Probiotic Cells in Fermented Milk Products

Encapsulation Materials & Methods Size of Capsules	Probiotic Cells	Product	Positive Effect	Reference
Gum arabic and inulin or maltodextrin, Spray drying (5.7-14.4 μm)	*Bifidobacterium lactis* BB-12	Lactose-free Greek-style yogurt	Probiotic viability was > 6.5 logCFU/g after 30 days of storage. Addition of the encapsulated cells increased the product's pH, firmness and adhesiveness.	Pinto *et al.*, 2019
Ca-alginate capsules, Extrusion method (528.8-646.1 μm)	*Lactobacillus casei* ATCC 393	Fermented milk	Higher viability compared to free cells after storage for 28 days in fermented milk (final levels 7.13 logCFU/g). Positive effect on sensorial characteristics of the new products.	Dimitrellou *et al.*, 2019a
Low methoxyl pectin, Freeze drying method (1000-1500 μm)	*Bifidobacterium breve* CICC 6182	Yogurt	Higher survival rate of encapsulated cells Improved yogurt's characteristics when added during the middle and late stages of yogurt fermentation.	Li *et al.*, 2019
Ca-alginate capsules with *Eleutherine americana* oligosaccharide extract, Extrusion method (1.47-1.59 mm)	*Lactobacillus plantarum* TISTR1465	Yogurt	Enhancement of probiotic antibacterial activity against enteropathogenic bacteria. Capsules were resistant to disintegration in simulated gastric juice (pH 2).	Phoem *et al.*, 2019
Ca-alginate capsules with plant extracts (moringa, fennel, sage, green tea)	*Lactobacillus plantarum* DSM 20205, *Pediococcus acidilactici* DSM 20238	Drinkable yogurt	Increase of cell survival due to encapsulation of additives compared to control capsules in simulated digestive fluids. Enhancement of stability of probiotic beads due to extracts during storage.	Shehata *et al.*, 2019

(Contd.)

Table 3: (*Contd.*)

Encapsulation Materials & Methods Size of Capsules	Probiotic Cells	Product	Positive Effect	Reference
Alginate–goats' milk–inulin (2.46-3.85 mm)	*Bifidobacterium animalis* subsp. *lactis* BB-12	Yogurt	Better survival of probiotic bacteria in the goat milk yogurt with addition of inulin stored over 28 days.	Prasanna and Charalampopoulos, 2019
Whey protein with maltodextrin and arrowroot starch, Extrusion, freeze-drying (315.3-687.09 µm)	*Lactobacillus plantarum*, *Weissela paramesenteroides*, *Enterococcus faecalis*, *Lactobacillus paraplantarum*	Yogurt	Significant increase of viable cells after freeze-drying compared to free cells at high bile salt concentrations and low acidity. Encapsulation with arrowroot starch and maltodextrin could allow viable probiotic bacteria to reach the large intestine.	Samedi and Charles, 2019
Gellan-caseinate mixture with milk protein concentrate, pH triggered gelation technique	*Lactobacillus paracasei* L26	Set yogurt	Better probiotic protection. Enhancement of qualitative characteristics of probiotic yogurt. Improvement of syneresis and viscosity during a 3 week storage at 4°C.	Moghaddas Kia *et al.*, 2018
Xanthan-chitosan-xanthan double layer particles, Extrusion method	*Bifidobacterium* BB01	Yogurt	Enhanced cell survival of encapsulated cells over free cells after 21 days of storage (final levels > 7 logCFU/g). Increased acidity but within suitable range for consumers' acceptance.	Chen *et al.*, 2017
Chitosan-alginate capsules, Extrusion method (100-120 µm)	*Streptococcus thermophilus*, *Lactobacillus delbrueckii*	Yogurt	Excellent performances of matrix and core-shell microcapsules. Protective effect of capsules in acidic environment.	De Prisco *et al.*, 2017

Method/Material	Microorganism	Product	Observations	Reference
Ca-alginate, Extrusion, spraying method (82–149.37 μm)	*Streptococcus thermophilus, Lactobacillus delbrueckii* subsp. *bulgaricus*	Yogurt	A 2 log increase in cell counts was achieved using encapsulated bacteria after spray drying compared to free cells.	Seth *et al.*, 2017
Xanthan-chitosan and xanthan-chitosan-xanthan hydrogels, Extrusion method	*Lactobacillus acidophilus*	Yogurt	Success in bacterial survival during storage for 21 days. Microcapsules showed higher viability in yogurt.	Shu *et al.*, 2017
Ca-alginate, Emulsion method (< 100 μm)	*Lactobacillus paracasei* A13, *Lactobacillus salivarius* subsp. *salivarius* CET 4063	Fermented milk	Microencapsulation conditions increased the resistance to the simulated digestion processes. No adversely effects on the sensory properties. Maintenance of high probiotic viability during fermented milk refrigerated storage.	Patrignani *et al.*, 2017
Alginate psyllium and alginate-gum arabic, Extrusion method (330 μm–1.5 mm)	*Enterococcus durans* IW3	Yogurt	Incorporation of psyllium and gum arabic resulted in improvement of alginate properties, enhancement of yogurt's storage stability at 4°C, endurance in gastro-intestinal conditions.	Nami *et al.*, 2017
Vegetal-Inulin and PVA, Emulsion, freeze-drying	*Bifidobacterium longum* LMG 13197	Yogurt	Encapsulated cells > 5 log units higher than unencapsulated. Protection of microparticles during exposure to simulated gastro-intestinal conditions.	Amakiri and Thantsha, 2016
Reconstituted skim milk, Spray drying method	*Lactobacillus casei* ATCC 393	Fermented milk	Higher cell counts during production and refrigerated storage for 28 days over the free cell system (final levels 7.98 logCFU/g). Overall high quality and acceptability of the new products.	Dimitrellou *et al.*, 2016a

(Contd.)

Table 3: (*Contd.*)

Encapsulation Materials & Methods Size of Capsules	Probiotic Cells	Product	Positive Effect	Reference
Re-assembled casein micelles or polysorbate-80, Emulsification method (89 nm (casein micelles) 7 nm (polysorbate-80))		Fat-free yogurt	Enrichment with vitamin D_3 increased the serum bioavailability of vitamin D_3 levels by 8 ng/mL. High protection during shelf-life. Improved rheological characteristic Pleasant taste and texture.	Levinson *et al.*, 2016
Alginate psyllium blend with inulin and fenugreek, Extrusion method (210-1100 μm)	*Enterococcus durans* 39C	Yogurt	Increase of prebiotic effect value. Enhancement of probiotic bacterial growth in the gastro-intestinal environment (> 47% compared to non-microencapsulated cells). High survival rate of viable probiotic cells during a 28 days storage time.	Haghshenas *et al.*, 2015
Ca-alginate with Thai herbal extracts (cashew flower, pennywort, yanang)	*Lactobacillus casei* 01, *Lactobacillus acidophilus* LA5, *Bifidobacterium lactis* Bb-12	Yogurt	Improvement in probiotic beads stability during storage. Positive correlation between increased probiotic growth and increased green tea concentration.	Chaikham, 2015
Coacervation- lyophilisation	*Lactobacillus acidophilus* Lac-04	Buffalo milk yogurt	Greater stability of microencapsulated cultures. Microcapsules presented a shelf-life around 120 days at 7°C.	Shoji *et al.*, 2013
Ca-alginate, Extrusion method (1.5-2.5 mm)	*Lactobacillus acidophilus* ATCC 4356	Yogurt	No substantially change of the yogurt's overall sensory properties. Encapsulation enhanced the survival of probiotic bacteria against an artificial human gastric digestive system.	Ortakci and Sert, 2012

properties to the matrix and therefore enhance the probiotic growth and survival in the GI environment. Noticeably, the use of psyllium or gum arabic in combination with alginates produces probiotic capsules of *Enterococcus durans* IW3 with improved survival ability at low pH, high bile salt concentration (0.5%), and long-time storage of yogurt (30 days), compared to alginate capsules and free cells. Additionally, capsules with psyllium lead to faster release of probiotic cells in simulated intestine fluid, but higher survival rate is recorded, due to the stimulating effect of psyllium on the probiotic bacteria (Nami *et al.*, 2017). In a similar study, an alginate-psyllium blend with fenugreek (F9) formulation was used for the encapsulation of *Enterococcus durans* 39C and the capsules were tested in yogurt production. They presented high encapsulation efficiency, high cell viability in digestive conditions and during storage of yogurt (up to one month), as well as increased release rates in colonic condition throughout a 12 h period (Haghshenas *et al.*, 2015). Following this trend, the alginate-based encapsulation of probiotics in combination with plant extracts has been also evaluated (Chaikham, 2015; Phoem *et al.*, 2019; Shehata *et al.*, 2019). Thus, the extracts of several plants, such as cashew flower (*Anacardium occidentale L.*), yanang (*Tiliacora triandra*), pennywort (*Centella asiatica*) green tea (*Camellia sinensis* L.), *Eleutherine americana*, fennel (*Foeniculum vulgare*), moringa (*Moringa oleifera*) and sage (*Salvia officinalis*) have been used with a final goal to increase probiotics survival. The use of extracts from cashew flower and green tea further enhance the stability of probiotics during yogurt refrigerated storage. In particular, higher survival rates has been reported for encapsulated *L. casei* 01 and *B. lactis* Bd-12 than *L. acidophilus* LA5 (Chaikham, 2015). Similarly, moringa and green tea extracts enhanced the stability of *Lactobacillus plantarum* DSM 20205 and *Pediococcus acidilactici* DSM 20238 beads in yogurt compared to the controls after storage at 4°C for 30 days (Shehata *et al.*, 2019). The use of *Eleutherine americana* oligosaccharide extract during alginate encapsulation of *L. plantarum* TISTR1465 improved cell survival after sequential exposure to simulated gastric and intestinal juices and refrigerated storage in yogurt, amending their antibacterial activity than the free cells. Moreover, the encapsulated cells caused less acidification during refrigerated storage than the free cells (Phoem *et al.*, 2019).

Prasanna and Charalampopoulos (2019) prepared a new encapsulating matrix for beneficial bacteria, like *Bifidobacterium animalis* subsp. *lactis* BB-12, using alginate, goat's milk and inulin, in order to be used in goat's milk-based probiotic fermented dairy products, avoiding the cross-contamination caused by using capsules based on cow's milk. These new capsules had more compact interior structural characteristics, provided better protection of probiotics under simulated GI conditions, increased cell survival rate in yogurt refrigerated storage for 28 days and resulted in slower post-acidification.

Xanthan-chitosan hydrogels have been also proposed for the microencapsulation of probiotics like *B. bifidum* BB01 and *L. acidophilus* (Chen *et al.*, 2017; Shu *et al.*, 2017). The application of probiotic capsules in yogurt extended its shelf-life both at 25° and 4°C, retaining high cell survival, while a less post-acidification was observed (Shu *et al.*, 2017). The use of double-layer microcapsules resulted in higher low-pH tolerance and bile salt resistance, whereas the single-layer microcapsules led to an improved release profile in simulated intestinal fluid (Chen *et al.*, 2017).

Biopolymer complexes and coacervates have been evaluated as microencapsulation shell materials for several functional components, including probiotic cells, with numerous applications in food matrices (Moschakis and Biliaderis, 2017). *L. paracasei* subsp. *paracasei* cells were microencapsulated in biopolymer complex coacervates, using whey protein isolate and gum arabic and provided improved survival during yogurt storage, as well as upon exposure to simulated gastric juice, with no effect on the rheological properties of the final product (Bosnea *et al.*, 2017). In a similar study, *L. acidophilus* was microencapsulated by complex coacervation, using casein and pectin, followed by freeze-drying and then applied in buffalo-milk yogurt production. Microencapsulation preserved the counts of *L. acidophilus* higher than 10^7 CFU/g at refrigerated conditions, but lacked cell protection at a pH similar to the human stomach. Lower values for post-acidification and greater stability were reported for yogurt prepared with microencapsulated cultures compared to the product produced by free cell culture (Shoji *et al.*, 2013).

Drying, especially freeze-drying and spray-drying, in the presence of coating material is another method that has been tested for the encapsulation of probiotic cells. Freeze-drying of several probiotics with arrowroot starch and whey as coating material produced capsules with increased survival under simulated GI conditions and yogurt storage for up to 90 days (Samedi and Charles, 2019). Three different spray-dried microcapsules of *B. lactis* BB-12 using gum arabic, inulin and maltodextrin as wall materials resulted in satisfactory probiotic viability (> 8 log CFU/g) throughout 30 days of storage at 4°C. Addition of the microcapsules (with gum arabic and inulin) in yogurt increased the product's pH, firmness and adhesiveness, while after 30 days of storage, probiotic cell counts were above 6.5 log CFU/g (Pinto *et al.*, 2019). Similar results were obtained by using spray-dried microencapsulated *L. casei* ATCC 393 cells (skim milk was used as wall material) under simulated GI conditions, as well as during fermented milk production and storage (Dimitrellou *et al.*, 2016a). The probiotic levels ranged above the minimum requirement for conferring a probiotic effect and the fermented milks has improved sensory characteristics.

Apart from the beneficial effect of encapsulation on probiotic cultures' viability, there is great evidence supporting the idea that microencapsulation of important nutraceuticals could increase their bioavailability, and thus can be addressed to consumers worldwide, in order to overcome health deficiencies (Table 2). In this vein, evaluation of fat-free yogurt enriched with encapsulated vitamin D_3 in casein matrix was assessed recently, and the new products improved the rheological characteristics and had a pleasant taste and texture. An increase in serum 23(OH)D (circulating form of Vitamin D_3) levels of 8 ng/mL was documented at the subjects, after consuming the yogurt daily for 14days (Levinson *et al.*, 2016).

Additionally, the common disorder of lactose intolerance has raised the need to develop new technologies to overcome this problem. Encapsulated lactase has been investigated for lactose hydrolysis in milk products (Zhang and Zhong, 2018). To achieve control of *in vivo* lactase release after the digestion process, emulsions composed of solid-in-oil-in-water were prepared and it was claimed that encapsulated lactase remained dispersed after incorporation and during storage of the milk products. On the other hand, free lactase powder supplemented into milk products resulted in 65% lactose hydrolysis after one week of refrigerated storage (Zhang and Zhong, 2018).

Moreover, encapsulation technology could meet the consumers' demand for utilizing natural preservatives over chemical substances. In this regard, essential oils have attracted considerable attention as biopreservatives, due to their inhibitory effects against food-borne pathogens and hence formulation of phytosome has been studied (Nazari *et al.*, 2019). The lipid-based phytosome is considered an appropriate delivery system for bioactive compounds that improves their stability, absorbance and antimicrobial activities during food processing and storage and also contributes in masking the off-flavor and the strong aroma that restrict the application of essential oils in the food industry (Ghanbarzadeh *et al.*, 2016). In this context, yogurt with encapsulated phytosome garlic essential oil had improved sensory characteristics and extended antimicrobial activity was recorded over samples enriched with non-encapsulated garlic oil (Nazari *et al.*, 2019).

Fruit pieces are usually added to commercial yogurt to provide a fruity taste. Following that trend, several studies were carried out using fruit pieces (strawberry, banana and apple) as immobilization support of probiotic cells, as well as commercial yogurt starter culture and then the immobilized cells were incorporated in yogurt and fermented milk (Sidira *et al.*, 2013; Dimitrellou *et al.*, 2019b), resulting in a positive effect on product quality, as confirmed by the sensory evaluation and the SPME GC-MS analysis of aroma-related compounds. Importantly, apple pieces retained the viability of probiotic cells during refrigerated storage at the essential concentration for providing health benefits. Probiotics immobilized in milk proteins were also incorporated in yogurt and fermented milk leading to products with higher protein content, improved quality characteristics and increased probiotic viability during processing and storage (Sidira *et al.*, 2017; Dimitrellou *et al.*, 2019b).

3.3   Applications in Ice-cream

Ice-cream is a popular dairy dessert made by a combination of components, such as milk (or skim milk), cream, sweeteners, stabilizers, emulsifiers, flavoring and coloring agents. It is also characterized as a frozen, aerated emulsion (oil in water), containing partially coalesced fat globules, air bubbles, ice crystals and unfrozen viscous serum (Goff, 1997; Marshall *et al.*, 2003). Ice-cream production involves pasteurization (at 80-85°C for 15 min) of the components mixture with continuous stirring and then rapid cooling down to 45-50°C. The cooled mixture requires aging at 4°C for 24-48 h and then is frozen to −18°C for hardening (Goff, 1997; Gharibzahedi *et al.*, 2018).

The composition of ice-cream, which includes milk proteins, fat and lactose, makes it a potential suitable carrier for probiotic cultures. In addition, ice-cream is a frozen product with high pH value (usually ranging 5.5-6.5), which may also contribute to improved viability of probiotic cultures during prolonged storage (Cruz *et al.*, 2009). In order to produce a probiotic ice-cream, each manufacturing step should be designed in a way as to ensure survival of probiotic cells and at least avoid negative effects (if improvement is not possible) on the physicochemical and sensory properties of the product (Stanton *et al.*, 2003).

Table 4 summarizes the main applications of encapsulated/immobilized probiotic cells in ice-cream products.

Homayouni *et al.* (2008) investigated the effect of microencapsulation on the survival of two probiotic strains (*L. casei* Lc01 and *B. lactis* BB-12) added in a synbiotic ice-cream, containing 1% resistant starch with free and encapsulated probiotics. The results indicated that encapsulation can significantly improve the survival of probiotic bacteria. When encapsulated probiotic bacteria in calcium alginate beads were used, cell survival rose by 30% during 6 months of storage at −20°C. Specifically, the final counts of *L. casei* and *B. lactis* free cells decreased by

Table 4: Application of Encapsulated/Immobilized Probiotic Cells in Ice-cream Products

Encapsulation Materials & Methods Size of Capsules	Probiotic Cells	Ice-cream	Positive Effect	Reference
Ca-alginate-chitosan, Extrusion method	*Lactobacillus rhamnosus* ASCC 290, *Lactobacillus casei* ATCC 334	Yellow mombin Ice-cream	Encapsulation efficiencies up to 79.5% were achieved. Microcapsules reduced cell loss in frozen storage and in simulated gastrointestinal conditions.	de Farias *et al.*, 2019
Ca-alginate and carrageenan (715, 727 µm)	*Lactobacillus acidophilus* ATTC 4356	Ice-cream	The encapsulated probiotic cell levels were up to 8.74 logCFU/mL after 120 days. During simulated gastrointestinal assay, the survival rate of encapsulated probiotic bacteria was higher than free cells.	Afzaal *et al.*, 2018
Ca-alginate starch capsules Emulsification (300-800 µm)	*Bifidobacterium longum* CFR815j	Ice-cream	Maintenance of Minimal Biological Values of bifidobacteria in ice-cream without affecting the sensory characteristics. Under low pH conditions, the encapsulated bacteria retained the viability of log 7.12 CFU/mL.	Kataria *et al.*, 2018
Konjac glucomannan hydrolysate (KGMH), Spray-drying	*Lactobacillus casei* 01	Ice-cream	Increasing the concentration of KGMH increased encapsulation efficiency (95.4%). Maintenance of cell viability during freezing at −18°C for 28 days.	Yanprapasiri *et al.*, 2018
Mixture of fatty acids and coating with molten chocolate, Spray-coating method (100-5000 µm)	*Lactobacillus rhamnosus* R0011, *Bifidobacterium longum* R0175	Ice-cream	Improved stability of encapsulated cells after storage for 6 months at -20°C. *Bifidobacterium* cells were more susceptible compare to *Lactobacillus* strain	Champagne *et al.*, 2015

Skim milk, glycose, maltodextrin or inulin and alginate, Spray-drying (1-1000 mm)	*Lactobacillus paracasei*	Modified ice-cream base formulations	Strain survived after the transit through the gastro-intestinal barrier. Encapsulated cells remained vital after rehydration.	Spigno *et al.,* 2015
Ca-alginate, Emulsification (14.25-21.35 μm)	*Lactobacillus casei* Lc-01, *Bifidobacterium lactis* BB-12	Synbiotic ice-cream	Viability ranged between 10^8-10^9 CFU/g at the end of three months storage of ice-cream. Good sensory evaluation with no marked off-flavour.	Homayouni *et al.,* 2008
Ca-alginate, Freeze-drying (0.064-1.17 mm)	*Lactobacillus acidophilus* LAMJLA1, Bifidobacteria spp. BBBDBB2	Frozen dairy dessert	Encapsulation improved the survival of cells. Encapsulated cells survived in acidic conditions (at pH 2.5).	Shah and Ravula, 2000

3.4 and 2.9 log, respectively, compared to 1.4 and 0.7 log in the encapsulated cells, while incorporation of encapsulated probiotics had no significant effect on the sensory properties. Similarly, encapsulation on alginates was proposed for maintaining a *B. longum* CFR815j cell population over 10^7 in the product (Kataria *et al.*, 2018) with no significant differences observed on chemical and organoleptic properties. Likewise, Afzaal *et al.* (2018) documented a significant improvement on the viability of *L. acidophilus* encapsulated on sodium alginate or carrageenan during ice-cream storage for 120 days at −20°C. Interestingly, a reduction of 3.75, 1.17 and 1.60 log CFU/mL was observed in free, encapsulated on sodium alginate or carrageenan cells, respectively. However, insertion of the encapsulated cells had a significant negative effect on the quality parameters and sensory characteristics of the products.

Spray-drying has been also used for the preparation of encapsulated probiotics for applications in ice-cream production. Spigno *et al.* (2015) investigated the possibility of producing spray-dried probiotic formulations for partial substitution of the commercial ice-cream bases. Two formulations were developed – one based on maltodextrins and another on inulin. However, encapsulation was possible only with inulin, but with a low process yield (51%) and high powder a_w (> 0.4) compared to the original formulation. Although a 82% death cell occurred during drying, the cells survived the ice-cream making process. In another study, Konjac Glucomannan Hydrolysate (KGMH) was used as wall material (Yanprapasiri *et al.*, 2018). KGMH (25% w/w) was able to protect probiotic cells from heat treatment during spray-drying and to maintain cell viability during freezing for 28 days. However, addition of 10% trehalose resulted in a significant survival rate decrease during spray-drying, due to osmotic stress, but cells remained viable when the probiotic powder was applied into ice-cream stored at −18°C for 28 days.

However, encapsulation of probiotics does not always result in improvement of cell viability. For example, Kailasapathy and Sultana (2003) produced ice-cream with alginate-encapsulated probiotics (*L. acidophius* and *B. lactis*) and concluded that probiotic microencapsulation did not significantly increase cell survival. Other researchers suggested that microencapsulation may increase probiotic viability, but this is rather a strain-dependent property (de Farias *et al.*, 2019). For example, the behavior evaluation of *L. rhamnosus* and *L. casei* added to ice-cream in free or encapsulated with alginate-chitosan form indicated that the encapsulated *L. rhamnosus* ASCC 290 resisted better at low temperatures, whereas the *L. casei* ATCC 334 showed greater survival during the encapsulation process and in the GI environment. Therefore, the best option for the preparation of functional yellow mombin ice-cream, taking into consideration all the cellular losses, is to use free *L. rhamnosus* ASCC 290 or encapsulated *L. casei* ATCC 334.

On the other hand, Shah and Ravula (2000) reported that the survival of probiotic bacteria in fermented frozen desserts improved with calcium alginate encapsulation.

4. Conclusion and Future Perspectives

Despite the abundance of methods proposed by several researchers for encapsulation/immobilization of probiotic cells, enzymes and functional ingredients for dairy foods, the technology has not yet been widely adopted by the industrial sector, mainly due to safety issues related to the suitability of the encapsulation/immobilization agents, confirmation of the stability and functionality of the probiotic cultures, enzymes and bioactive compounds and the lack of processes that can be readily scale-up. Ongoing research is focused on resolving the above concerns, as encapsulation/immobilization constitutes a successful way of protecting degradation of enzymes and functional substances and improving cell viability. The assessment of the industrial feasibility of encapsulation/immobilization technology is also a prerequisite for providing cost-effective, large-scale quantities of encapsulated/immobilized agents for specific commercial use.

Techniques commonly applied for encapsulation/immobilization of microbial cells, enzymes and functional constituents include attachment/adsorption to carrier surfaces, entrapment, cell co-aggregation, mechanical containment and encapsulation. For microencapsulation, methods like emulsion, extrusion, spray-drying, freeze-drying and coacervation are usually employed, while the most commonly used materials are carrageenan, alginate, modified starch, chitosan, xanthan, milk fat and milk or whey proteins.

Apart from incorporating probiotic cells into food matrices, which is definitely the most popular application, encapsulation technique is already examined for improving the nutritional value of dairy products. Indeed, many studies have focused on the insertion of encapsulated biologically active substances, like polyphenols, tocopherols, carotenoids or phytosterols that exist in minimum levels into dairies. Such bioactive compounds are well defined as natural antioxidants and their consumption has been linked with health benefits, like enhanced cardiovascular system. Fortification with encapsulated nutraceuticals is expected to lead to extended bioavailability during storage and improved bioavailability after consumption, as well as maintenance of their functionality, which is a very important issue, considering that specific nutrient deficiency can be accompanied by metabolic, skeletal, cardiovascular and immune-system-associated health problems (Hassanalilou *et al.*, 2018; Nargesi *et al.*, 2018). Likewise, enrichment of foods with encapsulated health-promoting substances or compounds that act as biopreservatives will broaden the encapsulation applications and lead to the production of novel food products.

In the near future, multiple delivery systems are expected to be developed, leading to complex nutritional matrices (for example co-encapsulation of probiotic cultures with functional food ingredients). However, more research is still required on developing encapsulated systems for controlled release and targeted-delivery of the functional ingredients, as well as on the selection of suitable food-grade purity encapsulation methods and immobilization supports that can be easily accepted by the consumers. In this vein, natural food carriers associated with health-benefits and able to act as delivery vehicles, such as fruits, nuts, cereals, etc, have a great potential.

References

Adhikari, K., Mustapha, A., Grun, I.U. and Fernando, L. (2000). Viability of microencapsulated bifidobacteria in set yogurt during refrigerated storage, *Journal of Dairy Science*, 83: 1946-1951.

Afzaal, M., Saeed, F., Arshad, M.U., Nadeem, M.T., Saeed, M. and Tufail, T. (2018). The effect of encapsulation on the stability of probiotic bacteria in ice-cream and simulated gastrointestinal conditions, *Probiotics and Antimicrobial Proteins*, 1-7.

Amakiri, A.C. and Thantsha, M.S. (2016). Survival of *Bifidobacterium longum* LMG 13197 microencapsulated in Vegetal or Vegetal-inulin matrix in simulated gastrointestinal fluids and yogurt, *Springer Plus*, **5**(1): 1343.

Amine, K.M., Champagne, C.P., Raymond, Y., St-Gelais, D., Britten, M., Fustier, P., .Salmien, S. and Lacroix, M. (2014). Survival of microencapsulated *Bifidobacterium longum* in Cheddar cheese during production and storage, *Food Control*, 37: 193-199.

Anal, A.K. and Singh, H. (2007). Recent advances in microencapsulation of probiotics for industrial applications and targeted delivery, *Trends in Food Science and Technology*, **18**(5): 240-251.

Azarnia, S., Lee, B., St-Gelais, D., Kilcawley, K. and Noroozi, E. (2011). Effect of free and encapsulated recombinant aminopeptidase on proteolytic indices and sensory characteristics of Cheddar cheese, *LWT – Food Science and Technology*, 44: 570-575.

Bagherpour, S., Alizadeh, A., Ghanbarzadeh, S., Mohammadi, M. and Hamishehkar, H. (2017). Preparation and characterization of betasitosterol-loaded nanostructured lipid carriers for butter enrichment, *Food Bioscience*, 20: 51-55.

Baró, L., Lara, F. and Plaza, J. (2017). Leche y derivados lácteos, pp. 21-43 *In:* A. Gil, R. Artacho and M.D. Ruiz (Eds.), *Tratado de Nutrición. Composición y calidad nutritiva de los alimentos (Tercera)*, Editorial Medica Panamericana.

Bedoya-Serna, C.M., Dacanal, G.C., Fernandes, A.M. and Pinho, S.C. (2018). Antifungal activity of nanoemulsions encapsulating oregano (*Origanum vulgare*) essential oil: *In vitro* study and application in Minas Padrão cheese, *Brazilian Journal of Microbiology*, 49: 929-935.

Bekatorou, A., Plessas, S. and Mallouchos, A. (2016). Cell immobilization technologies for applications in alcoholic beverages, pp. 933-955. *In*: M. Mishra (Ed.), *Handbook of Encapsulation and Controlled Release*, CRC Press, Boca Raton.

Bermúdez-Aguirre, D. and Barbosa-Cánovas, G.V. (2011). Quality of selected cheeses fortified with vegetable and animal sources of omega-3, *LWT – Food Science and Technology*, 44: 1577-1584.

Bosnea, L.A., Moschakis, T. and Biliaderis, C.G. (2017). Microencapsulated cells of *Lactobacillus paracasei* subsp. *paracasei* in biopolymer complex coacervates and their function in a yogurt matrix, *Food & Function*, **8**(2): 554-562.

Burgain, J., Gaiani, C., Linder, M. and Scher, J. (2011). Encapsulation of probiotic living cells: From laboratory scale to industrial applications, *Journal of Food Engineering*, **104**(4): 467-483.

Chaikham, P. (2015). Stability of probiotics encapsulated with Thai herbal extracts in fruit juices and yogurt during refrigerated storage, *Food Bioscience*, 12: 61-66.

Champagne, C., Lee, B. and Saucier, L. (2010). Immobilization of cells and enzymes for fermented dairy or meat products, pp. 345-365. *In*: N.J. Zuidam and V.A. Nedovic (Eds.), *Encapsulation Technologies for Active Food Ingredients and Food Processing*, Springer-Verlag, New York.

Champagne, C.P., Raymond, Y., Guertin, N. and Bélanger, G. (2015). Effects of storage conditions, microencapsulation and inclusion in chocolate particles on the stability of probiotic bacteria in ice-cream, *International Dairy Journal*, 47: 109-117.

Champagne, C.P. and Fustier, P. (2007). Microencapsulation for the improved delivery of bioactive compounds into foods, *Current Opinion in Biotechnology*, 18: 184-190.

Chandramouli, V., Kailasapathy, K., Peiris, P. and Jones, M. (2004). An improved method of microencapsulation and its evaluation to protect *Lactobacillus* spp. in simulated gastric conditions, *Journal of Microbiological Methods*, **56**(1): 27-35.

Chávarri, M., Marañón, I. and Villarán, M.C. (2012). Encapsulation technology to protect probiotic bacteria. *In*: E.C. Rigobelo (Ed.), *Probiotics*, InTech Open Publisher, Rijeka, DOI: 10.5772/50046

Chen, L., Yang, T., Song, Y., Shu, G. and Chen, H. (2017). Effect of xanthan-chitosan-xanthan double layer encapsulation on survival of *Bifidobacterium* BB01 in simulated gastrointestinal conditions, bile salt solution and yogurt, *LWT – Food Science and Technology*, 81: 274-280.

Chen, L.Y., Remondetto, G.E. and Subirade, M. (2006). Food protein-based materials as nutraceutical delivery systems, *Trends in Food Science and Technology*, 17: 272-283.

Chen, M.J. and Chen, K.N. (2007). Applications of probiotic encapsulation in dairy products, pp. 83-107. *In*: J.M. Lakkis (Ed.), *Encapsulation and Controlled Release Technologies in Food Systems*, Blackwell Publishing, USA.

Cruz, A.G., Antunes, A.E., Sousa, A.L.O., Faria, J.A. and Saad, S.M. (2009). Ice-cream as a probiotic food carrier, *Food Research International*, **42**(9): 1233-1239.

de Farias, T.G.S., Ladislau, H.F.L., Stamford, T.C.M., Medeiros, J.A.C., Soares, B.L.M., Arnaud, T.M.S. and Stamford, T.L.M. (2019). Viabilities of *Lactobacillus rhamnosus* ASCC 290 and *Lactobacillus casei* ATCC 334 (in free form or encapsulated with calcium alginate-chitosan) in yellow mombin ice-cream, *LWT – Food Science and Technology*, 100: 391-396.

De Prisco, A., van Valenberg, H.J., Fogliano, V. and Mauriello, G. (2017). Microencapsulated starter culture during yogurt manufacturing, effect on technological features, *Food and Bioprocess Technology*, **10**(10): 1767-1777.

Desai, K.G. and Park, H.J. (2005). Recent developments in microencapsulation of food ingredients, *Drying Technology*, 23: 1361-1394.

Dimantov, A., Kesselman, E. and Shamanic, E. (2004). Surface characterization and dissolution properties of high amylase core starch pectin coatings, *Food Hydrocolloids*, 18: 29-37.

Dimitrellou, D., Tsaousi, K., Kourkoutas, Y., Panas, P., Kanellaki, M. and Koutinas, A.A. (2008). Fermentation efficiency of thermally-dried immobilized kefir on casein as starter culture, *Process Biochemistry*, 43: 1323-1329.

Dimitrellou, D., Kourkoutas, Y., Koutinas, A.A. and Kanellaki, M. (2009). Thermally-dried immobilized kefir on casein as starter culture in dried whey cheese production, *Food Microbiology*, **26**(8): 809-820.

Dimitrellou, D., Kandylis, P. and Kourkoutas, Y. (2014a). Physicochemical and microbiological characteristics of probiotic yogurt produced with immobilized cells, *Journal of Biotechnology*, 185: S79.

Dimitrellou, D., Kandylis, P., Sidira, M., Koutinas, A.A. and Kourkoutas, Y. (2014b). Free and immobilized *Lactobacillus casei* ATCC 393 on whey protein as starter cultures for probiotic Feta-type cheese production, *Journal of Dairy Science*, 97: 4675-4685.

Dimitrellou, D., Kandylis, P., Kourkoutas, Y., Koutinas, A.A. and Kanellaki, M. (2015). Cheese production using kefir culture entrapped in milk proteins, *Applied Biochemistry and Biotechnology*, **176**(1): 213-230.

Dimitrellou, D., Kandylis, P., Petrović, T., Dimitrijević-Branković, S., Lević, S., Nedović, V. and Kourkoutas, Y. (2016a). Survival of spray dried microencapsulated *Lactobacillus casei* ATCC 393 in simulated gastrointestinal conditions and fermented milk, *LWT – Food Science and Technology*, 71: 169-174.

Dimitrellou, D., Kandylis, P. and Kourkoutas, Y. (2016b). Effect of cooling rate, freeze-drying, and storage on survival of free and immobilized *Lactobacillus casei* ATCC 393, *LWT – Food Science and Technology*, 69: 468-473.

Dimitrellou, D., Kandylis, P., Kourkoutas, Y. and Kanellaki, M. (2017). Novel probiotic whey cheese with immobilized lactobacilli on casein, *LWT – Food Science and Technology*, 86: 627-634.

Dimitrellou, D., Kandylis, P., Levic, S., Petrovic, T., Ivanovic, S., Nedovic, V. and Kourkoutas, Y. (2019a). Encapsulation of *Lactobacillus casei* ATCC 393 in alginate capsules for probiotic fermented milk production, *LWT – Food Science and Technology*, 116: 108501.

Dimitrellou, D., Kandylis, P. and Kourkoutas, Y. (2019b). Assessment of freeze-dried immobilized *Lactobacillus casei* as probiotic adjunct culture in yogurt, *Foods*, **8**(9): 374.

Dinakar, P. and Mistry, V.V. (1994). Growth and viability of *Bifidobacterium bifidum* in Cheddar cheese, *Journal of Dairy Science*, 77(10): 2854-2864.

El-Messery, T.M., El-Said, M.M. and Abdelkader Farahat, E.S. (2019). Production of functional processed cheese supplemented with nanoliposomes of mandarin peel extract, *Pakistan Journal of Biological Sciences*, 22: 247-256.

FAO/WHO (2002). Food and Agriculture Organization of the United Nations/World Health Organization, *Guidelines for the Evaluation of Probiotics in Food*, London, Ontario, Canada, April 30 and May 1, 2002.

FAO/WHO (2011). Codex Standard for Fermented Milks 243, pp. 1. *In: Codex Alimentarius Commission: Milk and Milk Products*, second edition, CODEX STAN 243-2003; World Health Organization & Food and Agriculture Organization of the United Nations: Rome, Italy.

Fenster, K., Freeburg, B., Hollard, C., Wong, C., Rønhave Laursen, R. and Ouwehand, A.C. (2019). The production and delivery of probiotics: A review of a practical approach, *Microorganisms*, **7**(3): 83.

Freeman, A. and Lilly, M.D. (1998). Effect of processing parameters on the feasibility and operational stability of immobilized viable microbial cells, *Enzyme and Microbial Technology*, **23**(5): 335-345.

Gardiner, G.E., Bouchier, P., O'Sullivan, E., Kelly, J., Collins, J.K., Fitzgerald, G., Paul Ross, R. and Stanton, C. (2002). A spray-dried culture for probiotic Cheddar cheese manufacture, *International Dairy Journal*, **12**(9): 749-756.

Genisheva, Z., Macedo, S., Mussatto, S., Teixeira, J.A. and Oliveira, J.M. (2012). Production of white wine by *Saccharomyces cerevisiae* immobilized on grape pomace, *Journal of the Institute of Brewing*, 118: 163-173.

Genisheva, Z., Mota, A., Mussatto, S., Oliveira, J.M. and Teixeira, J.A. (2014). Integrated continuous winemaking process involving sequential alcoholic and malolactic fermentation with immobilized cells, *Process Biochemistry*, 49: 1-9.

Ghanbarzadeh, B., Babazadeh, A. and Hamishehkar, H. (2016). Nano-phytosome as a potential food-grade delivery system, *Food Bioscience*, 15: 126-135.

Gharibzahedi, S.M.T., Koubaa, M., Barba, F.J., Greiner, R., George, S. and Roohinejad, S. (2018). Recent advances in the application of microbial transglutaminase crosslinking in cheese and ice-cream products: A review, *International Journal of Biological Macromolecules*, 107: 2364-2374.

Gobbetti, M., Corsetti, A., Smacchi, E., Zocchetti, A. and De Angelis, M. (1998). Production of Crescenza cheese by incorporation of bifidobacteria, *Journal of Dairy Science*, **81**(1): 37-47.

Goff, H.D. (1997). Colloidal aspects of ice-cream – A review, *International Dairy Journal*, 7(6-7): 363-373.

Górecka, E. and Jastrzębska, M. (2011). Immobilization techniques and biopolymer carriers, *Biotechnology and Food Science*, 75: 65-86.

Gryta, M. (2002). The assessment of microorganism growth in the membrane distillation system, *Desalination*, 142: 79-88.

Haghshenas, B., Nami, Y., Haghshenas, M., Barzegari, A., Sharifi, S., Radiah, D., Rosli, R. and Abdullah, N. (2015). Effect of addition of inulin and fenugreek on the survival of microencapsulated *Enterococcus durans* 39C in alginate-psyllium polymeric blends in simulated digestive system and yogurt, *Asian Journal of Pharmaceutical Sciences*, **10**(4): 350-361.

Hassanalilou, T., Khalili, L., Ghavamzadeh, S., Shokri, A., Payahoo, L. and Bishak, Y.K. (2018). Role of vitamin D deficiency in systemic lupus erythematosus incidence and aggravation, *Autoimmunity Highlights*, **9**(1): 1.

Heidebach, T., Forst, P. and Kulozik, U. (2009a). Microencapsulation of probiotic cells by means of rennet-gelation of milk proteins, *Food Hydrocolloids*, **23**(7): 1670-1677.

Heidebach, T., Forst, P. and Kulozik, U. (2009b). Transglutaminase-induced caseinate gelation for the microencapsulation of probiotic cells, *International Dairy Journal*, **19**(2): 77-84.

Homayouni, A., Azizi, A., Ehsani, M.R., Yarmand, M.S. and Razavi, S.H. (2008). Effect of microencapsulation and resistant starch on the probiotic survival and sensory properties of synbiotic ice-cream, *Food Chemistry*, **111**(1): 50-55.

Hyndman, C.L., Groboillot, A., Poncelet, D., Champagne, C. and Neufeld, R.J. (1993). Microencapsulation of *Lactococcus lactis* with cross-link gelatin membranes, *Journal of Chemical Technology and Biotechnology*, 56: 259-263.

Iravani, S., Korbekandi, H. and Mirmohammadi, S.V. (2015). Technology and potential applications of probiotic encapsulation in fermented milk products, *Journal of Food Science and Technology*, **52**(8): 4679-4696.

Jin, Y.L. and Speers, R.A. (1998). Flocculation of *Saccharomyces cerevisiae*, *Food Research International*, 31: 421-440.

Kailasapathy, K. and Lam, S.H. (2005). Application of encapsulated enzymes to accelerate cheese ripening, *International Dairy Journal*, 15: 929-939.

Kailasapathy, K. and Masondole, L. (2005). Survival of free and microencapsulated *Lactobacillus acidophilus* and *Bifidobacterium lactis* and their effect on texture of feta cheese *Australian Journal of Dairy Technology*, **60**(3): 252.

Kailasapathy, K. and Sultana, K. (2003). Survival and [beta]-D-galactosidase activity of encapsulated and free *Lactobacillus acidophilus* and *Bifidobacterium lactis* in ice-cream, *Australian Journal of Dairy Technology*, **58**(3): 223.

Kataria, A., Achi, S.C. and Halami, P.M. (2018). Effect of encapsulation on viability of *Bifidobacterium longum* CFR815j and physiochemical properties of ice-cream, *Indian Journal of Microbiology*, **58**(2): 248-251.

Katechaki, E., Panas, P., Kourkoutas, Y., Koliopoulos, D. and Koutinas, A.A. (2009). Thermally-dried free and immobilized kefir cells as starter culture in hard-type cheese production, *Bioresource Technology*, **100**(14): 3618-3624.

Khalifa, S.A., Omar, A.A. and Mohamed, A.H. (2017). The effect of substituting milk fat by peanut oil on the quality of white soft cheese, *International Journal of Dairy Science*, **12**(1): 28-40.

Kim, S.J., Cho, S.Y., Kim, S.H., Song, O.J., Shin, I.I.S., Cha, D.S. and Park, H.J. (2008). Effect of microencapsulation on viability and other characteristics in *Lactobacillus acidophilus* ATCC 43121, *LWT – Food Science and Technology*, **41**(3): 493-500.

Kourkoutas, Y., Bekatorou, A., Banat, I.M., Marchant, R. and Koutinas A.A. (2004). Immobilization technologies and support materials suitable in alcohol beverages production: A review, *Food Microbiology*, **21**(4): 377-397.

Kourkoutas, Y., Bosnea, L., Taboukos, S., Baras, C., Lambrou, D. and Kanellaki, M. (2006). Probiotic cheese production using *Lactobacillus casei* cells immobilized on fruit pieces, *Journal of Dairy Science*, 89: 1431-1451.

Krasaekoopt, W., Bhandari, B. and Deeth, H. (2004). The influence of coating materials on some properties of alginate beads and survivability of microencapsulated probiotic bacteria, *International Dairy Journal*, **14**(8): 737-743.

Kumari Narwal, R., Bhushan, B., Pal, A., Malhotra, S.P., Kumar, S. and Saharan, V. (2016). Inactivation thermodynamics and iso-kinetic profiling for evaluating operational suitability of milk clotting enzyme immobilized incomposite polymer matrix, *International Journal of Biological Macromolecules*, 91: 317-328.

Lebeau, T., Jouenne, T. and Junter, G.A. (1998). Diffusion of sugars and alcohols through composite membrane structures immobilizing viable yeast cells, *Enzyme and Microbial Technology*, 22: 434-438.

Levinson, Y., Ish-Shalom, S., Segal, E. and Livney, Y. D. (2016). Bioavailability, rheology and sensory evaluation of fat-free yogurt enriched with VD3 encapsulated in re-assembled casein micelles, *Food & Function*, 7: 1477.

Li, M., Jin, Y., Wang, Y., Meng, L., Zhang, N., Sun, Y., Hao, J. and Fu, Q. (2019). Preparation of *Bifidobacterium breve* encapsulated in low methoxyl pectin beads and its effects on yogurt quality, *Journal of Dairy Science*, 102: 1-12.

Liu, L., Chen, P., Zhao, W., Li, X., Wang, H. and Qu, X. (2017). Effect of microencapsulation with the Maillard reaction products of whey proteins and isomaltooligosaccharide on the survival rate of *Lactobacillus rhamnosus* in white brined cheese, *Food Control*, 79: 44-49.

Livney, Y.D. (2010). Milk proteins as vehicles for bioactives, *Current Opinion in Colloid and Interface Science*, **15**(1-2): 73-83.

Martins, S.C.S., Martins, C.M., Fiúza, L.M.C.C. and Santaella, S.T. (2013). Immobilization of microbial cells: A promising tool for treatment of toxic pollutants in industrial wastewater, *African Journal of Biotechnology*, 12: 4412-4418.

Mangione, M.R., Giacomazza, D., Bulone, D., Martorana, V., Cavallaro, G. and San Biagio, P.L. (2005). K+ and Na+ effects on the gelation properties of n-Carrageenan, *Biophysical Chemistry*, 113: 129-135.

Marshall, R.T., Goff, H.D. and Hartel, R.W. (2003). *Ice-cream*, New York: Springer.

Mitropoulou, G., Nedovic, V., Goyal, A. and Kourkoutas, Y. (2013). Immobilization technologies in probiotic food production, *Journal of Nutrition and Metabolism*, http://dx.doi.org/10.1155/2013/716861

Moghaddas Kia, E., Ghasempour, Z., Ghanbari, S., Pirmohammadi, R. and Ehsani, A. (2018). Development of probiotic yogurt by incorporation of milk protein concentrate (MPC) and microencapsulated *Lactobacillus paracasei* in gellan-caseinate mixture, *British Food Journal*, https://doi.org/10.1108/BFJ-12-2017-0668

Moschakis, T. and Biliaderis, C.G. (2017). Biopolymer-based coacervates: Structures, functionality and applications in food products, *Current Opinion in Colloid & Interface Science*, 28: 96-109.

Muñoz-Tébar, N., De la Vara, J.A., Ortiz de Elguea-Culebras, G., Cano, E.L., Molina, A., Carmona, M. and Berruga, M.I. (2019). Enrichment of sheep cheese with chia (*Salvia hispanica* L.) oil as a source of omega-3, *LWT – Food Science and Technology*, 108: 407-415.

Nami, Y., Haghshenas, B. and Yari Khosroushahi, A. (2017). Effect of psyllium and gum arabic biopolymers on the survival rate and storage stability in yogurt of *Enterococcus durans* IW 3 encapsulated in alginate, *Food Science and Nutrition*, **5**(3): 554-563.

Nargesi, S., Ghorbani, A., Shirzadpour, E., Mohamadpour, M., Mousavi, S.F. and Amraei, M. (2018). A systematic review and meta-analysis of the association between vitamin D deficiency and gestational diabetes mellitus, *Biomedical Research and Therapy*, 5: 2078-2095.

Nazari, M., Ghanbarzadeh, B., Kafil, H.S., Zeinali, M. and Hamishehkar, H. (2019). Garlic essential oil nanophytosomes as a natural food preservative: Its application in yogurt as food model, *Colloid and Interface Science Communications*, 30: 100176.

Nedovic, V.A., Obradovic, B., Leskosek-Cukalovic, I. and Vunjak-Novakovic, G. (2001). Immobilized yeast bioreactor systems for brewing-recent achievements, pp. 277-292. *In*: P. Thonart and M. Hofman (Eds.), *Engineering and Manufacturing for Biotechnology. Focus on Biotechnology*, vol. 4. Springer, Dordrecht.

Nejati, R., Gheisari, H.R., Hosseinzadeh, S. and Behbod, M. (2017). Viability of encapsulated *Lactobacillus acidophilus* (LA-5) in UF cheese and its survival under *in vitro* simulated gastrointestinal conditions, *International Journal of Dairy Technology*, **70**(1): 77-83.

Nejati, R., Gheisari, H., Hosseinzadeh, S. and Amin, H. (2011). Viability of encapsulated *Bifidobacterium lactis* (BB-12) in synbiotic UF cheese and its survival under *in vitro* simulated gastrointestinal conditions, *International Journal of Probiotics and Prebiotics*, **6**(3-4): 197-204.

Ningtyas, D.W., Bhandari, B., Bansal, N. and Prakash, S. (2019). The viability of probiotic *Lactobacillus rhamnosus* (non-encapsulated and encapsulated) in functional reduced-fat cream cheese and its textural properties during storage, *Food Control*, 100: 8-16.

Ortakci, F. and Sert, S. (2012). Stability of free and encapsulated *Lactobacillus acidophilus* ATCC 4356 in yogurt and in an artificial human gastric digestion system, *Journal of Dairy Science*, **95**(12): 6918-6925.

Oster, M. (2017). Trends, innovations and opportunities driving the global probiotics market, *Euromonitor International*. http://internationalprobiotics.org/wp-content/uploads/Trends-Innovations-and-Opportunities-Diving-the-Global-Probiotics-Market-Matthew-Oster.pdf

Oxley, J. (2016). Process-selection criteria, pp. 24-25. *In*: M. Mishra (Ed.), *Handbook of Encapsulation and Controlled Release*, CRC Press, Boca Raton.

Özer, B., Uzun, Y.S. and Kirmaci, H.A. (2008). Effect of microencapsulation on viability of *Lactobacillus acidophilus* LA-5 and *Bifidobacterium bifidum* BB-12 during Kasar cheese ripening, *International Journal of Dairy Technology*, **61**(3): 237-244.

Özer, B., Kirmaci, H.A., Şenel, E., Atamer, M. and Hayaloğlu, A. (2009). Improving the viability of *Bifidobacterium bifidum* BB-12 and *Lactobacillus acidophilus* LA-5 in white-brined cheese by microencapsulation, *International Dairy Journal*, **19**(1): 22-29.

Park, J.K. and Chang, H.N. (2000). Microencapsulation of microbial cells, *Biotechnology Advances*, 18: 303-319.

Patrignani, F., Siroli, L., Serrazanetti, D.I., Braschi, G., Betoret, E., Reinheimer, J.A. and Lanciotti, R. (2017). Microencapsulation of functional strains by high pressure homogenization for a potential use in fermented milk, *Food Research International*, 97: 250-257.

Peinado, R.A., Moreno, J.J., Maestre, O. and Mauricio, J.C. (2005). Use of a novel immobilization yeast system for winemaking, *Biotechnology Letters*, 27: 1421-1424.

Petrov, K.K., Petrova, P.M. and Beschkov, V.N. (2007). Improved immobilization of *Lactobacillus rhamnosus* ATCC 7469 in polyacrylamide gel, preventing cell leakage during lactic acid fermentation, *World Journal of Microbiology and Biotechnology*, 23: 423-428.

Phoem, A.N., Mayiding, A., Saedeh, F. and Permpoonpattana, P. (2019). Evaluation of *Lactobacillus plantarum* encapsulated with *Eleutherine americana* oligosaccharide extract as food additive in yogurt, *Brazilian Journal of Microbiology*, **50**(1): 237-246.

Picot, A. and Lacroix, C. (2004). Encapsulation of bifidobacteria in whey protein microcapsules and survival in simulated gastrointestinal conditions and in yogurt, *International Dairy Journal*, 14: 505-515.

Pinto, S.S., Fritzen-Freire, C.B., Dias, C.O. and Amboni, R.D. (2019). A potential technological application of probiotic microcapsules in lactose-free Greek-style yogurt, *International Dairy Journal*, 97: 131-138.

Prasanna, P.H. and Charalampopoulos, D. (2019). Encapsulation in an alginate – goats' milk-inulin matrix improves survival of probiotic *Bifidobacterium* in simulated gastrointestinal conditions and goats' milk yogurt, *International Journal of Dairy Technology*, 72(1): 132-141.

Prasanna, P.H.P. and Charalampopoulos, D. (2018). Encapsulation of *Bifidobacterium longum* in alginate-dairy matrices and survival in simulated gastrointestinal conditions, refrigeration, cow milk and goat milk, *Food Bioscience*, 21: 72-79.

Prevost, H. and Divies, C. (1987). Fresh fermented cheese production with continuous pre-fermented milk by a mixed culture of mesophilic lactic streptococci entrapped in Ca-Al ginate, *Biotechnology Letters*, 9(11): 789-794.

Rajam, R., Karthik, P., Parthasarathi, S., Joseph, G. and Anandharamakrishnan, C. (2012). Effect of whey protein–alginate wall systems on survival of microencapsulated *Lactobacillus plantarum* in simulated gastrointestinal conditions, *Journal of Functional Foods*, 4(4): 891-898.

Ramon-Portugal, F., Silva, S., Taillandier, P. and Strehaiano, P. (2003). Immobilized yeasts: Actual oenologic utilizations, *Wine Internet Technical Journal*, www.vinidea.net

Rathore, S., Desai, M.P., Liew, C.V., Chan, L.W. and Heng, P.W.S. (2013). Microencapsulation of microbial cells, *Journal of Food Engineering*, 116(2): 369-381.

Ray, S., Raychaudhuri, U. and Chakraborty, R. (2016). An overview of encapsulation of active compounds used in food products by drying technology, *Food Bioscience*, 13: 76-83.

Ré, M., Santana, M. and d'Avila, M. (2016). Encapsulation technologies for modifying food performance, pp. 643-684. *In*: M. Mishra (Ed.), *Handbook of Encapsulation and Controlled Release*, CRC Press, Boca Raton.

Samedi, L. and Charles, A.L. (2019). Viability of 4 probiotic bacteria microencapsulated with arrowroot starch in the simulated gastrointestinal tract (GIT) and yogurt. *Foods*, 8(5): 175.

Santacruz, S. and Castro, M. (2018). Viability of free and encapsulated *Lactobacillus acidophilus* incorporated to cassava starch edible films and its application to Manaba fresh white cheese, *LWT – Food Science and Technology*, 93: 570-572.

Santillo, A., Albenzio, M., Bevilacqua, A., Corbo, M.R. and Sevi, A. (2012). Encapsulation of probiotic bacteria in lamb rennet paste: Effects on the quality of Pecorino cheese, *Journal of Dairy Science*, 95(7): 3489-3500.

Seth, D., Mishra, H.N. and Deka, S.C. (2017). Effect of microencapsulation using extrusion technique on viability of bacterial cells during spray drying of sweetened yogurt. *International Journal of Biological Macromolecules*, 103: 802-807.

Shah, N.P. and Ravula, R.R. (2000). Microencapsulation of probiotic bacteria and their survival in frozen fermented dairy desserts, *Australian Journal of Dairy Technology*, 55(3): 139.

Shahidi, F. and Han, X.Q. (1993). Encapsulation of food ingredients, *Critical Reviews in Food Science and Nutrition*, 33: 501-547.

Shehata, M.G., Abd-Rabou, H.S. and El-Sohaimy, S.A. (2019). Plant extracts in probiotic encapsulation: Evaluation of their effects on strain survivability in juice and drinkable yogurt during storage and an *in vitro* gastrointestinal model, *Journal of Pure and Applied Microbiology*, 13(1): 609-617.

Shoji, A.S., Oliveira, A.C., Balieiro, J.C.D.C., Freitas, O.D., Thomazini, M., Heinemann, R.J.B., Okuro, C.S. and Fávaro-Trindade, C.S. (2013). Viability of *L. acidophilus* microcapsules and their application to buffalo milk yogurt, *Food and Bioproducts Processing*, **91**(2): 83-88.

Shori, A.B. (2015). The potential applications of probiotics on dairy and non-dairy foods focusing on viability during storage, *Biocatalysis and Agricultural Biotechnology*, **4**(4): 423-431.

Shori, A.B. (2017). Microencapsulation improved probiotics survival during gastric transit, *HAYATI Journal of Biosciences*, 24: 1-5.

Shu, G., He, Y., Chen, L., Song, Y., Meng, J. and Chen, H. (2017). Microencapsulation of *Lactobacillus acidophilus* by xanthan-chitosan and its stability in yogurt, *Polymers*, **9**(12): 733.

Sidira, M., Santarmaki, V., Kiourtzidis, M., Argyri, A.A., Papadopoulou, O.S., Chorianopoulos, N., Tassou, C., Kaloutsas, S., Galanis, A. and Kourkoutas, Y. (2017). Evaluation of immobilized *Lactobacillus plantarum* 2035 on whey protein as adjunct probiotic culture in yogurt production, *LWT – Food Science and Technology*, 75: 137-146.

Sidira, M., Saxami, G., Dimitrellou, D., Santarmaki, V., Galanis, A. and Kourkoutas, Y. (2013). Monitoring survival of *Lactobacillus casei* ATCC 393 in probiotic yogurt using an efficient molecular tool, *Journal of Dairy Science*, 96: 3369-3377.

Sipsas, V., Kolokythas, J., Kourkoutas, Y., Plessas, S., Nedovic, V.A. and Kanellaki, M. (2009). Comparative study of batch and continuous multi-stage fixed-bed tower (MFBT) bioreactor during winemaking using freeze-dried immobilized cells, *Journal of Food Engineering*, 90: 495-503.

Sodini, I., Boquien, C.Y., Corrieu, G. and Lacroix, C. (1997). Use of an immobilized cell bioreactor for the continuous inoculation of milk in fresh cheese manufacturing, *Journal of Industrial Microbiology and Biotechnology*, **18**(1): 56-61.

Spigno, G., Garrido, G.D., Guidesi, E. and Elli, M. (2015). Spray-drying encapsulation of probiotics for ice-cream application, *Chemical Engineering Transactions*, 43: 49-54.

Stanton, C., Desmond, C., Coakley, M., Collins, J.K., Fitzgerald, G. and Ross, P. (2003). Challenges facing development of probiotic-containing functional foods, pp. 27-58. *In:* G. Mazza (Ed.), *Handbook of Fermented Functional Foods*, Boca Raton: CRC Press.

Sultana, K., Godward, G., Reynolds, N., Arumugaswamy, R., Peiris, P. and Kailasapathy, K. (2000). Encapsulation of probiotic bacteria with alginate-starch and evaluation of survival in simulated gastrointestinal conditions and in yogurt, *International Journal of Food Microbiology*, 62: 47-55.

Sun, W. and Griffiths, M.W. (2000). Survival of bifidobacteria in yogurt and simulate gastric juice following immobilization in gellan-xanthan beads, *International Journal of Food Microbiology*, 61: 17-25.

Taillandier, P., Cazottes, M.L. and Strehaiano, P. (1994). Deacidification of grape musts by *Schizosaccharomyces* entrapped in alginate beads: A continuous-fluidised-bed process, *The Chemical Engineering Journal and the Biochemical Engineering Journal*, 55: B29-B33.

Truelstrup-Hansen, L., Allan-wojtas, P.M., Jin, Y.L. and Paulson, A.T. (2002). Survival of free and calcium-alginate microencapsulated *Bifidobacterium* spp. in simulated gastro-intestinal conditions, *Food Microbiology*, 19: 35-45.

Ullah, R., Nadeem, M., Imran, M., Taj Khan, I., Shahbaz, M. and Mahmud, A. (2018). Omega fatty acids, phenolic compounds, and lipolysis of Cheddar cheese supplemented with chia (*Salvia hispanica* L.) oil, *Journal of Food Processing and Preservation*, 42: 1-11.

Wu, Y. and Zhang, G. (2018). Synbiotic encapsulation of probiotic *Latobacillus plantarum* by alginate-arabinoxylan composite microspheres, *LWT – Food Science and Technology*, 93: 135-141.

Yanprapasiri, K., Lohsrithong, C., Setthachaimongkol, S., Mekkerdchoo, O. and Borompichaichartkul, C. (2018). Probiotic encapsulation by spray drying using konjac glucomannan hydrolysate as wall material and its application in ice-cream, *Italian Journal of Food Science*, 76: 36-40.

Yüksel-Bilsel, A. and Şahin-Yeşilçubuk, N. (2019). Production of probiotic kefir fortified with encapsulated structured lipids and investigation of matrix effects by means of oxidation and in vitro digestion studies, *Food Chemistry*, 296: 17-22.

Zacheus, O.M., Iivanainen, E.K., Nissinen, T.K., Lehtola, M.J. and Martikainen, P.J. (2000). Bacterial biofilm formation on polyvinyl chloride, polyethylene and stainless steel exposed to ozonated water, *Water Research*, 34: 63-70.

Zhang, Y. and Zhong, Q. (2018). Freeze-dried capsules prepared from emulsions with encapsulated lactase as a potential delivery system to control lactose hydrolysis in milk, *Food Chemistry*, 241: 397-402.

Zhao, X.-Q. and Bai, F.-W. (2009). Yeast flocculation: New story in fuel ethanol production, *Biotechnology Advances*, **27**(6): 849-856.

Zuidam, N.J. and Shimoni, E. (2010). Overview of microencapsulates for use in food products or processes and methods to make them, pp. 3-29. *In*: N. Zuidam and V. Nedovic (Eds.), *Encapsulation Technologies for Active Food Ingredients and Food Processing*, Springer, New York.

Encapsulation of Meat Product Ingredients and Influence on Product Quality

Slaviša Stajić[1]* and Dragan Vasilev[2]

[1] Department of Animal Source Food Technology, Faculty of Agriculture, University of Belgrade, Nemanjina 6, Belgrade, Serbia
[2] Department of Food Hygiene and Technology, Faculty of Veterinary Medicine, University of Belgrade, Bulevar oslobodjenja 18, Belgrade, Serbia

1. Introduction

For centuries meat has been considered important for optimal human growth and development (Pereira and Vicente, 2013). In the narrow sense, meat is defined as muscle tissue including interstitial fatty and connective tissues, blood and lymph vessels of animals (pig, poultry, ruminants, game). Generally speaking, meat is an important source of protein, fat, energy, essential amino acids, as well as micronutrients, such as iron, selenium, zinc and vitamin B_{12} (Zhang *et al.*, 2010; Pereira and Vicente, 2013).

In addition, meat also contains high water content (approximately 75 per cent in the muscle), and is therefore susceptible to various changes. These changes are induced by various internal and external factors, such as enzymes, microorganisms, oxygen, light, temperature. Processes that occur in meat over time include protein degradation, lipid oxidation and microflora development. As a consequence of these changes, meat deteriorates quickly, its quality degrades and meat becomes unsuitable or even unsafe for consumption.

Since meat deteriorated quickly and as it was not readily available in the early human communities, different preservation principles were developed to extend its sustainability. Drying, smoking, heat treatment and salting were the earliest treatments used in meat preservation. Combining muscle tissue with other edible animal tissues (fatty tissue, offals, skin tissue, etc.), with the addition of spices and application of one or more preservation treatments, has led to the emergence of different meat products.

---
*Corresponding author: stajic@agrif.bg.ac.rs

For centuries meat products were used in nutrition as a significant source of meat proteins, energy and other nutrients. Nowadays, the availability of fresh meat has increased significantly, new preservation procedures have been developed and conventional ones have been improved. Hence, in addition to good nutritional qualities, meat products are now more highly appreciated because of their sensory characteristics which depend on the protein/fat/water ratio, usage of non-meat ingredients (salt, nitrates, phosphates, etc.), preservation procedures (smoking, drying, fermentation, etc.) and their interactions. However, in the last decades of the 20th century, new scientific insights found that certain components of meat/meat products have an adverse effect on health (salt, nitrites, saturated fatty acids (SFA), etc.), while increase in the intake of some other components can facilitate the prevention and treatment of certain illnesses (antioxidants, n-3 polyunsaturated fatty acids (n-3 PUFA), minerals, vitamins, etc.). Concurrently with scientific discoveries, consumer awareness has also changed; therefore, nowadays people are showing greater interest in food that contains bioactive or functional components which will give additional benefits to their health status (Hygreeva *et al.*, 2014). Accordingly, demand for natural, organic and/or clean label meat products has also increased (Cho *et al.*, 2017). This creates room for the use of different ingredients containing components which do not originate from meat or are not present in sufficient amounts, and are not traditionally used, with the aim to suppress the deterioration of meat/meat products and/or provide physiological benefits to health. In line with that, numerous research was conducted with the aim of prolonging shelf-life and/or enhancing the functional value of meat and meat products by using natural or synthetic ingredients with antioxidant, antibacterial, antiviral and probiotic activity. To this end, different ingredients containing bioactive components, such as phytosterols, polyphenols, carotenoids, dietary lipids, probiotics, prebiotics and botanicals were used (Zhang *et al.*, 2010; Olmedilla-Alonso *et al.*, 2013; Hygreeva *et al.*, 2014).

These active and bioactive compounds undergo different reactions in meat systems, and therefore degrade during processing and storage. To prevent rapid degradation and undesirable reactions and maintain full functionality, these compounds should be stabilized or immobilized in meat systems (Table 1).

Emulsions, double emulsions and gel-like matrixes were used in stabilization and immobilization of oils rich with PUFA (Jimenez-Colmenero *et al.*, 2015; Serdaroğlu *et al.*, 2016; Stajić *et al.*, 2018a). Furthermore, double emulsions were used with the purpose of enclosing various aromas, bioactive compounds or sensitive food components (Serdaroğlu *et al.*, 2016). Different encapsulation techniques were used for the stabilization of synthetic or natural active ingredients which were added in animal feed to enhance the functional value of meat (Dunne *et al.*, 2011), extend meat's shelf-life (de Oliveira Monteschio *et al.*, 2017), and prolong raw and thermally processed ground meat deterioration (Kılıç *et al.* 2014; Pabast *et al.*, 2018) while in meat products, they act as replacements of additives and/or improve functional value (Muthukumarasamy and Holley, 2006; Lorenzo *et al.*, 2016).

2.   Encapsulation

In general, as presented in Table 1, activities aimed at improving the quality of meat/ meat products can be divided into direct and indirect ones. Animal diet represents an indirect activity which is a longer-lasting and not fully controlled process through

Table 1: An Overview of Immobilization Techniques for the Improvement of Meat/Meat Products Quality and Achieved Goals

Approach	Application	Active Component	Achieved Goals	Immobilization Technique
Indirect	Animal feed	Lipids and essential oils	Protection from ruminal biohydrogenation; FA profile meat improvement, prolong the shelf-life of meat	Spray-drying
Direct – meat and meat products	Row, freeze and thermal treated ground meats: patty-type, fresh sausages, etc.	Phosphates	Lipid oxidation reduction during storage	Spray-drying, fluid bed coating, spray-cooling
		Essential oils	Antioxidant and antimicrobial activity	Double emulsions, spray-drying, extrusion
		Fish oil, plant oils	Improvement of FA profile	Organogels, oleogels, spray-drying, external gelation
	Meat products: fermented sausages, emulsified sausages (e.g. frankfurters), etc.	Essential oils	Antioxidant and antimicrobial activity	Spray-drying
		Fish oil, plant oils	Improvement of FA profile	Organogels, oleogels, oil-bulking, double emulsions; spray-drying, extrusion, coacervation
		NaCl, ascorbic acid, acidulants	Delaying lipid rancidity and pH drop	Spray-drying, spray-cooling fluid bed coating, coacervation
		Probiotics	Protection sensitive bacteria; antimicrobial activity	Extrusion, emulsification, freeze-drying
		Volatile essential oils	Antimicrobial effect of active packaging systems	Nanoliposomes

FA: Fatty Acids

which meat quality can be improved (Živković *et al.*, 2013). However, direct addition of functional components into meat/meat batters (meat products) is a simpler and more controlled approach (Živković *et al.*, 2013).

3. Improving Meat Quality by Animal Diet – Indirect Approach

Many meat components, such as proteins, minerals, vitamins, conjugated linoleic acid, carnitine, carnosine, etc., have a positive effect on human health. However, meat lipids, due to the high content of SFA, have poorer nutritional qualities. Animal diet is the major factor influencing the composition and quality of muscle and fatty tissues, especially in monogastric animals (pigs, poultry and fish) since their organism absorbs fatty acids in their intact form (Živković *et al.*, 2013).

3.1 Improvement of the Fatty Acid Profile

In ruminants, due to ruminal biohydrogenation, compounds (such as unsaturated fatty acids (UFA)) are altered. Therefore, rumen protection technologies have emerged over the years and these include the encapsulation of UFA inside a microbial-resistant shell (Jenkins and Bridges Jr., 2007). The procedure implies the encapsulation of active compounds (e.g. oils rich in PUFA) in a pH sensitive matrix which remains intact at rumen pH, but gets broken down at the lower pH in the abomasum, thereby constituents were released for absorption in the small intestine (Dunne *et al.*, 2011; Scollan *et al.*, 2014).

According to literature data, ruminal protection of fatty acids by encapsulation has been achieved in several ways. One way implies the incorporation of fatty acids (or oils) within proteins (e.g. casein) cross-linked with formaldehyde which provides resistance to microbial degradation in rumen (Jenkins and Bridges Jr., 2007). Modification of this encapsulation procedure without the use of formaldehyde was also reported (Dunne *et al.*, 2011). Furthermore, encasing oils rich in PUFA within a sphere of high-melting-point SFA is another encapsulation procedure used for rumen-protected products (Jenkins and Bridges Jr., 2007).

Lipids encapsulated in formaldehyde-treated protein matrix were applied in the diet of ruminants in several research studies. Garrett *et al.* (1976) used encapsulated lipids supplement (70 per cent sunflower seed and 30 per cent soybean protected lipid – containing 21 per cent linoleic acid) in the feed of lambs and steers. The major changes were increases in the contents of linoleic and α-linolenic acids and decreases in the contents of myristic, palmitic, palmitoleic and oleic acids, in both fatty tissue and muscle lipids. Furthermore, they reported that approximately 18-25 per cent of the consumed linoleic acid (as encapsulated lipids) was stored in the body tissues of cattle.

Scollan *et al.* (2007) examined the effects of feeding an encapsulated lipids supplement (70 per cent soybean, 22 per cent linseed and 8 per cent sunflower-seed oils) on the fatty acid composition of beef *Longissimus thoracis* muscle and the associated subcutaneous adipose tissue. The results were similar to Garrett *et al.* (1976) – encapsulation provides a high level of protection from ruminal biohydrogenation, the meat had a lower content of total fat, and increases of linoleic and α-linolenic acids, and decreases of SFA contents were observed in the subcutaneous adipose tissue and muscle lipids. This increase of the PUFA/SFA ratio to 0.28 was also recorded, resulting in beef meat with healthier lipids content.

The limitation of the previous research was that it had no influence on long chain n-3 PUFA content, especially on the content of eicosapentaenoic (EPA) and docosahexaenoic acids (DHA), which are highly associated with numerous beneficial effects on human health (Zhang *et al.*, 2010). This can be overcome by using encapsulated fish oils, as reported by Dunne *et al.* (2011). In this research, fish oil was encapsulated by technology that was modified and formaldehyde is not used. Beef heifers were offered a ration, containing three different proportions of encapsulated fish oil fortified with vitamin E, and the fatty acid profile of neck muscle intramuscular lipids (neutral and polar lipids) were examined. The authors also stated that encapsulation provides considerable ruminal protection of dietary PUFA since the EPA and DHA content in meat increased three and two times, respectively. The authors concluded that by applying this technology of rumen protection, it is possible to achieve EPA and DHA concentrations which would allow muscle to be labeled as a 'source' of these fatty acids in human nutrition. Kitessa *et al.* (2001) also reported three times higher content of both EPA and DHA and twice higher content of linolenic acid in the *Longissimus* muscle of lambs fed with the addition of encapsulated (protein/ formaldehyde technology) tuna fish oil. Furthermore, significant incorporations of these fatty acids in omental and perirenal fat were observed.

In general, the results indicate a great potential of encapsulation as a technique for ruminal protection of dietary lipids, resulting in numerous commercial supplements which were used in some research. The reported differences were due to different sources of fatty acids (linseed, canola, fish oils), different muscles (neck, leg, loin) and fatty tissues (subcutaneous, omental, perirenal), which should be taken into account when considering the effects of the use of encapsulated dietary lipids.

3.2 Improvement of Oxidative Stability

A very important fact that should not be neglected is that the improvement in the fatty acid profile should not lead to intensive lipid oxidation and thus influence meat aroma and color, especially during storage and retail display (Dunne *et al.*, 2011).

The use of oil enriched with natural antioxidants (e.g. tocopherols) can suppress lipid oxidation in meat. In their research, Dunne *et al.* (2011) used encapsulated fish oil fortified with vitamin E and found that ground beef with the highest EPA and DHA contents (which corresponds to the highest dietary encapsulated oil intake) reached critical lipid oxidation levels (TBARS value – thiobarbituric acid reactive substances) at five to six days of simulated retail display, indicating that more vitamin E is required. However, although lipid oxidation is associated with color stability, it did not affect color stability.

In the research of de Oliveira Monteschio *et al.* (2017), encapsulated eugenol, thymol and vanillin mixture (commercial blend) was used alone or in combination with clove and/or rosemary essential oils, in the feed of heifers. The use of essential oils and/or their encapsulated active principles did not affect some of the observed parameters which are very important for meat quality, such as pH, fat thickness, marbling, muscle area (*Longissimus thoracis* was examined), water loss by thawing and dripping, and cooking loss during the aging period. Color parameters were not affected, except for redness (at 14 days of aging), which was lower in control than in other diets, indicating lower maintenance of red color during aging in the control feed. Meat from heifers fed with the use of a mixture of essential oils and their encapsulated active principles was more tender than control after 14 days of aging. Furthermore, the control feed showed the lowest antioxidant capacity and the highest lipid oxidation rate during 14 days of aging. The use of encapsulated eugenol, thymol and vanillin mixture and its mixture with clove essential oil in the feed of heifers resulted in meat with higher antioxidant activity and lower lipid oxidation rate during 14 days of aging. The results indicate a good potential for using encapsulated active compounds in animal feed to maintain and prolong the shelf-life of beef.

4. Improving Meat/Meat Products Quality – Direct Approach

Besides using different immobilized compounds in animal diet to improve meat quality and shelf-life, among which some can positively affect the immune and digestive system of animals and thus can be natural alternatives to antibiotics (de Oliveira Monteschio *et al.*, 2017), improving meat quality and prolonging meat stability during retail period can be achieved by direct application of immobilized compounds into meat systems.

4.1 Encapsulation of Commonly Used Non-meat Ingredients

Encapsulation technology can be used for the controlled release of commonly used compounds in meat systems (salt, phosphates, etc.) during the retail period, thus increasing and prolonging the stability (lipid, microbial...) of fresh meat, ground meats prepared for further thermal treatments (e.g. patties) or already thermal-treated ground meats.

Sodium chloride: Salt (sodium chloride) is one of the most common non-meat ingredients and probably the oldest non-meat ingredient used in meat processing. Apart from contributing to preservation and flavor, salt is very important for textural properties of meat products because it is critical to solubilization and extraction of salt-soluble proteins (Sebranek, 2009). On the other hand, salt lowers the freezing point and catalyzes lipid oxidation reactions in meat systems. One approach to overcoming this is the encapsulation of salt within layer(s) of hydrogenated vegetable oil, which melts at higher temperatures, prevents salt from interacting with proteins and water. This way, salt is released into the meat system slowly, or during the thermal processing, or when thermal processing is over. Various commercial products of encapsulated salt are offered for use in meat and fish processing, marinades, sauces and in seasoning mixtures. The use of encapsulated salts assumes that it is not required for protein extraction and preservation before thermal processing (Monahan and Troy, 1997).

Phosphates: Phosphates are commonly used ingredients in meat processing. They are used as blends of mono-, di- and poly-phosphates to improve the water-holding capacity (and thus the juiciness and tenderness of meat), but also to enhance color and lipid stability. Tripolyphosphates are more effective in prolonging lipid oxidation than di- or mono-phosphates, but they are hydrolyzed by phosphatases in the meat system before thermal processing (Sickler *et al.*, 2013). Thus, encapsulation protects phosphates from degradation by heat sensitive phosphatases active in raw meat or ground meat stored for some time before thermal processing (Sickler *et al.*, 2013).

In their research, Sickler *et al.* (2013) compared unencapsulated sodium tripolyphosphate (STP) and STP encapsulated with hydrogenated vegetable oil designed to release the phosphate once the temperature reaches 74°C, in ground turkey breast meat cold stored up to ten days. The cooking loss and pH were not influenced by encapsulation, which means that encapsulation would not alter the influence of STP on the water-holding capacity. The use of encapsulated STP significantly improved lipid stability during storage compared to unencapsulated STP, which probably increased pigment stability because encapsulated treatments were more red (higher a* values). The authors concluded that the results indicate a potential use of encapsulated STP in meat products, such as: uncured precooked ground turkey and chicken sausages and nuggets, beef patties, restructured roast beef, meat loaves, which have a delayed thermal processing step after incorporating ingredients. Similar to Sickler *et al.* (2013), Kılıç *et al.* (2014) reported that in general, the use of encapsulated phosphates (in hydrogenated vegetable oil) instead of unencapsulated can significantly improve the lipid stability of cooked ground chicken meat and ground beef without any significant adverse effects on cooking loss, color and pH values.

Some different results regarding the effect of encapsulated STP on lipid oxidation in cooked ground beef during the seven-day storage were reported by Bilecen and Kiliç (2019), who examined the encapsulation effect of STP and sodium pyrophosphate (SPP) and their combination. They confirmed the results of previous research, and found no effect of encapsulation of STT, and reported similar finds for SPP on cooking loss, pH and color parameters, but reported that only encapsulated SPP significantly reduced lipid oxidation during storage. The combination (1:1 ratio) of encapsulated STP and SPP also significantly reduced lipid oxidation, but treatment with this combination shows lower color stability (decrease of redness and increase of yellowness) during storage as compared to the one with encapsulated STP. The authors selected this combination for potential use in ready-to-eat meat product formulations, but pointed out that it may reduce the beneficial effect of encapsulated STP on cooking loss.

Since phosphates were encapsulated with hydrogenated vegetable oil, it is important to mention the research of Du and Claus (2015) and Claus *et al.* (2016), who concluded that the melting properties of wall material should allow the release of phosphates at higher temperatures. At these temperatures, the inactivation of phosphatises is more complete, reducing the phosphate degradation (by the heat-sensitive phosphatases) and allows phosphates' maximum effect in prolonging lipid oxidation.

Ascorbic/Erythorbic acids: In processing cured meat products, reducing compounds, such ascorbic acid, erythorbic acid and their sodium salts are commonly used. These compounds are very important in reactions of nitrite and myoglobin resulting in the formation of nitrosylmyoglobin, which is responsible for cured meat color (Sebranek, 2009). Acids are very reactive and can cause fast nitrite reduction and loss from curing mixture before nitrosylmyoglobin is formed, so that sodium salts are commonly used forms (Sebranek, 2009). Thus, the application of encapsulation techniques can be useful in providing protection and stability of ascorbic acid in cured meat systems. Comunian *et al.* (2014) encapsulated ascorbic acid by the coacervation technique (in gum arabic) and applied freeze-dried microcapsules in chicken frankfurter formulation. The results indicated that encapsulation provided the stability of ascorbic acid during processing and storage of frankfurters, and no significant differences in instrumental color, oxidative stability and sensory properties were found compared to the treatment containing sodium erythorbate.

Acidulants: Decreasing pH values in meat products have a preservative effect because lower external pH disturbs the homeostasis of different pathogens and spoilage bacteria (Leistner, 2000). However, due to fast pH drop and/or long time between mixing/staffing and thermal processing, this results in massive protein denaturation, which leads to texture deficiency and fat separation and in turn, to an unacceptable product (Barbut, 2005). Therefore, a slow and targeted release of acidulants, immediately before or during thermal processing, is recommended. This can also be achieved by their encapsulation within a layer of hydrogenated vegetable oil designed to melt (and thus release acid) at higher temperatures. Barbut (2005) reported that the addition of liquid lactic acid led to fast protein denaturation, clumping and moisture release of meat batters after thermal processing, while the use of encapsulated lactic acid (with hydrogenated vegetable oil designed to release acid at 51-55°C) resolved these deficiencies. Similar results were reported for encapsulated citric and gluconic acids. Cordray and Huffman (1985) reported that the use of encapsulated lactic acid and glucono-delta-lactone led to a more intense flavor and more desirable color of restructured heat processed meat products without an adverse effect on physico-chemical and sensory properties. The use of acidulants encapsulated (by spray-drying) in water soluble coatings (i.e. maltodextrin) and designed to be released after meat emulsion is formed (before thermal processing), so as not to alter the emulsion stability, was confirm by patents of Percel *et al.* (1985a, 1985b).

4.2 New/natural Antioxidants, Antimicrobials and Colorants

For several decades already, the meat processing industry has constantly been searching for natural alternatives to synthetic ingredients which were traditionally used, such as nitrates, pigments (i.e. cochineal carmine), butylated hydroxyanisole (BHA), butylated hydroxytoluene (BHT), etc.

Besides the encapsulation of commonly used ingredients (salt, phosphates, acidulants), which have been in use for several decades and has resulted in numerous commercial products, as of recently different encapsulation techniques have been used for immobilization of plant and herbal sourced bioactive compounds with antioxidant and antibacterial activity which exhibit a beneficial influence on human health. However, most of these bioactive compounds exhibit low water-solubility, strong off-flavors and are unstable in meat systems during processing and/or storage, which is why encapsulation is used to enable their protection as well as controlled and targeted release (Vinceković *et al.*, 2017).

In addition to spray-drying, other commonly used encapsulation techniques, applied to overcome restrictions of different compounds, are freeze drying, extrusion methods, molecular inclusion, coacervation, nanostructured lipid matrices or/and solvent evaporation (Gómez *et al.*, 2018).

In the research of Baldin *et al.* (2016), encapsulated (in maltodextrin by spray-drying) jabuticaba (*Myrciaria cauliflora*) extract was used in fresh pork sausage for providing red color pigment and antioxidant activity. The authors reported that encapsulated jabuticaba extract reduced lipid oxidation and provided pigment stability during 15 days of storage, but concentrations higher than 2 per cent in sausage formulation can impact sensory characteristics negatively. Moreover, the authors suggested the use of 2 per cent of encapsulated jabuticaba extract as a natural replacement for cochineal carmine in fresh sausages. Spray-drying was also used for encapsulation of propolis co-product extract (capsul® as wall material) to provide oxidative stability of burger meat during 28 days of freeze storage (Reis *et al.*, 2017). Encapsulated propolis co-product extract exhibits antioxidant activity as effective as sodium erythorbate and has potential usage as natural replacement for synthetic ingredients.

Wu *et al.* (2019) used myofibrillar proteins as carriers to encapsulate curcumin (lipophilic antioxidant) by the pH shifting method. The authors reported positive effects of myofibrillar proteins as carriers on the improvement of solubility and stability of curcumin-myofibrillar proteins nanocomplexes, which possess strong antioxidant activity and thus can improve the stability of marinated chicken meat during 12 days of storage.

Clove (*Syzygium aromaticum*, L.) essential oil is another potential natural alternative to chemical preservatives in meat processing due to its antimicrobial (against several pathogenic bacteria) and antioxidative activity (Radünz *et al.*, 2019). However, strong odor limits its usage. Encapsulation is suggested as a possible solution. Radünz *et al.* (2019) reported that encapsulated clove essential oil (extrusion, with sodium alginate and emulsifiers) showed low antioxidant activity but also strong antimicrobial inhibition. In burger-like meat products, after seven-day storage, encapsulated clove essential oil exhibited antimicrobial effect against *Staphylococcus aureus* in a similar way as would the presence of nitrite.

Spray-dried micronized encapsulated essential oil blend (citrus, onion, and garlic mixed in a sodium tripolyphosphate and potassium lactate base) was suggested as a potential alternative to potassium lactate in growth control of *Listeria monocytogenes* and *Salmonella* serovars in some ready-to-eat (RTE) meat products during storage (Casco *et al.*, 2015). However, the authors reported limitations regarding bacteria type and types of RTE meat products; namely, the encapsulated essential oil blend did not exhibit *Salmonella* reduction in chunked and formed cured ham and formed roast beef, but was as efficient as potassium lactate in emulsified poultry bologna. Furthermore, a potential alternative to potassium lactate in growth control of *Listeria monocytogenes* was reported in ham and roast beef products only. Also, sensory analysis was not performed, so the authors suggested consumer acceptability testing.

Healthier products can be obtained by replacing fatty tissue in meat products with oils with favorable fatty acid profile. However, due to high PUFA contents, these oils are susceptible to lipid oxidation. Different encapsulation techniques (described below) can be applied to delay or prevent lipid oxidation. Furthermore, incorporation of natural compounds with antioxidative activity in oils before encapsulation can provide more oxidative stability. While the replacement of 50 per cent of fatty tissue in burgers with encapsulated chia oil (by external ion gelation technique) decreases oxidative stability, the addition of rosemary in chia oil before encapsulation provides necessary oxidative stability during 120 days of storage, without impairing the sensory quality (Heck *et al.*, 2018).

4.3 Improvement of Fatty Acid Profile of Processed Meats

Given that scientists have established a link between the lower rates of heart disease among Greenland's Eskimos and their diet rich in n-3 PUFA (Nichols *et al.*, 2010), numerous research pointed out the benefit of these fatty acids, not only in the prevention of coronary heart disease, but also for other illnesses, such as cancer, inflammatory diseases, depression, diabetes type 2 (McAfee *et al.*, 2010; Zhang *et al.*, 2010). Furthermore, since the 1960s, SFA were marked for increasing risk of coronary heart disease and medical societies recommended lowering their intake, while new research indicated that the content of individual fatty acids, such as palmitic acid, is even greater than the total SFA (Kleber *et al.*, 2018). The amount of the fatty tissue and its fatty acid profile are of importance for sensory quality and stability of meat, especially for meat products (processed meat). Namely, color, odor, taste and texture are strongly dependent of total fat (fatty tissue) content, while SFAs affect the hardness of the fatty tissue and reduce its susceptibility to oxidation (Stajić *et al.*, 2018b). In general (with differences between the most commonly used animal species), oleic (C18:1) and palmitic (C16:0) are the most dominant in the fatty acids profile of meat and fatty tissue, while the n-3 PUFA content is generally low (Wood *et al.*, 2008). Furthermore, the contents of myristic acid (C14:0) and stearic acid (C18:0), which some research studies also correlated with coronary heart disease risk (McAfee *et al.*, 2010), are also important (Wood *et al.*, 2008). The total fat content in meat products can be very high, even exceeding 40 per cent, and which can be found in dry-fermented sausages. Meat and meat products are among the main sources of dietary fats, therefore the introduction of oils with a favorable fatty acids profile, as a replacement for fatty tissue, can help improve their nutritional properties. However, the created products must have the same or imperceptibly-altered sensory qualities. Because of their susceptibility to oxidation, oils rich in monounsaturated fatty acids (MUFA) and PUFA can be stabilized by emulsification and encapsulation before their implementation in meat systems (Delgado-Pando *et al.*, 2010; Josquin *et al.*, 2012). Literature data pointed at the oil-in-water emulsion system with soy protein isolates (SPI) as the earliest and the prevalent used method, while gel-like systems were noted in some new research (Stajić *et al.*, 2018b). Furthermore, different encapsulation techniques were used to increase the stability of oils, while further stabilization of encapsulated oils in gel-like matrix was reported in different types of meat products (e.g. fermented sausages, emulsified sausages, fresh sausage, etc.).

The effect of the use of immobilized oils on the quality of meat products depends on oil type and amount, immobilization technique and the production process and ingredients, which can, more or less, result in favorable lipids changes. Fermented sausages (dry and semi-dry) undergo the fermentation/drying/ripening processes and can be stored for several months in cold or room temperatures, either packed or unpacked. Emulsified sausages (e.g. frankfurters) after heat processing (70-75°C in thermal centre), are usually packed (vacuum or MAP) and cold-stored for several weeks. Ground meats (e.g. burgers, fresh sausages) can be frozen and stored for several months before heat processing.

Among meat products, fermented sausages have a very long tradition and the high fat content is very important for their sensory properties. Stajić *et al.* (2014) and Stajić *et al.* (2018b) used grapeseed and flaxseed oil respectively, encapsulated by electrostatic extrusion in calcium alginate, to partially replace pork backfat in dry-fermented sausage formulation. The oils were also added (to the same extent) as emulsion (with SPI) and alginate gel. The authors replaced 20 per cent of backfat with encapsulated grapeseed and flaxseed oil, with the average fat content of 40.3 per cent and 40.1 per cent (to gain about 5 per cent oil content in initially batch) respectively and with the average particle size of 714 and 1161 μm respectively (Fig. 1). In both studies, these encapsulated oils did not alter the production process to a large extent (weight loss, pH change, basic chemical composition), but they did lead to significantly lower hardness and chewiness compared to control and to the treatments with oils added as emulsion and alginate gel. The authors indicated the possibility that the large number of microspheres prevent meat pieces from binding firmly during fermentation and ripening. The influence on color was different and was attributed to the different color characteristics of oils. Regarding sensory characteristics, in both studies, treatments with encapsulated oils were, in general, perceived as acceptable. However, lower hardness and chewiness were perceived as less desirable by consumers/panelists. Encapsulation of oils (as other two treatments) provided stability to dry-fermented sausages during cold storage in vacuum.

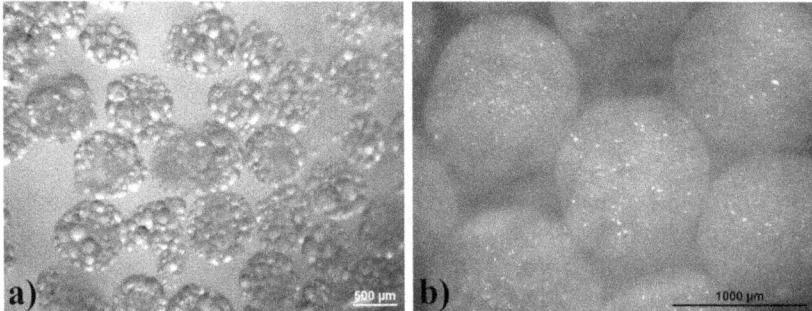

Fig. 1: Grapeseed oil (a) and flaxseed oil (b) encapsulated by electrostatic extrusion

Fish and flaxseed oils encapsulated by spray-drying (added as commercially-available ingredients) were used for partial replacement up to 30 per cent of pork backfat in fermented sausages formulation (Pelser *et al.*, 2007; Josquin *et al.*, 2012). Thus the nutritional properties improved significantly, with higher PUFA and lower SFA contents and lower n-6/n-3 rations. The use of encapsulated oils, despite higher PUFA contents, did not increase lipid oxidation during cold storage even in a modified atmosphere containing oxygen (Josquin *et al.*, 2012). Regarding lipid oxidation, fermented sausages with encapsulated fish oil (containing ascorbic acid) were found to be more stable than the ones with pre-emulsified (with SPI) fish oil (Josquin *et al.*, 2012). Contrary to Stajić *et al.* (2014) and Stajić *et al.* (2018b), the use of oils encapsulated by spray-drying led to firmer sausages without altering their sensory quality. Encapsulation masks the impact of sensory properties of fish oils (e.g. fishy taste), hence the sensory quality of fermented sausages with encapsulated fish oil was closest to the control treatment, unlike the sausages with liquid and pre-emulsified oil (with SPI). The results of this research pointed at encapsulation of fish oil as the best way to introduce this oil into fermented sausage formulation. The fish oil encapsulated by spray-drying (in maltodextrin, gum arabic and caseinate) and further stabilized in konjac glucomannan matrix was introduced in Spanish salchichón formulations to partially replace (up to 75 per cent) pork backfat (Lorenzo *et al.*, 2016). The increment of encapsulated fish oil (in konjac matrix) content led to harder and darker sausages, and also to a progressive increase in lipid oxidation, so that contrary to the previously-mentioned researches, encapsulation did not provide lipids protection. However, in the following research (Munekata *et al.*, 2017), oxidative stability was obtained by using natural antioxidants.

Frankfurters are worldwide popular emulsion-type and ready-to-serve meat products, made from pork and beef, as well as poultry meat. In the research of Salcedo-Sandoval *et al.* (2015), fish oil encapsulated in filled hydrogel particles (casein and pectin) was used as a replacement for backfat in low-fat pork frankfurters. The results indicated that the introduction of encapsulated fish oil into low-fat frankfurters will not alter technological properties – processing loss, purge loss, instrumental color and texture were similar (or even improved) to control treatment and moreover remain stable during the storage period. Encapsulation provides better oxidative stability as compared to oil-in-water emulsion. The possible explanation could be that encapsulated particles are more stable during thermal processing than conventional emulsions, and that casein and pectin are more effective at scavenging free radicals and chelating metals (Salcedo-Sandoval *et al.*, 2015). However, frankfurters with encapsulated fish oil were less oxidatively stable than the control treatment (only backfat), especially after 30 days of cold storage. Sensory properties were close to control and significantly better as compared to the treatment with emulsified fish oil. In the study of Domínguez *et al.* (2017), encapsulation (spray-drying; in maltodextrin, gum arabic and caseinate) of fish oil did not ensure its oxidative stability in pork frankfurters with partially replacement backfat, despite the use of BHT as antioxidant. Moreover, these frankfurters showed higher TBARs values, not only when compared to the control ones (only backfat), but also to frankfurters with pre-emulsified (with sodium caseinate) oil. The authors indicate that this may be due to high temperatures during the spry-drying process (and its length) which degrades BHT, or that antioxidant activity of BHT was limited by the encapsulation process.

Ground-type meat products, such as burgers (ground beef patty) and chicken nuggets, are widely consumed. Burgers and other ground beef patty-type products contain about 20-30 per cent (Heck *et al.*, 2019) of animal fat and the content of 20±2 per cent of fat was generally recommended for optimum palatability (Rust and Knipe, 2014). They can be sold fresh or frozen. The fat content was also important for the sensory quality (appearance, texture, aroma) and technological properties (cooking yield and shrinkage), and products with lower fat content had generally lower consumer acceptance (Rust and Knipe, 2014; Heck *et al.*, 2019). The encapsulated chia oil (by external ion gelation technique and 25 per cent% of oil content) was used for partial replacement of 50 per cent of backfat in beef burgers freeze-stored for 120 days. The modified burgers, besides having an improved fatty acid content, also had lower total fat contents, while instrumental color parameters were not affected, except lightness after cooking (higher values were measured). While 50 per cent replacement of backfat with encapsulated chia oil decreases oxidative stability, the addition of rosemary in chia oil before encapsulation provides the necessary oxidative stability during 120 days of storage, without impairing sensory quality, thus representing a new strategy to the introduction of oils with favorable fatty acid contents into cooked/ ready-to-cook meat products (Heck *et al.*, 2018; Heck *et al.*, 2019).

The demand for frozen breaded pre-fried ready-to-cook meat products, such as chicken nuggets, has increased in recent years (Pérez-Palacios *et al.*, 2018). Hence, these meat products have become interesting in the sense of creating functional food by changing the fatty acid profile. To improve the functional properties of chicken nuggets, spray-dried multilayered (lecithin, maltodextrin) microencapsulated fish oil (10 per cent oil content) was added at the level of 5 per cent into their formulation

(Pérez-Palacios *et al.*, 2018). The introduction of microencapsulated fish oil into chicken nuggets did not alter their sensory properties – consumers were not able to differentiate between the control treatment (without fish oil) and nuggets with microencapsulated fish oil, as shown in the results of the triangle test. Furthermore, as in the previously mentioned research, the encapsulation provides protection against lipid and protein oxidation, contrary to the nuggets with unencapsulated fish oil.

4.4 Probiotics

Improving the nutritional properties of meat products by using probiotics has been studied in the past several decades. Many fermented meat products, which do not require heat treatment during processing, such as dry-fermented sausages with probiotics, have been developed. The main requirements for the use of probiotics include their non-pathogenic nature, good resistance to conditions in the gastrointestinal tract, health benefits and a non-negative effect on sensory quality (Toldrá and Reig, 2011). Also, good adaptability to the conditions of dry-fermented sausages (low pH, high salt content, high nitrite content and low a_w values) is of great importance (Toldrá and Reig, 2011) because it is assumed that at least 6 log cfu/g of probiotics in the final product is required for them to have beneficial effects on human health (Muthukumarasamy and Holley, 2006). Again, encapsulation techniques could have the potential to provide protection to probiotics against such environmental and gastrointestinal conditions.

In the research of Muthukumarasamy and Holley (2006), the probiotic bacterium *Lactobacillus reuteri* was encapsulated by the extrusion and two-phase emulsion method with sodium alginate as a wall material and was added (as co-culture) in the amount of 7 cfu/g of meat batter (containing starter cultures *Pediococcus pentosaceus* and *Staphylococcus carnosus*), stuffed into fibrous casings and subjected to the fermentation/drying process. Both encapsulation techniques provide survival of *Lactobacillus reuteri* in sufficient numbers (> 6 log/g) until the end of the production process. The addition of encapsulated *Lactobacillus reuteri* did not influence pH and a_w values during production and in the final product. Furthermore, regarding sensory properties, no significant differences were found when comparing dry-fermented sausages containing encapsulated *Lactobacillus reuteri* with control (only starter cultures). According to the authors, this is probably the first study investigating the use of encapsulated probiotics in dry-fermented sausages, indicating that encapsulation techniques have great potential to provide a sufficient number of probiotics in this type of meat products.

In the next study, the same authors (Muthukumarasamy and Holley, 2007) encapsulated *Lactobacillus reuteri* and *Bifidobacterium longum*, separately or combined, by extrusion with sodium alginate as a wall material and added encapsulates to meat batter of dry-fermented sausages as a co-culture in the amount of 7 cfu/g. Regarding technological properties (pH and a_w values), the results confirmed those from the previous research. Encapsulation, as in the previous research, provided protection of probiotics during fermentation/drying process for both probiotics, but reduced their inhibitory activity against *E. coli* O157:H7. Therefore, the authors recommended a combination of unencapsulated and encapsulated forms to gain both viability and inhibitory activity of *Lactobacillus reuteri* against *E. coli* O157:H7 in dry-fermented sausages.

4.5   Active Packaging Systems

The packaging of fresh and processed meats (meat products) extends their shelf-life by reducing the influence of internal and external factors (oxygen, light, temperature, activity of enzymes and microorganisms) that cause meat/processed meat deterioration. Active packaging has become of interest in recent years because, in addition to decreasing meat/processed meats' decay, the developed active packaging technologies also reduce environmental pollution related to packaging (Wrona *et al.*, 2017). These technologies enable the incorporation of different active compounds into the packaging materials which, during the storage of meat/meat products, adsorb substances from the food or the environment (that cause deterioration) or are released into package environment to reduce the influence of factors that cause decay (McMillin, 2017). Thus, encapsulation could provide the protection of active compounds during their incorporation into packaging materials and also slow down and target their release during meat storage (Wrona *et al.*, 2017).

Several researches successfully applied encapsulated active compounds for developing active packaging for meat/meat products. Wrona *et al.* (2017) incorporated encapsulated green tea extract (crystalline microporous aluminosilicates as carrier) into polyethylene in two concentrations and packed minced pork meat in vacuum bags, covered by these materials under normal atmosphere. The encapsulation provides protection of green tea extract versus processing temperature during package material extrusion. Furthermore, active packaging increases the stability of minced meat up to nine days of storage. Minced meat packaged in active materials had lower met-myoglobin content (causes brown color), higher instrumental redness compared to control sample (no active packaging) at day nine of storage. The results of sensory evaluation correlated with the results of chemical and instrumental color analysis, while the aroma of experimental samples was assessed as fresh meat vs. vinegar in the control sample. However, active packaging with encapsulated green tea extract could not decrease the deterioration of minced meat after day nine and a higher concentration of the extract did not have a more pronounced impact during the entire storage period.

In the research of Pabast *et al.* (2018), nano-encapsulated *Satureja khuzestanica* essential oil (nanoliposomes, sonication method) was incorporated into chitosan-based material and used for coating the lamb meat that was cold stored for 20 days. Encapsulation enabled the controlled release of active compounds on meat surface during storage, while the coating with encapsulated *Satureja khuzestanica* essential oil significantly reduced microbial growth and provided stability during the entire storage period – the critical value of total viable counts (7 log CFU/g flesh) was not reached by the end of storage while in the control sample, it was reached after nine days. Moreover, significant reduction of lipid oxidation was also observed by the end of storage. Sensory quality (color and odor) was very stable throughout the whole storage period, which leads to the conclusion that chitosan coatings containing encapsulated Satureja essential oil have a potential for usage in active package development for the meat industry.

Another encapsulation technique was successfully developed and applied by Rajaei *et al.* (2017), to enhance meat's shelf-life. As in the previously mentioned research, in this research encapsulation also improved the effectiveness of essential oils. Clove essential oils (CEO) encapsulated by chitosan-myristic acid nanogel were used for coating beef meat cold-stored for 12 days. Nanogel-encapsulated CEO exhibited high antibacterial activity against *Salmonella enteric* and did not impair the color of beef meat during the storage period.

5. Other Techniques for Non-meat Ingredient Stabilization

As mentioned above, emulsions and double emulsions are able to stabilize some functional and bioactive compounds added to meat products with the aim to improve its nutritive value and provide functional properties. Emulsion type systems are mostly used as animal fat replacers, carrying oils rich in PUFA in order to improve the fatty acid profile of meat products and provide a health-promoting (functional) potential. Such an approach could be very challenging because of a possible adverse effect on sensory properties of products and their proneness to enhanced oxidation changes (Muguerza *et al.*, 2004; Vasilev *et al.*, 2010; Vasilev *et al.*, 2011; Grasso *et al.*, 2014).

Edible oils could be stabilized in emulsion-type systems by means of different strategies, including organogelation, oil-bulking, structured emulsions (hydrogelled and organogelled structured emulsions) (Jimenez-Colmenero *et al.*, 2015) and double emulsions (Serdaroğlu *et al.*, 2016).

5.1 Organogels

Organogels are bi-continuous systems which consist of gelators and non-polar solvents. The gelators tend to form fibrillary network structures that prevent the flow of non-polar phase. The most commonly used gelators could be divided into two groups: gelators that bind hydrogen, such as amino acids, amide and urea moieties, and carbohydrates, as well as gelators that do not bind hydrogen, including anthracene, anthraquinone and steroid-based molecules (Sahoo *et al.*, 2011). According to Esposito *et al.* (2018), organogelators could be divided into polymeric organic gelators (POGs) and low-molecular weight organogelators (LMOGs), and according to their origin, into synthetic molecules and biomolecules. Synthetic LMOGs include sorbitan monostearat, monopalmitate as well as glyceril fatty acid esters, while synthetic POGs are based on acrylic acid, sodium allyl or styrene sulfonate, polyethylene and polymethyl methacrylate, etc. As for biomolecules, lecithin is widely used, representing the group of phospholipids originating from soybean or egg yolk. Other biomolecules are phytosterols, sugars, carbohydrates, peptides and their derivates, as well as waxes.

The potential use of organogels in meat products has been studied by several authors. Barbut *et al.* (2016a) designed organogels consisting of canola oil as the solvent and ethylcellulose and sorbitan monostearate as gelators, in order to replace beef fat in meat batter. The organogel was prepared by mixing canola oil, ethylcelulose and sorbitan monostearate by means of overhead mechanical stirrer in a gravity convection oven, heated to 140°C. The whole process lasted 60 min (50 min to reach the target temperature, and 10 min of holding period), followed by chilling to the final temperature of 5°C. The organogel prepared in this manner, containing ethylcelulose in the concentration of 8 per cent, could be used as a beef-fat replacer in frankfurters without significantly affecting sensory properties. In higher concentrations of organogels, such as 10, 12 and 15 per cent in sausages, it resulted in higher hardness and less juiciness of the product. The same authors (Barbut *et al.*, 2016b) examined the potential use of organogel as fat replacer in breakfast sausage. The organogel was prepared by mixing canola oil with powdered gelators (10 per cent ethyl cellulose and 1.5 per cent sorbitan monostearate) at room temperature, followed by the same procedure described for frankfurters. In order to reduce lipid oxidation, the authors used butylated hydroxyl toluene (50 ppm) and rosemary oleoresin (0.6 per cent). As breakfast sausages belong to meat preparations intended for use after heat treatment, cooking loss is one of the important technological properties of such products. The use of organogels contributed to the retention of fat and water during heating, as well as to the enhancement of the texture, giving promising results in such type of meat products.

Soy proteins are widely used in the meat industry as a cheaper source of proteins, but they also show good gelating properties in organogels containing vegetable oils. Utrilla *et al*. (2014) used organogel consisting of olive oil, soy protein concentrate and water in the ratio of 10:1:8, respectively. The preparation process included mixing soy protein concentrate with hot water for two minutes followed by olive oil emulsification for another three minutes. Such obtained organogel was used as a substitute of high-fat pork meat (50 per cent fat) in a venison+pork Salchichon sausages. Salchichon is a type of fermented sausage, whose production includes the drying and ripening processes. From the technological point of view, it was possible to replace 15-55 per cent of fat pork in these sausages, but research into consumer acceptance showed that the sausages, where more than 25 per cent of pork meat was replaced, were not acceptable due to poor flavor and texture.

5.2   Oleogels

Oleogel is a type of organogel, which consists of two components, including liquid phase on the one hand, and the gelling agent on the other. The liquid phase could be polar, which is a characteristic of water (hydrogel), as well as non-polar, which includes oils (oleogel). The gelling agent forms a three-dimensional network which accounts for the immobilization of the liquid phase, giving stability to the whole system (Jimenez-Colmenero *et al*., 2015; Singh *et al*., 2017). The greatest advantage of oleogels in food processing is the possibility to obtain a stable system containing more than 90 per cent of oil by means of highly effective gelling agents. Such gelators include waxes, ethylcelulose, polymers, ceramides, phytosterol-based and carbohydrate-based ones (Patel *et al*., 2014; Singh *et al*.; 2017). Even more, oleogels could be relatively easy to obtain through a one-step process by mixing a gelling agent with oil under specific temperature and sharing conditions, or through a two-step process by drying (dehydration) of previously prepared oil-in-water emulsions (Jimenez-Colmenero *et al*., 2015).

The use of oleogels in meat products has recently been studied by several authors in order to replace animal fat in Bologna and Frankfurter-type sausages and patties, which are known as high-fat meat products.

Frankfurter-type sausages are high-fat emulsion-type cooked sausages, which often served as a model for the production of low-fat or PUFA-enriched meat products. But there are not many studies concerning the use of oleogels as fat replacers. Wolfer *et al*. (2018) used rice bran wax as the gelling agent and soybean oil as the liquid phase to replace the total amount of pork back-fat in frankfurter-type sausage, obtaining promising results. Oleogel-containing frankfurters showed a similar texture and aroma as the control, with the exception of lighter color and reduced flavor. The authors used two variations of soybean oil oleogels, where the first one contained 2.5

per cent rice bran wax, and the second 10 per cent rice bran wax, prepared in a steel bowl by mixing and heating to the temperature of 90°C for 2h. These oleogels were added to meat batter after reaching 4.4°C in the bowl chopper, followed by chopping to 13°C. There was also a third option, where oleogel with 2.5 per cent rice bran wax was later added to the batter in a bowl chopper, after reaching 10°C, followed by chopping to 13°C, in order to reduce the shear force acting on the oleogel during production. Such an approach helped to keep oleogel more intact, preventing it from association with the hydrophilic protein phase and being more similar to the control sausage, especially in color and texture. Even more, it showed that oleogels should not be treated exactly the same way as fatty tissue during production, but that they require a different approach in order to obtain optimal results. Another study concerning the use of oleogels in frankfurter-type sausages was conducted by Kouzounis *et al.* (2017), where monoglycerides and phytosterols served as gelators. These oleogels were prepared by addition of 20 per cent w/w monoglycerides and phytosterols in 1:1 and 3:1 ratio, in preheated sunflower oil (80°C), followed by stirring at 90°C for 60 min, homogenization in Ultra Turrax for 1 min and finally chilling at about 20°C. The 3:1 ratio was assessed as the optimal monoglycerides/phytosterols ratio for oleogel structuring, giving a stronger gel network in sunflower oil. This oleogel was used to replace 50 per cent of pork back-fat in frankfurters, where the product with acceptable sensory properties and improved fatty acid profile was obtained.

Bologna-type sausage is a finely comminuted cooked sausage from pork, usually containing up to 35 per cent animal fat. In order to produce a PUFA-enriched Bologna sausage, da Silva *et al.*, (2019) tried to replace 25, 50, 75 and 100 per cent animal fat by means of oleogel obtained by pork skin, water and sunflower oil. The pork skin served as the source for the gelling agent as the skin collagen was transformed into gelatin by previous cooking at 80°C for 40 min, followed by mixing with water and sunflower oil in the 1.5:1.5:1 ratio, respectively. The authors concluded that oleogel obtained in such a manner could replace up to 50 per cent of pork back-fat, without adverse effects on the sensory properties of Bologna sausage. The products where 75 and 100 per cent fat were replaced showed significantly lower redness and higher yellowness, as well as fewer defects in texture, aroma and overall acceptability.

The oleogel structuring is a promising solution for stabilization of PUFA-enriched patty-type products which can contain about 20-30 per cent of fat. Oh *et al.*, (2019) obtained oleogel by stabilizing canola oil with hydroxypropyl methylcellulose (HPMC), using it to replace 50 and 100 per cent of animal fat in meat patties. The oleogel preparation is somewhat more complex than the preceding ones and has several steps. The HPMC should first be dissolved in distilled water in the concentration of 1 per cent w/w and left overnight. The solution should be homogenized in a centrifuge at 11,000 rpm for 15 min, followed by freeze-drying, thus obtaining HPMC foams which should be ground and consequently added into the oil and shared by means of overhead stirrer at 400 rpm for 3 min. The mixture should be placed overnight in order to provide oil absorption which finally results in the formation of HPMC oleogel. The authors used canola oil as a PUFA source and obtained products with good sensory properties and oxidative stability, where products with a 50 per cent fat replacement with HPMC oleogel showed the best results.

5.3   Oil-bulking

Oil-bulking represents a process of incorporation of oils rich in polyunsaturated fatty acids into a gel-like matrix. Such oil-bulking systems could serve as a fatty tissue replacer in meat products. Polysaccharides possess the ability to form stable gels which could stabilize incorporated oil droplets. Konjac gum, alginate, inulin and dextrin are reported as good oil-bulking agents. The stability of such systems is based on the interactions between the oil molecules by means of hydrogen bonding of oil carbonyl groups and water, as well as carbohydrate molecules (Herrero *et al.*, 2014). The same authors reported that a stable oil-bulking system containing 55 per cent olive oil could be obtained through the following steps: first, 1 per cent sodium alginate, 1 per cent calcium sulfate, 0.75 per cent sodium pyrophosphate, 2.25 per cent dextrin and 40 per cent water were mixed in a homogenizer at 1500 rpm for 20s. The second step was the incorporation of olive oil in the matrix by slow adding and simultaneous mixing at the same speed. The other combination included the same process, but with 2.25 per cent inulin instead of dextrin as a bulking agent. Such oil-bulking systems possess good textural properties and could be used as fat replacers and at the same time for PUFA enrichment of meat products. Interestingly, the purpose of calcium sulphate and sodium pyrophosphate that were used in this study was to make the gelling process slower, which was particularly attributed to calcium salt. Some other studies (Vasilev *et al.*, 2016) indicate that calcium and potassium salts could also be successfully used as sodium chloride substitutes in functional fermented sausages enriched with inulin suspension as a fat replacer; so the overall approach should take into account, not only lowering saturated fat, but also lowering sodium in the design of healthier meat products.

Triki *et al.* (2013) investigated the influence of konjac-based olive oil-bulking system as a beef fat replacer in a Merguez sausage (fresh sausage from beef and lamb meat, spiced with chili pepper). Konjac gel was obtained by mixing 5 per cent konjac powder with 64.8 per cent water, including 3 min of mixing, 5 min of pause and another 3 min of mixing. Then 1 per cent of i-carrageenan was added followed by the addition of 20 per cent of olive oil and another 3 min of mixing. Along with this process, a starch gel was prepared by mixing 3 per cent corn starch powder with 16.2 per cent water. The starch gel was mixed with the above described konjac+carrageenan gel for another 3 min. After chilling to 10°C, 10 per cent calcium hydroxide was added to the mixture by gentle stirring. The oil-bulking system prepared in this manner was used to replace 75 and 100 per cent of beef fat in sausages (21.75 and 29 per cent in sausages). The sausages showed good sensory properties, oxidative stability and appropriate shelf-life.

Alginate was also successfully applied as a gelling agent to incorporate different plant oils into a gel-like matrix used as a fat replacer in different meat systems. Stajić *et al.* (2018a) used a commercial alginate mixture to prepare a gel-like matrix for linseed oil immobilization with alginate/oil/water ratio of 1/7/14 as a fat replacer in all-chicken frankfurters. They replaced 25 per cent and 50 per cent of chicken skin emulsion to provide approximately 2g and 4g of linseed oil in 100g of frankfurters which is about 50 and 100 per cent of the recommended daily intake of alpha-linolenic acid (ALA). Beside fatty acid profile improvement and an increase in yellowness (due to linseed oil characteristics), the modified frankfurters had similar physico-chemical characteristics as the control treatment, and were stable during the six-week cold storage and moreover, they had acceptable sensory properties. Alginate gel-like matrixes for plant oil (grapeseed and flaxseed oil) immobilization were also used (though in the modified alginate/oil/water ratio of 1/10/15) as back-fat replacement (at the level of 20 per cent) in dryfermented sausage formulation, as reported by Stajić *et al.* (2014)

and Stajić *et al.* (2018b). Moreover, in the research where grapeseed oil was used (Stajić *et al.*, 2014), the authors pointed out alginate gel matrix as the most promising immobilization technique among all other techniques that were experimented with (emulsion with soy protein isolate and encapsulation by electrostatic extrusion).

Inulin is a very promising gelator on the one hand because of its prebiotic and health-promoting properties, especially having in mind that some studies indicate a certain harmful potential of gums and carrageenans and on the other, because of its technological and sensory properties which make inulin a good fat replacer in meat products (Vasilev *et al.*, 2013; Vasilev *et al.*, 2017). An inulin-based oil-bulking system was designed by Glisic *et al.* (2019) in order to replace pork backfat in fermented sausages and thus enrich the product with a prebiotic (inulin) as well as PUFAs originated from linseed oil. The oil-bulking system consisted of two phases – the first included linseed pre-emulsion which was obtained with 15 per cent water, 20 per cent linseed oil and 3 per cent soybean lecithin mixed in a homogenizer until a thick paste consistency was obtained; the second phase was made of 35 per cent water, 25 per cent inulin powder and 2 per cent pork gelatine dissolved by heating (60°C) and homogenized for 2 min at medium speed. Then, the oil pre-emulsion was gradually added into the inulin suspension at room temperature, using a homogenizer at medium speed for about 3 min. The oil-bulking system obtained in this manner was chilled and consequently frozen before being used for sausage production. Such a system served to replace 64 per cent of pork back-fat (16 per cent in the sausage stuffing) in fermented sausages. The sausages were of good quality, but with reduced hardness to a certain degree, higher yellowness and they were also more susceptible to oxidation than the conventional sausages, which could be attributed to the PUFA-rich linseed oil.

5.4 Structured Emulsions

Structured emulsions consist of two or more phases, where one is dispersed into another in the form of small droplets and additionally structured – stabilized mainly by hydrogels and organogels. Because of the structuring process, the emulsions get a solid structure and could be successfully used as a fatty-tissue replacer in meat products (Jimenez-Colmenero *et al.*, 2015).

The use of hydrogels for emulsion structuring is described by Alejandre *et al.* (2019). The authors first prepared two phases: the oil phase consisted of 40 per cent canola oil, 0.05 per cent polysorbate 80 and 0.01 per cent BHT as antioxidant; the aqueous phase consisted of distilled water and 1.5 per cent and 3 per cent kappa carrageenan. These two phases were separately heated at 80°C and subsequently mixed and thoroughly homogenized. The authors also investigated the use of organogels for emulsion structuring, where a structured emulsion was prepared from 12 per cent ethylcellulose, 1.5 per cent or 3 per cent glycerol monostearate, 0.01 per cent BHT and canola oil, by heating in an oven to 140°C. The hydrogels and organogels prepared in this manner were used for beef-fat replacement in the emulsion type sausage. The modified sausages showed no significant difference in color and texture; moreover, they showed a lower oxidation level than the control, because of the presence of antioxidants. However, organogelled emulsions showed a better ability to stabilize meat batter than the hydrogelled emulsions which showed a bit less uniform microstructure and higher fat loss during heat treatment.

A study concerning the use of organogels for emulsion structuring in frankfurter sausages was conducted by Panagiotopoulou *et al.* (2016). The authors used a two-step approach: first, the preparation of an organogel, which was tested as a fat replacer itself, and second, the preparation of an organogel-in-water structured emulsion which also served as a fat substitute in frankfurters. The organogel was prepared from sunflower oil as a liquid phase, and phytosterol and γ-Oryzanol as a structuring agent. The structurants were used in two ratios – 30:70 and 60:40, in the amount of 10 per cent and 20 per cent (w/w) in sunflower oil. In order to obtain the organogel-in-water emulsion, such organogels were mixed with water at 80°C in a high shear dispenser Ultra Turrax for one minute at 14000 rpm. Tween 20 (1.7 per cent w/w) and xanthan gum (0.15 per cent w/w) served as emulsifying agents. As a pork back-fat substitute in frankfurters, the organogel containing phytosterol and γ-Oryzanol in the 60:40 ratio showed the best results, without affecting the sensory properties of the product, while the organogel with the 30:70 ratio was more appropriate for organogel-in-water emulsion stabilization.

5.5 Double Emulsions

Double emulsions represent multi-layered systems, where one layer represents oil-in-water emulsion and the other, water-in-oil emulsion. Such systems could serve as a carrier of functional ingredients in food, especially as animal fatty-tissue replacers in meat products. Serdaroğlu *et al.* (2016) described a two-step procedure: the inner water phase was obtained by distilled water and 0.6 per cent sodium chloride. Olive oil was prepared by mixing with 6.4 per cent emulsifier (polyglycerol polyricinoleate) at 50°C for 20 min. The emulsion was prepared by gradually introducing the water phase in the oil phase, and simultaneously mixing by high speed mixer (4400 rpm). The outer water phase was prepared by mixing 10 per cent sodium caseinate and 0.6 per cent sodium chloride at 50°C and 500 rpm for five minutes, followed by a one-hour stay in the water bath at 50°C. The phases prepared in this way were kept at 4°C for another 12 hin order to complete their fixation process before the final merging into a double emulsion. The final step included warming the phases at 50°C until they reached room temperature, followed by gradually pipetting the inner water-in-oil emulsion into the outer water phase and simultaneously mixing by 5200 rpm. The final double emulsion was stabilized by further emulsification for 10 min. This double emulsion was used as a fat replacer in a model system meat emulsion. The results showed appropriate technological properties, as well as oxidative stability, while the texture profile analysis showed significant changes, such as lower hardness, cohesiveness, gumminess and chewiness.

6. Conclusion and Future Perspectives

The stability of meat systems (raw meat, raw ground meat, processed meat) during storage and retail display are of great importance for the meat industry. Furthermore, with new scientific discoveries about the influence of meat/meat products on health, consumer awareness also changed and the design of natural, organic, clean label and meat products with a reduced negative impact or with bioactive or functional components also came in the focus of the meat industry. Immobilization techniques hence became a step in meat processing, enabling the preparation of different, primarily non-meat components (with bioactive compounds) for their optimal use. As these components are often in liquid form and given that powder is quite suitable for application in meat processing, different encapsulation techniques based on drying are becoming more interesting.

Taking into account all physico-chemical properties of different meat systems, active components, with proper selection of the encapsulation material and encapsulation techniques, new and/or natural ingredients for the meat industry can be produced in order to meet the ever complex demands of markets.

Acknowledgments

The study was supported by the Ministry of Education, Science and Technological Development of the Republic of Serbia - Contract numbers: 451-03-9/2021-14/200143 & 451-03-9/2021-14/200116

References

Alejandre, M., Astiasarán, I., Ansorena, D. and Barbut, S. (2019). Using canola oil hydrogels and organogels to reduce saturated animal fat in meat batters, *Food Research International*, 122: 129-136.

Baldin, J.C., Michelin, E.C., Polizer, Y.J., Rodrigues, I., de Godoy, S.H.S., Fregonesi, R.P., Pires, M.A., Carvalho, L.T., Fávaro-Trindade, C.S., de Lima, C.G., Fernandes, A.M. and Trindade, M.A. (2016). Microencapsulated jabuticaba (*Myrciaria cauliflora*) extract added to fresh sausage as natural dye with antioxidant and antimicrobial activity, *Meat Science*, 118: 15-21.

Barbut, S. (2005). Effects of chemical acidification and microbial fermentation on the rheological properties of meat products, *Meat Science*, 71: 397-401.

Barbut, S., Wood, J. and Marangoni, A. (2016a). Potential use of organogels to replace animal fat in comminuted meat products, *Meat Science*, 122: 155-162.

Barbut, S., Wood, J. and Marangoni, A. (2016b). Quality effects of using organogels in breakfast sausage, *Meat Science*, 122: 84-89.

Bilecen, D. and Kiliç, B. (2019). Determining the effects of encapsulated polyphosphates on quality parameters and oxidative stability of cooked ground beef during storage, *Food Science and Technology*, 39: 341-347.

Casco, G., Taylor, T.M. and Alvarado, C.Z. (2015). Evaluation of novel micronized encapsulated essential oil–containing phosphate and lactate blends for growth inhibition of *Listeria monocytogenes* and *Salmonella* on poultry bologna, pork ham, and roast beef ready-to-eat deli loaves, *Journal of Food Protection*, 78: 698-706.

Cho, M.G., Bae, S.M. and Jeong, J.Y. (2017). Egg shell and oyster shell powder as alternatives for synthetic phosphate: Effects on the quality of cooked ground pork products, *Korean Journal for Food Science of Animal Resources*, 37: 571-578.

Claus, J., Du, C. and Kılıç, B. (2016). Inhibition of lipid oxidation in ground turkey breasts by encapsulated polyphosphates as influenced by postmortem pH, *Meat Science*, 112: 129.

Comunian, T.A., Thomazini, M., Gambagorte, V.F., Trindade, M.A. and Favaro-Trindade, C.S. (2014). Effect of incorporating free or encapsulated ascorbic acid in chicken frankfurters on physicochemical and sensory stability, *Journal of Food Science and Engineering*, 4: 167-175.

Cordray, J.C. and Huffman, D.L. (1985). Restructured pork from hot processed sow meat: Effect of encapsulated food acids, *Journal of Food Protection*, 48: 965-968.

da Silva, S.L., Amaral, J.T., Ribeiro, M., Sebastião, E.E., Vargas, C., de Lima Franzen, F., Schneider, G., Lorenzo, J.M., Fries, L.L.M., Cichoski, A.J. and Campagnol, P.C.B. (2019). Fat replacement by oleogel rich in oleic acid and its impact on the technological, nutritional, oxidative, and sensory properties of Bologna-type sausages, *Meat Science*, 149: 141-148.

de Oliveira Monteschio, J., de Souza, K.A., Vital, A.C.P., Guerrero, A., Valero, M.V., Kempinski, E.M.B.C., Barcelos, V.C., Nascimento, K.F. and do Prado, I.N. (2017). Clove and rosemary essential oils and encapsuled active principles (eugenol, thymol and vanillin blend) on meat quality of feedlot-finished heifers, *Meat Science*, 130: 50-57.

Delgado-Pando, G., Cofrades, S., Ruiz-Capillas, C., Teresa Solas, M. and Jiménez-Colmenero, F. (2010). Healthier lipid combination oil-in-water emulsions prepared with various protein systems: An approach for development of functional meat products, *European Journal of Lipid Science and Technology*, 112: 791-801.

Domínguez, R., Pateiro, M., Agregán, R. and Lorenzo, J.M. (2017). Effect of the partial replacement of pork backfat by microencapsulated fish oil or mixed fish and olive oil on the quality of frankfurter type sausage, *Journal of Food Science and Technology*, 54: 26-37.

Du, C. and Claus, J.R. (2015). Inhibition of lipid oxidation in ground turkey with encapsulated phosphates as affected by meat age, phosphate type, and temperature release point, *Meat Science*, 101: 110. https://doi.org/10.1016/j.meatsci.2014.09.031

Dunne, P.G., Rogalski, J., Childs, S., Monahan, F.J., Kenny, D.A. and Moloney, A.P. (2011). Long chain n-3 polyunsaturated fatty acid concentration and color and lipid stability of muscle from heifers offered a ruminally protected fish oil supplement, *Journal of Agricultural and Food Chemistry*, 59: 5015-5025.

Esposito, C.L., Kirilov, P. and Roullin, V.G. (2018). Organogels, promising drug delivery systems: An update of state-of-the-art and recent applications, *Journal of Controlled Release*, 271: 1-20.

Garrett, W.N., Yang, Y.T., Dunkley, W.L. and Smith, L.M. (1976). Increasing the polyunsaturated fat content of beef and lamb, *Journal of Animal Science*, 42: 845-853.

Glisic, M., Baltic, M., Glisic, M., Trbovic, D., Jokanovic, M., Parunovic, N., Dimitrijevic, M., Suvajdzic, B., Boskovic, M. and Vasilev, D. (2019). Inulin-based emulsion-filled gel as a fat replacer in prebiotic- and PUFA-enriched dry fermented sausages, *International Journal of Food Science & Technology*, 54: 787-797.

Gómez, B., Barba, F.J., Domínguez, R., Putnik, P., Bursać-Kovačević, D., Pateiro, M., Toldrá, F. and Lorenzo, J.M. (2018). Microencapsulation of antioxidant compounds through innovative technologies and its specific application in meat processing, *Trends in Food Science & Technology*, 82: 135-147.

Grasso, S., Brunton, N.P., Lyng, J.G., Lalor, F. and Monahan, F.J. (2014). Healthy processed meat products – Regulatory, reformulation and consumer challenges, *Trends in Food Science & Technology*, 39: 4-17.

Heck, R.T., Fagundes, M.B., Cichoski, A.J., de Menezes, C.R., Barin, J.S., Lorenzo, J.M., Wagner, R. and Campagnol, P.C.B. (2019). Volatile compounds and sensory profile of burgers with 50% fat replacement by microparticles of chia oil enriched with rosemary, *Meat Science*, 148: 164-170.

Heck, R.T., Lucas, B.N., Santos, D.J.P.D., Pinton, M.B., Fagundes, M.B., de Araújo Etchepare, M., Cichoski, A.J., de Menezes, C.R., Barin, J.S., Wagner, R. and Campagnol, P.C.B. (2018). Oxidative stability of burgers containing chia oil microparticles enriched with rosemary by green-extraction techniques, *Meat Science*, 146: 147-153.

Herrero, A.M., Carmona, P., Jiménez-Colmenero, F. and Ruiz-Capillas, C. (2014). Polysaccharide gels as oil bulking agents: Technological and structural properties, *Food Hydrocolloids*, 36: 374-381.

Hygreeva, D., Pandey, M.C. and Radhakrishna, K. (2014). Potential applications of plant based derivatives as fat replacers, antioxidants and antimicrobials in fresh and processed meat products, *Meat Science*, 98: 47-57.

Jenkins, T.C. and Bridges Jr., W.C. (2007). Protection of fatty acids against ruminal biohydrogenation in cattle, *European Journal of Lipid Science and Technology*, 109: 778-789.

Jimenez-Colmenero, F., Salcedo-Sandoval, L., Bou, R., Cofrades, S., Herrero, A.M. and Ruiz-Capillas, C. (2015). Novel applications of oil-structuring methods as a strategy to improve the fat content of meat products, *Trends in Food Science & Technology*, 44: 177-188.

Josquin, N.M., Linssen, J.P.H. and Houben, J.H. (2012). Quality characteristics of Dutch-style fermented sausages manufactured with partial replacement of pork back-fat with pure, pre-emulsified or encapsulated fish oil, *Meat Science*, 90: 81-86.

Kılıç, B., Şimşek, A., Claus, J.R. and Atılgan, E. (2014). Encapsulated phosphates reduce lipid oxidation in both ground chicken and ground beef during raw and cooked meat storage with some influence on color, pH, and cooking loss, *Meat Science*, 97: 93-103.

Kitessa, S.M., Gulati, S.K., Ashes, J.R., Scott, T.W. and Fleck, E. (2001). Effect of feeding tuna oil supplement protected against hydrogenation in the rumen on growth and n-3 fatty acid content of lamb fat and muscle, *Australian Journal of Agricultural Research*, 52: 433-437.

Kleber, M.E., Delgado, G.E., Dawczynski, C., Lorkowski, S., März, W. and von Schacky, C. (2018). Saturated fatty acids and mortality in patients referred for coronary angiography – The *ludwigshafen* risk and cardiovascular health study, *Journal of Clinical Lipidology*, 12: 455-463. e453.

Kouzounis, D., Lazaridou, A. and Katsanidis, E. (2017). Partial replacement of animal fat by oleogels structured with monoglycerides and phytosterols in frankfurter sausages, *Meat Science*, 130: 38-46.

Leistner, L. (2000). Basic aspects of food preservation by hurdle technology, *International Journal of Food Microbiology*, 55: 181-186.

Lorenzo, J.M., Munekata, P.E.S., Pateiro, M., Campagnol, P.C.B. and Domínguez, R. (2016). Healthy Spanish salchichón enriched with encapsulated n-3 long chain fatty acids in konjac glucomannan matrix, *Food Research International*, 89: 289-295.

McAfee, A.J., McSorley, E.M., Cuskelly, G.J., Moss, B.W., Wallace, J.M.W., Bonham, M.P. and Fearon, A.M. (2010). Red meat consumption: An overview of the risks and benefits, *Meat Science*, 84: 1-13.

McMillin, K.W. (2017). Advancements in meat packaging, *Meat Science*, 132: 153-162.

Monahan, F.J. and Troy, D.J. (1997). Overcoming sensory problems in low fat and low salt products, pp. 257-281. *In:* A.M. Pearson & T.R. Dutson (Eds.), *Production and Processing of Healthy Meat, Poultry and Fish Products. Advances in Meat Research*, vol. 11, Springer, Boston, MA.

Muguerza, E., Gimeno, O., Ansorena, D. and Astiasarán, I. (2004). New formulations for healthier dry fermented sausages: A review, *Trends in Food Science & Technology*, 15: 452-457.

Munekata, P.E.S., Domínguez, R., Franco, D., Bermúdez, R., Trindade, M.A. and Lorenzo, J.M. (2017). Effect of natural antioxidants in Spanish salchichón elaborated with encapsulated n-3 long chain fatty acids in konjac glucomannan matrix, *Meat Science*, 124: 54-60.

Muthukumarasamy, P. and Holley, R.A. (2006). Microbiological and sensory quality of dry fermented sausages containing alginate-microencapsulated *Lactobacillus reuteri*, *International Journal of Food Microbiology*, 111: 164-169.

Muthukumarasamy, P. and Holley, R.A. (2007). Survival of *Escherichia coli* O157:H7 in dry fermented sausages containing micro-encapsulated probiotic lactic acid bacteria, *Food Microbiology*, 24: 82-88.

Nichols, P.D., Petrie, J. and Singh, S. (2010). Long-chain omega-3 oils – An update on sustainable sources, *Nutrients*, 2: 572-585.

Oh, I., Lee, J., Lee, H.G. and Lee, S. (2019). Feasibility of hydroxypropyl methylcellulose oleogel as an animal fat replacer for meat patties, *Food Research International*, in press, https://doi.org/10.1016/j.foodres.2019.01.012

Olmedilla-Alonso, B., Jiménez-Colmenero, F. and Sánchez-Muniz, F.J. (2013). Development and assessment of healthy properties of meat and meat products designed as functional foods, *Meat Science*, 95: 919-930.

Pabast, M., Shariatifar, N., Beikzadeh, S. and Jahed, G. (2018). Effects of chitosan coatings incorporating with free or nano-encapsulated Satureja plant essential oil on quality characteristics of lamb meat, *Food Control*, 91: 185-192.

Panagiotopoulou, E., Moschakis, T. and Katsanidis, E. (2016). Sunflower oil organogels and organogel-in-water emulsions (part II): Implementation in frankfurter sausages, *LWT – Food Science and Technology*, 73: 351-356.

Patel, A.R., Cludts, N., Sintang, M.D.B., Lesaffer, A. and Dewettinck, K. (2014). Edible oleogels based on water soluble food polymers: Preparation, characterization and potential application, *Food & Function*, 5: 2833-2841.

Pelser, W.M., Linssen, J.P.H., Legger, A. and Houben, J.H. (2007). Lipid oxidation in n-3 fatty acid enriched Dutch style fermented sausages, *Meat Science*, 75: 1-11.

Percel, P.J., Perkins, D.W. and Petricca, A.V. (1985a). Acidulated meat emulsion, U.S. Patent # 4,497,845.

Percel, P.J., Perkins, D.W. and Petricca, A.V. (1985b). Preparation of acidulated meat emulsions, U.S. Patent # 4,511,592.

Pereira, P.M.d.C.C. and Vicente, A.F.d.R.B. (2013). Meat nutritional composition and nutritive role in the human diet, *Meat Science*, 93: 586-592.

Pérez-Palacios, T., Ruiz-Carrascal, J., Jiménez-Martín, E., Solomando, J.C. and Antequera, T. (2018). Improving the lipid profile of ready-to-cook meat products by addition of omega-3 microcapsules: Effect on oxidation and sensory analysis, *Journal of the Science of Food and Agriculture*, 98: 5302-5312.

Radünz, M., da Trindade, M.L.M., Camargo, T.M., Radünz, A.L., Borges, C.D., Gandra, E.A. and Helbig, E. (2019). Antimicrobial and antioxidant activity of unencapsulated and encapsulated clove (*Syzygium aromaticum*, L.) essential oil, *Food Chemistry*, 276: 180-186.

Rajaei, A., Hadian, M., Mohsenifar, A., Rahmani-Cherati, T. and Tabatabaei, M. (2017). A coating based on clove essential oils encapsulated by chitosan-myristic acid nanogel efficiently enhanced the shelf-life of beef cutlets, *Food Packaging and Shelf-Life*, 14: 137-145.

Reis, A.S.d., Diedrich, C., Moura, C.d., Pereira, D., Almeida, J.d.F., Silva, L.D.d., Plata-Oviedo, M.S.V., Tavares, R.A.W. and Carpes, S.T. (2017). Physico-chemical characteristics of microencapsulated propolis co-product extract and its effect on storage stability of burger meat during storage at –15°C, *LWT – Food Science and Technology*, 76: 306-313.

Rust, R.E. and Knipe, C.L. (2014). Ethnic meat products in North America, pp. 555–557. *In:* M. Dikeman and C. Devine (Eds.), *Encyclopedia of Meat Sciences*, second edition, Academic Press, Oxford.

Sahoo, S., Kumar, N., Bhattacharya, C., Sagiri, S.S., Jain, K., Pal, K., Ray, S.S. and Nayak, B. (2011). Organogels: Properties and applications in drug delivery, *Designed Monomers and Polymers*, 14: 95-108.

Salcedo-Sandoval, L., Cofrades, S., Ruiz-Capillas, C. and Jiménez-Colmenero, F. (2015). Filled hydrogel particles as a delivery system for n-3 long chain PUFA in low-fat frankfurters: Consequences for product characteristics with special reference to lipid oxidation, *Meat Science*, 110: 160-168.

Scollan, N.D., Enser, M., Gulati, S.K., Richardson, I. and Wood, J.D. (2007). Effects of including a ruminally protected lipid supplement in the diet on the fatty acid composition of beef muscle, *British Journal of Nutrition*, 90: 709-716.

Scollan, N.D., Dannenberger, D., Nuernberg, K., Richardson, I., MacKintosh, S., Hocquette, J.-F. and Moloney, A.P. (2014). Enhancing the nutritional and health value of beef lipids and their relationship with meat quality, *Meat Science*, 97: 384-394.

Sebranek, J.G. (2009). Basic curing ingredients, pp. 1-23. *In:* R. Tarté (Ed.), *Ingredients in Meat Products: Properties, Functionality and Applications*, Springer New York, NY.

Serdaroğlu, M., Öztürk, B. and Urgu, M. (2016). Emulsion characteristics, chemical and textural properties of meat systems produced with double emulsions as beef fat replacers, *Meat Science*, 117: 187-195.

Sickler, M.L., Claus, J.R., Marriott, N.G., Eigel, W.N. and Wang, H. (2013). Antioxidative effects of encapsulated sodium tripolyphosphate and encapsulated sodium acid pyrophosphate in ground beef patties cooked immediately after antioxidant incorporation and stored, *Meat Science*, 94: 285-288.

Singh, A., Auzanneau, F.I. and Rogers, M.A. (2017). Advances in edible oleogel technologies – A decade in review, *Food Research International*, 97: 307-317.

Stajić, S., Živković, D., Tomović, V., Nedović, V., Perunović, M., Kovjanić, N., Lević, S. and Stanišić, N. (2014). The utilisation of grapeseed oil in improving the quality of dry fermented sausages, *International Journal of Food Science & Technology*, 49: 2356-2363.

Stajić, S., Stanišić, N., Tomasevic, I., Djekic, I., Ivanović, N. and Živković, D. (2018a). Use of linseed oil in improving the quality of chicken frankfurters, *Journal of Food Processing and Preservation*, 42: e13529.

Stajić, S., Stanišić, N., Lević, S., Tomović, V., Lilić, S., Vranić, D., Jokanović, M. and Živković, D. (2018b). Physico-chemical characteristics and sensory quality of dry fermented sausages with flaxseed oil preparations, *Polish Journal of Food and Nutrition Sciences*, 68: 367-375.

Toldrá, F. and Reig, M. (2011). Innovations for healthier processed meats, *Trends in Food Science & Technology*, 22: 517-522.

Triki, M., Herrero, A.M., Jiménez-Colmenero, F. and Ruiz-Capillas, C. (2013). Effect of preformed konjac gels, with and without olive oil, on the technological attributes and storage stability of merguez sausage, *Meat Science*, 93: 351-360.

Utrilla, M.C., García Ruiz, A. and Soriano, A. (2014). Effect of partial replacement of pork meat with an olive oil organogel on the physicochemical and sensory quality of dry-ripened venison sausages, *Meat Science*, 97: 575-582.

Vasilev, D., Đorđević, V., Karabasil, N., Dimitrijevic, M., Petrović, Z., Velebit, B. and Teodorović, V. (2017). Inulin as a prebiotic and fat replacer in meat products, *Theory and Practice of Meat Processing*, 2: 4-13.

Vasilev, D., Vuković, I., Saičić, S., Vasiljević, N., Milanović-Stevanović, M. and Tubić, M. (2010). The composition and significant changes in fats of functional fermented sausages, *Meat Technology*, 51: 27-35.

Vasilev, D., Vuković, I. and Saičić, S. (2011). Some quality parameters of functional fermented, cooked and liver sausages, *Meat Technology*, 52: 141-153.

Vasilev, D., Saičić, S. and Vasiljevic, N. (2013). Qualität und Nährwert von mit Inulin und Erbsenfasern als Fettgewebe-Ersatzstoffe horgestellten Rohwürsten, *Fleischwirtschaft*, 93: 123-127.

Vasilev, D., Jovetic, M., Vranić, D., Tomovic, V., Jokanović, M., Dimitrijevic, M., Karabasil, N. and Vasiljević, N. (2016). Qualität und Mikroflora von funktionellen Rohwürsten – Untersuchungen von Würsten, die mit KCl und CaCl2 als Kochsalz-Ersatzstoffe hergestellt und mit dem Probiotikum *L. casei* LC01 sowie einem Präbiotikum angereichert worden sind, *Fleischwirtschaft* , 96: 96-102.

Vinceković, M., Viskić, M., Jurić, S., Giacometti, J., Bursać Kovačević, D., Putnik, P., Donsì, F., Barba, F.J. and Režek Jambrak, A. (2017). Innovative technologies for encapsulation of Mediterranean plants extracts, *Trends in Food Science & Technology*, 69: 1-12.

Wolfer, T.L., Acevedo, N.C., Prusa, K.J., Sebranek, J.G. and Tarté, R. (2018). Replacement of pork fat in frankfurter-type sausages by soybean oil oleogels structured with rice bran wax, *Meat Science*, 145: 352-362.

Wood, J.D., Enser, M., Fisher, A.V., Nute, G.R., Sheard, P.R., Richardson, R.I., Hughes, S.I. and Whittington, F.M. (2008). Fat deposition, fatty acid composition and meat quality: A review, *Meat Science,* 78: 343-358.

Wrona, M., Nerín, C., Alfonso, M.J. and Caballero, M.Á. (2017). Antioxidant packaging with encapsulated green tea for fresh minced meat, *Innovative Food Science & Emerging Technologies*, 41: 307-313.

Wu, C., Li, L., Zhong, Q., Cai, R., Wang, P., Xu, X., Zhou, G., Han, M., Liu, Q., Hu, T. and Yin, T. (2019). Myofibrillar protein-curcumin nanocomplexes prepared at different ionic strengths to improve oxidative stability of marinated chicken meat products, *LWT – Food Science and Technology*, 99: 69-76.

Zhang, W., Xiao, S., Samaraweera, H., Lee, E.J. and Ahn, D.U. (2010). Improving functional value of meat products, *Meat Science*, 86: 15-31.

Živković, D., Stajić, S. and Stanišić, N. (2013). Possibilities for the production of meat and meat products with improved nutritional and functional value, pp. 188-199. *Proceedings of 10th International Symposium Modern Trends in Livestock Production*, Belgrade, Serbia.

Encapsulation of Food Supplements

Sudhanshu S. Behera[1,3] and Ramesh C. Ray[2]*

[1] Department of Fisheries and Animal Resource Development,
   Directorate of Fisheries, Government of Odisha, India
[2] Centre for Food Biology & Environment Studies, 1071/17, Jagamohan Nagar,
   Khandagiri PO, Bhubaneswar – 751030, Odisha, India
[3] National Institute of Technology, Raipur, Chhattisgarh, India

1. Introduction

Encapsulation is a technique for creating an external membrane or coating of one material over another one, applied for the protection and/or preservation of bioactive, volatile and easily degradable compounds from biochemical and thermal deterioration process (Saifullah *et al.*, 2019). Encapsulation of food supplements was first developed around 60 years ago as a technology for protecting the contents from the external environment. Encapsulation stabilizes the food supplements/ingredients, enhances their viability during processing and long-term storage of functional foods (Celli *et al.*, 2015) as well as preserves their metabolic activity before and after consumption. These food supplements/ingredients (bioactive components) include vitamins, lipids, peptides, fatty acids, antioxidants, minerals and living cells such as probiotics (Behera and Panda, 2020). Encapsulation improves and stabilizes the sensory properties of foods and helps in the homogenous distribution of food supplements/ingredients in the product (Aryee and Boye, 2015). Although different methods (spray-drying, emulsion, extrusion, liposomal, and lyophilization) and materials (polysaccharides, protein and lipids) are available, their selection will depend on the bioactive properties and the desired characteristics of the encapsulate and the final product (Celli *et al.*, 2015). Recently the food industry has shown growing interest in encapsulated functional foods due to increase in consumers' desire for a healthier lifestyle with enhanced nutritional and therapeutic values (Moumita *et al.*, 2018).

The present chapter provides an overview of the current research on classification of food supplements, used carrier materials, and market/market trends and methods/ techniques involved in the encapsulation of ingredients in functional foods/food products. The importance of encapsulation, as an efficient method of increasing complex properties, such as delayed release, quality, stability, thermal protection,

---

*Corresponding author: rc.ray6@gmail.com

viability and suitable sensorial profile as well as reviewing the newest achievements has been discussed.

2. Differences between Food Ingredients and Food Supplements

A food ingredient (functional food components, nutrients and bioactives) is any substance that is added to a food to achieve a desired effect. The term 'food ingredient' includes food additives, which are substances added to foods for specific technical and/or functional purposes during processing, storage and packaging (Cody *et al.*, 2012). The food supplements are also called dietary or nutritional supplements. It can be vitamins, minerals, amino acids, fatty acids and other substances delivered in the form of pills, tablets, capsules, liquid, etc. (www.foodingredientfacts.org).

3. Classification of Food Supplements and Food Supplements Encapsulation

The main advantage of encapsulation of food supplements in the food industries is high productivity of food additives/supplements and efficacy (Fig. 1).

3.1 Food Additives

There are six categories of food additives: preservatives, nutritional additives, flavoring agents, coloring agents, texturizing agents and miscellaneous agents (Branen and Haggerty, 1999).

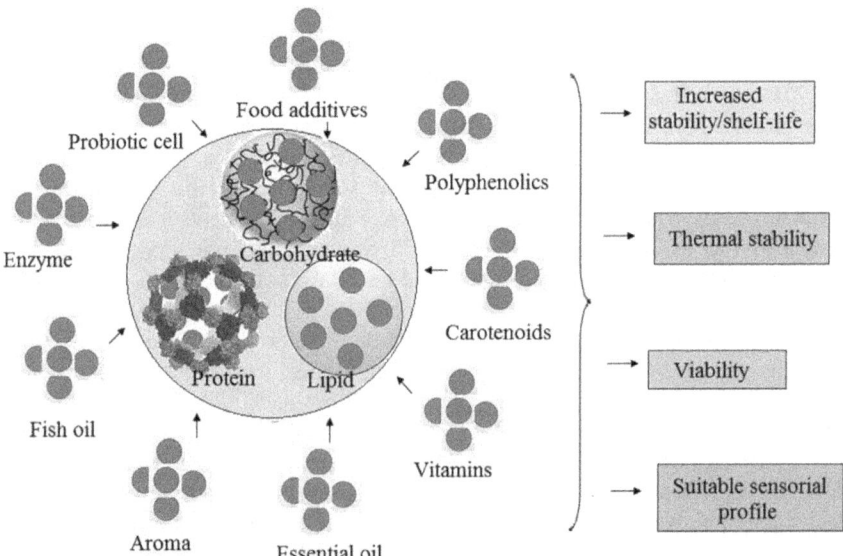

Fig. 1: Encapsulation of bioactive components with carrier biopolymers: A proposed model

3.1.1 Preservatives

Preservatives are a large group of food additives with diversified functionalities. There are basically three types of preservatives used in foods: antimicrobial, antioxidants and antibrowning agents. These additives are considered as preservatives by the European Scientific Committee for Foods (SCF) (Raju and Bawa, 2006; Granata *et al.*, 2018). Antimicrobials (benzoates, sorbates, propionates, parabens, apart from acidulants and sulfites) play an important role in extending the shelf-life of food (Branen and Haggerty, 1999; Raju and Bawa, 2006). Antioxidants (vitamin C, vitamin E, butylated hydroxyanisole (BHA), butylated hydroxytoluene (BHT)) are used to prevent autoxidation/oxidation of lipid or/and vitamin in food products (Branen and Haggerty, 1999; Granata *et al.*, 2018). The antibrowning agents (e.g. sulfites) are chemicals used to prevent both enzymatic and non-enzymatic browning and suppression of microbial growth and bleaching in food products, especially dried fruits or vegetables (Branen and Haggerty, 1999; Gonçalves *et al.*, 2016).

3.1.2 Nutritional Additives

The nutritional additives include vitamins, minerals and fibers that are commonly added to food supplements (Hundal *et al.*, 2019). Vitamins are commonly added to cereal products to restore nutritional loss and/or to enhance nutritive value of foods. For instance, addition of vitamin D in milk and B vitamins to bread has been associated with the prevention of nutritional deficiencies in the United States. Minerals, such as iron and iodine, have been used in preventing nutritional deficiencies in cereal products (Capozzi *et al.*, 2012). Fiber additives (e.g. cellulose, pectin and starch derivatives) have seen increased popularity in recent years. The fiber additives are not well defined, although they do have indirect nutritional value. However, fiber additives also provide improved texture in food products and are categorized as bulking agents, thickeners and/or stabilizers (Branen and Haggerty, 1999; Purhagen *et al.*, 2012).

3.1.3 Coloring Agents

Coloring agents (natural or synthetic substances) are commonly used as additives in food industry. Most of the coloring agents are used to improve the overall characteristics of the food. It also plays an important role in the sensory perception of the products (Ramesh and Muthuraman, 2018).

3.1.4 Flavoring Agents

Flavoring agents originate from plant or animal or synthetic sources and are used in greater number in foods. There are three major types of flavoring additives: sweeteners, natural and synthetic flavors and flavor enhancers. The most commonly used sweeteners are sucrose, glucose, fructose and lactose, with sucrose being most popular (Ramesh and Muthuraman, 2018). Artificial flavors are used to imitate the natural flavor of specific fruits, spices, cereals, beverages/juice, cookies, candies, etc. However, most artificial flavors are banned in the European Union (Ulloa, 2018).

3.1.5 Texturizing Agents

Texturizing agents, such as thickening, gelling, emulsifying, stabilizing agents are used in greater quantity in foods. These agents are used to add to or modify the overall texture/mouthfeel of food products. The emulsifiers and stabilizers are primary additives in this category (Karimi *et al.*, 2015). Emulsifiers include natural substances, such as lecithin and mono- and di- glycerides and several synthetic derivatives. The primary role of these agents is to allow flavors and oils to be dispersed throughout food products. Stabilizers include several natural gums (e.g. carrageenan) as well as natural and modified starches. They also are used to prevent evaporation and deterioration of volatile flavored oils (Branen and Haggerty, 1999; Karimi *et al.*, 2015).

3.1.6 Miscellaneous Agents

Several miscellaneous agents, such as acidulants (e.g. acetic acid, citric acid, malic acid, phosphoric acid, fumaric acid, and tartaric acid) have been used in many fruit-based products, representing more than 60 per cent of all food acids used. They act to reduce the pH, minimizing microbial growth and often enhancing the effect of weak acid preservatives (Raju and Bawa, 2006; Reddy *et al.*, 2016).

3.2   Nutraceuticals (Bioactive Polyphenols)

Phenolics (e.g. anthocyanins, phenolic acids, flavonoids, curcumin, catechins, quercetin and resveratrol) are an important group of bioactive compounds that are commonly used for the production of functional foods (Assadpour and Jafari, 2019). They can be found in different plant sources (e.g. cereal, fruits, vegetables and legumes) that are used as colorants in food products. Phenolics (natural colorants) can provide many health-promoting properties, including antioxidant, antimicrobial, anti-inflammatory, anticancer and antiallergic actions. However, they have low bioavailability and are high sensitive/chemically unstable to food-processing, storage conditions (temperature, oxygen and light) or in gastrointestinal tracts (pH, enzymes and nutrients) (Jia *et al.*, 2016). To improve the bioavailability/stability (at neutral pH and exposed to oxygen) of phenolic compounds, encapsulation seems to be an ideal process (Jia *et al.*, 2016; Assadpour and Jafari, 2019).

3.3   Carotenoids

Carotenoids (β-carotene, lycopene, astaxanthin, lutein and zeaxanthin) are another important group of natural pigments, used in the food industry (Liu *et al.*, 2018). They are sensitive to heat, oxidation and light due to their unsaturated chemical structures. Consequently, it is very important to prevent the degradation of carotenoids (oxidation or isomerisation), which decrease the nutritional properties, coloration and quality of final products. Since carotenoids are very expensive, some cost-effective techniques of encapsulation are desirable (Nedović *et al.*, 2013). Coronel-Aguilera and San Martín-González (2015) reported the encapsulation of β-carotene by spray-drying and fluidized bed coating with a solution of hydroxypropyl cellulose. The degradation of encapsulated β-carotene was shown to be prevented under spray-drying conditions (70°C and 7g/min) (Coronel-Aguilera and San Martín-González, 2015).

3.4   Vitamins

Vitamins occupy an important place in nutrition and participate in a variety of biological processes and provide health benefits to consumers (Gonnet *et al.*, 2010). The water-soluble and liposoluble vitamins (A, D, E and K) are mainly provided by foods. However, these molecules (vitamins) are sensitive to oxidation, thus encapsulation of vitamins is an appropriate means to preserve their properties during storage and enhance their physiological potencies (Gonnet *et al.*, 2010). Gonçalves *et al.* (2016) reported that microencapsulation may promote the stabilization of vitamin A.

3.5 Essential Oil

Essential oils (terpenes, phenols, alcohols, aldehydes, esters, and ketones) are slightly soluble in water and have a wide spectrum of biological activities, including antimicrobial activity against a wide range of food-borne pathogens (Nedović *et al.*, 2013). Oregano essential oil was incorporated in the inulin (15 per cent w/w) solutions using spray-drying conditions (120-190°C) and the essential oil was found successfully encapsulated in the system (Beirão-da-Costa *et al.*, 2013). Turasan *et al.* (2015) reported the encapsulation of rosemary essential oil (1, 8-cineole) and optimum formulation used for coating material (WP: whey protein concentrate and MD: maltodextrin) for the storage stability of microcapsules. The WP: MD ratio of 3:1 provided highest drying and encapsulation efficiency values and retention powers of essential oil in microcapsules found safe after 40 days of storage. Wen *et al.* (2016) reported that the cinnamon essential oil was fabricated with polyvinyl alcohol/β-cyclodextrin for effective food packaging/shelf-life of strawberry.

3.6 Flavor/Aroma and Fragrance

Flavor/aroma and fragrance play an important role in consumer satisfaction and consumption of foods. Flavor/fragrance stability in different foods has been of increasing interest because of its relationship with the quality and acceptability of foods (Chakraborty, 2017). Loss of flavor/aroma-producing compounds due to volatilization causes a reduction of flavor intensity and change in the typical food quality. Encapsulation of flavors inside a protective core prevents/limits rapid loss/degradation of these volatile ingredients prior to use in foods or beverages (Saifullah *et al.*, 2019).

3.7 Fish Oil

Fish oils have many dietary benefits (Behera, 2019), but due to their strong odors and rapid deterioration, their application in food formulations is limited (Ghorbanzade *et al.*, 2017). Encapsulation can be employed to prevent off-flavors of fish oil by preventing contact between light, oxygen or metal ions with fish oil. Moomand and Lim (2014) reported that the omega-3 fish oil was encapsulated in zein fibers using an electrospinning technique. The electrospun zein fibers provided a greater oxidative stability in comparison to non-encapsulated fish oil. Ghorbanzade *et al.* (2017) reported that yogurt fortified with nano-encapsulated fish oil had a higher DHA (docosahexaenoic acid) and EPA (eicosapentaenoic acid) contents than yogurt containing free fish oil. Di Giorgio *et al.* (2019) reported that the emulsification process (encapsulation) using soy protein was found suitable in the protection of fish oil from oxidative damage.

3.8 Enzyme Encapsulates

Encapsulation may be used to immobilize enzymes in food processing applications, such as fermentation and metabolite production processes (Agyei *et al.*, 2015). Gassara-Chatti *et al.* (2013) reported entrapment/encapsulation of ligninolytic enzyme (from *Phanerochaete chrysosporium*) in polyacrylamide/pectin-based hydrogels for its thermal stability and clarification of juice. The entrapment was found to increase the thermal stability of enzymes. In addition, the encapsulated enzymes reduced the phenolic content and proved efficient for juice clarification. Zhao *et al.* (2017) reported that immobilized enzyme (entrapped in alginate gel beads) prevented the problem of beer off-flavor (diacetyl) and reduced the beer maturation time, thus exhibiting a great potential application in beer brewing industry.

3.9   Encapsulation of Probiotic Living Cells

In the recent past, there has been a rising interest in producing functional foods containing encapsulated probiotic bacteria (Burgain *et al.*, 2011). Probiotics are one of the most promising ingredients for the production of functional foods or nutraceuticals. The probiotic cells that are present in food should survive in significant numbers (10^6-10^8 CFU/g), although the numbers vary from strain to strain (Martin *et al.*, 2013). Different probiotic cells have been encapsulated in different polymers/encapsulated materials for different purposes, such as increased shelf-life (in deep freezing using liquid nitrogen) and/or controlled release in gastrointestinal fluid (Behera *et al.*, 2018). Darukaradhya *et al.* (2013) studied the encapsulation of probiotic bacteria (*Lb. acidophilus* LAFTI L10 and *Bifidobacterium lactis* LAFTI B94) in calcium-alginate hydrogel and its effects in Cheddar cheese. The encapsulated probiotic bacteria survived better (10^7 CFU/g) than free probiotic bacteria (10^5 CFU/g) in Cheddar cheese during the long ripening period.

4.   Methods/Techniques of Encapsulation

Several encapsulation techniques (spray-drying, emulsification, extrusion, etc.) are used in food industry to make production processes more efficient (de Boer *et al.*, 2019). Some of these techniques that are important in the area of food supplements' encapsulation will be discussed below.

4.1   Spray-drying

Spray-drying is the most commonly (and oldest) used encapsulation techniques in the food industry. It is based on the transformation of a fluid into dry powder by spraying the liquid in a drying gas stream (de Boer *et al.*, 2019). The spray-drying can be utilized for encapsulation of a broad range of nutraceuticals, probiotics, flavors, enzymes and peptides. It is interesting to note that the drying process is very short (a few seconds), for which heat-sensitive ingredients can be encapsulated (Assadpour and Jafari, 2019). Moumita *et al.* (2018) studied the viability of probiotic bacteria and acceptability of dry shelf-stable synbiotic foods during long-term storage at room temperature. Synbiotic microcapsules were prepared by using *Pleurotus florida* extract as prebiotics and *Enterococcus faecium* as probiotic, through lyophilization and spray-drying methods. The results revealed that the spray dried form of fortification and survivability of probiotics were found more stable than lyophilized form.

4.2   Extrusion Methods

Extrusion methods are based on the dispersing of an aqueous solution of polymer (e.g. 0.6-3 wt per cent sodium alginate) and an active ingredient into droplets into a gelling bath (e.g. 0.05-1.5 per cent calcium chloride solution) (Nedović *et al.*, 2013). The formation of droplets occur in a controlled environment as opposed to spray-drying technique (Burgain *et al.*, 2011). Kim *et al.* (2017) reported that the probiotic *Lb. acidophilus* encapsulated with an electrostatic extrusion technique (ionic gelation of phytic acid and chitosan) showed increased survival rate (in simulated gastric fluid) and also provided protection against acid injury.

4.3   Emulsion Method

Emulsion method of encapsulation is based on the relationship between the dispersed/ discontinuous phase and dispersion/continuous phases. The dispersed phase, i.e. small volume of polymer slurry/cell is added to the large volume of continuous phase containing vegetable oils (corn, soy, sunflower and light paraffin oil). There are several types of emulsion: water/oil emulsions and oil/water emulsion; an alternative to these two is a water/oil/water double emulsion. The size of microcapsules depends on the reactor design and mixer geometry and can be controlled by varying the agitation speed and water/oil ratio (Nedović *et al.*, 2013).

4.4   Liposomal Entrapment

Liposome entrapment is a process of microencapsulation that has widely been used for encapsulation of food ingredients/supplements. It involves vesicles composed of one or more phospholipids bi-layers encapsulating a volume of aqueous media/ phase (Tan *et al.*, 2014). The mechanism of formation is based on the interactions, between phospholipids and water molecules. The polar head groups of phospholipids are exposed to the aqueous phases and hydrophobic hydrocarbon tails are forced to face each other in a bi-layer (da Silva Malheiros *et al.*, 2010). It was reported (Tan *et al.*, 2014) that encapsulation of carotenoids (lutein, β-carotene, lycopene and canthaxanthin) into liposomal membrane increases their storage stability, prevented lipid peroxidation and provided better *in vitro* release behavior in simulated gastrointestinal media.

4.5   Lyophilization

Lyophilization or freeze-drying is a low temperature dehydration process. It involves freezing the product by lowering the pressure and removing water by sublimation (ice to change directly from solid to vapor without passing through a liquid phase) (Vélez *et al.*, 2019). Vélez *et al.* (2019) studied the effect of lyophilization on conjugated linoleic acid loaded liposome with the aim of potential application in dairy products. The conjugated linoleic acid loaded liposiome resisted lyophilization conditions without cryoprotectants. Rodrigues *et al.* (2018) reported that the process of lyophilization was suitable for preserving the stability of microencapsulated jaboticaba (*Myrciaria cauliflora* Berg, popular fruit in Brazil) extract.

4.6   Other Encapsulation Techniques

Several other techniques, such as fluidized bed coating, coacervation, centrifugal suspension separation, cocrystallization, inclusion complexation and thermal gelation techniques have been involved in encapsulation of food supplements/ingredients (Kavitake *et al.*, 2018).

5. Carrier Materials for Food Supplements Encapsulation

5.1 Polysaccharides

Polysaccharides are most widely used for encapsulation in food applications. The most commonly used polysaccharides are starch as well as cellulose and their derivatives. Plant exudates and extracts, such as gum arabic, pectins and the marine extracts, such as alginate, are also present in foods and can be an efficient vehicle for encapsulating biomolecules. Xanthan and gellan gum are exopolysaccharides (EPS) produced by microorganisms and have the potential for development of efficient encapsulating systems (Nedović *et al.* 2013) (Table 1).

5.1.1 Starch and Starch Derivatives

Starch is a renewable biopolymer/polysaccharide widely distributed in nature. It is mainly composed of amylose (20-30 per cent) and highly branched amylopectin (70-80 per cent). The amylose/amylopectin component of starch is made of α-D-glucose units linked by α-1→4 and α-1→6 glycosidic bonds (Li *et al.*, 2016). It has been used since ages in food applications as a thickening, binding, sweetening and emulsifying agent (Mahmood *et al.*, 2017). Starch is slowly digestible or resistant to pancreatic amylases; thus it has a prebiotic effect which is of great interest as it is known to promote growth of intestinal microflora and subsequently induces health benefits within the body (Li *et al.*, 2016). Furthermore, selection of starch with smaller granule size, white in color and blended with a flavor exhibit attractive sensory characteristics for food applications (Li *et al.*, 2016). Li *et al.* (2016) investigated the survival potential of probiotic *Lb. plantarum* 299v microencapsulated in maize starch or partially hydrolyzed maize starches after acid, bile and heat treatment. The results showed that porous maize starch granules allowed a high probiotic-loading efficiency and enhanced protection against various stressful conditions compared to free cells. The amylose content of starch affects the physicochemical and functional properties of starch.

The starch is chemically modified to increase or alter certain functional attributes (Cruz-Benítez *et al.*, 2019). Cruz-Benítez *et al.* (2019) studied the potential of types of cassava starch (normal or waxy) and the dual chemical modification (acid hydrolysis and cross-linking process) with different amylase content for encapsulation of *Lb. pentosus*. The increase in the amylose content caused the formation of a structural network that facilitates the formation of microcapsules.

5.1.2 Cellulose and Cellulose Derivatives

Cellulose, the most abundant polysaccharide on Earth, belongs to the family of dietary fibers with strong hydrophilic (bulking) ability and is good for human intestinal health. Li *et al.* (2016a) reported the inherent strong hydrophibicity of cellulose for absorbing and conducive properties to high viability of probiotic cells. Cellulose cores-shell gels showed sustainable release of *Lb. plantarum* in simulated intestine fluid (viability lasting for 360min) with better rigid supporting property.

Cellulose derivatives, including ethyl cellulose, hydroxypropyal cellulose, carboxymethyl cellulose are widely used as carrier materials (Li *et al.*, 2016). Ethyl

Table 1: Carrier Materials and Some Properties Important for Food Supplements Encapsulation

Carrier Materials	Food Supplements	Properties	References
Starch and starch derivatives			
Starch	Probiotic *Lactobacillus plantarum*	High probiotic loading efficiency	Li *et al.*, 2016
Cassava starch	*Lactobacillus pentosus*	High probiotic loading efficiency	Cruz-Benitez *et al.*, 2019
Cellulose and cellulose derivatives			
Cellulose	*Lb. plantarum*	Sustainable release	Li *et al.*, 2016a
		Better rigid supporting	
EC	Polyphenols	Storage stability	Zheng *et al.*, 2011
Alginate			
Alginate hydrogels	Proteins encapsulation	Protect from aggregation and unfolding	Zhang *et al.*, 2016
		Increase encapsulation, retention and release	
Ca (II)–alginate beads	*Thymus serphyllum* L. aqueous extract (polyphenolic compounds)	Preservation of antioxidant activity	Stojanovic *et al.*, 2012
		Prevents thermal behavior under heating conditions	
Sodium alginate	Probiotics (*Staphylococcus succinus* and *Enterococcus fecium*) with prebiotics	Increase viability in simulated gastric conditions	Sathyabama and Vijayabharathi, 2014
Ca (II)–alginate beads	Bioactives (betacyanin and polyphenols)	Stability under processing conditions	Calvo *et al.*, 2019
Alginate-pectin microbeads	*Lactococcus lactis*	Best mechanical properties, and more stable	Bekhit *et al.*, 2016

(*Contd.*)

Table 1: (*Contd.*)

Carrier Materials	Food Supplements	Properties	References
Skim milk coated inulin-sodium alginate	*Lb. plantarum*	Better survival rate	Wang *et al.*, 2016
		Improve the probiotic functionality	
Alginate-starch microcapsules	*Lb. plantarum* and *Staphylococcus xylosus*	Enhance the stability of microcapsules	Bilenler *et al.*, 2017
Xanthan gum			
Xanthan and chitosan hydrogel	*Pediococcus acidilactici*	Increase the survival rates	Argin *et al.*, 2014
Xanthan, tara gum and xanthan-tara hydrogel	Purple Brazilian cherry (*Eugenia uniflora* L.) juice	Higher efficiency in encapsulation	Rutz *et al.*, 2013
Xanthan gum/gluten complexes	β-carotene	Improve bioaccessibility	Fu *et al.*, 2019
Gellan gum			
Gellan gum and sodium caseinate gel matrix	*Lactobacillus casei*	Higher survival of encapsulated LAB cells	Nag *et al.*, 2011
Whey protein isolate, Agar and gellan gum complex coacervates	Tuna oil	High encapsulation efficiency (95.8%)	Wang *et al.*, 2019
Gellan beads	Curcumin	Protect against oxidation/degradation	Ambebila *et al.*, 2019
Carrageenans (CGs)			
CGs/locust bean gum coated milk microspheres	*Lactobacillus bulgaricus*	Enhance survival rate	Shi *et al.*, 2013
CGs/CMC	*Lb. plantarum*	Prevention against adverse conditions	Dafe *et al.*, 2017
CGs	Cystatin (from almond)	Decreases activity	Siddiqui *et al.*, 2017
CGs	Dried milk powder	Reduction of surface fat	Foerster *et al.*, 2017

Encapsulation system	Compound	Property/Function	Reference
Konjac glucomannan			
Oxygenated KGM microspheres	Anthocyanins	Potential carrier of bioactive compounds	Lu et al., 2015
KGM matrix	Pork backfat	Reduction of fat content	Lorenzo et al., 2016
KGM hydrogel	*Lactobacillus acidophilus*	Enhance survival rate	Mu et al., 2018
Chitosan (CS)			
CMCS/zein complex	Vitamin D3	Controlled release property	Luo et al., 2012
		Improve the stability of labile nutrients	
CS-alginate beads	*Lb. plantarum*	Increases survival of the cells	Nualkaekul et al., 2012
EGCG/zein/chitosan NPs	EGCG and zein	Controlled release property	Liang et al., 2017
		Antioxidant activity	
CS-alginate-Eudragit (Eu) S100 NPs	*Lb. acidophilus* and *Lactobacillus rhamnosus*	Better viability	Ansari et al., 2017
CS-coated liposomes	SPH	Reduce storage loss	Li et al., 2015
Gelatin (GE)			
Gelatin	Curcumin	Food preservative	Gómez-Estaca et al., 2017
Gelatin-maltodextrin, and gum arabic mixture	Saffron	Retention of active component	Rajabi et al., 2015
Gelatin	EGCG	Retain antioxidant activity	Gómez-Mascaraque et al., 2015
Gelatin and gum arabic	Anthocyanins	High loading capacity	Shaddel et al., 2018
		More color retention capacity	

(Contd.)

Table 1: (*Contd.*)

Carrier Materials	Food Supplements	Properties	References
Gelatin-maltodextrin	Betalains	Higher yield	Castro-Muñoz *et al.*, 2015
		Maximum colorimetry values	
Milk protein			
Milk protein-PE and limonene	Fish oil	Prevents oxidation	Chen *et al.*, 2013
WPI	*Lb. plantarum* A17	Minimizing denaturation	Khem *et al.*, 2016

Table abbreviations: CGs: Carrageenans; WPI: Whey protein isolate; AG: Agar gum; GG: Gellan gum, CMC: Carboxymethyl cellulose; KGM: Konjac glucomannan matrix; CMCS: Carboxyethyl chitosan; EGCG: Epigallocatechin gallate; NPs: Nanoparticles; SPH: Salmon protein hydrolysates (SPH); CS: Chitosan; GA: gum arabic; GE: Gelatin; PE: phytosterol esters; WPI: Whey protein isolates; EC: Ethyl cellulose.

cellulose (EC) is a good choice of wall material due to its good chemical stability, film forming, biodegradability, etc. (Nedović *et al.*, 2013).

Polyphenols display highly antioxidative capability and prove to be major health-beneficial components in fruits and vegetables. Bayberry is an important dietary source of polyphenols (gallic acid, flavones and anthocyanins) among the bioactive compounds. Zheng *et al.* (2011) studied encapsulation of the polyphenolic extracts from bayberry by a phase separation method, using ethyl cellulose as a coating material. The microcapsules obtained were further characterized on the basis of antioxidant activity, polyphenols content, *in vitro* study and storage stability. The results revealed that the antioxidant activity of bayberry polyphenols was found to be effectively protected by microencapsulation. The storage stability of polyphenols remarkably improved by microencapsulation. Comparatively, the release rate of bayberry polyphenols increased significantly (87.37 per cent) under stimulated fluid (pH 8, 24h).

5.1.3 Alginate

Alginate is a natural polysaccharide/bioploymer derived from various species of algae (*kelp*, brown algae) and is composed of β-D-mannuronic and α-L-guluronic acids (Burgain *et al.*, 2011). Alginate hydrogels/beads are extensively used in cell encapsulation and calcium alginate (gelation occurs by cross-linking of Ca^{2+} ions) is preferred for encapsulating food supplements (Burgain *et al.*, 2011). Several studies have shown that the functional properties and biological activity of proteins can be enhanced by encapsulating them within hydrogels. Also, proteins encapsulated in polymer matrices may be protected from aggregation and unfolding. Moreover, these protein-loaded polymer particles can be designed to carry bioactive proteins to specific locations within the gastrointestinal tract (GIT) and release them at a controlled rate (Zhang *et al.*, 2016). Zhang *et al.* (2016) studied the encapsulation of whey protein in calcium-alginate beads. Protein-loaded beads were fabricated at three different pH values (pH 3, 5 and 7) to study the influence of protein-alginate electrostatic interactions on protein encapsulation, retention and release. The protein encapsulation and retention was highest at low pH, while protein release was highest at high pH.

Several studies have reported the use of hydrogel material for encapsulation of plant polyphenols to improve their functionality and stability in food products. Stojanovic *et al.* (2012) studied the encapsulation (using electrostatic extrusion) of *Thymus serphyllum* L. aqueous extract (containing polyphenolic compounds) within calcium alginate beads. The encapsulated *T. serphyllum* extract in calcium alginate can be used for preservation of extract's antioxidant activity and optimal thermal behavior under heating conditions, simulating the usual food processing. In recent years microencapsulation has also found to be a useful tool for stabilization of probiotic cells in functional food applications. It can enhance the viability of probiotic cells during processing, storage, consumption and during passage through the gastro-intestinal tract (Sathyabama and Vijayabharathi, 2014). Sathyabama and Vijayabharathi (2014) studied the co-encapsulation of two probiotics (*Staphylococcus succinus and Enterococcus fecium*) with prebiotics (*in vitro* fermentation, viz. sugar beet and chicory) and further encapsulated within sodium alginate (2 g/100 mL). The encapsulated cells were tested further for its ability to tolerate pH and bile and survival under gastrointestinal stress conditions. Also, they compared the storage viability of encapsulated cells against free cells. The results indicated an improvement in survival of co-encapsulated cells when exposed to acidic (pH 2.0-3.0) and bile (0.3, 0.6 and 0.8 g /100 mL) conditions. The encapsulated cells showed 98 per cent viability in simulated gastric environment. Viability was maintained throughout the storage period and found the range of 7.9-8.1 log CFU/mL for about a period of 30 days at 4°C.

The bioactive compounds from food waste can be extracted and used for the development of nutraceuticals and functional food additives. Calvo *et al.* (2019) studied the valuable bioactives, such as betacyanin and polyphenols of beetroot leaf and stem (a food waste), which were extracted and encapsulated in Ca (II)-alginate beads, using sucrose, gum arabic, guar gums and pectins as excipients. In addition, the guar gum can lead to Ca (II)-alginate beads with improved properties which constitute a promising enhancement in terms of stability under processing conditions for ingredients used in functional foods.

The process of microencapsulation can protect the cells for a longer period of time and also release the cells into the *in vivo* intestinal system. The encapsulation of probiotic cells (*Streptococcus phocae*) into alginate-chitosan gel capsules showed better survival (5.468 log cfu/mL) and remained stable when treated with simulated gastric and intestinal fluids (Kanmani *et al.*, 2011).

Bekhit *et al.* (2016) studied the composite hydrogel microspheres (alginate, pectin) containing *Lactococcus lactis*. The resultant microbeads (alginate/pectin: 75/25) were found to have best mechanical properties, were more stable and allowed the best release of nisin during the storage period. Moreover, the encapsulated alginate/pectin beads increased the viability of lactic acid bacteria (LAB) strains. The encapsulated *Lb. plantarum* in skim milk-coated inulin-sodium alginate (ISA) beads showed better survival rate as compared to free cells. Moreover, skim milk-coated ISA beads improved the probiotic functionality of *Lb. plantarum* and reduced the adverse action of freeze-drying conditions (Wang *et al.*, 2016). The effects of encapsulated (alginate-starch mixture) starter cultures (*Lb. plantarum* and *Staphylococcus xylosus*) on microbial and physicochemical properties of heat-treated sausages were investigated (Bilenler *et al.*, 2017). The results revealed that alginate-starch-based encapsulation technique enhanced the stability of microcapsules containing starter cultures in adverse conditions (70°C) (Bilenler *et al.*, 2017).

5.1.4 Xanthan Gum and Gellan Gum

Xanthan gum is an exopolysaccharide produced by microorganisms of the species *Xanthomonas*. Structurally, this gum is a heteropolysaccharide composed by units of β-D-glucose linked by link 1-4, containing side chains of β-D-mannose, 1,4-β-D-glucuronic acid and α-D-mannose residue (Rutz *et al.*, 2013). Xanthan gum and chitosan are two natural polymers that are capable of forming physically cross-linked hydrogels with reversible ionic linkages. Argin *et al.* (2014) studied the swelling and the release behaviors of xanthan and chitosan hydrogel system under the simulated gastrointestinal-tract conditions (pH 2.0, 37°C) in order to assess its potential as an enteric delivery system for the probiotic bacteria (*Pediococcus acidilactici*). The cell viability study showed that encapsulation increases the survival rates of cells against freeze-drying and in simulated gastric fluid.

Rutz *et al.* (2013) investigated the microencapsulation of the purple Brazilian cherry (*Eugenia uniflora* L.) juice by lyophilization as a drying method, using wall material, such as xanthan, tara gum and xanthan-tara hydrogel. The results revealed that the higher efficiency in encapsulation occurred with xanthan gum. Moreover, the hydrogel was most adequate for the release of carotenoids in gastric fluid simulation and intestinal fluid simulation with gradual release of the compound in water. Fu *et al.* (2019) studied the encapsulation of β-carotene stabilized with the biopolymer particles formed by wheat gluten/xanthan gum complexes. The resultant complex had good pH, salt, bioaccessibility (*in vitro*) and improved chemical stability during storage.

Gellan gum is an extracellular polysaccharide produced by bacterium *Pseudomonas elodea*. It contains a repeating unit composed of α-D-glucose, L-rhamnose and D-glucuronic acid (Prajapati *et al.*, 2013). *Lb. casei* cells were successfully entrapped into the gellan gum and sodium caseinate gel matrix by water-in-oil emulsion (Nag *et al.*, 2011). The survival of encapsulated LAB cells after 30min of incubation in simulated gastric fluid was claimed to be greater than that of free cells. The capsules also provided greater protection against bile salts (Nag *et al.*, 2011). Wang *et al.* (2019) used whey protein isolate, agar gum and gellan gum to create double-shell and multi-core microcapsules of tuna oil. The microcapsules exhibited low surface oil content (1.8 per cent), high encapsulation efficiency (95.8 per cent) and significantly enhanced oxidative stability. This research has proved that gellan beads significantly retard fast release of encapsulated material at pH 1.3 and achieve sustained release at pH 6.8. Ambebila *et al.* (2019) reported external gelation of curcumin-in-oil/gellan emulsions in order to entrap curcumin in gellan beads. An increased encapsulation yield (98 per cent) was claimed at higher gellan concentration (1 per cent w/v) due to the formation of more compact gel. The encapsulated curcumin offered protection against oxidation/degradation compared to free curcumin-in-oil solutions.

5.1.5 Carrageenans (CGs)

Carrageenans (CGs) are natural and hydrophilic polysaccharides extracted from numerous species of red seaweeds (Rhodophyta). Chemically, CGs are highly sulfated galactans, composed of D-galactose and 3, 6-anhydro-D-galactose repeating units (Chakraborty, 2017). CGs are primarily used in food industry as gelling, thickening and stabilizing agents (Siddiqui *et al.*, 2017). The encapsulation of probiotic cells in carrageenans has been performed by using extrusion and emulsion methods. Shi *et al.* (2013) encapsulated *Lb. bulgaricus* in CGs-locust bean gum-coated milk microspheres. The encapsulated *Lb. bulgaricus* showed better survival rate (1.5 log CFU/g for 2h) and stability in simulated gastro-intestinal conditions (2 g/100 mL bile salt solution; pH 2.5) compared with free cells. Dafe *et al.* (2017) reported that the carboxymethyl cellulose (CMC) and CGs blends could successfully protect *Lb. plantarum* cells against adverse conditions of the gastro-intestinal tract and bile salt solution. The cystatins are important regulatory proteins of plant and animal system and act as cysteine protease inhibitors. Siddiqui *et al.* (2017) studied the effect of CGs (a food additive) on cystatin purified from almond. The results revealed that CGs unfold cystatin, leading to decrease in its activity due to its inability to bind to proteases. Foerster *et al.* (2017) studied the surface fat formation on spray-dried milk powder through emulsion stabilization with CGs (0.3 per cent w/w). CGs improved the stability of fat globules and reduced the fat globule size.

5.1.6 Konjac Glucomannan (KGM)

Konjac glucomannan (KGM) is a natural water-soluble food-grade polysaccharide stored in tubers of *Amorphophallus konjac*. It is linked by β-D-glucose and β-D-mannose through β-1,4- or β-1,3-glycosidic linkages. It is widely accepted that KGM confers high viscosity, solubility and swelling in aqueous solutions, as well as specific biological functions (Behera and Ray, 2016).

Anthocyanins are sugar conjugates of flavonoids belonging to the class of phenolic compounds (Panda *et al.*, 2015) and represent essential group of nutraceuticals widely applied in the fortification of functional foods (Lu *et al.*, 2015; Panda *et al.*, 2015). Lu *et al.* (2015) investigated the absorption and release of anthocyanins by Fe^{3+} cross-linked oxygenated KGM microspheres. The results indicated that the oxygenated KGM microspheres can be potentially used as the carriers for targeted delivery of bioactive compounds in the intestine. The maximum anthocyanins absorption capacity was achieved at pH 3 and low salt concentration. They remained stable at pH 3 and were released from oxygenated KGM microspheres at pH above 5.

Commercial sausages are produced with high fat content (30 per cent). However, from a health point of view, excessive fat intake is not recommended. Lorenzo *et al.* (2016) reported the reduction of fat content (30 per cent) of Spanish *salchichon* when the pork backfat was replaced by konjac glucomannan matrix. The incorporation of microencapsulated fish oil in konjac gel increased the EPA and DHA contents and decreased the n-6/n-3 polyunsaturated fatty acids (PUFA) ratio as compared to the control samples. Mu *et al.* (2018) studied the encapsulation of *Lb. acidophilus* using KGM hydrogel. This probiotic was encapsulated in the hydrogel with a microencapsulation rate of 62.5 per cent.

5.1.7 Chitosan

Chitosan (CS) is a linear natural positive-charge (cationic) polysaccharide, which is obtained from chitin. It has been considered to be a versatile polymer for encapsulation and delivery of active ingredients (Luo *et al.*, 2012; Ansari *et al.*, 2017). Vitamin D is an essential nutrient for human health. It is one of the fat-soluble vitamins and consists of two major physiological forms, Vitamin D_2 (ergocalciferol) and vitamin D_3 (cholecalciferol) (Luo *et al.*, 2012). Luo *et al.* (2012) studied the zein (maize protein) nanoparticles coated with carboxyethyl chitosan (CMCS) to encapsulate vitamin D_3. The encapsulation of hydrophobic nutrients (vitamin D_3) in zein/CMCS complex nanoparticles would achieve the controlled release property and improve the stability of this labile nutrient.

The epigallocatechin gallate (EGCG) has been identified as a major tea polyphenol. In the presence of relatively high temperature, elevated oxygen concentrations and pH, EGCG is oxidized easily. Therefore, several approaches, including encapsulation, have emerged for its effective protection against degradation (Liang *et al.* 2017). Liang *et al.* (2017) studied the encapsulation of EGCG in zein/chitosan nanoparticles for controlled applications in the food system. Zein coating was found to greatly improve the controlled release property and antioxidant activity of encapsulated EGCG.

Chitosan has been used as a coating material for alginate beads and has been shown to increase the survival of probiotics in simulated gastric and intestinal juices. Nualkaekul *et al.* (2012) reported the effect of chitosan coated alginate beads in the survival of microencapsulated *Lb. plantarum* in simulated gastric solution. The survival of the cells in simulated gastric solution (pH 1.5) was improved in the case of the chitosan coated beads by 0.5-2 logs as compared to the uncoated beads.

The encapsulation of *Lb. casei* into calcium pectinate beads coated with chitosan provided capsules capable of delivering live probiotic bacteria into the intestine.

Bepeyeva *et al.* (2017) reported that the pectin-chitosan capsules provided protection to *Lb. casei* from the gastric acid (pH 2.0) and resulted in high levels of viable bacteria (9.6 log CFU/mL) released in the intestine. Eudragit (Eu) S100 is an anionic copolymer derived from metacrylic acid and methyl metacrylate (1:2). The microencapsulation of chitosan with calcium alginate-Eudragit (Eu) S100 nanoparticles is an efficient method for better viability of probiotic bacteria, *Lb. acidophilus* and *Lb. rhamnosus* (Ansari *et al.*, 2017).

The krill (*Euphausia superba*) oil is a rich source of eicosapentanoic acid (EPA) and docosahexanoic acid (DHA). Haider *et al.* (2017) reported encapsulation of krill oil in chitosan-tripolyphosphate nanoparticles (NPs). Chitosan NPs were successful in preventing the oxidation of krill oil. The results confirmed the suitability of the emulsion and electrostatic interaction-based method for the formation of oil-loaded chitosan NPs with greater encapsulation efficiency and loading capacity that will enhance their usage in food and pharmaceutical industry.

Atlantic salmon protein hydrolysates (SPH) (molecular weight < 1000 Da) have been shown to exhibit antidiabetic and antioxidative properties (Li *et al.*, 2015). Li *et al.* (2015) investigated encapsulation of SPH in chitosan-coated liposomes prepared from milk-fat globule membrane-derived phospholipids. The results revealed that the CH-coated liposomes showed reduced SPH loss during long-term storage at both 4 and 20°C.

5.2   Proteins

Proteins are also appropriate materials for encapsulation. Gelatin and whey proteins (milk) are some of the most commonly used proteins as wall materials.

5.2.1 Gelatin

Gelatin is obtained from partial hydrolysis of collagen which contains repeating sequences of glycine-proline- and hydroxyproline. It has been widely used for enhancing elasticity, stability and consistency of food products. Gelatin is commonly used as a food ingredient because it possesses unique gelation properties and is commercially available at a low cost (Gómez-Mascaraque *et al.*, 2015).

Curcumin is a polyphenol with a wide range of biological properties, including antioxidant and antimicrobial activity. However, its use as a food preservative is limited because of its insolubility in water (Gómez-Estaca *et al.*, 2017). The encapsulation of curcumin in gelatin greatly improved its antioxidant and antimicrobial properties, thus broadening its potential use as a food preservative (Gómez-Estaca *et al.*, 2017).

Saffron is a very sensitive spice and loses its active compounds during exposure to unfavorable environmental conditions. The encapsulation of saffron extract by various biopolymers (gelatin, maltodextrin and gum arabic) increased the retention of its active components as compared to free saffron (Rajabi *et al.*, 2015).

Green tea polyphenols are powerful antioxidants which have attracted great interest in the field of functional foods due to their numerous attributed health benefits. Epigallocatechin gallate (EGCG) is the most abundant and biologically active compound in green tea (Gómez-Mascaraque *et al.*, 2015). Gómez-Mascaraque *et al.* (2015) studied the electrosprayed gelatin as an edible carrier in the encapsulation of polyphenols (EGCG). The encapsulated EGCG retained its whole antioxidant activity compared to its free form.

Black raspberry (*Rubus occidentalis* L.) is a wild plant (grown mostly in the north of Iran) and it is a rich source of anthocyanins. It is an excellent source for producing red color for many foods (Shaddel *et al.*, 2018). Shaddel *et al.* (2018) studied the microencapsulation of potent natural food colorant (black raspberry anthocyanins) using wall materials, such as gelatin and gum arabic. It revealed that the encapsulation of anthocyanins improved their thermal and storage stability. The combination of gelatin, gum arabic and anthocyanin (1:1:1) was found to be most efficient for high color loading and retention during storage time.

Cactus pear (*Opuntia* app.) is a native fruit from America that grows in arid and semi-arid regions. Castro-Muñoz *et al.* (2015) reported spray-dried microcapsules of bioactive compounds (betalains) from purple cactus pear using a gelatin-maltodextrin composite. The encapsulation yield and the properties of the microcapsules were found to depend on the temperature and the gelatin-to-maltodextrin ratio. The best ratio (gelatin-to-maltodextrin) was found to be 2.5:7.5 with higher yield, good antioxidant activity and maximum colorimetric values related to the betalains content.

5.2.2 Milk Proteins

Milk proteins (whey protein isolate and sodium caseinate) are biopolymers/natural vehicles for bioactive substances and owing to their structural and physico-chemical properties, they are well adapted to several encapsulation purposes (Tavares *et al.*, 2014).

Fish oil is known to be a rich source of omega-3 long-chain polyunsaturated fatty acids, containing both DHA and EPA. Chen *et al.* (2013) studied the co-encapsulated microcapsules containing three lipophilic bioactive components, including fish oil, phytosterol esters (PE) and limonene using milk protein. The results revealed that the co-encapsulation of fish oil with PE effectively prevented polyunsaturated fatty acids (DHA and EPA) from oxidation.

Whey protein isolates (WPI) and/or in combination with carbohydrates has been reported to provide different degrees of protection of probiotic bacteria during spray-drying (Khem *et al.*, 2016a). Khem *et al.* (2016) reported that in comparison to other proteins and carbohydrates, WPI in its native state provided protection to probiotic, *Lb. plantarum* A17. This study showed that the optimum performance of encapsulated *Lb. plantarum* A17 cells is achieved by minimizing the degree of denaturation of WPI prior to spray-drying.

5.3 Lipids

Lipid materials, such as lecithin, fatty acids, fatty alcohols, waxes, glycerides and phospholipids have been used for food encapsulations. The important advantages of lipids as carriers in food encapsulation processes include ease of handling, stability and safety when used in food supplements/ingredients encapsulations (Nedović *et al.* 2013). Komaiko *et al.* (2016) suggested that sunflower phospholipids may be promising natural emulsifiers to deliver omega-3-fatty acids into food and beverages. Bryła *et al.* (2015) reported that lecithins (egg yolk, soybeans and sunflower) were used for the encapsulation of elderberry extract (into liposomes) for purposes of food enrichment.

6. Food Encapsulated Supplement Market and Market Trend

The global food encapsulation market size is projected to reach USD 14.1 billion by 2025, from USD 9.9 billion in 2020, recording a CAGR of 7.5 per cent during the forecast period. This market is mostly driven by the rising demand for encapsulated flavors from food manufacturers and due to increasing demand for fortified food products (http://www. marketsandmarkets.com/Market-Reports). Vitamins are functional ingredients that are used in food owing to their specific nutritional properties. The targeted and controlled release of vitamins often becomes important when added as a food ingredient. Thus, encapsulation is mostly adopted for vitamins for its targeted effect. Also, flavor manufacturers have been adopting this technology at a rapid speed, and hence, the vitamins segment was closely followed by flavors in terms of dominance in the global market (Food Encapsulation Market, Report code FB2207, March, 2020).

7. Bottlenecks

Despite many advantages of encapsulation technology in food industries, there are a few disadvantages also. The bottlenecks are discussed, in brief.

- The presence of foreign ingredients in the process of food encapsulation resulted in allergies/chronic diseases (stress and insecurity) in consumers. It can be managed through avoidance of problematic ingredients in the diet (Feehley *et al.*, 2019).
- The possible cross-reaction may occur between core and wall materials.
- Shelf-life of encapsulated food may be reduced due to the presence of hydrogroscopic food ingredients/supplements.
- The food encapsulation processing and production costs are high.
- More skilled workers and advance knowledge may be required to use this advanced and complex technology (Ozkan *et al.*, 2019).

8. Conclusion and Future Perspectives

The encapsulation technique provides an effective method to cover an array of active compounds with a protective wall material and thus, offer numerous advantages. It is also a potential method of blending bioactive components with the carrier biopolymers (polysaccharides, proteins and lipids) to obtain synergistic effects on the functional food products. To our knowledge, we feel the lack of sufficient research in the applications of such functional/bioactive ingredients in foods and the scientific validation of their technological and biological feasibility. Thus, future research should focus on overcoming the gap between conditions at research level and demands for large-scale applications, improving existing manufacturing technologies, choosing new processing conditions and new carrier materials that could be easily adopted at the industrial level.

References

Agyei, D., Shanbhag, B.K. and He, L. (2015). Enzyme engineering (immobilization) for food applications, pp. 213-235. *In:* Y. Yada and Rickey (Eds.), *Improving and Tailoring Enzymes for Food Quality and Functionality*, Woodhead Publishing, Cambridge.

Ambebila, E.N., Santamaría, E., Maestro, A., Gutiérrez, J.M. and González, C. (2019). Gellan hydrogels: Preparation, rheological characterization and application in encapsulation of curcumin, *Food Biophysics*, 14: 154-163.

Ansari, F., Pourjafar, H., Jodat, V., Sahebi, J. and Ataei, A. (2017). Effect of Eudragit S100 nanoparticles and alginate chitosan encapsulation on the viability of *Lactobacillus acidophilus* and *Lactobacillus rhamnosus*, *AMB Express*, 7: 144.

Argin, S., Kofinas, P. and Lo, Y.M. (2014). The cell release kinetics and the swelling behavior of physically cross-linked xanthan–chitosan hydrogels in simulated gastrointestinal conditions, *Food Hydrocolloids*, 40: 138-144.

Aryee, A.N. and Boye, J.I. (2015). Current and emerging trends in the formulation and manufacture of nutraceuticals and functional food products, pp. 1-52. *In:* J.I. Boye (Ed.), *Nutraceutical and Functional Food Processing Technology*, Wiley, London.

Assadpour, E. and Jafari, S.M. (2019). Advances in spray-drying encapsulation of food bioactive ingredients: From microcapsules to nanocapsules, *Annual Review of Food Science and Technology*, 10: 1-29.

Behera, S.S. and Ray, R.C. (2016). Konjac glucomannan, a promising polysaccharide of *Amorphophallus konjac* K. Koch in health care, *International Journal of Biological Macromolecules*, 92: 942-956.

Behera, S.S., Ray, R.C. and Zdolec, N. (2018). *Lactobacillus plantarum* with functional properties: An approach to increase safety and shelf-life of fermented foods, *BioMed Research International 2018*. https://doi.org/10.1155/2018/9361614

Behera, S.S. (2019). Dietary fish oil concentrates associated health benefits: A recent development of cardiovascular risk reduction, *Current Pharmaceutical Design*, 25: 4053-4062.

Behera, S.S. and Panda, S.K. (2020). Ethnic and industrial probiotic foods and beverages: Efficacy and acceptance, *Current Opinion in Food Science*, 32: 29-36.

Beirão-da-Costa, S., Duarte, C., Bourbon, A.I., Pinheiro, A.C., Januário, M.I.N., Vicente, A.A., Beirão-da-Costa, M.L. and Delgadillo, I. (2013). Inulin potential for encapsulation and controlled delivery of Oregano essential oil, *Food Hydrocolloids*, 33: 199-206.

Bekhit, M., Sánchez-González, L., Messaoud, G.B. and Desobry, S. (2016). Encapsulation of *Lactococcus lactis* subsp. *lactis* on alginate/pectin composite microbeads: Effect of matrix composition on bacterial survival and nisin release, *Journal of Food Engineering*, 180: 1-9.

Bepeyeva, A., de Barros, J.M., Albadran, H., Kakimov, A.K., Kakimova, Z.K., Charalampopoulos, D. and Khutoryanskiy, V.V. (2017). Encapsulation of *Lactobacillus casei* into calcium pectinate-chitosan beads for enteric delivery, *Journal of Food Science*, 82: 2954-2959.

Bilenler, T., Karabulut, I. and Candogan, K. (2017). Effects of encapsulated starter cultures on microbial and physicochemical properties of traditionally produced and heat treated sausages (sucuks), *LWT – Food Science and Technology*, 75: 425-433.

Branen, A.L. and Haggerty, R.J. (1999). Introduction to food additive, pp. 1-10. *In:* A. Larry Branen, P. Michael Davidson, Seppo Salminen and John H. Thorngate III (Eds.), *Food Additives*, Marcel Dekker, Inc, New York.

Bryła, A., Lewandowicz, G. and Juzwa, W. (2015). Encapsulation of elderberry extract into phospholipid nanoparticles, *Journal of Food Engineering*, 167: 189-195.

Burgain, J., Gaiani, C., Linder, M. and Scher, J. (2011). Encapsulation of probiotic living cells: From laboratory scale to industrial applications, *Journal of Food Engineering*, 104: 467-483.

Calvo, T.R.A., Santagapita, P.R. and Perullini, M. (2019). Functional and structural effects of hydrocolloids on Ca (II)-alginate beads containing bioactive compounds extracted from beetroot, *LWT – Food Science and Technology*, 111: 520-526.

Capozzi, V., Russo, P., Dueñas, M.T., López, P. and Spano, G. (2012). Lactic acid bacteria producing B-group vitamins: A great potential for functional cereals products, *Applied Microbiology and Biotechnology*, 96: 1383-1394.

Castro-Muñoz, R., Barragán-Huerta, B.E. and Yáñez-Fernández, J. (2015). Use of gelatin-maltodextrin composite as an encapsulation support for clarified juice from purple cactus pear (*Opuntia stricta*), *LWT – Food Science and Technology*, 62: 242-248.

Celli, G.B., Ghanem, A. and Brooks, M.S.L. (2015). Bioactive encapsulated powders for functional foods – A review of methods and current limitations, *Food and Bioprocess Technology*, 8: 1825-1837.

Chakraborty, S. (2017). Carrageenan for encapsulation and immobilization of flavor, fragrance, probiotics, and enzymes: A review, *Journal of Carbohydrate Chemistry*, 36: 1-19.

Chen, Q., McGillivray, D., Wen, J., Zhong, F. and Quek, S.Y. (2013). Co-encapsulation of fish oil with phytosterol esters and limonene by milk proteins, *Journal of Food Engineering*, 117: 505-512.

Cody, M.M., Gravani, R., Edge, M.S., Dooher, C. and White, C. (2012). International Food Information Council Foundation food and health survey, 2006–2010, food safety: A web-enabled survey, *Food Protection Trends*, 32: 309-326.

Coronel-Aguilera, C.P. and San Martín-González, M.F. (2015). Encapsulation of spray dried β-carotene emulsion by fluidized bed coating technology, *LWT – Food Science and Technology*, 62: 187-193.

Cruz-Benítez, M.M., Gómez-Aldapa, C.A., Castro-Rosas, J., Hernández-Hernández, E., Gómez-Hernández, E. and Fonseca-Florido, H.A. (2019). Effect of amylose content and chemical modification of cassava starch on the microencapsulation of *Lactobacillus pentosus*, *LWT – Food Science and Technology*, 105: 110-117.

da Silva Malheiros, P., Daroit, D.J. and Brandelli, A. (2010). Food applications of liposome-encapsulated antimicrobial peptides, *Trends in Food Science and Technology*, 21: 284-292.

Dafe, A., Etemadi, H., Zarredar, H. and Mahdavinia, G.R. (2017). Development of novel carboxymethyl cellulose/k-carrageenan blends as an enteric delivery vehicle for probiotic bacteria, *International Journal of Biological Macromolecules*, 97: 299-307.

Darukaradhya, J., Phillips, M. and Kailasapathy, K. (2013). Effect of encapsulation on the survival of probiotic bacteria in the presence of starter and non-starter lactic acid bacteria in Cheddar cheese over a 6-month ripening period, *International Journal of Fermented Foods*, 2: 63-76.

de Boer, F.Y., Imhof, A. and Velikov, K.P. (2019). Encapsulation of colorants by natural polymers for food applications, *Coloration Technology*; https://doi.org/10.1111/cote.12393

Di Giorgio, L., Salgado, P.R. and Mauri, A.N. (2019). Encapsulation of fish oil in soybean protein particles by emulsification and spray-drying, *Food Hydrocolloids*, 87: 891-901.

Feehley, T., Plunkett, C.H., Bao, R., Hong, S.M.C., Culleen, E., Belda-Ferre, P., Campbell, E., Aitoro, R., Nocerino, R., Paparo, L. and Andrade, J. (2019). Healthy infants harbor intestinal bacteria that protect against food allergy, *Nature Medicine*, 25: 448.

Foerster, M., Liu, C., Gengenbach, T., Woo, M.W. and Selomulya, C. (2017). Reduction of surface fat formation on spray-dried milk powders through emulsion stabilization with λ-carrageenan, *Food Hydrocolloids*, 70: 163-180.

Food Encapsulation Market, Report code FB2207, March 2020.

Fu, D., Deng, S., McClements, D.J., Zhou, L., Zou, L., Yi, J., Liu, C. and Liu, W. (2019). Encapsulation of β-carotene in wheat gluten nanoparticle-xanthan gum-stabilized pickering emulsions: Enhancement of carotenoid stability and bioaccessibility, *Food Hydrocolloids*, 89: 80-89.

Gassara-Chatti, F., Brar, S.K., Ajila, C.M., Verma, M., Tyagi, R.D. and Valéro, J.R. (2013). Encapsulation of ligninolytic enzymes and its application in clarification of juice, *Food Chemistry*, 137: 18-24.

Ghorbanzade, T., Jafari, S.M., Akhavan, S. and Hadavi, R. (2017). Nano-encapsulation of fish oil in nano-liposomes and its application in fortification of yogurt, *Food Chemistry*, 216: 146-152.

Gómez-Estaca, J., Balaguer, M.P., López-Carballo, G., Gavara, R. and Hernández-Muñoz, P. (2017). Improving antioxidant and antimicrobial properties of curcumin by means of encapsulation in gelatin through electrohydrodynamic atomization, *Food Hydrocolloids*, 70: 313-320.

Gómez-Mascaraque, L.G., Lagarón, J.M. and López-Rubio, A. (2015). Electrosprayed gelatin submicroparticles as edible carriers for the encapsulation of polyphenols of interest in functional foods, *Food Hydrocolloids*, 49: 42-52.

Gonçalves, A., Estevinho, B.N. and Rocha, F. (2016). Microencapsulation of vitamin A: A review, *Trends in Food Science & Technology*, 51: 76-87.

Gonnet, M., Lethuaut, L. and Boury, F. (2010). New trends in encapsulation of liposoluble vitamins, *Journal of Controlled Release*, 146: 276-290.

Granata, G., Stracquadanio, S., Leonardi, M., Napoli, E., Consoli, G.M.L., Cafiso, V., Stefani, S. and Geraci, C. (2018). Essential oils encapsulated in polymer-based nanocapsules as potential candidates for application in food preservation, *Food Chemistry*, 269: 286-292.

Haider, J., Majeed, H., Williams, P.A., Safdar, W. and Zhong, F. (2017). Formation of chitosan nanoparticles to encapsulate krill oil (*Euphausia superba*) for application as a dietary supplement, *Food Hydrocolloids*, 63: 27-34.

Hundal, J.S., Wadhwa, M. and Bakshi, M.P.S. (2019). Herbal feed additives containing essential oil: 1. Impact on the nutritional worth of complete feed *in vitro, Tropical Animal Health and Production*, 51: 1909-1917.

Jia, Z., Dumont, M.J. and Orsat, V. (2016). Encapsulation of phenolic compounds present in plants using protein matrices, *Food Bioscience*, 15: 87-104.

Kanmani, P., Kumar, R.S., Yuvaraj, N., Paari, K.A., Pattukumar, V. and Arul, V. (2011). Cryopreservation and microencapsulation of a probiotic in alginate-chitosan capsules improves survival in simulated gastrointestinal conditions, *Biotechnology and Bioprocess Engineering*, 16: 1106-1114.

Karimi, R., Azizi, M.H., Ghasemlou, M. and Vaziri, M. (2015). Application of inulin in cheese as prebiotic, fat replacer and texturizer: A review, *Carbohydrate Polymers*, 119: 85-100.

Kavitake, D., Kandasamy, S., Devi, P.B. and Shetty, P.H. (2018). Recent developments on encapsulation of lactic acid bacteria as potential starter culture in fermented foods – A review, *Food Bioscience*, 21: 34-44.

Khem, S., Bansal, V., Small, D.M. and May, B.K. (2016). Comparative influence of pH and heat on whey protein isolate in protecting *Lactobacillus plantarum* A17 during spray-drying, *Food Hydrocolloids*, 54: 162-169.

Khem, S., Small, D.M. and May, B.K. (2016a). The behavior of whey protein isolate in protecting *Lactobacillus plantarum*, *Food Chemistry*, 190: 717-723.

Kim, J.U., Kim, B., Shahbaz, H.M., Lee, S.H., Park, D. and Park, J. (2017). Encapsulation of probiotic *Lactobacillus acidophilus* by ionic gelation with electrostatic extrusion for enhancement of survival under simulated gastric conditions and during refrigerated storage, *International Journal of Food Science & Technology*, 52: 519-530.

Komaiko, J., Sastrosubroto, A. and McClements, D.J. (2016). Encapsulation of ω-3 fatty acids in nanoemulsion-based delivery systems fabricated from natural emulsifiers: Sunflower phospholipids, *Food Chemistry*, 203: 331-339.

Li, H., Turner, M.S. and Dhital, S. (2016). Encapsulation of *Lactobacillus plantarum* in porous maize starch, *LWT – Food Science and Technology*, 74: 542-549.

Li, W., Luo, X., Song, R., Zhu, Y., Li, B. and Liu, S. (2016a). Porous cellulose microgel particle: A fascinating host for the encapsulation, protection, and delivery of *Lactobacillus plantarum*, *Journal of Agricultural and Food Chemistry*, 64: 3430-3436.

Li, Z., Paulson, A.T. and Gill, T.A. (2015). Encapsulation of bioactive salmon protein hydrolysates with chitosan-coated liposomes, *Journal of Functional Foods*, 19: 733-743.

Liang, J., Yan, H., Wang, X., Zhou, Y., Gao, X., Puligundla, P. and Wan, X. (2017). Encapsulation of epigallocatechin gallate in zein/chitosan nanoparticles for controlled applications in food systems, *Food Chemistry*, 231: 19-24.

Liu, W., Wang, J., McClements, D.J. and Zou, L. (2018). Encapsulation of β-carotene-loaded oil droplets in caseinate/alginate microparticles: Enhancement of carotenoid stability and bioaccessibility, *Journal of Functional Foods*, 40: 527-535.

Lorenzo, J.M., Munekata, P.E.S., Pateiro, M., Campagnol, P.C.B. and Domínguez, R. (2016). Healthy Spanish salchichón enriched with encapsulated n-3 long chain fatty acids in konjac glucomannan matrix, *Food Research International*, 89: 289-295.

Lu, M., Li, Z., Liang, H., Shi, M., Zhao, L., Li, W., Chen, Y., Wu, J., Wang, S., Chen, X. and Yuan, Q. (2015). Controlled release of anthocyanins from oxidized konjac glucomannan microspheres stabilized by chitosan oligosaccharides, *Food Hydrocolloids*, 51: 476-485.

Luo, Y., Teng, Z. and Wang, Q. (2012). Development of zein nanoparticles coated with carboxymethyl chitosan for encapsulation and controlled release of vitamin D$_3$, *Journal of Agricultural and Food Chemistry*, 60: 836-843.

Mahmood, K., Kamilah, H., Shang, P.L., Sulaiman, S., Ariffin, F. and Alias, A.K. (2017). A review: Interaction of starch/non-starch hydrocolloid blending and the recent food applications, *Food Bioscience*, 19: 110-120.

Martin, M.J., Lara-Villoslada, F., Ruiz, M.A. and Morales, M.E. (2013). Effect of unmodified starch on viability of alginate-encapsulated *Lactobacillus fermentum* CECT5716, *LWT – Food Science and Technology*, 53: 480-486.

Moomand, K. and Lim, L.T. (2014). Oxidative stability of encapsulated fish oil in electrospun zein fibers, *Food Research International*, 62: 523-532.

Moumita, S., Das, B., Hasan, U. and Jayabalan, R. (2018). Effect of long-term storage on viability and acceptability of lyophilized and spray-dried synbiotic microcapsules in dry functional food formulations, *LWT – Food Science and Technology*, 96: 127-132.

Mu, R.J., Yuan, Y., Wang, L., Ni, Y., Li, M., Chen, H. and Pang, J. (2018). Microencapsulation of *Lactobacillus acidophilus* with konjac glucomannan hydrogel, *Food Hydrocolloids*, 76: 42-48.

Nag, A., Han, K.S. and Singh, H. (2011). Microencapsulation of probiotic bacteria using pH-induced gelation of sodium caseinate and gellan gum, *International Dairy Journal*, 21: 247-253.

Nedović, V., Kalušević, A., Manojlović, V., Petrović, T. and Bugarski, B. (2013). Encapsulation systems in the food industry, pp. 229-253. *In:* S. Yanniotis., P. Taoukis., N. Stoforos and V. Karathanos (Eds.), *Advances in Food Process Engineering Research and Applications*, Springer, Boston, MA.

Nualkaekul, S., Lenton, D., Cook, M.T., Khutoryanskiy, V.V. and Charalampopoulos, D. (2012). Chitosan coated alginate beads for the survival of microencapsulated *Lactobacillus plantarum* in pomegranate juice, *Carbohydrate Polymers*, 90: 1281-1287.

Ozkan, G., Franco, P., De Marco, I., Xiao, J. and Capanoglu, E. (2019). A review of microencapsulation methods for food antioxidants: Principles, advantages, drawbacks and applications, *Food Chemistry*, 272: 494-506.

Panda, S.K., Panda, S.H., Swain, M.R., Ray, R.C. and Kayitesi, E. (2015). Anthocyanin-rich sweet potato (*Ipomoea batatas* L.) beer: Technology, biochemical and sensory evaluation, *Journal of Food Processing and Preservation*, 39: 3040-3049.

Prajapati, V.D., Jani, G.K., Zala, B.S. and Khutliwala, T.A. (2013). An insight into the emerging exopolysaccharide gellan gum as a novel polymer, *Carbohydrate Polymers*, 93: 670-678.

Purhagen, J.K., Sjöö, M.E. and Eliasson, A.C. (2012). The anti-staling effect of pre-gelatinized flour and emulsifier in gluten-free bread, *European Food Research and Technology*, 235: 265-276.

Rajabi, H., Ghorbani, M., Jafari, S.M., Mahoonak, A.S. and Rajabzadeh, G. (2015). Retention of saffron bioactive components by spray-drying encapsulation using maltodextrin, gum arabic and gelatin as wall materials, *Food Hydrocolloids*, 51: 327-337.

Raju, P.S. and Bawa, A.S. (2006). Food additives in fruit processing, pp. 145-170. *In:* W.H. Hui (Ed.), *Handbook of Fruits and Fruit Processing*, Wily, London.

Ramesh, M. and Muthuraman, A. (2018). Flavoring and coloring agents: Health risks and potential problems, pp. 1-28. *In:* Alexandru Grumezescu and Alina Maria Holban (Eds.), *Natural and Artificial Flavoring Agents and Food Dyes*, Academic Press, London.

Reddy, C.K., Sivapriya, T.V.S., Kumar, U.A. and Ramalimgam, C. (2016). Optimization of food acidulant to enhance the organoleptic property in fruit jellies, *Journal of Food Processing and Technology*, 7; DOI:10.4172/2157-7110.1000635

Rodrigues, L.M., Dos Santos, S.S., Bergamasco, R.C. and Madrona, G.S. (2018). Jaboticaba byproduct encapsulation by lyophilization: pH and food application stability, *Journal of Food Process Engineering*, 41: 12639.

Rutz, J.K., Zambiazi, R.C., Borges, C.D., Krumreich, F.D., da Luz, S.R., Hartwig, N. and da Rosa, C.G. (2013). Microencapsulation of purple Brazilian cherry juice in xanthan, tara gums and xanthan-tara hydrogel matrixes, *Carbohydrate Polymers*, 98: 1256-1265.

Saifullah, M., Shishir, M.R.I., Ferdowsi, R., Rahman, M.R.T. and Van Vuong, Q. (2019). Micro and nano encapsulation, retention and controlled release of flavor and aroma compounds: A critical review, *Trends in Food Science & Technology*, 86: 230-251.

Sathyabama, S. and Vijayabharathi, R. (2014). Co-encapsulation of probiotics with prebiotics on alginate matrix and its effect on viability in simulated gastric environment, *LWT – Food Science and Technology*, 57: 419-425.

Shaddel, R., Hesari, J., Azadmard-Damirchi, S., Hamishehkar, H., Fathi-Achachlouei, B. and Huang, Q. (2018). Use of gelatin and gum arabic for encapsulation of black raspberry anthocyanins by complex coacervation *International Journal of Biological Macromolecules,* 107: 1800-1810.

Shi, L.E., Li, Z.H., Zhang, Z.L., Zhang, T.T., Yu, W.M., Zhou, M.L. and Tang, Z.X. (2013). Encapsulation of *Lactobacillus bulgaricus* in carrageenan-locust bean gum coated milk microspheres with double layer structure, *LWT – Food Science and Technology*, 54: 147-151.

Siddiqui, A.A., Feroz, A., Khaki, P.S.S. and Bano, B. (2017). Binding of λ-carrageenan (a food additive) to almond cystatin: An insight involving spectroscopic and thermodynamic approach, *International Journal of Biological Macromolecules*, 98: 684-690.

Stojanovic, R., Belscak-Cvitanovic, A., Manojlovic, V., Komes, D., Nedovic, V. and Bugarski, B. (2012). Encapsulation of thyme (*Thymus serpyllum* L.) aqueous extract in calcium alginate beads, *Journal of the Science of Food and Agriculture*, 92: 685-696.

Tan, C., Zhang, Y., Abbas, S., Feng, B., Zhang, X. and Xia, S. (2014). Modulation of the carotenoid bioaccessibility through liposomal encapsulation, *Colloids and Surfaces B: Biointerfaces*, 123: 692-700.

Tavares, G.M., Croguennec, T., Carvalho, A.F. and Bouhallab, S. (2014). Milk proteins as encapsulation devices and delivery vehicles: Applications and trends, *Trends in Food Science and Technology*, 37: 5-20.

Turasan, H., Sahin, S. and Sumnu, G. (2015). Encapsulation of rosemary essential oil, *LWT – Food Science and Technology*, 64: 112-119.

Ulloa, A.M. (2018). The aesthetic life of artificial flavors, *The Senses and Society*, 13: 60-74.

Vélez, M.A., Perotti, M.C., Hynes, E.R. and Gennaro, A.M. (2019). Effect of lyophilization on food grade liposomes loaded with conjugated linoleic acid, *Journal of Food Engineering*, 240: 199-206.

Wang, B., Adhikari, B. and Barrow, C. J. (2019). Highly stable spray dried tuna oil powders encapsulated in double shells of whey protein isolate-agar gum and gellan gum complex coacervates, *Powder Technology*, 358: 79-86.

Wang, L., Yu, X., Xu, H., Aguilar, Z.P. and Wei, H. (2016). Effect of skim milk coated inulin-alginate encapsulation beads on viability and gene expression of *Lactobacillus plantarum* during freeze-drying, *LWT – Food Science and Technology*, 68: 8-13.

Wen, P., Zhu, D.H., Wu, H., Zong, M.H., Jing, Y.R. and Han, S.Y. (2016). Encapsulation of cinnamon essential oil in electrospun nanofibrous film for active food packaging, *Food Control*, 59: 366-376.

Zhang, Z., Zhang, R., Zou, L. and McClements, D.J. (2016). Protein encapsulation in alginate hydrogel beads: Effect of pH on microgel stability, protein retention and protein release, *Food Hydrocolloids*, 58: 308-315.

Zhao, F., Wang, Q., Dong, J., Xian, M., Yu, J., Yin, H., Chang, Z., Mu, X., Hou, T. and Wang, J. (2017). Enzyme-inorganic nanoflowers/alginate microbeads: An enzyme immobilization system and its potential application, *Process Biochemistry*, 57: 87-94.

Zheng, L., Ding, Z., Zhang, M. and Sun, J. (2011). Microencapsulation of bayberry polyphenols by ethyl cellulose: Preparation and characterization, *Journal of Food Engineering*, 104: 89-95.

Nanoencapsulation in Food Technology

Spiros Paramithiotis* and Eleftherios H. Drosinos

Laboratory of Food Quality Control and Hygiene, Department of Food Science and Human Nutrition, Agricultural University of Athens, Iera Odos 75, GR-118 55 Athens, Greece

1. Introduction

Encapsulation is an approach originally employed to protect ingredients that are sensitive to processing or storage conditions. Indeed, the application of an extended range of encapsulated nutrients, processing aids and bioactive compounds has been reported in the literature (Gibbs *et al.*, 1999). Nanoencapsulation, i.e. encapsulation in a protective envelope of the nano scale, offers many advantages as well as new possibilities, mostly due to the larger surface to volume ratio. More accurately, nanoencapsulation improves solubility, biological activity and homogeneity of the dispersion. Moreover, in biomedical applications it improves bioavailability, controlled release and precision targeting (Rezaei *et al.*, 2019).

Depending on the properties of the food matrix and the nature of the encapsulated compound, a suitable nanocarrier has to be selected. The available nanocarriers may be divided into five classes (Assadpour and Jafari, 2018): (i) lipid based, which are prepared by oil, water and various solvents; (ii) naturally occurring, such as the nanostructures formed by casein, cyclodextrins, amylose etc; (iii) special-equipment demanding, such as nano spray-driers, electro-spinning/spraying etc; (iv) biopolymers, such as alginate or chitosan nano-gels, whey proteins nanotubes or nanofibrils etc, and (v) miscellaneous, such as inorganic ones, nanocrystals etc. Research has focused on the nanoencapsulation of processing aids, such as antimicrobials, flavors and pigments as well as on the nutritional enhancement of products through the incorporation of sensitive nutrients, such as vitamins and fatty acids, minerals, antioxidants, etc. A wide variety of nanoencapsulation approaches, conditions and formulations have been assessed *in vitro*. On the contrary, research is generally lacking in actual food systems. In the next paragraphs, the most recent research on the nanoencapsulation of

---

*Corresponding author: sdp@aua.gr

micronutrients, omega-3 polyunsaturated fatty acids, color additives and essential oils is summarized and critically discussed.

2. Micronutrients Nanoencapsulation

Minerals and vitamins are considered as micronutrients, i.e. nutrients that are essential in small amounts for the human body. However, deficiency of elements, such as calcium, magnesium, copper, iron, zinc, iodine and selenium seems to be quite common in both developed and developing countries for reasons associated either with the absence of animal products from the diet or with factors related to crop production (White and Broadley, 2009). In addition, vitamins are very sensitive to various processing steps, especially to heat treatment. For that purpose, alleviation strategies have been developed, mainly through crop improvement and food fortification. Regarding minerals, several studies report the increase of iron and zinc content of rice and wheat (Goto *et al.*, 1999; Masuda *et al.*, 2008, 2012; Lee *et al.*, 2009; Cakmak *et al.*, 2010; Kutman *et al.*, 2010; Johnson *et al.*, 2011; Ramzani *et al.*, 2016; Ciccolini *et al.*, 2017; Chattha *et al.*, 2017), calcium content of carrot, tomato, potato and baby leaf vegetables (Park *et al.*, 2005a, b; Kim *et al.*, 2006; Morris *et al.*, 2008; D'Imperio *et al.*, 2016), selenium content of lettuce, chickpea, pea, lentil and strawberries (Ramos *et al.*, 2010; Poblaciones *et al.*, 2013, 2014; Thavarajah *et al.*, 2015, 2017; Mimmo *et al.*, 2017) and iodine content of wheat and lettuce (Ren *et al.*, 2008; Voogt *et al.*, 2010). Similarly, biofortification of several plants, including tomato, bean, potato, rice maize, cassava, sweet potato, rapeseed and spinach with folate (Díaz de la Garza *et al.*, 2007; Blancquaert *et al.*, 2015; Ramírez Rivera *et al.*, 2016; Watanabe *et al.*, 2017; De Lepeleire *et al.*, 2018), provitamin A (De Moura *et al.*, 2015, 2016; Palmer *et al.*, 2016; Muzhingi *et al.*, 2017) vitamin B_1 (Dong *et al.*, 2015, 2016), vitamin B_6 (Raschke *et al.*, 2011; Li *et al.*, 2015a), vitamin E (Karunanandaa *et al.*, 2005; Raclaru *et al.*, 2006) etc. may also serve as examples of effective plant breeding and/or genetic modification. Alternatively, food fortification has been extensively considered and practiced. Minerals may be added directly in the form of their respective salts. In that case, issues such as solubility, stability and bioavailability as well as a possible effect on sensorial properties and texture along with economic considerations need to be properly addressed. On the other hand, encapsulation seems the only alternative regarding fortification with vitamins since protection from heat and oxidation is required.

Several studies currently exist in the preparation of nanoencapsulated minerals for food fortification or biomedical purposes (Table 1). Carriers, such as chitosan, alginate, whey protein, liposomes, nanoemulsions, etc. have been prepared by using a variety of approaches and effectively loaded with minerals. However, there is a significant lack of studies in actual food systems. Sharifi *et al.* (2013) loaded alginate nanoparticles with iron and zinc in the form of $FeCl_3$ and $ZnCl_2$, respectively and added them to ice cream. Their shape was characterized as smooth with a size range of 90-200 nm and a polydispersity index (PDI) of 0.1. Their addition in the ice cream had no statistically significant effect on the sensorial properties of the product. On the contrary, urine assay of the volunteers that consumed the ice cream revealed that more than 90 per cent of the loaded Fe/Zn was absorbed. Naveen and Kanum (2014) loaded phosphatidyl choline – cholesterol (70:30) nanoemulsion with iron and assessed bioavailability and toxicity in Wister rats. The shape of the nanoparticles was mostly

Table 1: Representative Recent Studies Assessing the Nanoencapsulation of Minerals

Loaded Material	NP System	Size (nm)	Polydispersity Index	Zeta Potential (mV)	Comment	Reference
Catechin/ catechin-Zn complex	β-chitosan	208-591	0.377-0.95	39.17-45.62	The complexes exhibited antibacterial activity against *E. coli* and *L. innocua*	Zhang *et al.*, 2016
Tea polyphenols/ tea polyphenols-Zn complex	β-chitosan	84.55	Nr	29.23	TP-Zn exhibited higher antioxidant activity than TP alone	Zhang and Zhao, 2015
Sodium selenite	CS-TPP	120-300	0.16-0.29	37–50	Zein coating decreased the selenite release from 85% to 30% within 4 h in PBS buffer and from 93% to 46% in simulated gastrointestinal conditions	Luo *et al.*, 2010
Methylseleninic acid, sodium selenite	CS-TPP	Nr	Nr	Nr	Nanoencapsulation increased selenium retention in cells and decreased cellular sensitivity and DNA damage	Zhang *et al.*, 2011
Iron (II) sulfate	Solid lipid NPs	300-500	Nr	–10 to +10	The potential of the examined formulations as a novel system for oral iron delivery was exhibited	Zariwala *et al.*, 2013
Iron (III) chloride, zinc chloride and manganese (II) chloride	CS-TPP	300	0.33	–5 to 25	The properties of the NPs indicated suitability for biomedical applications	Zahraei *et al.*, 2015
Ferrous fumarate	Water-in-oil-in-water (W/O/W) emulsion	600	0.35–0.40	–15 to 5	The efficiency, low cost and scale-up capacity of this approach were highlighted	Tang and Sivakumar, 2013
Sodium selenite	Guar gum NP	69–173	0.4	nr	Suitability for biomedical applications was indicated	Soumya *et al.*, 2013

Zinc chloride	Sodium alginate	200-230	0.4-1.0	-80 to -30	Alginate-zinc nanoparticles were presented as a reliable alternative to alginate nanoparticles cross-linked with polycations	Pistone *et al.*, 2015
Iron	Chitosan-iron casein succinylate ISC	700-1600	Nr	30-60	The developed formulation may be a useful oral delivery system of iron supplementation	Min *et al.*, 2016
Ferrous sulfate	Liposomes	150-200	Nr	Nr	The capacity of the microfluidization-based liposomal technology for a variety of water-soluble bioactive compounds for both food and nutraceutical applications was highlighted	Kosaraju *et al.*, 2006
Zinc chloride	whey protein	100	Nr	Nr	Applicability of another approach for effective delivery of sensitive micronutrients	Gulseren *et al.*, 2012

NP: Nanoparticle; *E.*: *Escherichia*; *L.*: *Listeria*; Nr: Not reported; TP: Tea polyphenols; PBS: Phosphate buffered saline; CS-TPP: Chitosan-tripolyphosphate spherical with an average size of 98 nm and zeta potential of -46.5mV. The *in vivo* bioavailability was characterized as appreciable and no toxic effects were detected.

Regarding nanoencapsulation of vitamins, it was achieved with a variety of methods (Table 2) including nanoemulsions (Ozturk *et al.*, 2014; Assadpour *et al.*, 2016; Campani *et al.*, 2016; Walia *et al.*, 2017), nanoliposomes (Li *et al.*, 2015b; Pezeshky *et al.*, 2016; Fathima *et al.*, 2016; Bochicchio *et al.*, 2016; Liu *et al.*, 2017), nanohydrogels (Tsuchido *et al.*, 2015; Peng *et al.*, 2016) and other nanocarriers based on a variety of materials, such as chitosan, cyclodextrin, poly-ε-caprolactone, sorbitan monooleate, etc. (Jimenez-Fernandez *et al.*, 2014; Heydari *et al.*, 2016; Pegoraro *et al.*, 2017). To the best of our knowledge no studies exist on the incorporation of nanoencapsulated vitamins in actual food systems.

3. Nanoencapsulation of Omega-3 Polyunsaturated Fatty Acids

A wide variety of omega-3 polyunsaturated fatty acids currently exists. Research has focused on alpha-linolenic acid (ALA), eicosapentaenoic acid (EPA) and docosahexaenoic acid (DHA) due to the health benefits that they exert (Pan *et al.*, 2012; Swanson *et al.*, 2012). The first is commonly found in plant oils while EPA and DHA in marine oils (Cholewski *et al.*, 2018). Due to their importance in human physiology, food supplementation has been extensively considered. However, certain issues, namely poor solubility, instability and fishy organoleptic character limit their application. Encapsulation is an effective way to address these issues and nanoencapsulation may present the already described benefits. Several studies are currently available on the nanoencapsulation of fish or plant oils or even purified polyunsaturated fatty acid (PUFA) preparations using nanoemulsions (Gulotta *et al.*, 2014; Walker *et al.*, 2015), nanoliposomes (Rasti *et al.*, 2012; Ghorbanzade *et al.*, 2017) or other nanocarriers, such as casein (Zimet *et al.*, 2011) or cyclodextrins (Choi *et al.*, 2010). In Table 3 representative recent studies are summarized.

However, as in the previous cases, there is a significant lack of studies in actual food systems. Ghorbanzade *et al.* (2017) and Zhong *et al.* (2018) studied yogurt fortification with fish oil. In the first study, fish oil was included into nanoliposomes consisting of sunflower oil and soy lecithin that were concomitantly incorporated into yogurt formulation. The size of the nanoliposomes was between 300 and 500 nm with PDI of 0.557; the encapsulation efficiency was 92.22 ± 0.19 per cent and around 70 per cent of the nanoliposomes were stable. Incorporation of nanoliposomes resulted in a significant reduction in acidity, syneresis and peroxide value. In addition, the product obtained after incorporation of nanoliposomes was organoleptically close to the control one and the DHA and EPA content was higher after 21 days of storage as compared to the control. Zhong *et al.* (2018) studied yogurt fortification with fish oil and γ-oryzanol nanoemulsified by Tween 80/Span 20. This fortification resulted in an organoleptically comparable product to plain yogurt. In addition, it resulted in attenuation of pH value and acidity increase and decreased apparent viscosity that occur during storage as compared to both plain yogurt and yogurt fortified with non-nanoemulsified fish oil and γ-oryzanol. Moreover, nanoemulsification resulted in higher DHA and EPA content and lower peroxide values throughout the storage as compared to the yogurt fortified with non-nanoemulsified fish oil and γ-oryzanol. Finally, it affected syneresis throughout storage. More accurately, it resulted in the highest syneresis during the first week but the lowest during the last week of storage.

Table 2: Representative Recent Studies Assessing the Nanoencapsulation of Vitamins

Loaded Vitamin	NP System	Size (nm)	Polydispersity Index	Zeta Potential (mV)	Comment	Reference
D3	Oil in water nanoemulsion based on fish oil and tween 20	300-450	Nr	Nr	Promising approach for efficient vitamin D delivery	Walia *et al.*, 2017
Coenzyme Q10 and vitamin E acetate	Semisolid preparations based on poly-ε-caprolactone, sorbitan monooleate and gellan gum	130-370	Nr	Nr	Promising anti-edematogenic, anti-inflammatory and antioxidant activities demonstrated in an animal model of UVB irradiation-induced inflammation	Pegoraro *et al.*, 2017
D3	Nanostructured lipid carriers	< 200	0.18-0.28	−30 to −40	The lipid carriers were able to protect the vitamin D_3 in simulated gastric conditions, but released more than 90% after 8 h of digestion in simulated intestinal fluid	Park *et al.*, 2017
Folic acid	Water in oil nanoemulsions encapsulated into maltodextrin-whey protein.	< 100	Nr	Nr	Very effective approach for folic acid delivery was introduced	Assadpour *et al.*, 2016
B12, tocopherol and ergocalciferol	Nanoliposomal vectors based on L-α-phosphatidylcholine from egg yolk, Cholesterol and α-phosphatidyl - DL glycerol sodium salt from egg yolk lecithin	40-51	Nr	Nr	Ten days stability of the nanovesicles at environmental conditions was reported	Bochicchio *et al.*, 2016

(Contd.)

Table 2: (*Contd.*)

Loaded Vitamin	NP System	Size (nm)	Polydispersity Index	Zeta Potential (mV)	Comment	Reference
C	Nanocarriers based on chitosan, cyclodextrin and tripolyphosphate	< 300	0.34–0.38	34-38	The reported approach seems a promising tool for oral administration of water soluble compounds	Jimenez-Fernandez et al., 2014
C	Hydrogel based on bovine serum albumin and citrus peel pectin	120-150	< 0.06	−25.90 to 0.79	Potential delivery system based on food grade and natural biopolymers was presented	Peng et al., 2016
B2	β-cyclodextrin	100-150	Nr	10.9-11.3	The cationic β-cyclodextrin polymers with high guanidine content exhibited slower vitamin release compared to the other formulations examined	Heydari et al., 2016
B2	Solid lipid NPs based on fully hydrogenated canola oil	104 ± 5.7	Nr	Nr	The employed technique was able to produce small particles with significant encapsulation efficiency	Couto et al., 2017
A palmitate	Nanoliposomes based on lecithin and cholesterol	76-115	Nr	Nr	The low encapsulation efficiency of this approach was highlighted	Pezeshky et al., 2016
K1	Nanoemulsion based on α-tocopherol, ethanol and Tween 80	100.6 ± 6.6 to 263.0 ± 6.3	Nr	−34.90 ± 15.39 to −15.08 ± 1.69	This study presented nanoemulsions as an alternative to topical delivery of vitamin K1	Campani et al., 2016

NP: Nanoparticle; Nr: not reported

Table 3: Representative Recent Studies Assessing the Nanoencapsulation of Fatty Acids

Loaded Fatty Acid	NP System	Size (nm)	Polydispersity Index	Zeta Potential (mV)	Comment	Reference
DHA	Various combinations of maltodextrin, mannitol, acacia gum, methyl cellulose, sodium alginate, sodium caseinate, starch, skim milk	283.8 ± 1.23 to 385 ± 1.91	Nr	−24.3 to −28.5	The optimized formulation consisting of microalgae oil (21.45%), maltodextrin (19.27%), acacia gum (8.56%), methyl cellulose (5.35%), sodium alginate (3.21%), starch (9.63%), sodium caseinate (21.41%) skim milk (9.63%) exhibited improved shelf life, bioavailability, oxidative stability and memory enhancement effects	Singh *et al.*, 2018
Antarctic Krill (*Euphausia superba*) oil	Chitosan	100–300	0.199	+ 26.6 to + 37.7	The efficacy of the proposed methodology to prevent krill oil oxidation was exhibited.	Haider *et al.*, 2017
DHA	lipids (Precirol ATO5®, Miglyol-812®) and Tween 60®	211 ± 8 to 323 ± 41 depending on DHA concentration	Nr	−28	Nanoencapsulation enhanced DHA antibacterial activity against *H. pylori*. The activity was associated with changes in the cell membrane. No cytotoxic affect to human epithelial cells was observed at the bactericidal concentrations	Seabra *et al.*, 2017

(Contd.)

Table 3: (*Contd.*)

Loaded Fatty Acid	NP System	Size (nm)	Polydispersity Index	Zeta Potential (mV)	Comment	Reference
Cod liver oil	PVA nanofibers	Fiber diameter of 172 ± 44 to 252 ± 68 (WPI emulsifier); fiber diameter of 163 ± 45 to 120 ± 59 and bead diameter of 295 ± 42 to 784 ± 422 (FPH emulsifier)	Nr	Nr	PVA nanofibers exhibited higher physical stability with smaller droplet size with WPI as emulsifier. However, poor oxidative stability was observed most probably due to the presence of traces of metals in PVA	Garcia-Moreno *et al.*, 2016
Fish oil	β-cyclodextrin, PCL	250–700	0.5–0.7	−31 to −26	Aggregation of nanoparticles and concomitant fish oil retarded release were observed when high amounts of core material were used in the nanocomplex. PCL retarded fish oil release more effectively	Choi *et al.*, 2010
Fish oil extracted from Caspian Sea sprat	Oil-in-water nanoemulsions with WPI as emulsifier	66 ± 8.48 to 703 ± 36.06 depending on WPI:FO WR, ultrasound time and pH value	Nr	−37.5 ± 0.2 to +4.3 ± 0.2 depending on the pH value	The reported results may be useful in designing nanoemulsions as fish oil delivery systems especially when high energy sonication is considered	Nejadmansouri *et al.*, 2016
EPA and DHA	Nanoliposomes composed of DPPC, DL-α-tocopherol and preservatives	73.4 ± 1.03 to 107.2 ± 1.8 depending on the formulation	< 0.2	Nr	Use of α-tocopherol resulted in stabilization of physical properties such as mean particle size, size distribution and encapsulation efficiency	Sahari *et al.*, 2016

| Pomegranate seed oil | Zein alone or crosslinked with citric acid | 246 ± 6.0 to 442.6 ± 25.7 | 0.308 ± 0.008 to 0.402 ± 0.03 | Nr | Crosslinking of zein particles with citric acid resulted in the development of smaller particles, higher mass yield and higher encapsulation efficiency. Improved heat resistance was observed when crosslinking took place before desolvation with water | Tavakolipour *et al.*, 2015 |

NP: Nanoparticle; Nr: Not reported; DHA: Docosahexaenoic acid; *H: Helicobacter*; PVA: Polyvinyl alcohol; WPI: Whey protein isolate; FPH: Fish protein hydrolysate; PCL: Polycaprolactone; FO: Fish oil; WR: Weight ratio; EPA: Eicosapentaenoic acid; DPPC: (1,2-dipalmitoyl-sn-glycero-3-phosphocholine); SLN: Solid lipid nanoparticles

4. Nanoencapsulation of Color Additives

According to the US Food and Drug Administration (FDA), any substance that imparts color to food, drug, cosmetic or human body is defined as a color additive (https://www. fda.gov/Food/IngredientsPackagingLabeling/FoodAdditivesIngredients/ucm488219. htm). In food, color additives may be used as a standardization aid, to impart color to colorless foods as well as to modify or enhance naturally occurring color. For that purpose, a wide variety of naturally occurring pigments have been employed, aimed at replacing their synthetic counterparts, mostly for reasons of safety. From a chemical perspective, the vast majority of food colorants may either be polyphenols (e.g. curcumin), flavonoids (e.g. anthocyanins) or tetraterpenoids (e.g. carotenoids). Apart from the coloring activity, an extended range of biological activities, including antimicrobial and antioxidant ones, has also been revealed (Rao and Rao, 2007; Vilchez *et al*., 2011; Li *et al*., 2014; Raposo *et al*., 2015; Yousuf *et al*., 2016; Li *et al*., 2017; Sathasivam and Ki, 2018; Gorzynik-Debicka *et al*., 2018). However, their poor water solubility (with the exception of anthocyanins) and chemical instability, especially during food processing, limit their use and dictate the use of encapsulation.

Several studies are currently available on the nanoencapsulation of pigments, such as lutein (Lacatusu *et al*., 2013), crocin (Mehrnia *et al*., 2016), anthocyanins (Ravanfar *et al*., 2016), lycopene (Nerome *et al*., 2013), β-carotene (de Freitas Zompero *et al*., 2015), astaxanthin (Li *et al*., 2016), curcumin (Almeida *et al*., 2018) etc. In Table 4, recent studies on nanoencapsulation of pigments are presented. On the contrary, studies referring to application in actual food systems are rather limited. Saiz-Abajo *et al*. (2013) encapsulated β-carotene into casein micelles, achieving average diameter of ca. 80 nm, zeta potential of ca. –34 mV, while encapsulation efficiency and yield were 100±1 per cent and 60±5 per cent, respectively. Nanoencapsulation effectively protected β-carotene during heat treatment. Indeed, the amount of free β-carotene that was degraded during heating at 80°C for 8 h, pasteurization and sterilization were reported at 83.5 per cent, 67.9 per cent and 75.9 per cent, respectively. On the contrary, the nanoencapsulated compound was reduced by 30.9 per cent, 35.7 per cent and 51.0 per cent, respectively. Similar results were obtained by high hydrostatic pressure processing; the nanoencapsulated β-carotene was reduced by 8.6 per cent as compared to 28.1 per cent of the control. Finally, nanoencapsulated and free β-carotene were incorporated into a cookie recipe that included baking at 180°C for 20 min; a reduction of 41.5 per cent and 66.7 per cent was reported, respectively. Almeida *et al*. (2018) encapsulated curcumin into polyvinylpyrrolidone; the obtained nanoparticles had diameters ranging from 200 to 712 nm and PDI of 0.050±0.009. The biological activities of the nanoencapsulated curcumin along with a commercial water-dispersible preparation and a powder formulation containing 65 per cent curcumin were comparatively assessed. The latter presented the highest antioxidant and anti-inflammatory activities followed by the nanoencapsulated and the water-dispersible formulations. Regarding cytotoxicity, it depended upon the cell-line that was employed; however, the powder and the nanoencapsulated formulations were most effective. As far as the antimicrobial activity was concerned, all formulations, with the exception of the powder one, exhibited antimicrobial activity against *Pseudomonas aeruginosa* and *Morganella morganii*. Interestingly, Gram positive bacteria were more sensitive, especially to the powder formulation. Then, these curcumin preparations were incorporated into yogurt. Slight changes in nutritional composition but significant ones in color were reported. Finally, no cytotoxic activity was observed.

Table 4: Representative Recent Studies Assessing the Nanoencapsulation of Pigments

Loaded Pigment	NP System	Size (nm)	Polydispersity Index	Zeta Potential (mV)	Comment	Reference
Lutein	LNCs consisting of GS, CW, FO, corn oil, Tween 80, lecithin and Poloxamer 407	Diameter 167.5±0.793-387.4±10.24	0.172±0.016-0.419±0.006	−34.5±4.61 - −27.4±0.68	An effective *in vitro* antioxidant capacity was revealed; lutein-LNCs scavenged up to 98% of oxygen free radicals	Lacatusu *et al.*, 2013
Astaxanthin	SLNs consisting of SA, GMS, GDS, soybean oil and Tween 20	SA-SLNs: ~170; GMS-SLNs: ~210; GDS-SLNs: ~250; optimized: 163.9±16.9–173.2±9.2	Optimized: 0.142±0.071 – 0.201±0.018	SA-SLNs: ~−30; GMS-SLNs: ~−17; GDS-SLNs: ~−20; optimized: −37.7±3.6 - −33.5±3.8	Nanoencapsulation provided significantly improved stability of astaxanthin compared to the non-encapsulated compound, at 4 and 25 °C without significant modification of the particle size. In addition, prolonged release was observed in simulated gastric and intestinal juices	Li *et al.*, 2016
Crocin	Water in oil nanoemulsion consisting of olive oil, Span 80 and PGPR	with Span 80: 17±2 - 40±5; with PGPR: 9.7±3-21±4	Nr	Nr	The smallest droplet size was observed at SWR 100% and increased with decreasing SWR. Increase of shear viscosity was also observed by increasing SWR	Mehmia *et al.*, 2016
Astaxanthin	LNCs consisting of Precirol® ATO 5, sunflower oil, Tween® 80 and Poloxamer 407	60±7	0.33±0.09	−25.5±0.7	LNCs were stable for 1 month at 4 °C. Nanoencapsulated astaxanthin exhibited higher antioxidant capacity than non-encapsulated	Rodriguez-Ruiz *et al.*, 2018

(Contd.)

Table 4: (*Contd.*)

Loaded Pigment	NP System	Size (nm)	Polydispersity Index	Zeta Potential (mV)	Comment	Reference
Astaxanthin	Nanospheres consisting of PCPLC	68.3±0.35 to 312±5.83	Nr	−30.2	Astaxanthin could not be effectively encapsulated in EC or PB4. The encapsulation efficiency and loading of PCPLC were determined at 98% and 40%, respectively. Minimal heat degradation of the encapsulated astaxanthin was exhibited after 2 h at 70 °C (w/w); nearly complete degradation of the unencapsulated astaxanthin was observed at the same conditions	Tachaprutinun *et al.*, 2009
Carotenoids (lycopene, β-carotene, lutein, canthaxanthin)	Liposomes coated by chitosan	10-15	≤ 0.3	9.3 ± 0.2	The molecular structure of the carotenoid affected the encapsulating and retaining ability of the chitosan liposomes; β-carotene and lutein were more effectively protected than lycopene and canthaxanthin.	Tan *et al.*, 2016

LNCs: Lipid nanocarriers; nr: not reported; GS: Glycerol stearate; CW: Carnauba wax; FO: Fish oil; SLNs: Solid lipid nanoparticles; GMS: Glycerol monostearate; GDS: Glycerol distearates; SA: Stearic acid; PGPR: Polyglycerol polyrecioleate; SWR: Surfactant to water ratio; PCPLC: Poly(ethylene oxide)-4-methoxycinnamoylphthaloylchitosan; EC: Ethylcellulose; PB4: Poly(vinylalcohol-co-vinyl-4-methoxycinnamate)

5. Nanoencapsulation of Essential Oils

The antimicrobial activity of essential oils has been extensively studied *in vitro*, in model food systems and in actual food systems (examples of recent reviews are Fierascu *et al.*, 2018; Hu *et al.*, 2018; Pisoschi *et al.*, 2018; Sharifi-Rad *et al.*, 2018a, b; Stringaro *et al.*, 2018; Vasconcelos *et al.*, 2018). Despite effectiveness, their use in food is hindered by the effect on the organoleptic properties and their reduced activity compared to the respective *in vitro* that occurs due to their poor solubility in aqueous systems and their hydrophobic interactions with other food components. These disadvantages may be effectively addressed by encapsulation in general and nanoencapsulation in particular. Nanoencapsulation and nanoemulsification of essential oils have been the epicenter of intensive study aimed at both applications in the food industry and biomedical ones (examples of recent review articles are Bilia *et al.*, 2014; Ferreira and Nunes, 2019). In Table 5, a few representative recent studies assessing the nanoencapsulation of essential oils are presented. Regarding the application in actual food systems, nanoemulsification has been extensively considered, either as a dipping solution or through incorporation in edible coatings.

Nanoemulsification of essential oils, mostly into chitosan-based edible coatings, has been extensively considered. It seems an attractive way to reduce quality deterioration originating either from physicochemical reasons, such as moisture loss and oxidation and at the same time to suppress growth of microorganisms that may contaminate fresh products during minimal processing. Moreover, coating may serve as a carrier of compounds that may offer additional functions, such as antimicrobial activity; chitosan is usually selected to form the coating as it exerts antimicrobial activity by itself. Thus, effectiveness of a variety of essential oils to prolong shelf-life of fresh-cut products has been assessed. Kim *et al.* (2013) studied the ability of lemongrass essential oil incorporated into a carnauba wax-based coating to inhibit growth of *Salmonella* Typhimurium and *Escherichia coli* O157:H7 and at the same time preserve important quality parameters of plums. The coating effectively inhibited growth of both pathogens during storage at 4 and 25°C in an essential oil concentration-dependent manner. Flavor, fracturability and glossiness of the plums were not negatively affected by the coating.

Severino *et al.* (2014a, b) studied the effectiveness of edible coatings consisting of carvacrol, bergamot, lemon or mandarin essential oils nanoemulsified into native or modified chitosan when combined with ozonated water, UV-C and γ-ray treatments. More accurately, Severino *et al.* (2014a) inoculated broccoli florets with a cocktail of 5 *Listeria monocytogenes* strains (serotypes 1/2a and 1/2b) at 6 logCFU/g, applied the aforementioned treatments and stored the product at 4°C for 13 days. It was reported that the most effective coating was the one containing 0.05 per cent (w/v) mandarin essential oil, which resulted in a reduction of the population by 1.46 logCFU/g after six days of storage. The most effective combination was the one of gamma irradiation with essential oil nanoemulsified coating which restricted the growth of the pathogen to less than 1 logCFU/g during storage. Severino *et al.* (2014b) inoculated green beans with 6 logCFU/g *L. innocua* and studied the inactivation potential of ozonated water, UV-C and γ-ray treatments combined with modified chitosan-based coating containing mandarin essential oil during storage at 4°C for 15 days. In this case, the most effective combination was the one between UV-C and coating, which resulted in a 3 logCFU/g reduction of *L. innocua* population. Furthermore, *L. innocua* population

Table 5: Representative Recent Studies Assessing the Nanoencapsulation of Essential Oils

Loaded Essential Oil	NP System	Size (nm)	Polydispersity Index	Zeta Potential (mV)	Comment	Reference
Thyme oil	Sodium caseinate and soy lecithin	82.5 ± 3.0 to 125.5 ± 2.4 depending on the formulation	0.22 ± 0.01 to 0.27 ± 0.01 depending on the formulation	Nr	Nanoencapsulation resulted in more pronounced initial reduction of *S.* Enteritidis, *L. monocytogenes* and *E. coli* populations in both TSB broth and 2 per cent reduced fat milk at 21°C than free thyme oil. However, the difference between nanoencapsulated and free thyme oil after 72h was strain and concentration dependent	Xue *et al.*, 2015
Lippia sidoides essential oil	Chitosan plus cashew gum, chichá gum and angico gum	17 to 800 nm	Nr	− 40 to + 30	The effectiveness of the proposed approach was exhibited. In addition, the average encapsulating efficiency was 62 per cent, with cashew and chichá gum affecting positively loading capacity	Paula *et al.*, 2017
Salvia officinalis essential oil	Nanoemulsion with Tween 80 and Span 80	222.3 ± 2.1	Nr	Nr	Nanoemulsion decreased the MIC of the non-nanoemulsified sage oil by two times regarding *Sh. dysenteriae* and by four times regarding *S. Typhi* and *E. coli*. Destruction of the cell membrane is the proposed mode of action.	Moghimi *et al.*, 2016

(Contd.)

Thymol, eugenol	Nanoemulsion with LAE and lecithin	55 and 75 for thymol and eugenol, respectively	Nr	ca. + 40	The combination of lecithin and LAE improved the physical properties of the nanoemulsions but failed to improve antimicrobial activity, especially against Gram-negative bacteria.	Ma *et al.*, 2016
Eugenol	Nanoemulsion with gum arabic and lecithin	103.6 ± 7.5	<0.2	-50 to -40	MIC and MBC against *L. monocytogenes* were strain dependent. No such trend was revealed regarding *S.* Enteritidis	Hu *et al.*, 2016
Origanum majorana, *Syzgium aromaticum*, *Cinnamomum zeylanicum* essential oils	Alginate/clay nanocomposite films	Na	Na	Na	The antimicrobial activity of the examined essential oils against *E. coli*, *St. aureus*, and *L. monocytogenes* was retained throughout a 12 days' storage period when they were included in the nanocomposite films. The activity dependent upon the amount of the incorporated essential oil	Alboofetileh *et al.*, 2014
Eugenia caryophyllata essential oil	chitosan nanoparticles	ca. 300	Nr	ca. 30	Controlled release over 56 days was observed. Improvement of the antimould activity against *A. niger* compared to the nonencapsulated essential oil was reported	Hasheminejad *et al.*, 2019

LAE: Lauryl arginate; Nr: Not reported; Na: Not applicable; *S.*: *Salmonella*; *L.*: *Listeria*; *E.*: *Escherichia*; *St.*: *Staphylococcus*; *Sh.*: *Shigella*; *A.*: *Aspergillus* remained constant throughout storage. In addition, no loss of firmness or color was reported. The effectiveness of the afore-mentioned essential oils when combined with modified atmosphere packaging (MAP) and gamma irradiation to inactivate *E. coli* O157:H7 and *S.* Typhimurium that were inoculated at 6 logCFU/g on green bean during storage at 4°C for 13 days, was assessed by Severino *et al.* (2015). The most effective combination was γ-irradiation treatment, coating and MAP storage which resulted in reduction of pathogenic bacteria population to undetectable levels throughout storage.

Preservation of rucola color and texture as well as inhibition of microbial growth, especially during the first days of storage (4°C for three days and then 8°C for 21 days) upon coating with chitosan containing nanoemulsified lemon essential oil, was reported by Sessa *et al.* (2015). Salvia-Trujillo *et al.* (2015) studied *E. coli* inactivation, microbiological stability and the evolution of physicochemical quality parameters, such as color and firmness, of Fuji apple pieces during storage at 4°C. More accurately, the pathogen was inoculated at 6 logCFU/g onto apple pieces, which were then covered with nanoemulsion-based alginate edible coatings in which lemongrass essential oil was nanoemulsified. The pieces were placed in trays, sealed with polypropylene film and placed at 4°C. The presented results highlighted the advantage of nanoemulsification over conventional emulsification in incorporation of the essential oil, regarding both safety and quality maintenance. More accurately, more effective inactivation of *E. coli* as well as indigenous microbiota over conventional emulsification during storage was observed; on the other hand, no significant changes in quality parameters were observed. Tastan *et al.* (2017) inoculated fresh-cut cucumber slices with 7 logCFU/g *E. coli* and investigated the decontamination efficacy of pulsed light combined with chitosan-based edible coating containing carvacrol nanoemulsions. The antimicrobial coating resulted in the reduction of the population by less than 1 logCFU/g. Similarly, pulsed light treatment resulted in a reduction by 0.6-2.8 logCFU/g, depending on the dose. The effectiveness of the combined treatment was highlighted since the reduction of the population reached 5.9 logCFU/g with the combination of 3 per cent carvacrol nanoemulsion with a pulsed light dose of 12 J/cm^2. Finally, Artiga-Artigas *et al.* (2017) developed an alginate-based coating containing mandarin fiber and oregano essential oil, applied it to low-fat cut cheese and studied the effect on texture, water vapor resistance, color, inoculated *Staphylococcus aureus* as well as native microbiota during storage at 4°C for 15 days. Coated cheese presented lower whiteness index (WI). No significant changes in WI during storage were observed when oregano essential oil was applied at concentrations above 2.0 per cent. On the contrary, when it was applied at less than 1.5 per cent, an increase in WI was observed until the 13th day. Textural analysis, hardness, elasticity, cohesiveness and gumminess in the coated cheese were comparable irrespective of the amount of essential oil incorporated but lower than uncoated samples, especially hardness and gumminess. During storage, elasticity and cohesiveness remained unaltered; hardness and gumminess of the uncoated samples were reduced to the respective levels of coated ones and adhesiveness was reduced in all samples but more intensively in uncoated ones. Water vapor resistance was higher in the uncoated samples followed by the coating containing 2.5 per cent essential oil. The lowest water vapor resistance was observed in the coated samples with 2.0 and 1.5 per cent essential oil. From a microbiological perspective, essential oil restricted growth of psychrophilic bacteria, yeasts and molds as well as the inoculated *St. aureus*.

Dipping into essential oil nanoemulsions has also been extensively studied as a processing step aimed at decontaminating mostly fresh products. Landry *et al.* (2014,

2015) examined the effectiveness of carvacrol nanoemulsion against *S*. Enteritidis and *E. coli* O157:H7 that were inoculated on to *mung* bean, alfalfa seeds, radish seeds and broccoli seeds. Soaking of the former three into the nanoemulsion containing ≥ 4000 ppm carvacrol for more than 30min resulted in complete inactivation of the pathogens when they were inoculated at 2 or 3 logCFU/g. Moreover, the pathogens could not be detected after sprouting. On the contrary, complete inactivation was not observed when the pathogens were inoculated at 5 or 8 logCFU/g. In that case, a reduction of the population by ca. 2-3 logCFU/g was evident. As far as broccoli seeds are concerned, the ineffectiveness of this approach was exhibited since the reduction of the pathogens population was ca. 1 logCFU/g and the pathogens were always detected after sprouting. This was attributed to the increased glucosinolate, isothiocyanate and sulforaphane content of broccoli seeds and possible interactions that may destabilize the nanoemulsion or the reactive hydroxyl group of carvacrol. Similar results were reported by Landry *et al*. (2016). In that study, the effectiveness of carvacrol nanoemulsion acidified with 50 mM acetic or levulinic acid to inactivate a cocktail of *S*. Enteritidis inoculated on *mung* beans and broccoli seeds was assessed. Complete inactivation of the pathogen was observed only at the inoculum level of 4 logCFU/g. At higher inoculum levels, a reduction of the population by 1-5 logCFU/g, depending on the treatment, was reported. Ruengvisesh *et al*. (2015) encapsulated eugenol and carvacrol into surfactant micelles consisting of Tween 20, sodium dodecyl sulfate (SDS) and CytoGuard®LA 20 (CG20) and examined the decontamination efficiency of spinach leaf surfaces inoculated with ca. 7 logCFU/g of *E. coli* O157:H7 and *S*. Saintpaul. Regarding both pathogens, the lowest minimum inhibitory concentration (MIC) and minimum bactericidal concentration (MBC) were obtained by SDS plus eugenol or carvacrol while the highest values were observed by Tween 20 plus eugenol or carvacrol. Application of eugenol-loaded nanomicelles on spinach surface resulted in reduction of the population to 2.2-4.9 logCFU/cm^2 and to 2.2-4.8 logCFU/cm^2 regarding *E. coli* O157:H7 and *S*. Saintpaul, respectively, independent of the application method, i.e. spraying or immersion.

The MIC and MBC of oregano essential oil nanoemulsion against *L. monocytogenes*, *S*. Typhimurium and *E. coli* O157:H7 were determined by Bhargava *et al*. (2015) at 0.0625 per cent with the exception of *E. coli* O157:H7 MBC that was determined at 0.125 per cent. Then, lettuce was inoculated with 7 logCFU/g of the pathogens, dipped into nanoemulsions of 0.05 per cent and 0.1 per cent oregano essential oil for 1min and stored at 4°C for 72h. The population of all three pathogens was reduced by 2.31 to 3.57 logCFU/g with *E. coli* O157:H7 being the most sensitive and *S*. Typhimurium the least. Only marginal differences were observed between the different oregano oil concentrations. Nanoemulsified oregano essential oil was also studied by Moraes-Lovison *et al*. (2017). In that study, a nanoemulsion was developed through the phase inversion temperature method. The MIC and MBC, using freshly prepared nanoemulsions for *St. aureus*, were reported at 0.56 ± 0.06 and 0.90 mg/mL, respectively, and for *E. coli* at 0.60 and 3.32 ± 0.58, respectively. Then, chicken pate was inoculated with 10^4-10^7 CFU/g of the pathogens and stored at 4°C for eight days. The population of the pathogens remained constant throughout storage. On the contrary, the addition of pure oregano essential oil (0.6 g/kg) or the nanoemulsified preparation (12 g/kg) resulted only in marginal population reduction. Ozogul *et al*. (2017) studied the antimicrobial effect of rosemary (*Rosmarinus officinalis*), laurel (*Laurus nobilis*), sage (*Salvia officinalis*) and thyme (*Thymus vulgaris*) essential oil nanoemulsions on rainbow trout (*Oncorhynchus mykiss*) during ice storage for 24 days. Only marginal differences in the mesophilic, psychrotrophic and Enterobacteriaceae population between the untreated samples and the ones treated with the essential oil preparations were observed. The combination of acidic electrolyzed water containing ≤ 4 mg/L free available chlorine and carvacrol nanoemulsion to control the autochthonous microbiota of shredded cabbage was assessed by Sow *et al.* (2017). Only marginal reduction, i.e. less than 1 logCFU/g, of total aerobic mesophilic, total aerobic psychrophilic and yeast and mould counts was achieved. Hill *et al.* (2017) encapsulated cinnamon bark extract (CBE) into chitosan-co-poly-N-isopropylacrylamide (chitosan-PNIPAAM), beta-cyclodextrin (BCD) and poly-D,L-lactide-co-glycolide (PLGA) and applied their aqueous solution on the surface of organic romaine lettuce (*Lactuca sativa* L. var. longifolia) that had been previously inoculated with 4.5-6.5 logCFU/g of *L. monocytogenes*. The product was stored at 5 and 10°C and the population of the pathogen was monitored regularly. The PNIPAAM-chitosan-CBE formulation offered growth inhibition throughout storage; the population of the pathogen remained at the inoculation level at both temperatures. This was also the case when the BCD-CBE was applied. However, when BCD-CBE was applied, growth of the pathogen during the initial days of storage was observed. The effectiveness of the PNIPAAM-chitosan-CBE formulation was attributed to the initial high level of CBE release followed by a steady gradual one that characterizes this specific material (Hill and Gomes, 2014). The most effective combination was further assessed for the effect on color, chlorophyll and carotenoid contents as well as the organoleptic properties. Color, in terms of L*, a* and b* parameters was not affected by the application of nanoencapsulated CBE. On the contrary, total chlorophyll and carotenoid content was significantly affected, depending on the amount of CBE and chitosan-PNIPAAM nanoparticles used. Regarding the organoleptic evaluation of the product, no significant differences between the evaluated quality characteristics were observed.

Regarding other food products, Pan *et al.* (2014) encapsulated thymol in sodium caseinate and studied the activity against *L. monocytogenes* in milk of different fat concentration (1.1-1.8 per cent). The size of the nanocapsules was less than 140 nm and got slightly decreased during 30 days' storage at 21°C. The zeta potential was significantly affected by the pH value and the incorporation of thymol. Regarding the antilisterial activity, encapsulation resulted in a more rapid reduction of the pathogen population that was less effective at higher fat levels.

6. Conclusion and Future Perspectives

Nanoencapsulation of processing aids, such as antimicrobial compounds, flavors and pigments as well as sensitive nutrients, such as vitamins, fatty acids and antioxidants has been studied to some extent. The results obtained from *in vitro* experimentation have exhibited the flexibility of the available methodological approaches and have improved our knowledge regarding the factors that affect nanoencapsulation quality as well as a series of adjustments that may improve the process. On the contrary, the effect that the complexity of actual food systems may have on the stability of the nanostrucuture and the biological activity of the encapsulated compounds is yet to be thoroughly assessed. Due to the increased significance that these applications have, both from technological advancement and nutritional perspectives, it is expected that research will focus on addressing such issues in the near future.

References

Alboofetileh, M., Rezaei, M., Hosseini, H. and Abdollahi, M. (2014). Antimicrobial activity of alginate/clay nanocomposite films enriched with essential oils against three common foodborne pathogens, *Food Control*, 36: 1-7.

Almeida, H.H.S., Barros, L., Barreira, J.C.M., Calhelha, R.C., Heleno, S.A., Sayer, C., Miranda, C.G., Leimann, F.V., Barreiro, M.F. and Ferreira, I.C.F.R. (2018). Bioactive evaluation and application of different formulations of the natural colorant curcumin (E100) in a hydrophilic matrix (yogurt), *Food Chemistry*, 261: 224-232.

Artiga-Artigas, M., Acevedo-Fani, A. and Martín-Belloso, O. (2017). Improving the shelf-life of low-fat cut cheese using nanoemulsion based edible coatings containing oregano essential oil and mandarin fiber, *Food Control*, 76: 1-12.

Assadpour, E., Maghsoudlou, Y., Jafari, S.-M., Ghorbani, M. and Aalami, M. (2016). Optimization of folic acid nano-emulsification and encapsulation by maltodextrin-whey protein double emulsions, *International Journal of Biological Macromolecules*, 86: 197-207.

Assadpour, E. and Jafari, S.M. (2018). A systematic review on encapsulation of food bioactive ingredients and nutraceuticals by various nanocarriers, *Critical Reviews in Food Science and Nutrition*, 8: 1-47.

Bhargava, K., Conti, D.S., da Rocha, S.R. and Zhang, Y. (2015). Application of an oregano oil nanoemulsion to the control of foodborne bacteria on fresh lettuce, *Food Microbiology*, 47: 69-73.

Bilia, A.R., Guccione, C., Isacchi, B., Righeschi, C., Firenzuoli, F. and Bergonzi, M.C. (2014). Essential oils loaded in nanosystems: A developing strategy for a successful therapeutic approach, *Evidence-based Complementary and Alternative Medicine*, 2014: 651593.

Blancquaert, D., Van Daele, J., Strobbe, S., Kiekens, F., Storozhenko, S., De Steur, H., Gellynck, X., Lambert, W., Stove, C. and Van Der Straeten, D. (2015). Improving folate (vitamin B$_9$) stability in biofortified rice through metabolic engineering, *Nature Biotechnology*, 33: 1076-1078.

Bochicchio, S., Barba, A.A., Grassi, G. and Lamberti, G. (2016). Vitamin delivery: Carriers based on nanoliposomes produced via ultrasonic irradiation, *LWT – Food Science and Technology*, 69: 9-16.

Cakmak, I., Pfeiffer, W.H. and Mcclafferty, B. (2010). Biofortification of durum wheat with zinc and iron, *Cereal Chemistry*, 87: 10-20.

Campani, V., Biondi, M., Mayol, L., Cilurzo, F., Pitaro, M. and De Rosa, G. (2016). Development of nanoemulsions for topical delivery of vitamin K$_1$, *International Journal of Pharmaceutics*, 511: 170-177.

Chattha, M.U., Hassan, M.U., Khan, I., Chattha, M.B., Mahmood, A., Chattha, M.U., Nawaz, M., Subhani, M.N., Kharal, M. and Khan, S. (2017). Biofortification of wheat cultivars to combat zinc deficiency, *Frontiers in Plant Science*, 8: 281.

Choi, M.-J., Ruktanonchai, U., Min, S.-G., Chun, J.-Y. and Soottitantawat, A. (2010). Physical characteristics of fish oil encapsulated by β-cyclodextrin using an aggregation method or polycaprolactone using an emulsion–diffusion method, *Food Chemistry*, 119: 1694-1703.

Cholewski, M., Tomczykowa, M. and Tomczyk, M. (2018). A comprehensive review of chemistry, sources and bioavailability of omega-3 fatty acids, *Nutrients*, 10: 1662.

Ciccolini, V., Pellegrino, E., Coccina, A., Fiaschi, A.I., Cerretani, D., Sgherri, C., Quartacci, M.F. and Ercoli, L. (2017). Biofortification with iron and zinc improves nutritional and nutraceutical properties of common wheat flour and bread, *Journal of Agricultural and Food Chemistry*, 65: 5443-5452.

Couto, R., Alvarez, V. and Temelli, F. (2017). Encapsulation of Vitamin B$_2$ in solid lipid nanoparticles using supercritical CO$_2$, *The Journal of Supercritical Fluids*, 120: 432-442.

de Freitas Zompero, R.H., López-Rubio, A., de Pinho, S.C., Lagaron, J.M. and de la Torre, L.G. (2015). Hybrid encapsulation structures based on β-carotene-loaded nanoliposomes within electrospun fibers, *Colloids and Surfaces B: Biointerfaces*, 134: 475-482.

De Lepeleire, J., Strobbe, S., Verstraete, J., Blancquaert, D., Ambach, L., Visser, R.G.F., Stove, C. and Van Der Straeten, D. (2018). Folate biofortification of potato by tuber-specific expression of four folate biosynthesis genes, *Molecular Plant*, 11: 175-188.

De Moura, F.F., Miloff, A. and Boy, E. (2015). Retention of provitamin A carotenoids in staple crops targeted for biofortification in Africa: Cassava, maize and sweet potato, *Critical Reviews in Food Science and Nutrition*, 55: 1246-1269.

De Moura, F.F., Moursi, M., Donahue Angel, M., Angeles-Agdeppa, I., Atmarita, A., Gironella, G.M., Muslimatun, S. and Carriquiry, A. (2016). Biofortified β-carotene rice improves vitamin A intake and reduces the prevalence of inadequacy among women and young children in a simulated analysis in Bangladesh, Indonesia, and the Philippines, *The American Journal of Clinical Nutrition*, 104: 769-775.

Díaz de la Garza, R.I., Gregory, J.F. and Hanson, A.D. (2007). Folate biofortification of tomato fruit, *Proceedings of the National Academy of Sciences of the USA*, 104: 4218-4222.

D'Imperio, M., Renna, M., Cardinali, A., Buttaro, D., Serio, F. and Santamaria, P. (2016). Calcium biofortification and bioaccessibility in soilless 'baby leaf' vegetable production, *Food Chemistry*, 213: 149-156.

Dong, W., Stockwell, V.O. and Goyer, A. (2015). Enhancement of thiamin content in *Arabidopsis thaliana* by metabolic engineering, *Plant and Cell Physiology*, 56: 2285-2296.

Dong, W., Thomas, N., Ronald, P.C. and Goyer, A. (2016). Overexpression of thiamin biosynthesis genes in rice increases leaf and unpolished grain thiamin content but not resistance to *Xanthomonas oryzae* pv. *Oryzae, Frontiers in Plant Science*, 7: 616.

Fathima, S.J., Fathima, I., Abhishek, V. and Khanum, F. (2016). Phosphatidylcholine, an edible carrier for nanoencapsulation of unstable thiamine, *Food Chemistry*, 197: 562-570.

Ferreira, C.D. and Nunes, I.L. (2019). Oil nanoencapsulation: Development, application, and incorporation into the food market, *Nanoscale Research Letters*, 14: 9.

Fierascu, I., Dinu-Pirvu, C.E., Fierascu, R.C., Velescu, B.S., Anuta, V., Ortan, A. and Jinga, V. (2018). Phytochemical profile and biological activities of *Satureja hortensis* L.: A review of the last decade, *Molecules*, 23: E2458.

Garcia-Moreno, P.J., Stephansen, K., van der Kruijs, J., Guadix, A., Guadix, E.M., Chronakis, I.S. and Jacobsen, C. (2016). Encapsulation of fish oil in nanofibers by emulsion electrospinning: Physical characterization and oxidative stability, *Journal of Food Engineering*, 183: 39-49.

Ghorbanzade, T., Jafari, S.M., Akhavan, S. and Hadavi, R. (2017). Nano-encapsulation of fish oil in nanoliposomes and its application in fortification of yogurt, *Food Chemistry*, 216: 146-152.

Gibbs, B.F., Kermasha, S., Alli, I. and Mulligan, C.N. (1999). Encapsulation in the food industry: A review, *International Journal of Food Sciences and Nutrition*, 50: 213-224.

Gorzynik-Debicka, M., Przychodzen, P., Cappello, F., Kuban-Jankowska, A., Gammazza, A.M., Knap, N., Wozniak, M. and Gorska-Ponikowska, M. (2018). Potential health benefits of olive oil and plant polyphenols, *International Journal of Molecular Sciences*, 19: 686.

Goto, F., Yoshihara, T., Shigemoto, N., Toki, S. and Takaiwa, F. (1999). Iron fortification of rice seed by the soybean ferritin gene, *Nature Biotechnology*, 17: 282-286.

Gulotta, A., Saberi, A.H., Nicoli, M.C. and McClements, D.J. (2014). Nanoemulsion-based delivery systems for polyunsaturated (ω-3) oils: Formation using a spontaneous emulsification method, *Journal of Agricultural and Food Chemistry*, 62: 1720-1725.

Gulseren, İ., Fang, Y. and Corredig, M. (2012). Zinc incorporation capacity of whey protein nanoparticles prepared with desolvation with ethanol, *Food Chemistry*, 135: 770-774.

Haider, J., Majeed, H., Williams, P.A., Safdar, W. and Zhong, F. (2017). Formation of chitosan nanoparticles to encapsulate krill oil (*Euphausia superba*) for application as a dietary supplement, *Food Hydrocolloids*, 63: 27-34.

Hasheminejad, N., Khodaiyan, F. and Safari, M. (2019). Improving the antifungal activity of clove essential oil encapsulated by chitosan nanoparticles, *Food Chemistry*, 275: 113-122.

Heydari, A., Doostan, F., Khoshnood, H. and Sheibani, H. (2016). Water-soluble cationic poly (β-cyclodextrin-co-guanidine) as a controlled vitamin B_2 delivery carrier, *RSC Advances*, 6: 33267-33278.

Hill, L.E. and Gomes, C.L. (2014). Characterization of temperature and pH-responsive poly-Nisopropylacrylamide-co-polymer nanoparticles for the release of antimicrobials, *Materials Research Express*, 1: 35405.

Hill, L.E., Oliveira, D.A., Hills, K., Giacobassi, C., Johnson, J., Summerlin, H., Taylor, T.M. and Gomes, C.L. (2017). A comparative study of natural antimicrobial delivery systems for microbial safety and quality of fresh-cut lettuce, *Journal of Food Science*, 82: 1132-1141.

Hu, Q., Gerhard, H., Upadhyaya, I., Venkitanarayanan, K. and Luo, Y. (2016). Antimicrobial eugenol nanoemulsion prepared by gum arabic and lecithin and evaluation of drying technologies, *International Journal of Biological Macromolecules*, 87: 130-140.

Hu, Q., Zhou, M. and Wei, S. (2018). Progress on the antimicrobial activity research of clove oil and eugenol in the food antisepsis field, *Journal of Food Science*, 83: 1476-1483.

Jimenez-Fernandez, E., Ruyra, A., Roher, N., Zuasti, E., Infante, C. and Fernández-Díaz, C. (2014). Nanoparticles as a novel delivery system for vitamin C administration in aquaculture, *Aquaculture*, 432: 426-433.

Johnson, A.T., Kyriacou, B., Callahan, D.L., Carruthers, L., Stangoulis, J., Lombi, E. and Tester M. (2011). Constitutive overexpression of the OsNAS gene family reveals single-gene strategies for effective iron- and zinc-biofortification of rice endosperm, *PLoS One*, 6: e24476.

Karunanandaa, B., Qi, Q., Hao, M., Baszis, S.R., Jensen, P.K., Wong, Y.H., Jiang, J., Venkatramesh, M., Gruys, K.J., Moshiri, F., Post-Beittenmiller, D., Weiss, J.D. and Valentin, H.E. (2005). Metabolically engineered oilseed crops with enhanced seed tocopherol, *Metabolic Engineering*, 7: 384–400.

Kim, C.K., Han, J.S., Lee, H.S., Oh, J.Y., Shigaki, T., Park, S.H. and Hirschi, K. (2006). Expression of an *Arabidopsis* CAX2 variant in potato tubers increases calcium levels with no accumulation of manganese, *Plant Cell Reports*, 25: 1226-1232.

Kim, I.-H., Lee, H., Kim, J.E., Song, K.B., Lee, Y.S., Chung, D.S. and Min, S.C. (2013). Plum coatings of lemongrass oil-incorporating carnauba wax-based nanoemulsion, *Journal of Food Science*, 78: E1551-E1559.

Kosaraju, S.L., Tran, C. and Lawrence, A. (2006). Liposomal delivery systems for encapsulation of ferrous sulfate: Preparation and characterization, *Journal of Liposome Research*, 16: 347-358.

Kutman, U.B., Yildiz, B., Ozturk, L. and Cakmak, I. (2010). Biofortification of durum wheat with zinc through soil and foliar applications of nitrogen, *Cereal Chemistry*, 87: 1-9.

Lacatusu, I., Mitrea, E., Badea, N., Stan, R., Oprea, O. and Meghea, A. (2013). Lipid nanoparticles based on omega-3 fatty acids as effective carriers for lutein delivery. Preparation and in vitro characterization studies, *Journal of Functional Foods*, 5: 1260-1269.

Landry, K.S., Chang, Y., McClements, D.J. and McLandsborough, L. (2014). Effectiveness of a novel spontaneous carvacrol nanoemulsion against *Salmonella enterica* Enteritidis Enteritidis and *Escherichia coli* O157:H7 on contaminated mung bean and alfalfa seeds, *International Journal of Food Microbiology*, 187: 15-21.

Landry, K.S., Micheli, S., McClements, D.J. and McLandsborough, L. (2015). Effectiveness of a spontaneous carvacrol nanoemulsion against *Salmonella enterica* Enteritidis and *Escherichia coli* O157:H7 on contaminated broccoli and radish seeds, *Food Microbiology*, 51: 10-17.

Landry, K.S., Komaiko, J., Wong, D.E., Xu, T., McClements, D.J. and McLandsborough, L. (2016). Inactivation of *Salmonella* on sprouting seeds using a spontaneous carvacrol nanoemulsion acidified with organic acids, *Journal of Food Protection*, 79: 1115-1126.

Lee, S., Jeon, U.S., Lee, S.J., Kim, Y.K., Persson, D.O., Husted, S., Schjørring, J.K., Kakei, Y., Masuda, H., Nishizawa, N.K. and An, G. (2009). Iron fortification of rice seeds through activation of the nicotianamine synthase gene, *Proceedings of the National Academy of Science of the USA*, 106: 22014-22019.

Li, A.-N., Li, S., Zhang, Y.-J., Xu, X.-R., Chen, Y.-M. and Li, H.-B. (2014). Resources and biological activities of natural polyphenols, *Nutrients*, 6: 6020-6047.

Li, D., Wang, P., Luo, Y., Zhao, M. and Chen, F. (2017). Health benefits of anthocyanins and molecular mechanisms: Update from recent decade, *Critical Reviews in Food Science and Nutrition*, 57: 1729-1741.

Li, K.T., Moulin, M., Mangel, N., Albersen, M., Verhoeven-Duif, N.M., Ma, Q., Zhang, P., Fitzpatrick, T.B., Gruissem, W. and Vanderschuren, H. (2015a). Increased bioavailable vitamin B_6 in field-grown transgenic cassava for dietary sufficiency, *Nature Biotechnology*, 33: 1029-1032.

Li, M., Zahi, M.R., Yuan, Q., Tian, F. and Liang, H. (2016). Preparation and stability of astaxanthin solid lipid nanoparticles based on stearic acid, *European Journal of Lipid Science and Technology*, 118: 592-602.

Li, T., Yang, S., Liu, W., Liu, C., Liu, W., Zheng, H., Zhou, W. and Tong, G. (2015b). Preparation and characterization of nanoscale complex liposomes containing medium-chain fatty acids and vitamin C, *International Journal of Food Properties*, 18: 113-124.

Liu, W., Tian, M., Kong, Y., Lu, J., Li, N. and Han, J. (2017). Multilayered vitamin C nanoliposomes by self assembly of alginate and chitosan: Long-term stability and feasibility application in mandarin juice, *LWT – Food Science and Technology*, 75: 608-615.

Luo, Y., Zhang, B., Cheng, W.H. and Wang, Q. (2010). Preparation, characterization and evaluation of selenite-loaded chitosan/TPP nanoparticles with or without zein coating, *Carbohydrate Polymer*, 82: 942-951.

Ma, Q., Davidson, P.M. and Zhong, Q. (2016). Nanoemulsions of thymol and eugenol co-emulsified by lauric arginate and lecithin, *Food Chemistry*, 206: 167-173.

Masuda, H., Suzuki, M., Morikawa, K.C., Kobayashi, T., Nakanishi, H., Takahashi, M., Higuchi, K., Nakanishi, H. and Nishizawa, N.K. (2008). Increase in iron and zinc concentrations in rice grains via the introduction of barley genes involved in phytosiderophore synthesis, *Rice*, 1: 100-108.

Masuda, H., Ishimaru, Y., Aung, M.S., Kobayashi, T., Kakei, Y., Takahashi, M., Higuchi, K., Nakanishi, H. and Nishizawa, N.K. (2012). Iron biofortification in rice by the introduction of multiple genes involved in iron nutrition, *Scientific Reports*, 2: 1-6.

Mehrnia, M.-A., Jafari, S.-M., Makhmal-Zadeh, B.S. and Maghsoudlou, Y. (2016). Crocin loaded nano-emulsions: Factors affecting emulsion properties in spontaneous emulsification, *International Journal of Biological Macromolecules*, 84: 261–267.

Mimmo, T., Tiziani, R., Valentinuzzi, F., Lucini, L., Nicoletto, C., Sambo, P., Scampicchio, M., Pii, Y. and Cesco, S. (2017). Selenium biofortification in *Fragaria ananassa*: Implications on strawberry fruits quality, content of bioactive health beneficial compounds and metabolomic profile, *Frontiers in Plant Science*, 8: 1887.

Min, K.A., Cho, J.H., Song, Y.K. and Kim, C.K. (2016). Iron casein succinylate-chitosan coacervate for the liquid oral delivery of iron with bioavailability and stability enhancement, *Archives of Pharmacal Research*, 39: 94-102.

Moghimi, R., Aliahmadi, A., McClements, D.J. and Rafati, H. (2016). Investigations of the effectiveness of nanoemulsions from sage oil as antibacterial agents on some food borne pathogens, *LWT – Food Science and Technology*, 71: 69-76.

Moraes-Lovison, M., Marostegan, L.F.P., Peres, M.S., Menezes, I.F., Ghiraldi, M., Rodrigues, R.A.F., Fernandes, A.M. and Pinho, S.C. (2017). Nanoemulsions encapsulating oregano essential oil: Production, stability, antibacterial activity and incorporation in chicken pate. *LWT – Food Science and Technology*, 77: 233-240.

Morris, J., Hawthorne, K.M., Hotze, T., Abrams, S.A. and Hirschi, K.D. (2008). Nutritional impact of elevated calcium transport activity in carrots, *Proceedings of the National Academy of Sciences of the USA*, 105: 1431-1435.

Muzhingi, T., Palacios-Rojas, N., Miranda, A., Cabrera, M.L., Yeum, K.J. and Tang, G. (2017). Genetic variation of carotenoids, vitamin E and phenolic compounds in Provitamin A biofortified maize, *Journal of the Science of Food and Agriculture*, 97: 793-801.

Naveen, S. and Kanum, F. (2014). Characterization and evaluation of iron nano-emulsion prepared by high speed homogenization, *International Journal of Biology, Pharmacy and Allied Sciences*, 3: 45-55.

Nejadmansouri, M., Hosseini, S.M.H., Niakosari, M., Yousefi, G.H. and Golmakan, M.T. (2016). Physicochemical properties and storage stability of ultrasound-mediated WPI-stabilized fish oil nanoemulsions, *Food Hydrocolloids*, 61: 801-811.

Nerome, H., Machmudah, S., Wahyudiono, Fukuzato, R., Higashiura, T., Youn, Y.-S., Lee, Y.-W. and Goto, M. (2013). Nanoparticle formation of lycopene/β-cyclodextrin inclusion complex using supercritical antisolvent precipitation, *Journal of Supercritical Fluids*, 83: 97-103.

Ozogul, Y., Yuvka, İ., Ucar, Y., Durmus, M., Kösker, A.R., Öz, M. and Ozogul, F. (2017). Evaluation of effects of nanoemulsion based on herb essential oils (rosemary, laurel, thyme and sage) on sensory, chemical and microbiological quality of rainbow trout (*Oncorhynchus mykiss*) fillets during ice storage, *LWT – Food Science and Technology*, 75: 677-684.

Ozturk, B., Argin, S., Ozilgen, M. and McClements, D.J. (2014). Formation and stabilization of nanoemulsion-based vitamin E delivery systems using natural surfactants: Quillaja saponin and lecithin, *Journal of Food Engineering*, 142: 57-63.

Palmer, A.C., Healy, K., Barffour, M.A., Siamusantu, W., Chileshe, J., Schulze, K.J., West, K.P. Jr. and Labrique, A.B. (2016). Provitamin A carotenoid-biofortified maize consumption increases pupillary responsiveness among Zambian children in a randomized controlled trial, *Journal of Nutrition*, 146: 2551-2558.

Pan, A., Chen, M., Chowdhury, R., Wu, J.H.Y., Sun, Q., Campos, H., Mozaffarian, D. and Hu, F.B. (2012). α-Linolenic acid and risk of cardiovascular disease: A systematic review and meta-analysis, *The American Journal of Clinical Nutrition*, 96: 1262-1273.

Pan, K., Chen, H., Davidson, P.M. and Zhong, Q. (2014). Thymol nanoencapsulated by sodium caseinate: Physical and antilisterial properties, *Journal of Agricultural and Food Chemistry*, 62: 1649-1657.

Park, S.H., Kang, T.S., Kim, C.K., Han, J.S., Kim, S., Smith, R.H., Pike, L.M. and Hirschi, K.D. (2005a). Genetic manipulation for enhancing calcium content in potato tuber, *Journal of Agricultural and Food Chemistry*, 53: 5598-5603.

Park, S., Cheng, N.H., Pittman, J.K., Yoo, K.S., Park, J., Smith, R.H. and Hirschi, K.D. (2005b). Increased calcium levels and prolonged shelf life in tomatoes expressing *Arabidopsis* Hþ/Ca2þ transporters, *Plant Physiology*, 139: 1194-1206.

Park, S.J., Garcia, C.V., Shin, G.H. and Kim, J.T. (2017). Development of nanostructured lipid carriers for the encapsulation and controlled release of vitamin D_3, *Food Chemistry*, 225: 213-219.

Paula, H.C.B., Oliveira, E.F., Carneiro, M.J.M. and de Paula, R.C.M. (2017). Matrix effect on the spray drying nanoencapsulation of *Lippia sidoides* essential oil in chitosan-native gum blends, *Planta Medica*, 83: 392-397.

Pegoraro, N.S., Barbieri, A.V., Camponogara, C., Mattiazzi, J., Brum, E.S., Marchiori, M.C.L., Oliveira, S.M. and Cruz, L. (2017). Nanoencapsulation of coenzyme Q10 and vitamin E acetate protects against UVB radiation-induced skin injury in mice, *Colloids and Surfaces B: Biointerfaces*, 150: 32-40.

Peng, H., Chen, S., Luo, M., Ning, F., Zhu, X. and Xiong, H. (2016). Preparation and self-assembly mechanism of bovine serum albumin–citrus peel pectin conjugated hydrogel: A potential delivery system for vitamin C, *Journal of Agricultural and Food Chemistry*, 64: 7377-7384.

Pezeshky, A., Ghanbarzadeh, B., Hamishehkar, H., Moghadam, M. and Babazadeh, A. (2016). Vitamin A palmitate-bearing nanoliposomes: Preparation and characterization, *Food Bioscience*, 13: 49-55.

Pisoschi, A.M., Pop, A., Georgescu, C., Turcuş, V., Olah, N.K. and Mathe, E. (2018). An overview of natural antimicrobials role in food, *European Journal of Medicinal Chemistry*, 143: 922-935.

Pistone, S., Qoragllu, D., Smistad, G. and Hiorth, M. (2015). Formulation and preparation of stable crosslinked alginate-zinc nanoparticles in the presence of a monovalent salt, *Soft Matter*, 11: 5765-5774.

Poblaciones, M.J., Rodrigo, S., Santamaria, O., Chen, Y. and McGrath, S.P. (2014). Selenium accumulation and speciation in biofortified chickpea (*Cicer arietinum* L.) under Mediterranean conditions, *Journal of the Science of Food and Agriculture*, 94: 1101-1106.

Poblaciones, M.J., Rodrigo, S.M. and Santamaria, O. (2013). Evaluation of the potential of peas (*Pisum sativum* L.) to be used in selenium biofortification programs under Mediterranean conditions, *Biological Trace Element Research*, 151: 132-137.

Raclaru, M., Gruber, J., Kumar, R., Sadre, R., Lühs, W., Zarhloul, M., Friedt, W., Frentzen, M. and Weier, D. (2006). Increase of the tocochromanol content in transgenic *Brassica napus* seeds by overexpression of key enzymes involved in prenylquinone biosynthesis, *Molecular Breeding*, 18: 93-107.

Ramírez Rivera, N.G., García-Salinas, C., Aragão, F.J. and Díaz de la Garza, R.I. (2016). Metabolic engineering of folate and its precursors in Mexican common bean (*Phaseolus vulgaris* L.), *Plant Biotechnology Journal*, 14: 2021-2032.

Ramos, S.J., Faquin, V., Guilherme, L.R.G., Castro, E.M., Avila, F.W., Carvalho, G.S., Bastos C.E.A. and Oliveira, C. (2010). Selenium biofortification and antioxidant activity in lettuce plants fed with selenate and selenite, *Plant Soil and Environment*, 56: 584-588.

Ramzani, P.M., Khalid, M., Naveed, M., Ahmad, R. and Shahid, M. (2016). Iron biofortification of wheat grains through integrated use of organic and chemical fertilizers in pH affected calcareous soil, *Plant Physiology and Biochemistry*, 104: 284-293.

Rao, A.V. and Rao, L.G. (2007). Carotenoids and human health, *Pharmacological Research*, 55: 207-216.

Raposo, M.F.D.J., Morais, A.M.M.B.D. and Morais, R.M.S.C.D. (2015). Carotenoids from marine microalgae: A valuable natural source for the prevention of chronic diseases, *Marine Drugs*, 13: 5128-5155.

Raschke, M., Boycheva, S., Crevecoeur, M., Nunes-Nesi, A., Witt, S., Fernie, A. R., Amrhein N. and Fitzpatrick, T.B. (2011). Enhanced levels of vitamin B-6 increase aerial organ size and positively affect stress tolerance in *Arabidopsis, The Plant Journal*, 66: 414-432.

Rasti, B., Jinap, S., Mozafari, M.R. and Yazid, A.M. (2012). Comparative study of the oxidative and physical stability of liposomal and nanoliposomal polyunsaturated fatty acids prepared with conventional and Mozafari methods, *Food Chemistry*, 135: 2761-2770.

Ravanfar, R., Tamaddon, A.M., Niakousari, M. and Moein, M.R. (2016). Preservation of anthocyanins in solid lipid nanoparticles: Optimization of a microemulsion dilution method using the Placket–Burman and Box–Behnken designs, *Food Chemistry*, 199: 573-580.

Ren, Q., Fan, F., Zhang, Z., Zheng, X. and DeLong, G.R. (2008). An environmental approach to correcting iodine deficiency: Supplementing iodine in soil by iodination of irrigation water in remote areas, *Journal of Trace Elements in Medicine and Biology*, 22: 1-8.

Rezaei, A., Fathi. M. and Jafari, S.M. (2019). Nanoencapsulation of hydrophobic and low-soluble food bioactive compounds within different nanocarriers, *Food Hydrocolloids*, 88: 146-162.

Rodriguez-Ruiz, V., Salatti-Dorado, J.Á., Barzegari, A., Nicolas-Boluda, A., Houaoui, A., Caballo, C., Caballero-Casero, N., Sicilia, D., Venegas, J.B., Pauthe, E., Omidi, Y., Letourneur, D., Rubio, S., Gueguen, V. and Pavon-Djavid, G. (2018). Astaxanthin-loaded nanostructured lipid carriers for preservation of antioxidant activity, *Molecules*, 23: 2601.

Ruengvisesh, S., Loquercio, A., Castell-Perez, E. and Taylor, T.M. (2015). Inhibition of bacterial pathogens in medium and on spinach leaf surfaces using plant-derived antimicrobials loaded in surfactant micelles, *Journal of Food Science*, 80: M2522-M2529.

Sahari, M.A., Moghimi, H.R., Hadian, Z., Barzegar, M. and Mohammadi, A. (2016). Improved physical stability of docosahexaenoic acid and eicosapentaenoic acid encapsulated using nanoliposome containing a-tocopherol, *International Journal of Food Science and Technology*, 51: 1075-1086.

Saiz-Abajo, M.-J., González-Ferrero, C., Moreno-Ruiz, A., Romo-Hualde, A. and González-Navarro, C.J. (2013). Thermal protection of β-carotene in re-assembled casein micelles during different processing technologies applied in food industry, *Food Chemistry*, 138: 1581-1587.

Salvia-Trujillo, L., Rojas-Graü, M.A., Soliva-Fortuny, R. and Martín-Belloso, O. (2015).

Use of antimicrobial nanoemulsions as edible coatings: Impact on safety and quality attributes of fresh-cut Fuji apples, *Postharvest Biology and Technology*, 105: 8-16.

Sathasivam, R. and Ki, J.-S. (2018). A review of the biological activities of microalgal carotenoids and their potential use in healthcare and cosmetic industries, *Marine Drugs*, 16: 26.

Seabra, C.L., Nunes, C., Gomez-Lazaro, M., Correia, M., Machado, J.C., Gonçalves, I.C., Reis, C.A., Reis, S. and Martins, M.C.L. (2017). Docosahexaenoic acid loaded lipid nanoparticles with bactericidal activity against *Helicobacter pylori*, *International Journal of Pharmaceutics*, 519: 128-137.

Sessa, M., Ferrari, G. and Donsì, F. (2015). Novel edible coating containing essential oil nanoemulsions to prolong the shelf life of vegetable products, *Chemical Engineering Transactions*, 43: 55-60.

Severino, R., Vu, K.D., Donsì, F., Salmieri, S., Ferrari, G. and Lacroix, M. (2014a). Antimicrobial effects of different combined non-thermal treatments against *Listeria monocytogenes* in broccoli florets, *Journal of Food Engineering*, 124: 1-10.

Severino, R., Vu, K.D., Donsì, F., Salmieri, S., Ferrari, G. and Lacroix, M. (2014b). Antibacterial and physical effects of modified chitosan based-coating containing nanoemulsion of mandarin essential oil and three non-thermal treatments against *Listeria innocua* in green beans, *International Journal of Food Microbiology*, 191: 82-88.

Severino, R., Ferrari, G., Vu, K.D., Donsì, F., Salmieri, S. and Lacroix, M. (2015). Antimicrobial effects of modified chitosan based coating containing nanoemulsion of essential oils, modified atmosphere packaging and gamma irradiation against *Escherichia coli* O157:H7 and *Salmonella* typhimurium on green beans, *Food Control*, 50: 215-222.

Sharifi, A., Golestan, L. and Sharifzadeh Baei, M. (2013). Studying the enrichment of ice cream with alginate nanoparticles including Fe and Zn salts, *Journal of Nanoparticles*, ID no. 754385.

Sharifi-Rad, M., Mnayer, D., Morais-Braga, M.F.B., Carneiro, J.N.P., Bezerra, C.F., Coutinho, H.D.M., Salehi, B., Martorell, M., Del Mar Contreras, M., Soltani-Nejad, A., Uribe, Y.A.H., Yousaf, Z., Iriti, M. and Sharifi-Rad, J. (2018a). *Echinacea* plants as antioxidant and antibacterial agents: From traditional medicine to biotechnological applications, *Phytotherapy Research*, 32: 1653-1663.

Sharifi-Rad, M., Nazaruk, J., Polito, L., Morais-Braga, M.F.B., Rocha, J.E., Coutinho, H.D.M., Salehi, B., Tabanelli, G., Montanari, C., Del Mar Contreras, M., Yousaf, Z., Setzer, W.N., Verma, D.R., Martorell, M., Sureda, A. and Sharifi-Rad, J. (2018b). *Matricaria* genus as a source of antimicrobial agents: From farm to pharmacy and food applications, *Microbiological Research*, 215: 76-88.

Singh, H., Kumar, C., Singh, N., Paul, S. and Jain, S.K. (2018). Nanoencapsulation of docosahexaenoic acid (DHA) using a combination of food grade polymeric wall materials and its application for improvement in bioavailability and oxidative stability, *Food & Function*, 9: 2213-2227.

Soumya, R.S., Vineetha, V.P., Reshma, P.L. and Raghu, K.G. (2013). Preparation and characterization of selenium incorporated guar gum nanoparticle and its interaction with H9c2 cells, *PLoS One*, 8: e74411.

Sow, L.C., Tirtawinata, F., Yang, H., Shao, Q. and Wang, S. (2017). Carvacrol nanoemulsion combined with acid electrolysed water to inactivate bacteria, yeast *in vitro* and native microflora on shredded cabbages, *Food Control*, 76: 88-95.

Stringaro, A., Colone, M. and Angiolella, L. (2018). Antioxidant, antifungal, antibiofilm, and cytotoxic activities of *Mentha* spp. essential oils, *Medicines (Basel)*, 5(4): E112.

Swanson, D., Block, R. and Mousa, S.A. (2012). Omega-3 fatty acids EPA and DHA: Health benefits throughout life, *Advances in Nutrition*, 3: 1-7.

Tachaprutinun, A., Udomsup, T., Luadthong, C. and Wanichwecharungruang, S. (2009). Preventing the thermal degradation of astaxanthin through nanoencapsulation, *International Journal of Pharmaceutics*. 374: 119-124.

Tan, C., Feng, B., Zhang, X., Xia, W. and Xia, S. (2016). Biopolymer-coated liposomes by electrostatic adsorption of chitosan (chitosomes) as novel delivery systems for carotenoids, *Food Hydrocolloids*, 52: 774-784.

Tang, S.Y. and Sivakumar, M. (2013). A novel and facile liquid whistle hydrodynamic cavitation reactor to produce submicron multiple emulsions, *AIChE Journal*, 59: 155-167.

Tastan, O., Pataro, G., Donsi, F., Ferrari, G. and Baysal, T. (2017). Decontamination of fresh-cut cucumber slices by a combination of a modified chitosan coating containing carvacrol nanoemulsions and pulsed light, *International Journal of Food Microbiology*, 260: 75-80.

Tavakolipour, H., Bagheri, L. and Madadlou, A. (2015). Pomegranate seed oil-loaded particles of the zein cross-linked with citric acid, *Journal of Food Process Engineering*, 38: 49-56.

Thavarajah, D., Thavarajah, P., Vial, E., Gebhardt, M., Lacher, C., Kumar, S. and Combs G.F. (2015). Will selenium increase lentil (*Lens culinaris* Medik) yield and seed quality? *Frontiers in Plant Science*, 6: 356.

Thavarajah, D., Abare, A., Mapa, I., Coyne, C.J., Thavarajah, P. and Kumar, S. (2017). Selecting lentil accessions for global selenium biofortification, *Plants (Basel)*, 6(3): E34.

Tsuchido, Y., Sasaki, Y., Sawada, S.-I. and Akiyoshi, K. (2015). Protein nanogelation using vitamin B6-bearing pullulan: Effect of zinc ions, *Polymer Journal*, 47: 201-205.

Vasconcelos, N.G., Croda, J. and Simionatto, S. (2018). Antibacterial mechanisms of cinnamon and its constituents: A review, *Microbial Pathogenesis*, 120: 198-203.

Vilchez, C., Forjan, E., Cuaresma, M., Bedmar, F., Garbayo, I. and Vega, J.M. (2011). Marine carotenoids: Biological functions and commercial applications, *Marine Drugs*, 9: 319-333.

Voogt, W., Holwerda, H.T. and Khodabaks, R. (2010). Biofortification of lettuce (*Lactuca sativa* L.) with iodine: The effect of iodine form and concentration in the nutrient solution on growth, development and iodine uptake of lettuce grown in water culture, *Journal of the Science of Food and Agriculture*, 90: 906-913.

Walia, N., Dasgupta, N., Ranjan, S., Chen, L. and Ramalingam, C. (2017). Fish oil-based vitamin D nanoencapsulation by ultrasonication and bioaccessibility analysis in simulated gastro-intestinal tract, *Ultrasonics Sonochemistry*, 39: 623-635.

Walker, R.M., Decker, E.A. and McClements, D.J. (2015). Physical and oxidative stability of fish oil nanoemulsions produced by spontaneous emulsification: Effect of surfactant concentration and particle size, *Journal of Food Engineering*, 164: 10-20.

Watanabe, S., Ohtani, Y., Tatsukami, Y., Aoki, W., Amemiya, T., Sukekiyo, Y., Kubokawa, S. and Ueda, M. (2017). Folate biofortification in hydroponically cultivated spinach by the addition of phenylalanine, *Journal of Agricultural and Food Chemistry*, 65: 4605-4610.

White, P.J. and Broadley, M.R. (2009). Biofortification of crops with seven mineral elements often lacking in human diets-iron, zinc, copper, calcium, magnesium, selenium and iodine, *New Phytologist*, 182: 49-84.

Xue, J., Davidson, P.M. and Zhong, Q. (2015). Antimicrobial activity of thyme oil co-nanoemulsified with sodium caseinate and lecithin, *International Journal of Food Microbiology*, 210: 1-8.

Yousuf, B., Gul, K., Wani, A.A. and Singh, P. (2016). Health benefits of anthocyanins and their encapsulation for potential use in food systems: A review, *Critical Reviews in Food Science and Nutrition*, 56: 2223-2230.

Zahraei, M., Monshi, A., Shahbazi-Gahrouei, D., Amirnasr, M., Behdadfara, B. and Rostami, M. (2015). Synthesis and characterization of chitosan coated manganese zinc ferrite nanoparticles as MRI contrast agents, *Journal of Nanostructures*, 5: 77-86.

Zariwala, M.G., Elsaid, N., Jackson, T.L., Corral López, F., Farnaud, S., Somavarapu, S. and Renshaw, D. (2013). A novel approach to oral iron delivery using ferrous sulphate loaded solid lipid nanoparticles, *International Journal of Pharmaceutics*, 456: 400-407.

Zhang, H. and Zhao, Y. (2015). Preparation, characterization and evaluation of tea polyphenol-Zn complex loaded β-chitosan nanoparticles, *Food Hydrocolloids*, 48: 260-273.

Zhang, H., Jung, J. and Zhao, Y. (2016). Preparation, characterization and evaluation of antibacterial activity of catechins and catechins–Zn complex loaded β-chitosan nanoparticles of different particle sizes, *Carbohydrate Polymers*, 137: 82-91.

Zhang, S., Luo, Y., Zeng, H., Wang, Q., Tian, F., Song, J. and Cheng, W.H. (2011). Encapsulation of selenium in chitosan nanoparticles improves selenium availability and protects cells from selenium induced DNA damage response, *Journal of Nutritional Biochemistry*, 22: 1137-1142.

Zhong, J., Yang, R., Cao, X., Liu, X. and Qin, X. (2018). Improved physicochemical properties of yogurt fortified with fish oil/-oryzanol by nanoemulsion technology, *Molecules*, 23: E56.

Zimet, P., Rosenberg, D. and Livney, Y.D. (2011). Re-assembled casein micelles and casein nanoparticles as nano-vehicles for ω-3 polyunsaturated fatty acids, *Food Hydrocolloids*, 25: 1270-1276.

Safety, Standards and Regulations of Nanotechnology in Food

Milad Rouhi[1], Leila Nateghi[2], Mostafa Aghamirzaei[3,4]*, Shima Yousefi[5], Farhad Garavand[6,7]*, Samira Feyzi[8], Majid Nooshkam[8] and Seid Mahdi Jafari[9]

[1] Department of Food Science and Technology, School of Nutrition Sciences and Food Technology, Research Center for Environmental Determinants of Health (RCEDH), Health Institute, Kermanshah University of Medical Sciences, Kermanshah, Iran
[2] Department of Food Science & Technology, College of Agriculture, Varamin-Pishva Branch, Islamic Azad University, Varamin, Iran
[3] Department of Food Science and Technology, Faculty of Agriculture, University of Tabriz, Tabriz 5166616471, Islamic Republic of Iran
[4] Vice-Chancellor for Food and Drug Administration, Healthcare Network of Fardis, Alborz University of Medical Sciences, Karaj, 3176794107, Islamic Republic of Iran
[5] College of Food Science and Technology, Science and Research Branch, Islamic Azad University, Tehran, Iran
[6] Department of Food Science & Engineering, University of Tehran, Karaj, Iran
[7] Department of Food Chemistry & Technology, Teagasc Food Research Centre, Moorepark, Fermoy, Co. Cork, Ireland
[8] Department of Food Science and Technology, Faculty of Agriculture, Ferdowsi University of Mashhad (FUM), Mashhad, Iran
[9] Department of Food Materials and Process Design Engineering, Gorgan University of Agricultural Sciences and Natural Resources, Gorgan, Iran

1. An Overview on Nanotechnology Applications in Food Industry

Nanotechnology offers a vast number of opportunities for the agro-food industry. Nanotechnology could be beneficial for improving pesticides, enriching animal feeds and smart sensors for diagnosis of animal disease. Development of nanostructures leads to the improvement of security, safety and quality of food products (Busquets, 2017). Food industry is seeking novel, effective and cheaper tools to manufacture and store food products and nanotechnology can play an important role in food processing

*Corresponding authors: aghamirzaei.ma88@gmail.com;
farhadgaravand@ut.ac.ir, farhad.garavand@teagasc.ie

(Dasgupta *et al.*, 2015). Nanomaterials and nanotechnology-based processes are increasingly being used within the food technology. Some of the most important applications of nanotechnology in the food-processing area are shelf-life extenders (Chaudhry *et al.*, 2008, 2017), nanofiltration (Zhou *et al.*, 2015), liquid membrane filtrations (Garavand and Madadlou, 2014), enzyme immobilization (Bai *et al.*, 2006), texture improvement (Graveland-Bikker and De Kruif, 2006), color/flavor/aroma enhancers (Freiberger *et al.*, 2015) and meat replacements (Manski *et al.*, 2007).

In addition, food-processing could benefit from improvements in food safety and market value, preserving food quality during storage and reducing food waste via real-time detection and control of possible chemical and microbiological hazards (e.g. pesticides, heavy metals, antibiotics and pathogens) thanks to the application of nanotechnology. The development of nanosensors can be useful for fast detection of environmental changes during food storage (i.e. temperature, relative humidity and oxygen exposure), physicochemical changes of food (i.e. color and pH), the presence of chemical hazards (e.g. nitrite or sulfite), or microbial contaminations.

Food packaging is one of the most attractive areas for using nanotechnology in the food industry (Rouhi *et al.*, 2017; Mirzaei-Mohkam *et al.*, 2019). Nanostructured antimicrobials and food-contact materials are used in active packaging to extend shelf-life and to preserve food organoleptic properties, respectively (Kariminejad *et al.*, 2018). Moreover, intelligent packaging systems have been designed to act as nanosensors that inform the consumers on the quality or freshness of the sealed food products during storage or transportation (Cerqueira *et al.*, 2014; Dasgupta *et al.*, 2015). They would serve as 'noses' that can detect chemicals released during food spoilage (Sozer and Kokini, 2009). Advantages of these fluidic sensors are their small dimensions and ability to detect relevant compounds rapidly in a small volume (few microliters), which has already led to widespread applications (e.g. medical, biological and chemical analysis) (Vo-Dinh *et al.*, 2001).

Using nanotechnology in food products can, also, improve their nutritional value. Incorporation of nanosized bioactive compounds, such as antioxidants, vitamins, polyphenols and omega-3 fatty acids increases their bioavailability. In addition, development of nanoencapsulation technologies offers possible solutions to improve the bioavailability of many functional compounds by enhancing their absorption in the gastrointestinal tract, which is critical for their activity (Chau *et al.*, 2007).

2. Market, Health Concerns and Public Consent

The economic gains and market differentiation are some of the benefits of nanotechnology applications in new food products (Neethirajan and Jayas, 2011; Handford *et al.*, 2014; Amenta *et al.*, 2015). Today, the European Union (EU) and the US Department of Agriculture (USDA) consider nanotechnology as a potentially major tool in solving key societal challenges, such as food security and safety, nutrition and environmental protection (EC, 2014; USDA, 2015). A considerable number of nanotechnology products are expected to reach the market in the next few years.

Despite the new market opportunities and the potential benefits, nanotechnology application to a sensitive area wherein food industry is facing some technological, societal and regulatory issues (Chaudhry *et al.*, 2017). The recent failure of genetically- modified foods due to public concerns for their real and perceived risks could be repeated for the nanotechnology modified foods (Williams *et al.*, 2004). The increase in public awareness and in building consumer confidence call for addressing of some important questions regarding the potential benefits of nanotechnology application in food industry for consumers, environment and/or industry. The toxicity of nanoparticles is dependent on their composition, structure and the exposure route and intervals (Oberdürster, 2000; Sayes *et al.*, 2006a, 2006b; Wick *et al.*, 2007; Breysse *et al.*, 2013; Pedata *et al.*, 2015). The dietary consumption of nanoparticles, mainly of TiO_2 and mixed silicates, is estimated around 10^{12} particles/person per day in developed countries (Oberdörster, 2004). The use of specific products, such as salad dressing with the nano-TiO_2 whitening agent, can result in more than 40-fold increase in the daily average intake (Oberdörster, 2004).

The public preconceived fears due to the poorly understood fate of nanoparticles in biological systems are the main factor that can limit the speed and scope of nanotechnology application. However, taking into account the wide use of natural and synthetic food additives and supplements, one might expect that nanotechnology in food industry can, as well, be accepted by public. Some studies regarding the awareness and acceptance of consumers of food products developed by using nanotechnology give an insight into the public opinion. Handford *et al.* (2014) analyzed the awareness and attitudes of the emerging use of nanotechnology in the agro-food sector in Ireland. The results showed a low awareness of nanotechnology applications in agro-food and respondents were neither positive nor negative towards these applications (Handford *et al.*, 2014).

Some surveys in the EU concluded that although consumer opinion is conducive to other applications of nanotechnology, it may not be entirely positive to its applications to food (Bieberstein *et al.*, 2012; Epp, 2017). The study conducted by Bieberstein *et al.* (2012) in France and Germany regarding food nano-fortifications has shown that consumers are reluctant to accept nanotechnology application in food (Bieberstein *et al.*, 2012).

Conflicting results were obtained by Siegrist *et al.* (2007), who investigated effective factors on acceptability of nanotechnology in food products, either foods or nano-packaging, in the German-speaking part of Switzerland. In general, participants were hesitant to buy nanotechnology-based products. However, nanotechnology in packaging was perceived as more beneficial than nanotechnology applied directly to foods (Siegrist *et al.*, 2007). Similarly, Zhou (2013) reported that US consumers were more willing to accept nanotechnology when they became aware of the benefits added to the food product (Zhou, 2013).

The study by Roosen *et al.* (2015) on the trust and acceptance of consumers to purchase food products developed with nanotechnology in Canada and Germany showed real concern for consumers. However, it was unclear if the consumers' concerns were related to the lack of information regarding nanotechnology in general or lack of knowledge about the uses of nanotechnology in the food industry specifically (Roosen *et al.*, 2015). In contrast to these results, the study on American adults' awareness and attitudes toward nanotechnology and synthetic biology showed that although the majority of respondents had little or no knowledge of nanotechnology, they expressed positive expectations when told about the potential benefits and risks of the technology (HRA, 2008).

Therefore, to minimize the public confusion and anxieties and to increase the awareness, understanding and acceptability of nanotechnology incorporation in food sector, the scientists, governmental entities and agro-food industries should cooperate in clarifying potential benefits and risks of its application and consumption, while implementing the legislative framework in this area.

3.   Interaction of Nanomaterials with Cells

Due to application of nanomaterials in food technology and the other industries, the human body is exposed to a multitude of nanoparticles every day (Powell *et al.*, 2010). Therefore, understanding the behavior of nanoparticles in the passage through the human body, entrance and possible disruption of the cellular functions is very important. Nanomaterials chemical composition, size, porosity, density and interaction with biological system could affect their bioavailability and possible toxicity (Oberdörster *et al.*, 2005a; Nel *et al.*, 2006).

3.1   Cellular and Extracellular Transfer Mechanisms of Nanoparticles

Nanoparticles are transported through cells and extracellular matrix (i.e. transcellular and paracellular transport) (Forth *et al.*, 1987; Innes and Ogden, 1999; Jung *et al.*, 2000). There are three main types of cellular uptake/transport mechanisms of nanoparticles: (1) endocytosis, specially pinocytosis and receptor-(clathrin-)mediated endocytosis, (2) energy-dependent selectively active transport by membrane proteins and (3) passive transport via facilitated diffusion of molecules through selective transmembrane protein channels (Fig. 1) (Innes and Ogden, 1999; Asati *et al.*, 2010; Cho *et al.*, 2010; Holpuch *et al.*, 2010). It has been observed that the uptake of nanoparticles occurs mainly by endocytosis in epithelial cells (Axson *et al.* 2015), while pinocytosis is one of the main mechanisms of nanoparticle transport (Shukla *et al.*, 2005; Cai *et al.*, 2007; Ryman-Rasmussen *et al.*, 2007; Buono *et al.*, 2009; Lund *et al.*, 2011).

The potential transport processes of nanoparticles are probably dependent on their properties, such as size, shape, surface charge, concentration and reactivity. Nanomaterials with positively-charged surfaces are transported by interaction on the negatively-charged tight intercellular filaments (van der Lubben *et al.*, 2001).

3.2   Cytotoxicity of Nanoparticles

Potential cytotoxicity effects of nanoparticles after their entrance into cells depend on their size, surface charge and concentration of the nanoparticle, type of encountered cell, as well as tissue environmental conditions (Asati *et al.*, 2010). It has been shown that metallic nanoparticles cause more cellular damage and faster toxic reactions than micron and/or bigger size materials due to their high surface area (Kroll *et al.*, 2009). Current toxicological procedures could not exactly recognize the risks derived from most nanomaterials (Stone *et al.*, 2010) and many cytotoxic mechanisms are currently unknown. However, some studies have shown nanoparticles could pose a risk to human health (Gurr *et al.*, 2005; Oberdörster *et al.*, 2005b; Buzea *et al.*, 2007; Ivask *et al.*, 2014).

Nanoparticle cytotoxicity results from negative effects on cell membrane and disruption of intracellular process through chemical reactions. Release of metal ions from metallic nanoparticles causes cellular damage through oxidative stress (Dick *et al.*, 2003; Barillet *et al.*, 2010) due to production of free radical species through

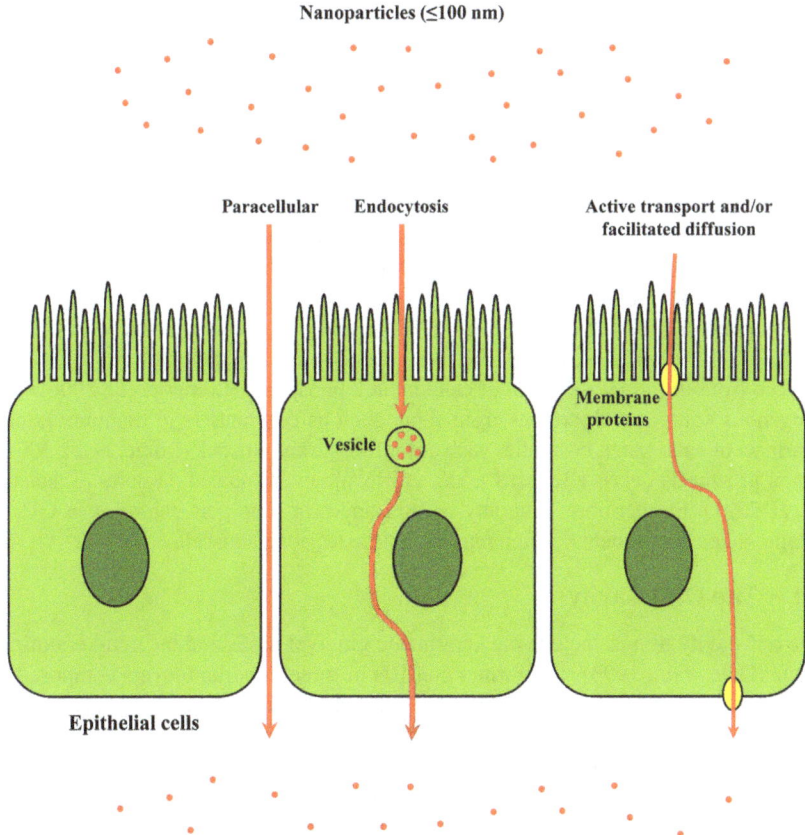

Fig. 1: Transcellular and paracellular transfer pathways of nanoparticles in the epithelial cells

the Haber–Weiss cycle and Fenton reaction (Buzea *et al.*, 2007). The main targets of intracellular toxicity are mitochondria (Hussain *et al.*, 2005), actin cytoskeleton (Soenen *et al.*, 2009) and nucleus (Prasad *et al.*, 2013).

Some studies on the application of super magnetic iron oxide nanomaterials as contrast agents in magnetic resonance imaging (MRI) have shown that these nanoparticles could induce tumor formation, DNA damage, changes in gene expression profiles, oxidative stress, disruption in iron homeostasis and signaling pathways and impairment of cell cycle regulation (Singh *et al.*, 2010).

However, some studies have shown that plasmatic homeostasis proteins, such as ceruloplasmin and metallothionein, could reduce the cytotoxicity of silver nanoparticles (Miyayama *et al.*, 2014; Liu *et al.*, 2017). Metal nanoparticles can interact with such proteins and form a so-called corona which in turn determines their cellular outcome (Fleischer and Payne, 2014; Sasidharan *et al.*, 2015). Moreover, Ge *et al.* (2011) found an interaction of single-wall carbon nanotubes with human serum proteins. Cytotoxicity assays with human acute monocytic leukemia cell line and human umbilical vein endothelial cells revealed that these protein-coated nanotubes resulted in reduced cytotoxicity (Ge *et al.*, 2011).

4. The Gastrointestinal Fate of Nanoparticles

Gastrointestinal tract (GIT) is responsible for digestion of foods, absorption of nutrients and removal of waste materials. The GIT of humans with large surface area (200 m^2) has the greatest uptake potential of body systems (Helander and Fändriks, 2014). Dissolution, separation, interaction, decomposition, adsorption, absorption into the body and bioavailability of nanoparticles could be affected by different roles and chemical environments of various membranes and sections of GIT (Best *et al.*, 2015). In addition to peristaltic-derived pressure up to 150 mm Hg, pH changes in GIT can affect nanoparticle physicochemical properties (Bellmann *et al.*, 2015; Fröhlich and Fröhlich, 2016).

In general, GIT is exposed to endogenous and exogenous nanoparticles (Lomer *et al.*, 2002; Powell *et al.*, 2007). The endogenous sources of nanoparticles in GIT are derived from intestinal secretion of calcium and phosphate (Lomer *et al.*, 2004). The exogenous sources are particles from food, such as colorants (*e.g.* titanium oxide), pharmaceuticals, water, cosmetics such as toothpaste and lipstick (Lomer *et al.*, 2004), dental prosthesis debris (Ballestri *et al.*, 2001) and even inhaled particles (Takenaka *et al.*, 2001). Furthermore, a variety of substances in food gets degraded in GIT to smaller-sized components, which may include sizes at a nanoscale.

4.1   The Oral Cavity

The oral cavity tissues include the keratinized gingival tissue and the non-keratinized tissue (Dale *et al.*, 1990). The former consists of three important layers composed of epithelial, fibroblastic and macrophage cells. The highly keratinized gingival epithelial cells that cover the surface of gums are the first to get in contact with nanoparticles. Strong intercellular junctions between these cells provide resistance to mechanical forces, enzymatic damage and barrier against nanoparticles (Presland and Dale, 2000).

Below the epithelial and fibroblastic cells, there is a layer of macrophage cells that act as defending mechanisms against foreign bodies, such as microorganisms and nanomaterials. Macrophage cells protect the body via identifying, engulfing and destroying foreign nanomaterials through phagocytosis (Kimball *et al.*, 2006). In addition, in the event of infection or foreign body penetration, these cells act as an immune-system stimulator, resulting in increasing the influx of immune cells to the tissue and breaking down the foreign body, increasing the permeability of the blood vessels to remove the pathogen and/or nanomaterials.

The oral cavity environment is mildly acidic (pH= 6.5-6.8) with some enzymes (Aframian *et al.*, 2006). Saliva contains many innate or acquired (antibodies) defense mechanisms. Antimicrobial proteins including lysozyme, lactoperoxidase system, lactoferrin, and high weight molecular glycoproteins are among the innate or unspecific immunological factors (van Nieuw Amerongen *et al.*, 2004; Andrade *et al.*, 2014). Lysozyme degrades peptidoglycan or biopolymer nanoparticles, such as chitosan used in oral health care products. Lactoperoxidase accelerates the breakdown rate of both organic and inorganic materials, such as metal-based nanoparticles through reactions, such as Fenton's reaction releasing metal ions (Kohler and Jenzer, 1989). In the study by Pinheiro *et al.* (2016), lactoferrin and saliva enzymes were used as stabilizing agents of curcumin nanoemulsions against acid digestion in the stomach. However, results showed less stability and breakdown of nanoemulsions in the duodenum and other parts of the gastrointestinal tract (Pinheiro *et al.*, 2016).

Kästner *et al.* (2017) investigated *in vitro* digestion of poly(acrylic acid) stabilized silver nanoparticles (with a radius of 3.1nm) in the absence and presence of food components including oil, starch, skimmed milk powder and a mixture of these three components. The particles were stable at the basic pH, but aggregated at acidic pH. Therefore, as expected, aggregation occurred when the primary nanoparticles were incubated in the artificial saliva at pH 6.4 under stirring for 5min at 37°C. In the presence of food components, results showed that except for milk and food mixture, most of the primary nanoparticles immediately aggregate in the saliva. However, the non-aggregated section of nanoparticles in all cases showed constant radii in saliva, like they were in the absence of food (Kästner *et al.*, 2017). Therefore, the type of food could affect the nanoparticle size and aggregation in the presence of artificial saliva and its fate in the gastrointestinal tract. However, an artificial saliva solution could not be a true oral cavity model for solid foods due to the lack of mechanical mastication. On the other hand, once the sample is liquid, the mastication is very poor, the residence time in the mouth is very short and saliva enzymes would not act.

Generally, as a result of little paracellular transport in the keratinized cells and saliva flow, the bioavailability and uptake of inorganic nanomaterials into the body through the oral mucosa is very insignificant (Lendenmann *et al.*, 2000; Garrod and Chidgey, 2008; Best *et al.*, 2015). However, lipid nanoparticles show greater absorption in the oral mucosa, resulting in their accumulation and potential immune activation and toxicity in the human body (Kotta *et al.*, 2012; McClements, 2013).

4.2   The Esophagus

The esophagus is a tubular tissue with four layers: mucosa, submucosa, muscularis and tunica adventitia (Harper, 2001). The epithelial cell layer and mucosa constitute the most inner layer of the esophageal lumen in contact with foods or nanoparticles. A lot of inhaled nanoparticles can move up through the trachea to hypopharynx with the help of mucociliary cells and then enter the stomach through the esophagus (Takenaka *et al.*, 2001; Braakhuis *et al.*, 2015). It seems that smaller absorptive area and shorter contact time in the esophagus compared to the other area of the gastrointestinal tract cause low nanoparticles bioavailability and uptake.

4.3   The Stomach

Environment of the stomach is highly acidic with a pH ranging from 1.2 to 5.0 (Gray, 1981; McConnell *et al.*, 2008), with a high concentration of proteases enzymes (Gray, 1981). It is important to understand the effects of strong pH on properties and behavior of nanomaterials. Studies have shown conflicting results in changes of the nanoparticles' shape, size, aggregation, bioavailability and cytotoxicity due to highly acidic pH, as described in the following sentences.

If complete dissolution of nanoparticles occurs in GIT, the specific characteristics related to the nanoscale no longer exist and exposure and hazard can be assessed, based on the conventional chemical form of the substance. The release of many types of metal ions due to acid erosion of metalliferous nanoparticles could result in oxidative stresses and highly cytotoxic effects (Sato and Donaldson, 1998; Nel *et al.*, 2006; Banavi *et al.*, 2020). Seok *et al.* (2013) reported high bioavailability and cytotoxicity of zinc oxide nanoparticles through the release of zinc ions under stomach environmental conditions (Seok *et al.*, 2013). Peters *et al.* (2012) showed that SiO_2 nanoparticles can form large agglomerates in human stomach conditions due to the formation of chloride bridges between nanoparticles. However, at subsequent intestinal digestion stage, the nano-sized silica particles reappeared again, even in smaller sizes and higher amounts than in the mouth digestion stage (Peters *et al.*, 2012). The addition of surfactant-like bile juices at the start of the intestine digestion stage may have a large impact on the stability and availability of nano-sized silica particles (DePalma *et al.*, 1966; Goodwin, 2009).

Similar reactions with chloride from gastric acid have been reported (Rogers *et al.*, 2012; Jiang *et al.*, 2018). Rogers *et al.* (2012) reported that the exposure of citrate-stabilized silver nanoparticles to the *in vitro* synthetic model of the human stomach fluid (pH 1.5) for 1h resulted in change in zeta potential, rapid aggregation of nanoparticles, partial silver oxidation in acidic media and reaction with chloride ions to form chemically-modified AgCl particles (Rogers *et al.*, 2012). In addition, Jiang *et al.* (2018) found that the polyvinylpyrrolidone (PVP)-coated silver nanoparticles, after incubation in the medium that mimics human gastric fluid, were aggregated and precipitated as AgCl salts and showed decreased cellular uptake by Caco-2 cells and cytotoxicity (Jiang *et al.*, 2018).

The bioavailability of some nanomaterials is decreased by the stomach acidic environment. Wick *et al.* (2007) investigated the effect of acidic pH on the aggregation of carbon nanotubes. Results showed a strange reversal of the classical model of bioavailability-related toxicology. Carbon nanotubes dispersed freely at neutral pH with a high bioavailability, but little toxicity. However, at lower pH, they were agglomerated, exhibiting in turn a lower bioavailability and greater cytotoxicity (Wick *et al.*, 2007). Similarly, silver nanoparticles aggregated and thus reduced their bioavailability under stomach acidic conditions (Mwilu *et al.*, 2013). It was observed that the smaller particles had higher affinity towards aggregation, whereas bigger particles maintained their individual particle shape. However, some studies have shown that some nanostructured materials remain stable under gastric conditions. For instance, Mackie *et al.* (2019) found low flocculation and coalescence of emulsions stabilized by cellulose nanocrystal in simulated gastric conditions (Mackie *et al.*, 2019).

The presence of metal or metal ion-coordinating moieties, such as sulfhydryl, carboxylate and amine groups in food components may also influence the dissolution and precipitation of silver from silver nanoparticle (Rogers *et al.*, 2012). In a study by Kästner *et al.* (2017), the effect of artificial gastric juice digestion (pH= 2) at 37°C for 2h on the colloidal stability of 3.1nm silver nanoparticles with and without food additives was investigated. After gastric digestion without food, the size of non-aggregated section of nanoparticles decreased to 2.5nm and the remaining particles formed aggregates with a mean radius of 9nm. Also in the presence of starch and oil, the radii decreased in the stomach to 2.5nm and 2.2nm, and the aggregates' radii increased to 10nm and 9nm, respectively, whereas no radius reduction and only very small aggregates of 6nm radius were observed in the stomach in the presence of both milk and food mixture (Kästner *et al.*, 2017). The stomach can absorb amino acids and small molecules. However, the main absorption and chemical digestion take place in the small intestine (Kästner *et al.*, 2017).

4.4   Small Intestine

The small intestine is formed of the duodenum, jejunum and ileum regions. Most of the chemical digestion through GIT occurs in the duodenum with pH value of 6-7

(Gray, 1981; Powell *et al.*, 2010). The enhancement of absorptive surface is due to the presence of villi and enterocytes. Small microparticles cross the epithelial cells by endocytosis. Larger nanoparticles and microparticles can also cross the intestinal epithelium through transcytosis by M-cell-uptake. Moreover, nanoparticles can form the gaps in villi apical zone through a membrane disruption process and then enter inside (Elder *et al.*, 2009). Also, the weak cell junctions in disease conditions result in increasing the nanoparticle paracellular uptake (Powell *et al.*, 2010).

The exposure of duodenum endothelium to nanoparticles could cause oxidative stress, DNA damage and inflammation (Ahamed *et al.*, 2010) and even increase the autoimmune activation of macrophage, resulting in Crohn's disease (Lomer *et al.*, 2002). However, Mackie *et al.* (2019) exposed gastric digested cellulose nanocrystal emulsions to the murine intestinal mucosa and demonstrated that the intestinal mucus layer entrapped the cellulose nanocrystals and prevented them from reaching the underlying epithelium (Mackie *et al.*, 2019).

Ileum is a highly-sensitive contractile tissue with supporting sympathetic nerve system which controls the contraction wave of the ileum and determines the flow of digested foods. This long region with >2m length and >20m^2 surface area makes the greatest contact surface through GIT for uptake of nutrients (Helander and Fändriks, 2014). However, this property is thought to cause the greatest bioavailability and uptake of nanoparticle in GIT.

Yao *et al.* (2015) determined the influence of size and concentration of gold nanoparticles on their absorption, accumulation and cytotoxicity by Caco-2 cells as the *in vitro* model of human small intestine. Results showed that by decreasing the mean particle diameter of gold nanoparticles from 100 to 50 to 15nm, their absorption rate by the Caco-2 cells increased, but their cellular accumulation decreased. Moreover, gold nanoparticles accumulation caused depolarization of mitochondria membranes, representing cytotoxicity in the Caco-2 cells (Yao *et al.*, 2015).

Changes in conditions of different parts of small intestine may alter nanoparticles' physicochemical properties, translocation across biological barriers, distribution and cellular/molecular responses in the body. Some nanomaterials, such as zinc that dissolves in the stomach acids, could pose a huge risk to the ileum health when used frequently. Studies have shown that some metal ions, such as zinc and cobalt interfere with this natural contraction process and even spasm of the ileum (Schnieden and Small, 1971). In contrast, many nanomaterials, such as silver nanoparticles, were re-dispersed by increasing pH in the small intestine, especially in the ileum (pH of 7-8), which increases their bioavailability to the epithelium, allowing them to reach the circulatory blood (Walczak *et al.*, 2012; Mwilu *et al.*, 2013). Contradictory results for various nanoparticles represent the knowledge gaps in this area and raise a major hurdle in the estimation of the nanoparticles' safety in food and drinks (Chaudhry *et al.*, 2008).

In addition to the intestinal conditions, food components have been observed to affect the particle size, aggregation and adherence of nanoparticles to Caco-2 cells during *in vitro* studies. In a study by Sinnecker *et al.* (2014), differences in adherence of carboxylated polystyrene nanoparticles to the Caco-2 cells were observed and they were correlated with the nature of corona coating. Presence of proteins, such as bovine serum albumin and casein, resulted in decreased adherence, whereas meat extract had no effect and intestinal fluids increased the adherence of engineered nanoparticles (Sinnecker *et al.*, 2014).

In agreement with these food-dependent results, Peters *et al.* (2012) observed changes in the presence, dissolution, agglomeration and release of nano-silica particles from different food products subjected to *in vitro* digestion in different sections of GIT (Peters *et al.*, 2012).

4.5   Colon

The colon removes water and remaining nutrients from the digested and indigestible food (Gray, 1981). Moreover, thousands of bacterial species exist in colon and play important roles in improving digestive and immunological functions, including biosynthesis of short-chain fatty acid butyrate and vitamin B_{12}, reducing the toxicity of some drugs and toxins, maturation and education of immune system and antagonistic effects on pathogenic bacteria (Zoetendal *et al.*, 2004; Manson *et al.*, 2008; Walk and Young, 2008; Young, 2012; Bergin and Witzmann, 2013).

Several studies have investigated the colonic absorption of nanoparticles. Aggregated or agglomerated nanoparticles may be absorbed less than well-dispersed ones, or in some cases may pass through the body without being absorbed. In other words, individual nanoparticles will most probably be transported across the intestinal epithelial layer into the bloodstream. Sergent *et al.* (2012) reported that SiO_2-25 nm and SiO_2-100 nm nanoparticles had little cytotoxic and genotoxic effects on HT-29 cells after a 24h exposure. However, cytotoxicity and genotoxicity showed an inverse dose-dependent relationship to SiO_2-100 nm nanoparticles (Sergent *et al.*, 2012). Similarly, Pelka *et al.* (2009) showed that <100 nm platinum nanoparticles at the concentration of 1000 ng/cm^2 enter the HT29 (human colon carcinoma) cells after incubation for 24h. Results showed an inverse relationship between cytotoxicity in the colon's epithelial cells and platinum nanoparticle size. The toxicity mechanisms were not as expected through the free radical generation and oxidative stress of metal ions, but due to other unknown toxicity mechanisms (Pelka *et al.*, 2009).

The absorption of nanoparticles in colon is expected to be less pronounced than in small intestine. However, some nanoparticles with antimicrobial properties in addition to extending shelf-life of foods could change the balance of colonic microflora. Taylor *et al.* (2015) investigated the phenotypic response of the gut microbial community to ZnO, CeO_2 and TiO_2 nanoparticles' ingestion as a model microbial community within colon. Results indicated that the nanoparticle-exposed community showed significant differences as compared to the unexposed one in multiple phenotypic traits, including short-chain fatty acid production, hydrophobicity, sugar content of the extracellular polymeric substance and electrophoretic mobility. The nanoparticles, especially TiO_2, caused nonlethal but significant changes to the phenotype of the native bacterial population, which may be related to the development of inflammatory diseases in the colon, such as ulcerative colitis (Taylor *et al.*, 2015).

4.6   Access to Systemic Blood

Nanoparticles can translocate into the bloodstream and access each organ by crossing the epithelium (Hansson, 2012). Stone *et al.* (2010) showed that once nanoparticles are transported into the systemic blood, they can accumulate in many organs of the body, including the liver, bone, nervous tissues and even brain (Stone *et al.*, 2010). Silicon oxide nanoparticles, carbon–iron nanotubes, gold and silver nanoparticles have shown a higher affinity towards accumulation in the liver and spleen in animal models (Balasubramanian *et al.*, 2010; Hasezaki *et al.*, 2011; De Jong *et al.*, 2013).

In the study by van der Zande *et al.* (2012), the 28-day oral exposure of rats to <20 nm non-coated, <15 nm PVP-coated silver nanoparticles, $AgNO_3$, or only carrier solution was investigated. Dissection was performed on day 29 after a wash-out period of up to two months. Silver was present in all the examined organs with the highest levels in the liver and spleen for all silver treatments. Silver concentrations in organs were highly correlated with the amount of Ag^+ in the silver nanoparticle suspension, indicating that mainly Ag^+, and to much lesser extent silver nanoparticles, passed the intestinal barrier in the exposed rats. Results also showed that in all groups, silver was cleared from most organs after two months of washing out, but remarkably not from brain and testis. Detection of silver nanoparticles in silver nanoparticle-exposed rats and also remarkably in $AgNO_3$-exposed rats, demonstrated the *de novo* formation of nanoparticles from Ag^+ *in vivo*. Although the accumulation of silver nanoparticles in different organs was observed, no hepatotoxicity or immunotoxicity effects were observed. Oral exposure to silver nanoparticles appears to be very similar to exposure to silver salts (van der Zande *et al.*, 2012).

The accumulation of nanoparticles could result in direct inflammation of the host tissue and indirect cellular damage through immune system activation (Kreyling *et al.*, 2009; Farrera and Fadeel, 2015). The ability of nanoparticles to accumulate in different cells and tissues causes damage to the function of affected organs via the immune response against these materials as foreign bodies. For example, silver nanoparticles, even at low doses, cause immunotoxic response (De Jong *et al.*, 2013). Immune activation in response to metal oxide nanoparticles or carbon nanomaterials, occurs through a process of antibody-mediated cellular activation. Moreover, accumulation and exposure of nanoparticles correlate with the development of chronic autoimmune diseases, such as Crohn's disease, as discussed previously, through activation of immune responses (Lomer *et al.*, 2002).

5. Regulations of Nanomaterials in Food by International Regulatory Systems

Although some national and international organizations have recommended strategies for risk assessment of nanomaterials, no comprehensive and detailed instructions have been currently developed for the safety assessment of nanomaterials applied in foods (SCCP, 2007; SCENIHR, 2009). The reasons behind this are following: (1) lack of precise and accurate quantitative analytical methods of nanomaterials, (2) lack of evaluation methods of their safety, and (3) insufficient information about the potential hazards posed by nanomaterials.

Available conventional analytical methods for identifying the hazards of different chemicals were developed for molecular form of materials based on their chemical and physical properties. However, the ability of these methods may not be sufficient to predict the health effects of nanoscale materials as a result of their different chemical and physical behavior within the biological systems (FDA, 2007; Dingman, 2008; Miller *et al.*, 2008; Chaudhry and Castle, 2011; Cushen *et al.*, 2012).

Some analytical methods, such as microscopy and chemical ones, are currently used to detect nanomaterials and characterize their quality and quantity under certain conditions (Tiede *et al.*, 2008; Barlow *et al.*, 2009). However, many difficulties cause imprecise and inaccurate results obtained by these methods, as follows (Barlow *et al.*, 2009):

1. Diverse nanomaterials with different properties cannot be analyzed using only a single method, but a combination of techniques.
2. Food matrix could affect the properties of the nanomaterials, for instance, interactions of nanomaterials with cell constituents and/or other materials in food may result in false positive or negative results. Also insufficient content of nanomaterials in food may be below sensitivity of conventionally-used experimental techniques.
3. The engineered nanomaterials may not be differentiated from the natural similar nanostructures.
4. The limited number of standardized reference nanomaterials.
5. Losses during preparatory steps of some chemical analytical methods.

Considering these limitations in the analysis of nanomaterials in food, at present, the consumers must rely on the information provided by the industry to know about the presence, type and concentration of nanomaterials used in the relevant food products (Barlow *et al.*, 2009). Moreover, international organizations have not provided the highest allowed concentration levels of any nanomaterials in foods due to the lack of reliable and sensitive analytical methods and therefore, have only provided key general principles and potential risk assessments, resulting from nanoscience and nanotechnologies in food.

5.1 World Health Organization (WHO) and Food and Agricultural Organization of the United Nations (FAO)

Codex Alimentarius Commission, established by WHO and FAO of the United Nations, has no special standard references for food incorporating nanomaterials or nanoscale food additives. WHO noted that nanomaterials, before application in food and food processing, should be assessed for the potential health and environmental risks, similar to all new materials (WHO, 2008).

Due to the accelerated global development and interest in the use of nanotechnology, and concerns and limited knowledge on the potential food safety issues, FAO and WHO organized a meeting in 2010 on 'FAO/WHO Expert Meeting on the Application of Nanotechnologies in the Food and Agriculture Sectors: Potential Food Safety Implications' to provide information and decisions on the existing and projected applications of nanotechnologies and nanomaterials in food technology, methodologies to evaluate the potential risk of nanomaterials in food, potential food safety and human health risks and address the transparent and constructive dialogues on nanotechnology among stakeholders (FAO/WHO, 2010).

The FAO/WHO Expert Meeting stated some recommendations and future plans, including:

• Addressing internationally recognized and clear definitions relating to processes, materials, products and applications of nanotechnologies by the Codex Alimentarius Commission.
• Characterizing the interaction between nanoparticles and other substances in the food matrix.
• Considering the whole life cycle of engineered nanomaterials in the environment because of probable indirect human exposure.
• Using a tiered risk assessment approach for application of nanotechnologies to food and feed, consisting of an initial screening level, characterization of a

material, classification of nanomaterials based on their toxicity and evaluation of dose-response relationships.

• Developing innovative and interdisciplinary researches, validating testing methods to reduce uncertainties in risk assessment, improve risk-assessment strategies for the application of nanotechnologies in food and feed and address specific data gaps in different areas.

• Commencing to make collaborative decisions in the application of nanotechnologies in agriculture and food with a broad range of stakeholders, aiming to increase the awareness of policy-makers, analysts, R&D funding agencies and all stakeholders.

• Providing guidance, training and capacity-building resources for governments to engage stakeholders and increase the public understanding of the potential risks and benefits of nanoscience and nanotechnologies in food and agriculture.

To summarize and analyze the information of previous expert meeting and determine possible courses of action to be followed by FAO and WHO, they published a report on 'state of the art on the initiatives and activities relevant to risk assessment and risk management of nanotechnologies in the food and agriculture sectors'. This technical paper presents relevant activities at the national and regional level and by international governmental and nongovernmental organizations, as well as risk management of nanotechnologies in the food and agriculture sector (FAO/WHO, 2013).

5.2   European Commission (EC) and European Food Safety Authority (EFSA)

In February 2008, the European Commission (EC) released a recommendation to the Member States to adopt a Code of Conduct for Nanosciences and Nanotechnologies research to investigate some effects of nanotechnology on human health and the environment (EFSA, 2011). This Code is complementary to legislation and provides the Member States, employers, research funders, researchers and, more generally, all individuals and civil society organizations involved or interested in nanosciences and nanotechnologies research, with guidelines favoring a responsible and open approach to nanoscience and nanotechnology (N&N) research in the community.

The EC Communication on the regulatory aspects of nanomaterials was held in June 2008 to cover the potential health, safety and environmental risks in relation to nanomaterials (Lidén, 2011). It concluded that as a critical factor for implementation and legislation, knowledge on the characterization of nanomaterials, their hazards, exposure, risk assessment and risk management should be improved. The EC activities are coordinated with international partners and stakeholders in appropriate forums, such as the Organization for Economic Co-operation and Development (OECD) and the International Organization for Standardization (ISO). In addition, responsible authorities and agencies should carefully control the market and pre-market to monitor risks related to nanomaterials. Furthermore, the EC Communication on 'the second regulatory review on nanomaterials' in October 2012, had a greater emphasis on the technical challenges and the requirements for implementing the case-by-case risk assessment (EC, 2012).

On the other hand, in response to public concerns on the effects of nanotechnology and nanomaterial on nutritional value or bioavailability of foods, EFSA recommended that the safety of nanotechnology applications in food and feed needs to be evaluated

(EFSA, 2008). EFSA, following the EC request, reported a scientific opinion of the scientific committee on the potential risks arising from nanoscience and nanotechnologies on food and feed safety in February 2009. According to the opinion, the physiochemical properties of materials may be changed by nanomaterials, much more different and/or intense than the dissolved and micro/macro-sized of the same material. These effects occur as a result of high surface-to-mass ratio and surface reactivity of the nanomaterials that could pose potential health and environmental risks (Barlow *et al.*, 2009).

EFSA concluded that a case-by-case risk assessment of nanomaterials is probably applicable and useful. In this approach, current conventional toxicity-testing could be performed as a suitable starting point. However, uncertainties in this area increase as a result of currently limited information in previously-mentioned areas, such as difficulties to detect, characterize and measure the nanomaterials in food and their toxicokinetics and toxicology. Moreover, there is limited information in the spread of nanomaterial used in food and feed area and public health exposure to them. Therefore, EFSA recommended the development of methods for detection, characterization and measurement of nanomaterials in food and biological tissues, as well as the determination of public and environmental exposure to food-derived nanomaterials and their toxicity (Barlow *et al.*, 2009).

5.3   US Food and Drug Administration (FDA)

FDA is a member agency in the National Nanotechnology Initiative (NNI), a federal research and development program established to coordinate the multi-agency efforts in nanoscale science, engineering and technology. In August 2006, FDA formed an internal Nanotechnology Task Force to determine ways for addressing any knowledge or policy gaps in safety evaluation and regulation of products containing nanomaterials.

The initial report of the Task Force was published to support the development of regulatory pathways for nanotechnology products. The report provided scientific knowledge and issues of the potential effects of nanomaterials relevant to FDA's regulation of products, as well as regulatory policy issues in this area. According to this report, the available information does not suggest that all nanoscale materials are hazardous as compared to the same materials with non-nanoscale dimensions. The report also commented that products with nanomaterials make regulatory challenges similar to those using other emerging technologies. Therefore, the uncertain nature of the science highlights the need for rapid development of a transparent, consistent and predictable regulatory pathway for these products (FDA, 2007). Afterwards, FDA published three guidance documents for industry: in June 2014, relating to the regulatory status of nanotechnology application in food ingredients and food contact substances (FDA, 2014a); the use of nanomaterials in food for animals (FDA, 2015), and the safety of nanomaterials in cosmetic products (FDA, 2014b).

5.4   Food Standards in Australia and New Zealand (FSANZ)

In Australia and New Zealand, food nano-additives and nano-ingredients are regulated under the Food Standards Code by FSANZ. This authority ceases to regulate nanomaterials soluble in water or oil in the final food, or in the gastrointestinal tract. Conversely, nanoscale or microscale materials insoluble in water and oil and non-biodegradable ones may require additional regulatory consideration (Fletcher and Bartholomaeus, 2011).

FSANZ highlights the need for risk assessments to ensure that small particles in food are as safe as larger ones. In this approach, FSANZ examines a wide range of scientific evidence to establish the safe consumption of foods or food ingredients. Moreover, premarket safety assessment and approval are inevitable for any new food substances before they can be legally commercialized (FSANZ, 2008). Therefore, FSANZ also investigates any new types of engineered nanoparticles for application in food to ensure the protection and safety and public health (FSANZ, 2008).

To improve the strategies, share experiences and information, and ensure a consistent regulatory response; FSANZ engages with international food regulatory partners, other key national regulatory agencies, industry, non-governmental organizations and the public through various forums including fact sheets, web videos and presentations at international conferences and workshops. As a result of rapid development in nanotechnologies, FSANZ has an ongoing program of work to review and revise, where necessary, existing arrangements for the regulation of nanotechnologies in food, such as monitoring the rapidly developing scientific literature (Fletcher and Bartholomaeus, 2011).

6. Conclusion and Future Perspectives

Currently, increasing types of foods using nanotechnology are produced, making their way into the market. Therefore, the public will increasingly be the consumer of these products, most of whom are without awareness of potential benefits or risks of nanomaterials. To increase the public awareness of nanotechnology application in food products, the regulators, scientists and stakeholders should collaborate to determine the environmental and biological fates of different nanoparticles, develop validated testing methods, address specific data gaps and set out the legislative framework in this area.

References

Aframian, D., Davidowitz, T. and Benoliel, R. (2006). The distribution of oral mucosal pH values in healthy saliva secretors, *Oral Diseases*, **12**(4): 420-423.

Ahamed, M., AlSalhi, M.S. and Siddiqui, M. (2010). Silver nanoparticle applications and human health. *Clinica Chimica Acta*, **411**(23-24): 1841-1848.

Amenta, V., Aschberger, K., Arena, M., Bouwmeester, H., Moniz, F.B., Brandhoff, P., Gottardo, S., Marvin, H.J., Mech, A. and Pesudo, L.Q. (2015). Regulatory aspects of nanotechnology in the agri/feed/food sector in EU and non-EU countries, *Regulatory Toxicology and Pharmacology*, **73**(1): 463-476.

Andrade, F.B.D., Oliveira, J.C.D., Yoshie, M.T., Guimarães, B.M., Gonçalves, R.B. and Schwarcz, W.D. (2014). Antimicrobial activity and synergism of lactoferrin and lysozyme against cariogenic microorganisms, *Brazilian Dental Journal*, **25**(2): 165-169.

Asati, A., Santra, S., Kaittanis, C. and Perez, J.M. (2010). Surface-charge-dependent cell localization and cytotoxicity of cerium oxide nanoparticles, *ACS Nano*, **4**(9): 5321-5331.

Axson, J.L., Stark, D.I., Bondy, A.L., Capracotta, S.S., Maynard, A.D., Philbert, M.A., Bergin, I.L. and Ault, A.P. (2015). Rapid kinetics of size and pH-dependent dissolution and aggregation of silver nanoparticles in simulated gastric fluid, *The Journal of Physical Chemistry C*, **119**(35): 20632-20641.

Bai, Y.-X., Li, Y.-F., Yang, Y. and Yi, L.-X. (2006). Covalent immobilization of triacylglycerol lipase onto functionalized nanoscale SiO_2 spheres, *Process Biochemistry*, **41**(4): 770-777.

Balasubramanian, S.K., Jittiwat, J., Manikandan, J., Ong, C.-N., Liya, E.Y. and Ong, W.-Y. (2010). Biodistribution of gold nanoparticles and gene expression changes in the liver and spleen after intravenous administration in rats, *Biomaterials*, **31**(8): 2034-2042.

Ballestri, M., Baraldi, A., Gatti, A.M., Furci, L., Bagni, A., Loria, P., Rapanà, R.M., Carulli, N. and Albertazzi, A. (2001). Liver and kidney foreign bodies granulomatosis in a patient with malocclusion, bruxism, and worn dental prostheses, *Gastroenterology*, **121**(5): 1234-1238.

Banavi, P., Sadeghi, E., Garavand, F., Heydari, M. and Rouhi, M. (2020). Release behavior of metals from tin-lined copper cookware into food simulants during cooking and cold storage, *Environmental Science and Pollution Research*: 1-11.

Barillet, S., Simon-Deckers, A., Herlin-Boime, N., Mayne-L'Hermite, M., Reynaud, C., Cassio, D., Gouget, B. and Carrière, M. (2010). Toxicological consequences of TiO_2, SiC nanoparticles and multi-walled carbon nanotubes exposure in several mammalian cell types: An *in vitro* study, *Journal of nanoparticle Research,* **12**(1): 61-73.

Barlow, S., Chesson, A., Collins, J.D., Flynn, A., Hardy, A., Jany, K.-D., Knaap, A., Kuiper, H., Larsen, J.C. and Le Neindre, P. (2009). The potential risks arising from nanoscience and nanotechnologies on food and feed safety, *EFSA Journal*, **7**(3).

Bellmann, S., Carlander, D., Fasano, A., Momcilovic, D., Scimeca, J.A., Waldman, W.J., Gombau, L., Tsytsikova, L., Canady, R. and Pereira, D.I. (2015). Mammalian gastrointestinal tract parameters modulating the integrity, surface properties, and absorption of food-relevant nanomaterials, *Wiley Interdisciplinary Reviews: Nanomedicine and Nanobiotechnology*, **7**(5): 609-622.

Bergin, I.L. and Witzmann, F.A. (2013). Nanoparticle toxicity by the gastrointestinal route: Evidence and knowledge gaps, *International Journal of Biomedical Nanoscience and Nanotechnology*, **3**(1-2).

Best, M., Phillips, G., Fowler, C., Rowland, J. and Elsom, J. (2015). Characterisation and cytotoxic screening of metal oxide nanoparticles putative of interest to oral healthcare formulations in non-keratinised human oral mucosa cells *in vitro*, *Toxicology in vitro*, **30**(1): 402-411.

Bieberstein, A., Roosen, J., Marette, S., Blanchemanche, S. and Vandermoere, F. (2012). Consumer choices for nano-food and nano-packaging in France and Germany, *European Review of Agricultural Economics*, **40**(1): 73-94.

Braakhuis, H.M., Kloet, S.K., Kezic, S., Kuper, F., Park, M.V., Bellmann, S., van der Zande, M., Le Gac, S., Krystek, P. and Peters, R.J. (2015). Progress and future of *in vitro* models to study translocation of nanoparticles, *Archives of Toxicology*, **89**(9): 1469-1495.

Breysse, P.N., Delfino, R.J., Dominici, F., Elder, A.C., Frampton, M.W., Froines, J.R., Geyh, A.S., Godleski, J.J., Gold, D.R. and Hopke, P.K. (2013). US EPA particulate matter research centers: Summary of research results for 2005-2011, *Air Quality, Atmosphere & Health*, **6**(2): 333-355.

Buono, C., Anzinger, J.J., Amar, M. and Kruth, H.S. (2009). Fluorescent pegylated nanoparticles demonstrate fluid-phase pinocytosis by macrophages in mouse atherosclerotic lesions, *The Journal of Clinical Investigation*, **119**(5): 1373-1381.

Busquets, R. (2017). *Emerging Nanotechnologies in Food Science*, Elsevier.

Buzea, C., Pacheco, I.I. and Robbie, K. (2007). Nanomaterials and nanoparticles: Sources and toxicity, *Biointerphases*, **2**(4): MR17-MR71.

Cai, Y., Liu, Y., Yan, W., Hu, Q., Tao, J., Zhang, M., Shi, Z. and Tang, R. (2007). Role of hydroxyapatite nanoparticle size in bone cell proliferation, *Journal of Materials Chemistry*, **17**(36): 3780-3787.

Cerqueira, M.A., Pinheiro, A.C., Silva, H.D., Ramos, P.E., Azevedo, M.A., Flores-Lopez, M.L., Rivera, M.C., Bourbon, A.I., Ramos, O.L. and Vicente, A.A. (2014). Design of bio-nanosystems for oral delivery of functional compounds, *Food Engineering Reviews*, **6**(1-2): 1-19.

Chau, C.-F., Wu, S.-H. and Yen, G.-C. (2007). The development of regulations for food nanotechnology, *Trends in Food Science & Technology*, **18**(5): 269-280.

Chaudhry, Q., Scotter, M., Blackburn, J., Ross, B., Boxall, A., Castle, L., Aitken, R. and Watkins, R. (2008). Applications and implications of nanotechnologies for the food sector, *Food Additives and Contaminants*, **25**(3): 241-258.

Chaudhry, Q. and Castle, L. (2011). Food applications of nanotechnologies: An overview of opportunities and challenges for developing countries, *Trends in Food Science & Technology*, **22**(11): 595-603.

Chaudhry, Q., Castle, L. and Watkins, R. (2017). *Nanotechnologies in Food*, Royal Society of Chemistry.

Cho, E.C., Au, L., Zhang, Q. and Xia, Y. (2010). The effects of size, shape, and surface functional group of gold nanostructures on their adsorption and internalization by cells, *Small*, **6**(4): 517-522.

Cushen, M., Kerry, J., Morris, M., Cruz-Romero, M. and Cummins, E. (2012). Nanotechnologies in the food industry – Recent developments, risks and regulation, *Trends in Food Science & Technology*, **24**(1): 30-46.

Dale, B.A., Salonen, J. and Jones, A.H. (1990). New approaches and concepts in the study of differentiation of oral epithelia, *Critical Reviews in Oral Biology & Medicine*, **1**(3): 167-190.

Dasgupta, N., Ranjan, S., Mundekkad, D., Ramalingam, C., Shanker, R. and Kumar, A. (2015). Nanotechnology in agro-food: From field to plate, *Food Research International*, 69: 381-400.

De Jong, W.H., Van Der Ven, L.T., Sleijffers, A., Park, M.V., Jansen, E.H., Van Loveren, H. and Vandebriel, R.J. (2013). Systemic and immunotoxicity of silver nanoparticles in an intravenous 28 days repeated dose toxicity study in rats, *Biomaterials*, **34**(33): 8333-8343.

DePalma, R.G., Hubay, C.A. and Levey, S. (1966). The micellar properties of bile, *JAMA*, **195**(11): 943-945.

Dick, C.A., Brown, D.M., Donaldson, K. and Stone, V. (2003). The role of free radicals in the toxic and inflammatory effects of four different ultrafine particle types, *Inhalation Toxicology*, **15**(1): 39-52.

Dingman, J. (2008). Guest Commentary: Nanotechnology: Its Impact on Food Safety, *Journal of Environmental Health*, **70**(6): 47-50.

EC (2012). *Second Regulatory Review on Nanomaterials*, COM, European Commission.

EC (2014). *The EU Framework Programme for Research and Innovation*, European Commission.

EFSA (2008). *Nanotechnology*, European Food Safety Authority.

EFSA (2011). Guidance on the risk assessment of the application of nanoscience and nanotechnologies in the food and feed chain, *EFSA Journal*, **9**(5): 2140.

Elder, A., Vidyasagar, S. and DeLouise, L. (2009). Physicochemical factors that affect metal and metal oxide nanoparticle passage across epithelial barriers, *Wiley Interdisciplinary Reviews: Nanomedicine and Nanobiotechnology*, **1**(4): 434-450.

Epp, A. (2017). *Nanoview – Perception of Nanotechnology by the German Population and Media Coverage*, Paper read at 1st joint symposium on nanotechnology, BfR. 20.

FAO/WHO (2010). FAO/WHO expert Meeting on the *Application of Nanotechnologies in the Food and Agriculture Sectors: Potential Food Safety Implications.* Meeting Report, Food and Agriculture Organization/World Health Organization.

FAO/WHO (2013). *State-of-the-art on the Initiatives and Activities Relevant to Risk Assessment and Risk Management of Nanotechnologies in the Food and Agriculture Sectors.* FAO/WHO technical paper, Food and Agriculture Organization/World Health Organization.

Farrera, C. and Fadeel, B. (2015). It takes two to tango: Understanding the interactions between engineered nanomaterials and the immune system, *European Journal of Pharmaceutics and Biopharmaceutics*, 95: 3-12.

FDA (2007). *Nanotechnology: A Report of the US Food and Drug Administration Nanotechnology Task Force*, Washington DC, Food and Drug Administration.

FDA (2014a). *Guidance for Industry: Assessing the Effects of Significant Manufacturing Process Changes, Including Emerging Technologies, on the Safety and Regulatory Status of Food Ingredients and Food Contact Substances, Including Food Ingredients that are Color Additives*, Food and Drug Administration.

FDA (2014b). *Guidance for Industry: Safety of Nanomaterials in Cosmetic Products* [final], Food and Drug Administration.

FDA (2015). *Guidance for Industry: Use of Nanomaterials in Food for Animals* [draft], Food and Drug Administration.

Fleischer, C.C. and Payne, C.K. (2014). Nanoparticle–cell interactions: Molecular structure of the protein corona and cellular outcomes, *Accounts of Chemical Research*, 47(8): 2651-2659.

Fletcher, N. and Bartholomaeus, A. (2011). Regulation of nanotechnologies in food in Australia and New Zealand, *International Food Risk Analysis Journal*, 1(2): 33-40.

Forth, W., Henschler, D. and Rummel, W. (1987). *Pharmacology and Toxicology* for students of medicine, veterinary medicine, pharmacy, chemistry, biology as well as for physicians, veterinarians and pharmacists [assisted by 35 authors] [German].

Freiberger, E.B., Kaufmann, K.C., Bona, E., de Araújo, P.H.H., Sayer, C., Leimann, F.V. and Gonçalves, O.H. (2015). Encapsulation of roasted coffee oil in biocompatible nanoparticles, *LWT – Food Science and Technology*, 64(1): 381-389.

Fröhlich, E. and Fröhlich, E. (2016). Cytotoxicity of nanoparticles contained in food on intestinal cells and the gut microbiota, *International Journal of Molecular Sciences*, 17(4): 509.

FSANZ (2008). Small Particles, Nanotechnology and Food, *Food Standards*, Australia, New Zealand.

Garavand, F. and Madadlou, A. (2014). Recovery of phenolic compounds from effluents by a microemulsion liquid membrane (MLM) extractor, *Colloids and Surfaces A: Physicochemical and Engineering Aspects*, 443: 303-310.

Garrod, D. and Chidgey, M. (2008). Desmosome structure, composition and function, *Biochimica et Biophysica Acta (BBA) – Biomembranes*, 1778(3): 572-587.

Ge, C., Du, J., Zhao, L., Wang, L., Liu, Y., Li, D., Yang, Y., Zhou, R., Zhao, Y. and Chai, Z. (2011). Binding of blood proteins to carbon nanotubes reduces cytotoxicity, *Proceedings of the National Academy of Sciences*, 108(41): 16968-16973.

Goodwin, J. (2009). *Colloids and Interfaces with Surfactants and Polymers*, John Wiley & Sons.

Graveland-Bikker, J. and De Kruif, C. (2006). Unique milk protein based nanotubes: Food and nanotechnology meet, *Trends in Food Science & Technology*, 17(5): 196-203.

Gray, H. (1981). *Anatomy Descriptive and Surgical: 'Gray's Anatomy'*, Classics of Medicine Library.

Gurr, J.-R., Wang, A.S., Chen, C.-H. and Jan, K.-Y. (2005). Ultrafine titanium dioxide particles in the absence of photoactivation can induce oxidative damage to human bronchial epithelial cells, *Toxicology*, **213**(1-2): 66-73.

Handford, C.E., Dean, M., Henchion, M., Spence, M., Elliott, C.T. and Campbell, K. (2014). Implications of nanotechnology for the agri-food industry: Opportunities, benefits and risks, *Trends in Food Science & Technology*, **40**(2): 226-241.

Hansson, G.C. (2012). Role of mucus layers in gut infection and inflammation, *Current Opinion in Microbiology*, **15**(1): 57-62.

Harper, D. (2001). Online Etymology Dictionary.

Hasezaki, T., Isoda, K., Kondoh, M., Tsutsumi, Y. and Yagi, K. (2011). Hepatotoxicity of silica nanoparticles with a diameter of 100 nm. *Die Pharmazie – An International Journal of Pharmaceutical Sciences*, **66**(9): 698-703.

Helander, H.F. and Fändriks, L. (2014). Surface area of the digestive tract–revisited, *Scandinavian Journal of Gastroenterology*, **49**(6): 681-689.

Holpuch, A.S., Hummel, G.J., Tong, M., Seghi, G.A., Pei, P., Ma, P., Mumper, R.J. and Mallery, S.R. (2010). Nanoparticles for local drug delivery to the oral mucosa: Proof of principle studies, *Pharmaceutical Research*, **27**(7): 1224-1236.

HRA (2008). *Awareness of and Attitudes Toward Nanotechnology and Synthetic Biology*, Peter D. Hart Research Associates, Inc.

Hussain, S., Hess, K., Gearhart, J., Geiss, K. and Schlager, J. (2005). *In vitro* toxicity of nanoparticles in BRL 3A rat liver cells, *Toxicology in vitro*, **19**(7): 975-983.

Innes, N. and Ogden, G. (1999). A technique for the study of endocytosis in human oral epithelial cells, *Archives of Oral Biology*, **44**(6): 519-523.

Ivask, A., Kurvet, I., Kasemets, K., Blinova, I., Aruoja, V., Suppi, S., Vija, H., Käkinen, A., Titma, T., Heinlaan, M., Visnapuu, M., Koller, D., Kisand, V. and Kahru, A. (2014). Size-dependent toxicity of silver nanoparticles to bacteria, yeast, algae, crustaceans and mammalian cells *in vitro*, *PLoS One*, **9**(7): e102108.

Jiang, X., Wu, Y., Gray, P., Zheng, J., Cao, G., Zhang, H., Zhang, X., Boudreau, M., Croley, T.R. and Chen, C. (2018). Influence of gastrointestinal environment on free radical generation of silver nanoparticles and implications for their cytotoxicity, *NanoImpact*, 10: 144-152.

Jung, T., Kamm, W., Breitenbach, A., Kaiserling, E., Xiao, J. and Kissel, T. (2000). Biodegradable nanoparticles for oral delivery of peptides: Is there a role for polymers to affect mucosal uptake? *European Journal of Pharmaceutics and Biopharmaceutics*, **50**(1): 147-160.

Kariminejad, M., Sadeghi, E., Rouhi, M., Mohammadi, R., Askari, F., Taghizadeh, M. and Moradi, S. (2018). The effect of nano-SiO_2 on the physicochemical and structural properties of gelatin-polyvinyl alcohol composite films, *Journal of Food Process Engineering*, **41**(6): e12817.

Kästner, C., Lichtenstein, D., Lampen, A. and Thünemann, A.F. (2017). Monitoring the fate of small silver nanoparticles during artificial digestion, *Colloids and Surfaces A: Physicochemical and Engineering Aspects*, 526: 76-81.

Kimball, J.R., Nittayananta, W., Klausner, M., Chung, W.O. and Dale, B.A. (2006). Antimicrobial barrier of an *in vitro* oral epithelial model, *Archives of Oral Biology*, **51**(9): 775-783.

Kohler, H. and Jenzer, H. (1989). Interaction of lactoperoxidase with hydrogen peroxide: Formation of enzyme intermediates and generation of free radicals, *Free Radical Biology and Medicine*, **6**(3): 323-339.

Kotta, S., Khan, A.W., Pramod, K., Ansari, S.H., Sharma, R.K. and Ali, J. (2012). Exploring oral nanoemulsions for bioavailability enhancement of poorly water-soluble drugs, *Expert Opinion on Drug Delivery*, **9**(5): 585-598.

Kreyling, W.G., Semmler-Behnke, M., Seitz, J., Scymczak, W., Wenk, A., Mayer, P., Takenaka, S. and Oberdörster, G. (2009). Size dependence of the translocation of inhaled iridium and carbon nanoparticle aggregates from the lung of rats to the blood and secondary target organs, *Inhalation Toxicology*, **21**(suppl. 1): 55-60.

Kroll, A., Pillukat, M.H., Hahn, D. and Schnekenburger, J. (2009). Current *in vitro* methods in nanoparticle risk assessment: Limitations and challenges, *European Journal of Pharmaceutics and Biopharmaceutics*, **72**(2): 370-377.

Lendenmann, U., Grogan, J. and Oppenheim, F. (2000). Saliva and dental pellicle – A review, *Advances in Dental Research*, **14**(1): 22-28.

Lidén, G. (2011). The European Commission tries to define nanomaterials, *Annals of Occupational Hygiene*, **55**(1): 1-5.

Liu, W., Worms, I.A., Herlin-Boime, N., Truffier-Boutry, D., Michaud-Soret, I., Mintz, E., Vidaud, C. and Rollin-Genetet, F. (2017). Interaction of silver nanoparticles with metallothionein and ceruloplasmin: Impact on metal substitution by Ag (i), corona formation and enzymatic activity, *Nanoscale*, **9**(19): 6581-6594.

Lomer, M.C., Thompson, R.P. and Powell, J.J. (2002). Fine and ultrafine particles of the diet: Influence on the mucosal immune response and association with Crohn's disease, *Proceedings of the Nutrition Society*, **61**(1): 123-130.

Lomer, M.C., Hutchinson, C., Volkert, S., Greenfield, S.M., Catterall, A., Thompson, R.P. and Powell, J.J. (2004). Dietary sources of inorganic microparticles and their intake in healthy subjects and patients with Crohn's disease, *British Journal of Nutrition*, **92**(6): 947-955.

Lund, T., Callaghan, M.F., Williams, P., Turmaine, M., Bachmann, C., Rademacher, T., Roitt, I.M. and Bayford, R. (2011). The influence of ligand organization on the rate of uptake of gold nanoparticles by colorectal cancer cells, *Biomaterials*, **32**(36): 9776-9784.

Mackie, A., Gourcy, S., Rigby, N., Moffat, J., Capron, I. and Bajka, B. (2019). The fate of cellulose nanocrystal stabilised emulsions after simulated gastrointestinal digestion and exposure to intestinal mucosa, *Nanoscale*, **11**(6): 2991-2998.

Manski, J.M., van der Goot, A.J. and Boom, R.M. (2007). Formation of fibrous materials from dense calcium caseinate dispersions, *Biomacromolecules*, **8**(4): 1271-1279.

Manson, J.M., Rauch, M. and Gilmore, M.S. (2008). The commensal microbiology of the gastrointestinal tract, *GI Microbiota and Regulation of the Immune System*, 15-28. Springer,

McClements, D.J. (2013). Edible lipid nanoparticles: Digestion, absorption and potential toxicity, *Progress in Lipid Research*, **52**(4): 409-423.

McConnell, E.L., Basit, A.W. and Murdan, S. (2008). Measurements of rat and mouse gastrointestinal pH, fluid and lymphoid tissue, and implications for *in-vivo* experiments, *Journal of Pharmacy and Pharmacology*, **60**(1): 63-70.

Miller, G., Lowrey, N. and Senjen, R. (2008). *Out of the Laboratory and on to Our Plates: Nanotechnology in Food & Agriculture*, Friends of the Earth.

Mirzaei-Mohkam, A., Garavand, F., Dehnad, D., Keramat, J. and Nasirpour, A. (2019). Optimisation, antioxidant attributes, stability and release behaviour of carboxymethyl cellulose films incorporated with nanoencapsulated vitamin E, *Progress in Organic Coatings*, 134: 333-341.

Miyayama, T., Arai, Y., Suzuki, N. and Hirano, S. (2014). Cellular distribution and behavior of metallothionein in mammalian cells following exposure to silver nanoparticles and silver ions, *Yakugaku zasshi: Journal of the Pharmaceutical Society of Japan*, **134**(6): 723-729.

Mwilu, S.K., El Badawy, A.M., Bradham, K., Nelson, C., Thomas, D., Scheckel, K.G., Tolaymat, T., Ma, L. and Rogers, K.R. (2013). Changes in silver nanoparticles exposed to human synthetic stomach fluid: Effects of particle size and surface chemistry, *Science of the Total Environment*, 447: 90-98.

Neethirajan, S. and Jayas, D.S. (2011). Nanotechnology for the food and bioprocessing industries, *Food and Bioprocess Technology*, 4(1): 39-47.

Nel, A., Xia, T., Mädler, L. and Li, N. (2006). Toxic potential of materials at the nanolevel, *Science*, 311(5761): 622-627.

Oberdörster, E. (2004). Manufactured nanomaterials (fullerenes, C60) induce oxidative stress in the brain of juvenile largemouth bass. *Environmental Health Perspectives*, 112(10): 1058-1062.

Oberdörster, G., Maynard, A., Donaldson, K., Castranova, V., Fitzpatrick, J., Ausman, K., Carter, J., Karn, B., Kreyling, W. and Lai, D. (2005a). Principles for characterizing the potential human health effects from exposure to nanomaterials: Elements of a screening strategy, *Particle and fibre Toxicology*, 2(1): 8.

Oberdörster, G., Oberdörster, E. and Oberdörster, J. (2005b). Nanotoxicology: An emerging discipline evolving from studies of ultrafine particles, *Environmental Health Perspectives*, 113(7): 823-839.

Oberdürster, G. (2000). Toxicology of ultrafine particles: *In vivo* studies, *Philosophical Transactions of the Royal Society of London. Series A: Mathematical, Physical and Engineering Sciences*, 358(1775): 2719-2740.

Pedata, P., Stoeger, T., Zimmermann, R., Peters, A., Oberdörster, G. and D'Anna, A. (2015). Are we forgetting the smallest, sub 10 nm combustion generated particles? *Particle and Fibre Toxicology*, 12(1): 34.

Pelka, J., Gehrke, H., Esselen, M., Türk, M., Crone, M., Bräse, S., Muller, T., Blank, H., Send, W. and Zibat, V. (2009). Cellular uptake of platinum nanoparticles in human colon carcinoma cells and their impact on cellular redox systems and DNA integrity, *Chemical Research in Toxicology*, 22(4): 649-659.

Peters, R., Kramer, E., Oomen, A.G., Herrera Rivera, Z.E., Oegema, G., Tromp, P.C., Fokkink, R., Rietveld, A., Marvin, H.J. and Weigel, S. (2012). Presence of nano-sized silica during in vitro digestion of foods containing silica as a food additive, *ACS Nano*, 6(3): 2441-2451.

Pinheiro, A.C., Coimbra, M.A. and Vicente, A.A. (2016). *In vitro* behaviour of curcumin nanoemulsions stabilized by biopolymer emulsifiers – Effect of interfacial composition, *Food Hydrocolloids*, 52: 460-467.

Powell, J.J., Thoree, V. and Pele, L.C. (2007). Dietary microparticles and their impact on tolerance and immune responsiveness of the gastrointestinal tract, *British Journal of Nutrition*, 98(S1): S59-S63.

Powell, J.J., Faria, N., Thomas-McKay, E. and Pele, L.C. (2010). Origin and fate of dietary nanoparticles and microparticles in the gastrointestinal tract, *Journal of Autoimmunity*, 34(3): J226-J233.

Prasad, R.Y., Wallace, K., Daniel, K.M., Tennant, A.H., Zucker, R.M., Strickland, J., Dreher, K., Kligerman, A.D., Blackman, C.F. and DeMarini, D.M. (2013). Effect of Treatment Media on the Agglomeration of Titanium Dioxide Nanoparticles: Impact on Genotoxicity, Cellular Interaction, and Cell Cycle, *ACS Nano* 7(3): 1929-1942.

Presland, R.B. and Dale, B.A. (2000). Epithelial structural proteins of the skin and oral cavity: Function in health and disease, *Critical Reviews in Oral Biology & Medicine*, 11(4): 383-408.

Rogers, K.R., Bradham, K., Tolaymat, T., Thomas, D.J., Hartmann, T., Ma, L. and Williams, A. (2012). Alterations in physical state of silver nanoparticles exposed to synthetic human stomach fluid, *Science of the Total Environment*, 420: 334-339.

Roosen, J., Bieberstein, A., Blanchemanche, S., Goddard, E., Marette, S. and Vandermoere,

F. (2015). Trust and willingness to pay for nanotechnology food, *Food Policy*, 52: 75-83.

Rouhi, M., Razavi, S.H. and Mousavi, S.M. (2017). Optimization of crosslinked poly (vinyl alcohol) nanocomposite films for mechanical properties, *Materials Science and Engineering: C*, 71: 1052-1063.

Ryman-Rasmussen, J.P., Riviere, J.E. and Monteiro-Riviere, N.A. (2007). Surface coatings determine cytotoxicity and irritation potential of quantum dot nanoparticles in epidermal keratinocytes, *Journal of Investigative Dermatology*, **127**(1): 143-153.

Sasidharan, A., Riviere, J.E. and Monteiro-Riviere, N.A. (2015). Gold and silver nanoparticle interactions with human proteins: Impact and implications in biocorona formation, *Journal of Materials Chemistry B*, **3**(10): 2075-2082.

Sato, Q.Z.Y.K.K. and Donaldson, K.N.N.K.K. (1998). Differences in the extent of inflammation caused by intratracheal exposure to three ultrafine metals: Role of free radicals, *Journal of Toxicology and Environmental Health Part A*, **53**(6): 423-438.

Sayes, C.M., Liang, F., Hudson, J.L., Mendez, J., Guo, W., Beach, J.M., Moore, V.C., Doyle, C.D., West, J.L. and Billups, W.E. (2006a). Functionalization density dependence of single-walled carbon nanotubes cytotoxicity *in vitro*, *Toxicology Letters*, **161**(2): 135-142.

Sayes, C.M., Wahi, R., Kurian, P.A., Liu, Y., West, J.L., Ausman, K.D., Warheit, D.B. and Colvin, V.L. (2006b). Correlating nanoscale titania structure with toxicity: A cytotoxicity and inflammatory response study with human dermal fibroblasts and human lung epithelial cells, *Toxicological Sciences*, **92**(1): 174-185.

SCCP (2007). Preliminary Opinion on Safety of Nanomaterials in Cosmetic Products. Scientific Committee on Consumer Products.

SCENIHR (2009). Risk Assessment of Products of Nanotechnologies, Scientific Committee on Emerging and Newly Identified Health Risks.

Schnieden, H. and Small, R. (1971). Spasmolytic effects of cadmium and zinc ions upon the guinea-pig isolated ileum preparation, *British Journal of Pharmacology*, **41**(3): 488-499.

Seok, S.H., Cho, W.S., Park, J.S., Na, Y., Jang, A., Kim, H., Cho, Y., Kim, T., You, J.R. and Ko, S. (2013). Rat pancreatitis produced by 13-week administration of zinc oxide nanoparticles: Biopersistence of nanoparticles and possible solutions, *Journal of Applied Toxicology*, **33**(10): 1089-1096.

Sergent, J.-A., Paget, V. and Chevillard, S. (2012). Toxicity and genotoxicity of nano-SiO_2 on human epithelial intestinal HT-29 cell line, *Annals of Occupational Hygiene*, **56**(5): 622-630.

Shukla, R., Bansal, V., Chaudhary, M., Basu, A., Bhonde, R.R. and Sastry, M. (2005). Biocompatibility of gold nanoparticles and their endocytotic fate inside the cellular compartment: A microscopic overview, *Langmuir*, **21**(23): 10644-10654.

Siegrist, M., Cousin, M.-E., Kastenholz, H. and Wiek, A. (2007). Public acceptance of nanotechnology foods and food packaging: The influence of affect and trust, *Appetite*, **49**(2): 459-466.

Singh, N., Jenkins, G.J., Asadi, R. and Doak, S.H. (2010). Potential toxicity of superparamagnetic iron oxide nanoparticles (SPION), *Nano Reviews*, **1**(1): 5358.

Sinnecker, H., Ramaker, K. and Frey, A. (2014). Coating with luminal gut-constituents alters adherence of nanoparticles to intestinal epithelial cells, *Beilstein Journal of Nanotechnology*, **5**(1): 2308-2315.

Soenen, S.J.H., Illyes, E., Vercauteren, D., Braeckmans, K., Majer, Z., De Smedt, S.C. and De Cuyper, M. (2009). The role of nanoparticle concentration-dependent induction of cellular stress in the internalization of non-toxic cationic magnetoliposomes, *Biomaterials*, **30**(36): 6803-6813.

Sozer, N. and Kokini, J.L. (2009). Nanotechnology and its applications in the food sector, *Trends in Biotechnology*, **27**(2): 82-89.

Stone, V., Nowack, B., Baun, A., van den Brink, N., von der Kammer, F., Dusinska, M., Handy, R., Hankin, S., Hassellöv, M. and Joner, E. (2010). Nanomaterials for environmental studies: Classification, reference material issues, and strategies for physico-chemical characterization, *Science of the Total Environment*, **408**(7): 1745-1754.

Takenaka, S., Karg, E., Roth, C., Schulz, H., Ziesenis, A., Heinzmann, U., Schramel, P. and Heyder, J. (2001). Pulmonary and systemic distribution of inhaled ultrafine silver particles in rats, *Environmental Health Perspectives*, **109**(suppl. 4): 547-551.

Taylor, A.A., Marcus, I.M., Guysi, R.L. and Walker, S.L. (2015). Metal oxide nanoparticles induce minimal phenotypic changes in a model colon gut microbiota, *Environmental Engineering Science*, **32**(7): 602-612.

Tiede, K., Boxall, A.B., Tear, S.P., Lewis, J., David, H. and Hassellöv, M. (2008). Detection and characterization of engineered nanoparticles in food and the environment, *Food Additives and Contaminants*, **25**(7): 795-821.

USDA (2015). USDA Awards $3.8 Million in Grants for Nanotechnology Research.

van der Lubben, I.M., Verhoef, J.C., Borchard, G. and Junginger, H.E. (2001). Chitosan and its derivatives in mucosal drug and vaccine delivery, *European Journal of Pharmaceutical Sciences*, **14**(3): 201-207.

van der Zande, M., Vandebriel, R.J., Van Doren, E., Kramer, E., Herrera Rivera, Z., Serrano-Rojero, C.S., Gremmer, E.R., Mast, J., Peters, R.J. and Hollman, P.C. (2012). Distribution, elimination, and toxicity of silver nanoparticles and silver ions in rats after 28-day oral exposure, *ACS Nano*, **6**(8): 7427-7442.

van Nieuw Amerongen, A., Bolscher, J.G. and Veerman, E.C. (2004). Salivary proteins: Protective and diagnostic value in cariology? *Caries Research*, **38**(3): 247-253.

Vo-Dinh, T., Cullum, B.M. and Stokes, D.L. (2001). Nanosensors and biochips: Frontiers in biomolecular diagnostics, *Sensors and Actuators B: Chemical*, **74**(1-3): 2-11.

Walczak, A.P., Fokkink, R., Peters, R., Tromp, P., Herrera Rivera, Z.E., Rietjens, I.M., Hendriksen, P.J. and Bouwmeester, H. (2012). Behaviour of silver nanoparticles and silver ions in an *in vitro* human gastrointestinal digestion model, *Nanotoxicology*, **7**(7): 1198-1210.

Walk, S.T. and Young, V.B. (2008). Emerging insights into antibiotic-associated diarrhea and Clostridium difficile infection through the lens of microbial ecology, *Interdisciplinary Perspectives on Infectious Diseases*, 2008.

WHO (2008). Nanotechnology. International Food Safety Authorities Network (INFOSAN) Information Note No. 01/2008 (Rev 1. March 2008), 2008 [cited 9 September 2008], World Health Organization.

Wick, P., Manser, P., Limbach, L.K., Dettlaff-Weglikowska, U., Krumeich, F., Roth, S., Stark, W.J. and Bruinink, A. (2007). The degree and kind of agglomeration affect carbon nanotube cytotoxicity, *Toxicology Letters*, **168**(2): 121-131.

Williams, P., Stirling, E. and Keynes, N. (2004). Food fears: A national survey on the attitudes of australian adults about the safety and quality of food. *Asia Pacific Journal of Clinical Nutrition*, **13**(1): 32-39.

Yao, M., He, L., McClements, D.J. and Xiao, H. (2015). Uptake of gold nanoparticles by intestinal epithelial cells: Impact of particle size on their absorption, accumulation, and toxicity, *Journal of Agricultural and Food Chemistry*, **63**(36): 8044-8049.

Young, V.B. (2012). The intestinal microbiota in health and disease, *Current Opinion in Gastroenterology*, **28**(1): 63.

Zhou, D., Zhu, L., Fu, Y., Zhu, M. and Xue, L. (2015). Development of lower cost seawater desalination processes using nanofiltration technologies – A review, *Desalination*, 376: 109-116.

Zhou, G. (2013). Nanotechnology in the food system: Consumer acceptance and willingness to pay. Theses and Dissertations – Agricultural Economics, 1-122.

Zoetendal, E.G., Collier, C.T., Koike, S., Mackie, R.I. and Gaskins, H.R. (2004). Molecular ecological analysis of the gastrointestinal microbiota: A review, *The Journal of Nutrition*, **134**(2): 465-472.

Index

For Product Safety Concerns and Information please contact our EU
representative GPSR@taylorandfrancis.com
Taylor & Francis Verlag GmbH, Kaufingerstraße 24, 80331 München, Germany